T0213543

Communications in Computer and Information Science 700

Commenced Publication in 2007
Founding and Former Series Editors:
Alfredo Cuzzocrea, Xiaoyong Du, Orhun Kara, Ting Liu, Dominik Ślęzak,
and Xiaokang Yang

More information about this series at http://www.springer.com/series/7899

Vladimir M. Vishnevskiy · Konstantin E. Samouylov
Dmitry V. Kozyrev (Eds.)

Distributed Computer and Communication Networks

20th International Conference, DCCN 2017
Moscow, Russia, September 25–29, 2017
Proceedings

 Springer

Editors
Vladimir M. Vishnevskiy
V.A. Trapeznikov Institute of Control
  Sciences
Russian Academy of Sciences
Moscow
Russia

Dmitry V. Kozyrev
V.A. Trapeznikov Institute of Control
  Sciences
Russian Academy of Sciences
Moscow
Russia

Konstantin E. Samouylov
RUDN University
Moscow
Russia

ISSN 1865-0929   ISSN 1865-0937 (electronic)
Communications in Computer and Information Science
ISBN 978-3-319-66835-2   ISBN 978-3-319-66836-9 (eBook)
DOI 10.1007/978-3-319-66836-9

Library of Congress Control Number: 2017952879

Printed on acid-free paper

This Springer imprint is published by Springer Nature
The registered company is Springer International Publishing AG
The registered company address is: Gewerbestrasse 11, 6330 Cham, Switzerland

Preface

This volume contains a collection of revised selected full papers presented at the 20th International Conference on Distributed Computer and Communication Networks (DCCN2017), held in Moscow, Russia, September 25–29, 2017.

The conference constituted a continuation of the traditional international conferences of the DCCN series, which have taken place in Bulgaria (Sofia, 1995, 2005, 2006, 2008, 2009, 2014), Israel (Tel Aviv, 1996, 1997, 1999, 2001), and Russia (Moscow, 1998, 2000, 2003, 2007, 2010, 2011, 2013, 2015, 2016) in the last 20 years. The main idea of the conference is to provide a platform and forum for researchers and developers from academia and industry from various countries working in the area of theory and applications of distributed computer and communication networks, mathematical modeling, methods of control, and optimization of distributed systems, by offering them a unique opportunity to share their views as well as discuss developments and pursue collaboration in this area. The content of this volume is related to the following subjects:

1. Computer networks architecture and topology: control and management, design, optimization, routing, resource allocation
2. Analytical modeling and simulation, performance, and QoS evaluation of info-communication systems
3. Centimeter and millimeter wave wireless network technologies: local networks and 4G/5G cellular networks
4. RFID and sensor networks
5. Distributed information systems applications: Internet of Things, Big Data, Intelligent Transport Systems
6. Distributed systems and cloud computing, software-defined networks, virtualization
7. Stochastic and statistical methods in modeling of information systems
8. Queueing theory and reliability theory in computer network applications
9. Unmanned aircrafts and high-altitude platform stations (HAPS): control, communication, applications

The DCCN 2017 conference gathered 176 submissions from authors from 15 different countries. From these, 132 high-quality papers in English were accepted and presented during the conference, 39 of which were recommended by session chairs and selected by the Program Committee for the Springer proceedings.

All the papers selected for the proceedings are given in the form presented by the authors. These papers are of interest to everyone working in the field of computer and communication networks.

We thank all the authors for their interest in DCCN, the members of the Program Committee for their contributions, and the reviewers for their peer-reviewing efforts.

September 2017

Vladimir Vishnevskiy
Konstantin Samouylov

Organization

DCCN 2017 was jointly organized by the Russian Academy of Sciences (RAS), the V.A. Trapeznikov Institute of Control Sciences of RAS (ICS RAS), the Peoples' Friendship University of Russia (RUDN), the National Research Tomsk State University, and the Institute of Information and Communication Technologies of the Bulgarian Academy of Sciences (IICT BAS).

International Program Committee

V.M. Vishnevskiy (Chair)	ICS RAS, Russia
K.E. Samouylov (Co-chair)	RUDN University, Russia
A.M. Gortsev (Co-chair)	Tomsk State University, Russia
A.M. Andronov	Transport and Telecommunication Institute, Latvia
L.I. Abrosimov	Moscow Power Engineering Institute, Russia
Mo Adda	University of Portsmouth, UK
T.I. Aliev	ITMO University, Russia
A.S. Bugaev	Moscow Institute of Physics and Technology, Russia
T. Czachorski	Institute of informatics of the Polish Academy of Sciences, Poland
A.N. Dudin	Belarusian State University, Belarus
V.V. Devyatkov	Bauman Moscow State Technical University, Russia
D. Deng	National Changhua University of Education, Taiwan
M.A. Fedotkin	State University of Nizhni Novgorod, Russia
E. Gelenbe	Imperial College London, UK
A. Gelman	IEEE Communications Society, USA
D. Grace	York University, UK
J. Kolodziej	Cracow University of Technology, Poland
G. Kotsis	Johannes Kepler University Linz, Austria
U. Krieger	University of Bamberg, Germany
A. Krishnamoorthy	Cochin University of Science and Technology, India
A.E. Koucheryavy	Bonch-Bruevich Saint-Petersburg State University of Telecommunications, Russia
Ye.A. Koucheryavy	Tampere University of Technology, Finland
L. Lakatos	Budapest University, Hungary
E. Levner	Holon Institute of Technology, Israel
S.D. Margenov	Institute of Information and Communication Technologies of the Bulgarian Academy of Sciences, Bulgaria
G.K. Mishkoy	Academy of Sciences of Moldova, Moldavia
E.V. Morozov	Institute of Applied Mathematical Research of the Karelian Research Centre RAS, Russia

A.A. Nazarov	Tomsk State University, Russia
P. Nikitin	University of Washington, USA
S.A. Nikitov	Institute of Radio Engineering and Electronics of RAS, Russia
D.A. Novikov	ICS RAS, Russia
M. Pagano	Pisa University, Italy
V.V. Rykov	Gubkin Russian State University of Oil and Gas, Russia
L.A. Sevastianov	RUDN University, Russia
P. Stanchev	Kettering University, USA
S.N. Stepanov	Moscow Technical University of Communication and Informatics, Russia
H. Tijms	Vrije Universiteit Amsterdam, Netherlands
V.A. Vasenin	Lomonosov Moscow State University, Russia
S.N. Vasiliev	ICS RAS, Russia
S.I. Vinitsky	Joint Institute for Nuclear Research, Russia
E. Yakubov	Holon Institute of Technology, Israel
Yu.P. Zaychenko	Kyiv Polytechnic Institute, Ukraine

Organizing Committee

V.M. Vishnevskiy (Chair)	ICS RAS, Russia
K.E. Samouylov (Vice Chair)	RUDN University, Russia
A.N. Nazarov (Vice Chair)	Tomsk State University, Russia
A.S. Mandel	ICS RAS, Russia
S.P. Moiseeva	Tomsk State University, Russia
Yu.V. Gaidamaka	RUDN University, Russia
D.V. Kozyrev	RUDN University and ICS RAS, Russia
A.A. Larionov	ICS RAS, Russia
O.V. Semenova	ICS RAS, Russia
R.E. Ivanov	ICS RAS, Russia
T. Atanasova	IICT BAS, Bulgaria
V.M. Vorobyev	R&D Company "Information and Networking Technologies", Russia
S.N. Kupriyakhina	ICS RAS, Russia

Organizers and Partners

Organizers

Russian Academy of Sciences
RUDN University
V.A. Trapeznikov Institute of Control Sciences of RAS
National Research Tomsk State University

Institute of Information and Communication Technologies of the Bulgarian Academy of Sciences
Research and Development Company "Information and Networking Technologies"

Support

Information support was provided by the IEEE Russia Section. Financial support was provided by the Russian Foundation for Basic Research. The conference was held in the framework of the RUDN University Competiteveness Enhancement Program "5-100."

Contents

Modeling the Process of Dynamic Resource Sharing Between LTE and NB-IoT Services

Vyacheslav Begishev[1,2,3,4], Andrey Samuylov[1,2,3,4(✉)],
Dmitri Moltchanov[1,2,3,4], and Konstantin Samouylov[1,2,3,4]

[1] Tampere University of Technology, Tampere, Finland
[2] Department of Applied Probability and Informatics,
Peoples' Friendship University of Russia (RUDN University),
6 Miklukho-Maklaya Street, Moscow 117198, Russia
{begishev_vo,samuylov_ak,molchanov_da,samuylov_ke}@rudn.university
[3] Institute of Informatics Problems,
Federal Research Center "Computer Science and Control"
of the Russian Academy of Sciences, 44-2 Vavilov Street, Moscow 119333, Russia
[4] Department of Electronics and Communications Engineering,
Tampere University of Technology, 10 Korkeakoulunkatu, 33720 Tampere, Finland
{andrey.samuylov,dmitri.moltchanov}@tut.fi

Abstract. The Internet of Things (IoT) undergoes fundamental changes, expanding its infrastructure with more advanced and mobile devices. As the IoT develops, the existing cellular communication technologies often do not provide sufficient coverage while modern IoT terminals are often expensive and characterized by a short battery life. To address these issues, in Release 13 (LTE Advanced Pro) published in 2016, 3GPP consortium has proposed the Narrow-Band IoT (NB-IoT) technology as an efficient way to provide a wide range of new capabilities and services in a wireless cellular network. Having specified three operational regimes, 3GPP did not provide guidelines on the way resource sharing has to be done between LTE and NB-IoT traffic. In this paper, the in-band NB-IoT service model is presented, where a certain amount of LTE radio resources are exclusively allocated to LTE and NB-IoT users while the rest are shared between them. We analyze the proposed system for performance metrics of interest including NB-IoT and LTE session drop probabilities and resource utilization.

Keywords: Internet of Things · NB-IoT (Narrow Band IoT) · Analytical model · Session drop probability · Resource sharing

1 Introduction

Low power consumption, affordable sensor prices, high radio transmission distance and support for a large number of connections are the basic requirements for an Internet of Things (IoT) based wireless network. With these requirements in mind, a new type of wireless networks, low-power WANs (LPWAN), have been

© Springer International Publishing AG 2017
V.M. Vishnevskiy et al. (Eds.): DCCN 2017, CCIS 700, pp. 1–12, 2017.
DOI: 10.1007/978-3-319-66836-9_1

proposed in the past. The most successful examples include GPRS/EDGE-based GSM, LTE-M, LoRaWAN, SIGFOX [1]. The development of these technologies was initiated and supported by several leading bodies including IEEE, ETSI and 3GPP. However, due to technological imperfections none of these solutions have gained the widespread popularity.

The recently standardized in 3GPP Rel. 13 Narrow-band IoT (NB-IoT) technology is expected fulfill the strict requirements on energy efficiency, LTE compatibility, and simplicity of operation [2,3]. This new technology can be relatively easily implemented in existing LTE cellular networks, adding a few new categories of devices designed specifically to reduce the complexity of endpoints. To achieve the required efficiency, the width of the wireless channel was reduced to 180 kHz, which allowed gaining the signal by 20 dB. These changes significantly increase the effective transmission distance for low-power devices and increase their autonomy up to 10 years or more depending on the application.

In this paper, we investigate the resource sharing strategies between NB-IoT and LTE sessions in an LTE cell. First, we formalize an NB-IoT model for in-band LTE operation, where NB-IoT traffic coexists in the same band with LTE sessions. We consider the dynamic RBs sharing for LTE and NB-IoT traffic with minimal resource guarantees to both. Introducing the minimum resource guarantees into dynamic resource sharing scheme allows to alleviate the problem of LTE and NB-IoT performance degradation of fully dynamic case while the nature of dynamic resource for the rest of resources still enables the high degree of statistical multiplexing of sessions. The performance metrics of interest are the probability of drop of NB-IoT and LTE sessions. The proposed model allows to determines the dynamic capacity of a LTE cell with respect to both types of traffic and can be further used to optimize LTE BS equipment for various operational conditions.

The rest of the paper is organized as follows. In Sect. 2 we introduce the system model. The system is analyzed is Sect. 3. Numerical results are illustrated in Sect. 4. Conclusions are drawn in the last section.

2   System Model

We consider a cell of circular coverage area with a base station located at the center. LTE users are provided with a single service, e.g., voice telephony or streaming video. There are C resource blocks (RB) in the system. Each resource block is further divided into c basic channels channels. Resource allocation for both LTE and NB-IoT sessions is performed in terms of the number of basic channels. However, an LTE session may only requests resources in terms of multiple resources blocks, i.e., multiples of c. To provide the service LTE users are reserved R_L channels and for NB-IoT R_N channels. Then $C_L = C - R_N$ channels are available for LTE service users and $C_N = C - R_L$ channels are available for NB-IoT.

Suppose that b radio channels are required to transmit a data block of any type from NB-IoT. In order to transmit the current number of data blocks, radio

Table 1. Notation used in this paper.

Parameter	Definition
C	Number of basic channels in LTE cell
c	Number of basic channels in a single RB
λ	Intensity of NB-IoT session arrivals
θ	Mean NB-IoT session volume in bits
ν	Intensity of LTE session arrivals
$1/\mu$	Mean LTE session duration
$\rho = \lambda\theta$	NB-IoT offered traffic load
$a = \nu/\mu$	LTE offered traffic load
R_N	Number of basic channels reserved for NB-IoT
R_L	Number of basic channels reserved for LTE
$c(m)$	Number of channels currently occupied by NB-IoT
b	Minimal number of channels for NB-IoT session
d	Number of channels required by LTE session
M	Maximal number of NB-IoT sessions in RB
$p(m, n)$	Probability of having m NB-IoT and n LTE sessions
$G(\mathcal{X})$	Normalization constant
p_N	NB-IoT session drop probability
p_L	LTE session drop probability
$\mathcal{B}_N, \mathcal{B}_L$	Blocking sets of NB-IoT/LTE sessions
$E[b_N], E[b_L]$	Mean number of occupied channels by NB-IoT/LTE
$E[b_{NL}]$	Mean number of occupied channels by NB-IoT and LTE

resources are allocated by fixed size bands over c radio channels. Then $M = \lfloor c/b \rfloor$ determines the maximum number of data blocks that can be simultaneously transmitted on the same resource range. Let $S = \lfloor C_N/c \rfloor$ be the maximum number of fixed size resource ranges that can be allocated in the cell for NB-IoT maintenance. We assume that the incoming stream of NB-IoT requests for transmission of data blocks is Poisson with the intensity λ $[c^{-l}]$ and the length of the block is distributed exponentially with the average θ [bit] (see Fig. 1).

To provide the LTE service, d radio channels are required. It is assumed that the incoming flow of requests for LTE users is Poisson flow with the intensity ν $[c^{-l}]$ and the time of granting the LTE service is distributed exponentially with an average $1/\mu$ [s]. Let $a = \nu/\mu$ be the intensity of the proposed load generated by users of the LTE service. The notation used in this paper is summarized in the Table 1.

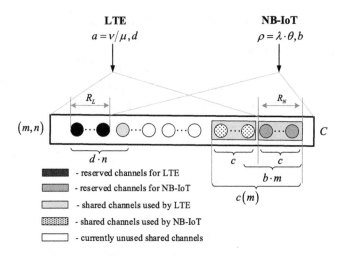

Fig. 1. The resource allocation process in LTE cell

3 Performance Analysis

Let $m(t)$, $t > 0$ and $n(t)$, $t > 0$, define the number of active NB-IoT and LTE sessions, respectively. Thus, the system state of an LTE cell serving both NH-IoT and LTE sessions can be described by the two dimensional stochastic process, $(m(t), n(t), t > 0)$ over the following state space

$$\mathcal{X} = \{m \geq 0, n \geq 0 : nd \leq C - R_L, c(m) \leq C - R_L, nd + c(m) \leq C\}, \quad (1)$$

where $c(m) = c \lceil bm/M \rceil$.

For the proposed resource sharing scheme, the blocking subsets for NB-IoT and LTE users are represented as

$$\begin{aligned}
\mathcal{B}_N &= \{(m, n) \in \mathcal{X} : c(m + 1) > C - \max(nd, R_L)\}, \\
\mathcal{B}_L &= \{(m, n) \in \mathcal{X} : (n + 1)d > C - \max(c(m), R_N)\}.
\end{aligned} \quad (2)$$

Observe that $(m(t), n(t), t > 0)$ is stationary Markov process. Under the Markovian assumption there exists a stationary probability distribution $p(m, n), (m, n) \in \mathcal{X}$. Then by the Kolmogorov criterion, $(m(t), n(t))$ is a reversible stochastic process implying that the stationary probabilities $p(m, n), (m, n) \in \mathcal{X}$, satisfy the following system

$$\begin{cases}
p(m, n)(c(m)/\theta) = p(m - 1, n)\lambda, \ m > 0, (m, n) \in \mathcal{X}, \\
p(m, n)\mu n = p(m, n - 1)\nu, \ n > 0, (m, n) \in \mathcal{X}.
\end{cases} \quad (3)$$

Solving (3) the stationary probability distribution of the given Markov process $(m(t), n(t))$ has the following form

$$p(m, n) = G^{-1}(\mathcal{X}) \left(\frac{\rho}{Mb}\right)^m \left(\prod_{i=1}^{m} \left\lceil \frac{i}{M} \right\rceil\right)^{-1} \frac{a^n}{n!}, \ (m, n) \in \mathcal{X}, \quad (4)$$

where

$$G(\mathcal{X}) = \sum_{(m,n)\in\mathcal{X}} \left(\frac{\rho}{Mb}\right)^m \left(\prod_{i=1}^{m}\left\lceil\frac{i}{M}\right\rceil\right)^{-1} \frac{a^n}{n!}, \tag{5}$$

is the normalizing constant [8].

Consider now LTE and NB-IoT session drop probabilities. To reduce the complexity of the calculations we introduce the recurrent relationship of non-standardized probabilities of macrostates of the system in the following form

$$\mathcal{X} = \bigcup_{s=0}^{S} \mathcal{X}_s, \tag{6}$$

where $\mathcal{X}_s = \{(m,n) \in \mathcal{X} : c(m) = sc\}$.

Using (6) the blocking subspaces for LTE sessions can be written as

$$\mathcal{B}_L = \bigcup_{s=0}^{S} \{\mathcal{X} : (n+1)d > C - c(m) \cup (n+1)d > C - R_N\}$$

$$= \bigcup_{s=0}^{S} \{\mathcal{X} : (n+1)d > C - \max(c(m), R_N), c(m) = sc\}$$

$$= \bigcup_{s=0}^{S} \left\{\mathcal{X} : n+1 > \left\lfloor\frac{C - \max(sc, R_N)}{d}\right\rfloor, \left\lceil\frac{mb}{c}\right\rceil = s\right\}$$

$$= \bigcup_{s=0}^{S} \left\{\mathcal{X} : n = \left\lfloor\frac{C - \max(sc, R_N)}{d}\right\rfloor, \left\lceil\frac{mb}{c}\right\rceil = s\right\}$$

$$= \left\{\mathcal{X} : n = \left\lfloor\frac{C - R_N}{d}\right\rfloor, m = 0\right\} \bigcup$$

$$\bigcup_{s=1}^{S} \left\{\mathcal{X} : n = \left\lfloor\frac{C - \max(sc, R_N)}{d}\right\rfloor, s < \frac{mb}{c} < s\right\}. \tag{7}$$

The LTE session drop probability is then

$$p_L = p\left(0, \left\lfloor\frac{C - R_N}{d}\right\rfloor\right) + \sum_{s=1}^{\lfloor\frac{C-R_L}{c}\rfloor} \sum_{m=[s-1]M+1}^{sM} p\left(m, \left\lfloor\frac{C - \max[sc, R_N]}{d}\right\rfloor\right). \tag{8}$$

Similarly, the subset for NB-IoT session blocking is

$$
\begin{aligned}
\mathcal{B}_N &= \{\mathcal{X} : c(m+1) > C - nd \cup c(m+1) > C - R_L\} \\
&= \{\mathcal{X} : c(m+1) > C - \max(nd, R_L), c(m) = sc\} \\
&= \bigcup_{s=0}^{S} \{\mathcal{X} : nd > C - \max(c(m), R_L), c(m) = sc\} \\
&= \bigcup_{s=0}^{S} \left\{\mathcal{X} : n - 1 \ge \left\lfloor \frac{C - \max(sc, R_L)}{d} \right\rfloor, \left\lceil \frac{mb}{c} \right\rceil c = sc\right\} \\
&= \bigcup_{s=0}^{S} \left\{\mathcal{X} : n \ge \left\lfloor \frac{C - \max(sc, R_L)}{d} \right\rfloor, m = \frac{sc}{b}\right\}.
\end{aligned}
\tag{9}
$$

Using $\mathcal{B}_N$ the NB-IoT session drop probability is

$$
p_N = \sum_{n=0}^{\lfloor \frac{R_L}{d} \rfloor} p\left(\left\lceil \frac{C - R_L}{c} \right\rceil M, n\right) + \sum_{s=\lfloor \frac{R_N}{c} \rfloor}^{\lfloor \frac{C-R_L}{c} \rfloor - 1} \sum_{n=\lfloor \frac{C-(s+1)c}{d} \rfloor + 1}^{\lfloor \frac{C-\max(sc, R_N)}{d} \rfloor} p(sM, n).
\tag{10}
$$

The mean number of basic channels occupied by NB-IoT sessions is

$$
E[b_N] = M \sum_{m=0}^{\lfloor \frac{C-R_L}{b} \rfloor} \sum_{n=0}^{\left\lfloor \frac{C-\max\left(\lceil \frac{m}{M} \rceil c, R_N\right)}{d} \right\rfloor} \left\lceil \frac{m}{M} \right\rceil p(m, n).
\tag{11}
$$

The mean number of basic channels occupied by LTE sessions is

$$
E[b_L] = d \sum_{m=0}^{\lfloor \frac{C-R_L}{b} \rfloor} \sum_{n=0}^{\left\lfloor \frac{C-\max\left(\lceil \frac{m}{M} \rceil c, R_N\right)}{d} \right\rfloor} n\, p(m, n).
\tag{12}
$$

The radio resource utilization coefficient is now given by

$$
U = \frac{E[b_N] + E[b_L]}{C}.
\tag{13}
$$

4   Numerical Assessment

In this section, we numerically analyze the performance of an LTE cell simultaneously serving both NB-IoT and LTE sessions. As an example of LTE service we will consider streaming video. NB-IoT service is represented by a simple telemetry transfer. The system parameters used in further analysis are summarized in the Table 2.

To keep the results realistic and easily interpretable, instead of parameterizing the traffic models with certain arrival rates we present the results with respect to the number of NB-IoT and LTE devices having certain identical intersession time. Owning to the large number of considered devices the superposition theorem applies leading to the Poisson nature of traffic arrival rates.

Table 2. System parameters used in numerical analysis.

Parameter	Definition	Value
C	Number of basic channels in LTE cell	100
c	Number of basic channels in a RB	4
R_N	Number of channels reserved for NB-IoT	$[0, 1, ..., 100]$
R_L	Number of channels reserved for LTE	$[0, 1, ..., 100]$
b	Number of channels for NB-IoT session	1
d	Number of channels for LTE session	4
θ	Mean NB-IoT session volume	100 kbit
$1/\mu$	Mean LTE session duration	10 s
λ	Intensity of NB-IoT session arrivals	1 per min
ν	Intensity of LTE session arrivals	0.2 per min

4.1 Fully Dynamic Resource Sharing

Consider first the case when no actions are taken by the network operator to isolate LTE and NB-IoT traffic. This corresponds to the fully dynamic resource sharing scheme when no resources are reserved for both LTE and NB-IoT traffic, i.e., $R_N = R_H = 0$.

Figure 2 illustrates resource utilization of the fully dynamic resource sharing scheme as a function of the number of NB-IoT devices for several value of LTE users. Expectedly, when both types of traffic are not limited in the resource usage and are allowed to fully share the available bandwidth, resource utilization may theoretically achieve 100%. This behavior is observed in Fig. 2 where all the

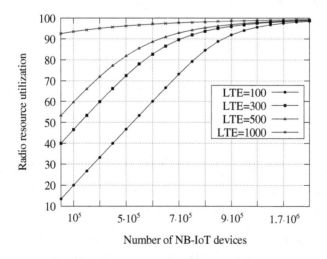

Fig. 2. Resource utilization for fully dynamics resource sharing scheme.

curves eventually approach 100% as the number of NB-IoT devices increases. In addition, we would like to highlight that the number of supported users could achieve rather large values.

While fully dynamic scheme is definitely favorable solution with respect to resource utilization, performance of particular types of traffic can be compromised especially in case of severe load mismatch. The reason is complete lack of isolation of traffic types from each other. Figure 3 illustrates the NB-IoT drop probability for two values of NB-IoT arrival intensities as a function of the LTE session intensities for two types of LTE session volumes, $d = 4$ and $d = 8$. Expectedly, as one may observe, the NB-IoT session drop probability increases as the number of LTE users grow. However, the qualitative behavior is different for different values of LTE session volumes. Logically, for rather small number of LTE users (up to approximately 400 for $d = 8$) the packet drop probability for $3e5$ NB-IoT sensors is higher than that for $1e5$ sensors. However, as the number of LTE sessions increases further, higher losses are experienced in the system with $3e5$ NB-IoT sensors. The reasons is the complex interplay induced by the specifics of resource scheduling, where the whole RB is allocated to upon arrival of a new Nb-IoT session. The same behavior is observed for $d = 4$.

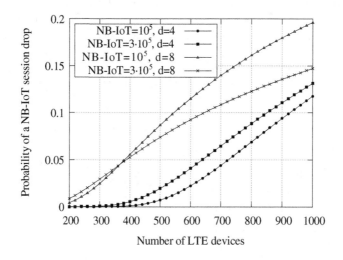

Fig. 3. NB-IoT drop probability as a function of LTE users, $R_H = R_N = 0$.

Analyzing the data presented in Fig. 3 further, we note the drastic difference between NB-IoT session drop probabilities for the same number of NB-IoT sensors but different session volumes requested by LTE users. Indeed, the curves corresponding to $d = 4$ are always below than those corresponding to $d = 8$. The difference could approach two orders of magnitude. Notice that prolonging the curves in Fig. 3 increasing the number of LTE users even further, the curve corresponding for $d = 8$ and $3e5$ NB-IoT sensors eventually crosses the one corresponding to $d = 8$ and $3e5$ NB-IoT sensors. However, this interesting effect

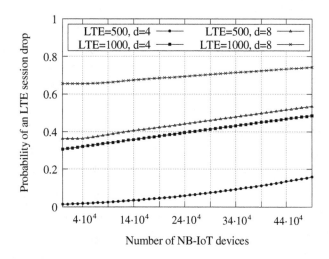

Fig. 4. LTE drop probability as a function of NB-IoT sensors, $R_H = R_N = 0$.

is of no practical importance as the Nb-IoT sensor drop probability is rather high. Finally, we note that for any $d > 1$ NB-IoT session drop probability never approaches 1.

Consider now the effect of the number of NB-IoT sensors on the LTE session performance for different values of LTE session volumes d, shown in Fig. 4. Analyzing the presented data, one may note rather smooth behavior of drop probability values. The reason is that extremely small fraction of the total offered traffic load generated by an individual NB-IoT sensor. Furthermore, notice the difference between curves corresponding to the same offered traffic load, 1000 LTE users with $d = 4$ and 500 LTE users and $d = 8$. This implies, that under the same offered traffic load from LTE better performance is provided when LTE session volumes as smaller.

4.2 Dynamic Resource Sharing with Reservations

The fully dynamic resource sharing scheme is characterized by excellent resource utilization. However, at the same time, it does not allow to provide any kind of performance guarantees in terms of NB-IoT and/or LTE session drop probabilities. Although for any LTE session volume, d, greater than one the NB-IoT session drop probability never approaches 1, this only happens in impractical regime of the system. We now proceed studying the case when some resources are exclusively allocated to NB-IoT and LTE traffic types while the rest are dynamically shared between them.

Figure 5 illustrates the resource utilization for dynamics resource sharing scheme with reservation, for different values of γ, where γ determined the amount of resources exclusively allocated to LTE and measured in percentage. For comparison purposes resource utilization for fully dynamic case is also illustrated

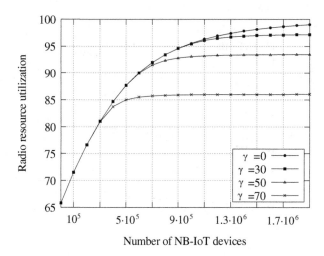

Fig. 5. Resource utilization of dynamic resource sharing scheme with reservations.

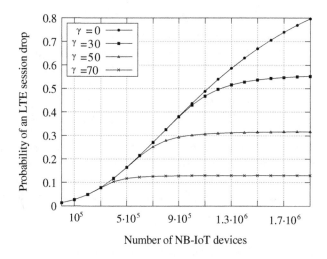

Fig. 6. LTE drop probability as a function of NB-IoT users.

and marked with $\gamma = 0$. As one may observe, the higher the value of γ the lower the resource utilization with fully dynamic case providing the upper bound. As expected, this is the price to pay when introducing a degree of isolation between traffic types. However, observe that when $\gamma = 30\%$ the curve closely follow the one corresponding to fully dynamic case implying that the careful choice of γ may still provide good resource utilization.

Consider now the gain provided by the dynamic sharing with reservation in terms of LTE session drop probability shown in Fig. 6. As one may observe, increasing the number of Nb-IoT sensors the fully dynamic scheme again provides the upper bound in the LTE session drop probability. Increasing the amount of resources exclusively reserved for LTE we drastically decrease the LTE session drop probability starting from some number of the number of NB-IoT sensors. Once again, the gain could be as high as two order of magnitude. What is more important is that exclusive assignment of resources to LTE users allows to keep the LTE session drop probability at the constant level inducing isolation between traffic types.

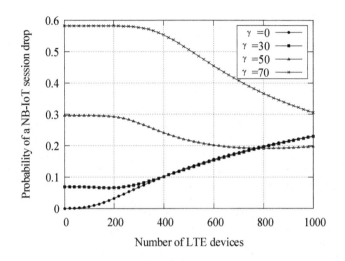

Fig. 7. NB-IoT drop probability as a function of LTE users.

Finally, let us analyze the behavior of the NB-IoT session drop probability as a function of the number of LTE users for $1e6$ number pf NB-IoT users, shown in Fig. 7. As one may observe, the NB-IoT session drop probability is heavily affected by the amount of resources exclusively allocated to LTE. When the number of LTE users is low the considered probability stays at the level dictated by the γ. Then, as the number of LTE users increases, LTE traffic start to affect the NB-IoT session drop probability. Eventually, for large number of LTE users the drop probability converges to a constant value again dictated by the current exclusive allocation of resources to NB-IoT traffic.

5   Conclusions

In this paper, we have proposed a mathematical model for in-band operational mode of NB-IoT technology, where the LTE resources are shared between NB-IoT and LTE services. To ensure the minimum performance guarantees and enable a degree of isolation between services we have assumed a certain minimum allocation exclusively provided to NB-IoT and LTE services. The rest of the resources are assumed to be dynamically shared between two session types. Using the proposed model we have derived session drop probability for both services, the average system throughput, and the amount of resources occupied by session types.

Our numerical results demonstrate that tuning the amount of reserved resources one may provide the required performance guarantees for both types of traffic. The proposed model can be used by vendors specializing in the provision of telecommunications services to determine the optimal resource sharing strategy for in-band operational mode of NB-IoT.

Acknowledgments. The publication was financially supported by the Ministry of Education and Science of the Russian Federation (the Agreement number 02.a03.21.0008) and by RFBR (research projects No. 16-07-00766, 17-07-00142).

References

1. Petrov, V., Mikhaylov, K., Moltchanov, D., Andreev, S., Fodor, G., Torsner, J., Yanikomeroglu, H., Juntti, M., Koucheryavy, Y.: When IoT keeps people in the loop: a path towards a new global utility. arXiv preprint arXiv:1703.00541 (2017)
2. TR.45.820: Cellular system support for ultra-low complexity and low throughput internet of things (CIoT). Technical report, Release 13, 3GPP (2016)
3. Petrov, V., Samuylov, A., Begishev, V., Moltchanov, D., Andreev, S., Samouylov, K., Koucheryavy, Y.: Vehicle-based relay assistance for opportunistic crowdsensing over narrowband IoT (NB-IoT). IEEE Internet Things J. (2017)
4. 3GPP: Cellular system support for ultra-low complexity and low throughput Internet of Things (CIoT). 3GPP TR 45.820/r13, December 2015
5. 3GPP: Standardization of NB-IoT completed, June 2016. http://www.3gpp.org/news-events/3gpp-news/1785-nb_iot_complete
6. Cardone, G., Corradi, A., Foschini, L., Ianniello, R.: ParticipAct: a large-scale crowdsensing platform. IEEE Trans. Emerg. Top. Comput. **4**, 21–32 (2015)
7. Margelis, G., Piechocki, R., Kaleshi, D., Thomas, P.: Low throughput networks for the IoT: lessons learned from industrial implementations, In: 2015 IEEE 2nd World Forum on Internet of Things (WF-IoT), pp. 181–186. IEEE (2015)
8. Borodakiy, V.Y., Buturlin, I.A., Gudkova, I.A., Samouylov, K.E.: Modelling and analysing a dynamic resource allocation scheme for M2M traffic in LTE networks. In: Balandin, S., Andreev, S., Koucheryavy, Y. (eds.) NEW2AN/ruSMART 2013. LNCS, vol. 8121, pp. 420–426. Springer, Heidelberg (2013). doi:10.1007/978-3-642-40316-3_37

Nonparametric Analysis of Extremes on Web Graphs: PageRank Versus Max-Linear Model

Natalia M. Markovich[1], Maxim Ryzhov[1], and Udo R. Krieger[2(✉)]

[1] V.A. Trapeznikov Institute of Control Sciences,
Russian Academy of Sciences, Profsoyuznaya Str. 65, 117997 Moscow, Russia
markovic@ipu.rssi.ru
[2] Fakultät WIAI, Otto-Friedrich-Universität,
An der Weberei 5, 96047 Bamberg, Germany
udo.krieger@ieee.org

Abstract. We analyze the cluster structure in large networks by means of clusters of exceedances regarding the influence characteristics of nodes. As the latter characteristics we use PageRank and the Max-Linear model and compare their distributions and dependence structure. Due to the heaviness of tail and dependence of PageRank and Max-Linear model observations, the influence indices appear by clusters or conglomerates of nodes grouped around influential nodes. The mean size of such clusters is determined by a so called extremal index. It is related to the tail index that indicates the heaviness of the distribution tail. We consider graphs of Web pages and partition them into clusters of nodes by their influence.

Keywords: Web graph · PageRank · Max-Linear model · Extremal index · Tail index

1 Introduction

The evaluation of the influence of nodes in a Web graph $G = (V, E)$ is an important problem of Web identification. PageRank (PR) and in-degree are the most popular indices of such influence. By Google's definition [2] PR is the rank of a Web page p_i. It is determined by

$$R(p_i) = c \sum_{p_j \in N(p_i)} \frac{R(p_j)}{D_j} + (1 - c)\, q_i, \qquad i = 1, ..., n, \tag{1}$$

where $N(p_i)$ is the set of pages that link to p_i (in-degree), D_j is the number of outgoing links of page p_j (out-degree), and $c \in (0, 1)$ is a damping factor. $q = (q_1, q_2, ..., q_n)$ is a personalization probability vector such that $q_i \geq 0$ and $\sum_{i=1}^{n} q_i = 1$ holds, e.g. a uniform distribution $q_i = 1/n$, and n is the total number of pages p_i or corresponding nodes $i \in V$ of the Web graph G. The definition is simplified omitting the term relating to dangling nodes.

© Springer International Publishing AG 2017
V.M. Vishnevskiy et al. (Eds.): DCCN 2017, CCIS 700, pp. 13–26, 2017.
DOI: 10.1007/978-3-319-66836-9_2

On the other hand, PR of a random page $i \in V$ can be considered as a weighted branching process

$$R_i = \sum_{j=1}^{N_i} A_j \, R_i^{(j)} + Q_i, \quad i = 1, ..., n, \tag{2}$$

denoting $R_i = R(p_i)$, $A_j =^d c/D_j$, $Q_i = (1 - c) \, q_i$, [8,17]. Here, $\{R_i^{(j)}, j = 1, ..., N_i\}$ denotes the ranks of N_i nodes j with links outgoing to the node i. '$=^d$' denotes the equality in distribution. Moreover, PR was considered in [14] as an autoregressive process with random coefficients $\{A_j\}$ and a random depth N_i of dependence.

As an alternative to PR we use a Max-Linear Model (MLM), [5]. The MLM may be determined by the substitution of sums in (2) by maxima

$$R_i = \bigvee_{j=1}^{N_i} A_j R_i^{(j)} \vee Q_i, \quad i = 1, ..., n.$$

Such a model is practically useful when a largest rank of the most influential follower of a node is only available. In this case, (2) is not applicable.

Our first objective is to compare PR and the MLM by the tail and extremal indices. The tail index shows the heaviness of the distribution tail of the rank variable R_i. The reciprocal of the extremal index approximates the mean cluster size of ranks. We determine the *cluster* around a node of interest as a conglomerate of nodes connected to this node such that at least one node in the conglomerate has a rank that exceeds a sufficiently high threshold u.

Our second objective is a clustering of networks by evaluating the extremal indices of nodes. A node $i \in V = \{1, ..., n\}$ is considered as a root of a branching tree and its extremal index θ_i is estimated by samples of ranks of its followers. Since θ_i is the dependence measure around that node, the visualization of clusters of the network may be done by circles with diameter $1/\theta_i$ around each node. In [4,10] the clustering of nodes or the associated graph partition is proposed in terms of disconnected or weakly connected communities of nodes using samples of node indices. In this paper we develop a corresponding stochastic approach.

Usually, in- and out-degrees of nodes, i.e. the number of incoming and outgoing links of a node, are measured. They can be modelled by regularly varying distributions. The distribution function $F(x)$ is called regularly varying of tail index $\alpha > 0$ if $1 - F(x) \sim x^{-\alpha} \ell(x)$ as $x \to \infty$, where $\ell(x)$ is a slowly varying function, i.e. $\lim_{x \to \infty} \ell(tx)/\ell(x) = 1$, $\forall t > 0$, holds. In real-world networks $\alpha \in (1, 3)$ is observed, [3].

A term of the PR process that dominates its tail (i.e., one that has a smallest tail index α_{min}) may determine the cluster structure of the network controlled by the extremal index θ. We compare nonparametric estimates of α_{min} and θ for PR and the MLM by a study of a real network.

The paper is organized as follows. In Sect. 2 a theoretical basis of our study is given. We propose an adaptation of the blocks estimator of the extremal index to a Web graph modelled by Thorny Branching Trees. In Sect. 3 we then compare the tail and extremal indices of PR and MLM by a study of a Web graph sample. Finally, we present some conclusions on our new nonparametric analysis approach.

2 Theoretical Foundation of Web Graph Modeling

We consider a Web graph $G = (V, E)$ with a sample $\{R_n, n \geq 1\}$ of a random rank variable, [8,13–15,17].

Definition 1 *([9], p. 53). The stationary sequence $\{R_n, n \geq 1\}$ is said to have extremal index $\theta \in [0, 1]$ if for each $0 < \tau < \infty$ there is a sequence of real numbers $u_n = u_n(\tau)$ such that it holds*

$$\lim_{n \to \infty} n(1 - F(u_n)) = \tau, \quad \lim_{n \to \infty} P\{M_n \leq u_n\} = e^{-\tau \theta}, \tag{3}$$

where $M_n = \max\{R_1, ..., R_n\} = \bigvee_{j=1}^{n} R_j$ is used.

For independent r.v.s R_j $\theta = 1$ holds, but the converse is not true. $\theta \approx 0$ implies a strong dependence. As $1/\theta$ approximates the mean cluster size, $\theta = 0$ implies that the maximum M_n likely does not exceed a sufficiently high threshold u.

The practical significance of θ is that it determines the distribution of a first hitting time. This is the minimal time required to reach a sufficiently important node with a high rank [13]. The extremal index evaluates the mean first hitting time to find a subset of nodes with highest ranks in a network. This result helps to compare sampling random walks that are used to gather information about nodes and ranking algorithms like PR and the MLM.

It is a problem to get analytical formulae for θ of a PR process when the distributions of its components and their dependence are unknown.

For a given personalization vector $q_i = 1/n, 1 \leq i \leq n = |V|$, the scale-free PR $R_i^{(n)} = nR_i$ of a node i can be computed iteratively [17] by

$$\widehat{R}_i^{(n,0)} = 1, \quad \widehat{R}_i^{(n,k)} = \sum_{j \to i} \frac{c}{D_j} \widehat{R}_j^{(n,k-1)} + (1 - c), \quad k > 0, \tag{4}$$

until the difference between two consecutive iterations will be small enough. Here, $j \to i$ implies that node j links to node i, i.e. $(j, i) \in E$. To calculate the corresponding MLM values $\{X_i\}$ one can insert ranks obtained by (4) into

$$X_i = \bigvee_{j=1}^{N_i} \frac{c}{D_j} R_i^{(j)} \vee (1 - c), \quad i = 1, ..., n. \tag{5}$$

The stationary regularly varying distribution of PR R_i is derived in [8,17] under slightly different assumptions. Considering (2) and assuming that all r.v.s in the triple $(N_i, A_j R_i^{(j)}, Q_i)$ are mutually independent and that $\{N_i\}$, $\{A_j R_i^{(j)}\}$, $\{Q_i\}$ are sequences of iid regularly varying r.v.s, it is derived that the stationary distribution of R_i is regularly varying with $\alpha = \min\{\alpha_N, \alpha_{AR}, \alpha_Q\}$, i.e. with the minimal tail index among the tail indices of all components in the triple. The same is proved in [14] under more relaxed conditions, i.e. $\{A_j R_i^{(j)}\}$ are assumed to be iid regularly varying r.v.s. It is derived therein that PR and MLM have the same tail and extremal indices.

An open question is whether the same tail and extremal indices are preserved for PR and the MLM in case that the ranks of followers of a node are dependent due to possible links among those followers. We check it for Web graphs by a nonparametric estimation of the tail and extremal indices.

2.1 Tail Index Estimation

Let $\{R_n, n \geq 1\}$ be a stationary sequence of r.v.s. of node ranks. To estimate the tail index α of these ranks, we use Hill's estimator [7] and the SRCEN estimator [16]. Hill's estimator is determined by

$$\widehat{\alpha}(n, k) = \left(\frac{1}{k} \sum_{i=1}^{k} \ln R_{(n-i+1)} - \ln R_{(n-k)} \right)^{-1}, \tag{6}$$

where $k \in \mathbb{N}$, $1 \leq k < n$, is the number of largest order statistics of $\{R_n\}$. It is the most popular estimator and it may be applied for $\alpha > 0$ and iid data.

The SRCEN estimator may be applied for $0 < \alpha < 2$. It is determined by

$$\widehat{\alpha}(n, b) = 2[n/b^2] \ln(b) / \sum_{i=1}^{[n/b^2]} \xi_i(b) \tag{7}$$

where $\xi_i(b) = \ln \left(\sum_{j=(i-1)b^2+1}^{ib^2} R_j^2 \right) - 1/b \sum_{k=1}^{b} \ln \left(\sum_{j=(k-1)b^2+(k-1)b+1}^{(k-1)b^2+kb} R_j^2 \right)$, and $[\cdot]$ denotes the integer part. We chop the data $\{R_1, ..., R_n\}$ into non-overlapping blocks of size b^2, e.g., $b = [n^{1/3}]$.

By a simulation it was shown that Hill is better than SRCEN for many cases of iid series, whereas SRCEN overcomes Hill for dependent data, [16].

2.2 Extremal Index Estimation by the Blocks Estimator

Regarding a graph structure we will use the blocks estimator [1] as the most appropriate one. Then a cluster is defined as a block of data $\{R_i\}$, where at least one observation exceeds a threshold u. The estimator states as follows

$$\widehat{\theta} = \frac{n \sum_{j=1}^{k} 1 \left(M_{(j-1)r, jr} > u \right)}{rk \sum_{i=1}^{n} 1 \left(R_i > u \right)}, \tag{8}$$

where $M_{i,j} = \max\{R_{i+1}, ..., R_j\}$, k is the number of blocks, $r = [n/k]$ is the number observations in the block, and $1(\cdot)$ is the indicator of an event.

In [15] a modification of the blocks estimator is proposed for Web graphs, where generations of followers of a root node in the branching tree are considered

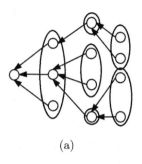
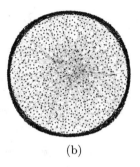

(a)   (b)

Fig. 1. Generations of followers of the root node in the branching tree used as blocks (1(a)), and graph of the Berkeley-Stanford dataset (1(b)).

as blocks (see Fig. 1(a)). The intuition behind (8) is that the blocks should not be overlapping. Typically, a node may be present in several generations due to loops, see Fig. 1(b). As the blocks of the generations are not equal-sized, one can use for a given threshold level u the ratio

$$\widehat{\theta}(u) = C(u)/N(u) \tag{9}$$

instead of (8), where $N(u)$ is the number of exceedances over u and $C(u)$ is the number of clusters. u is the most sensitive parameter of (9). It may be found visually corresponding to a stability interval of the plot $(u, \widehat{\theta}(u))$. For big data such as Web graphs we may select u by bootstrap methods, [12].

The same argument concerns the Hill' and SRCEN tools, where we have to select the number of the largest order statistics k and the block size b, respectively. For this purpose one may also use bootstrap methods, [12].

2.3   Bootstrap Method

In the following we briefly describe the bootstrap method to evaluate k in (6), [12].

Algorithm 1.

1. *Generate B re-samples $\{R_1^*, ..., R_{n_1}^*\}$ of size $n_1 < n$ with replacement from the original observations $\{R_i, 1 \leq i \leq n\}$, where n_1 is defined as*

$$n_1 = n^{\beta_b}, \qquad 0 < \beta_b < 1.$$

The number of the largest order statistics $k_1 \in \{1, ..., n_1 - 1\}$ corresponding to any re-sample relates to k and n by

$$k = k_1 \left(\frac{n}{n_1}\right)^{\alpha_b}, \qquad 0 < \alpha_b < 1. \tag{10}$$

2. *Estimate B values $\widehat{\alpha}_{BS}(n_1, k_1, b)$ of the tail index by each $b \in \{1, \ldots, B\}$ of these B re-samples.*
3. *Calculate the mean squared error (MSE) by these re-samples,*

$$MSE(n_1, k_1) = (bias(n_1, k_1))^2 + \widehat{var}(n_1, k_1), \qquad (11)$$

where the bias and the variance are determined by the following terms

$$bias(n_1, k_1) = \widehat{\alpha}_{BS}(n_1, k_1) - \widehat{\alpha}(n, k) = \frac{1}{B}\sum_{b=1}^{B}\widehat{\alpha}_{BS}(n_1, k_1, b) - \widehat{\alpha}(n, k),$$

$$\widehat{var}(n_1, k_1) = \frac{1}{B-1}\sum_{b=1}^{B}\left(\frac{1}{B}\sum_{b=1}^{B}\widehat{\alpha}_{BS}(n_1, k_1, b) - \widehat{\alpha}_{BS}(n_1, k_1, b)\right)^2,$$

for a tail index estimate $\widehat{\alpha}(n, k)$ in (6) and find a minimal $MSE(n_1, k_1)$ among different $k_1 \in \{1, ..., n_1 - 1\}$.
4. *Using the obtained k_1, find the optimal k by (10) and then the corresponding estimate $\widehat{\alpha}(n, k)$ by (6).*

Replacing k and k_1 in Algorithm 1 by b and b_1, respectively, one can estimate the parameter b in (7) in the same way as the parameter k. In [6] it is recommended to choose $\alpha_b = 2/3$ and $\beta_b = 1/2$ for Hill's estimator. This selection leads to a bootstrap estimate of the MSE that is asymptotically close to the real MSE. To our best knowledge, the optimal values of the bootstrap parameters α_b and β_b are not obtained yet regarding SRCEN. But in this case we shall use the same values, too.

The same bootstrap algorithm can be applied to estimate u in (9), where k and k_1 in (10) may be interpreted as the total numbers of exceedances in the sample and in the re-sample, respectively. Then one can find u corresponding to the selected k and determine the estimate of the extremal index $\widehat{\theta}(u)$. In this case the values α_b and β_b are not precisely known due to the lack of theory and we may take $\alpha_b = 2/3$ and $\beta_b = 1/2$ as well. It is a subject of our future research to derive these values α_b and β_b by theoretical arguments.

3   Comparison of PageRank and the Max-Linear Model

We study the Web graph of the Berkeley-Stanford dataset in which nodes represent Web pages and edges represent hyperlinks between those pages, [11]. The graph contains 685230 nodes and 7600595 edges, [10]. We calculate PR and the MLM of each node by (4) and (5) with $c = 0.85$. The scatter plot in Fig. 2(a) shows the presence of outliers and, hence, the heavy-tailed distributions of PR and MLM.

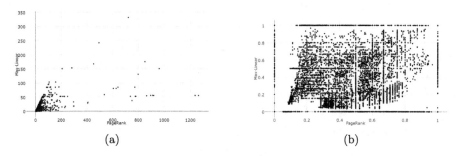

(a)   (b)

Fig. 2. Scatter plots of the MLM versus PR for 150000 nodes (2(a)) and extremal indices of PR versus the MLM for 685230 nodes (2(b)).

3.1 Tail Index Estimation

We estimate the tail index by PR and the MLM values that are obtained from the underlying datasets by the estimators (6) and (7) (see Fig. 3). Usually, the tail index value is taken according to a stability interval of the Hill's plot $(k, \widehat{\alpha}(n, k))$ regarding k. In the same way one can find the stability interval of the plot $(b, \widehat{\alpha}(n, b))$ of the SCREN estimator regarding b. Since the plots may have several stability intervals, we apply the bootstrap method with the number of bootstrap re-samples $B = 300$ and obtain the Hill's estimate equal to 1.081 and 1.052, and the SRCEN estimate equal to 1.3 and 1 for PR and MLM, respectively. Similar values can be obtained considering the first stability intervals from the left of the plots. Regarding the MLM, the values are closer for both estimators since the block-maxima used for the estimation in this case belong to the distribution tail in the same way as for the Hill' estimator that uses only the largest order statistics. As the tail index of PR and MLM are close to 1, this outcome implies that their distributions are likely regularly varying with infinite variance.

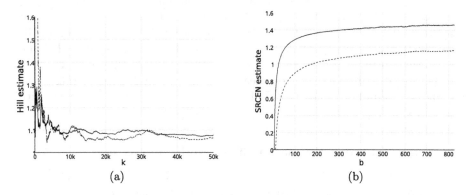

(a)   (b)

Fig. 3. Tail index estimation by Hill's estimator (3(a)), and the SRCEN estimator (3(b)): PR (solid line), MLM (dashed line).

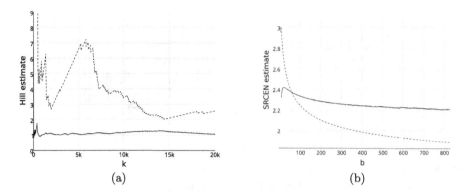

Fig. 4. Tail index estimation of the in- and out-degrees by Hill's estimator (4(a)), and the SRCEN estimator (4(b)): in-degree (solid line), out-degree (dashed line).

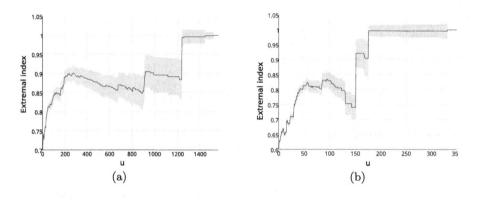

Fig. 5. Extremal index estimates by (9) for PR (5(a)) and the MLM (5(b)).

Moreover, we estimate the tail indices of the in- and out-degrees (see Fig. 4) and calculate their bootstrap values 1.026 and 2.730 for Hill's estimate and 2.2373 and 2.2650 for the SCREN estimate, respectively. The tail index of the in-degree is close to one which is a similarity regarding the tail index of PR and the MLM. This result implies that the distribution of the in-degree has a heavier tail than the distribution of the out-degree. Hence, the in-degree determines the heaviness of tail of PR and the MLM. This outcome is in the agreement with the results of [14, 17].

3.2   Extremal Index Estimation of All Nodes in a Graph

To estimate the extremal index θ of the whole dataset, (then $1/\theta$ implies the mean cluster size over the whole network,) we select first generations of followers of each node as blocks. To avoid the overlapping of blocks, we copy the same sample 300 times and select blocks in such a way that each node belongs to only one generation. In Fig. 5 the blocks estimates of PR and the MLM averaged over

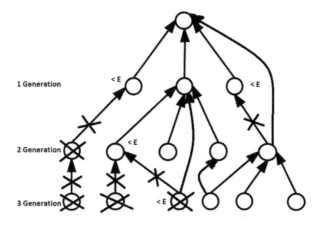

Fig. 6. Example of the truncation of the branching tree of a Web graph for a given ε.

300 samples are shown with the standard deviations. From the stability intervals we select an extremal index of about $\widehat{\theta} \in [0.85, 0.9]$ for PR and $\widehat{\theta} \in [0.8, 0.85]$ for MLM.

3.3 Extremal Index Estimation of an Individual Node

We estimate also the extremal index θ of each node. We consider a node as a root of the corresponding branching tree and generations of its followers as blocks. As the branching tree of a node can be very large, we propose the following truncation of the tree, see Fig. 6. Starting from the root, we take into consideration only a limited number of descendants using the following rule. If

$$\frac{c^{k-1}\widehat{R}_{j_k}}{\prod_{m=1}^{k-1} D_{j_m}\widehat{R}_i} < \varepsilon, \qquad 0 < \varepsilon < 1, \tag{12}$$

then the kth node will be included in the truncated graph but not its descendants. As a node may belong to different generations due to loops, some descendants may be preserved in the truncated graph as a member of the generations nearest to the root. The term on the left-hand side of (12) is arising by recursive replacements in (4) instead of $\widehat{R}_j^{(n,k-1)}$:

$$\widehat{R}_i = \sum_{j_1 \to i} \sum_{j_2 \to j_1} \cdots \sum_{j_k \to j_{k-1}} \frac{c}{D_{j_k}} \cdot \frac{c^{k-1}}{D_{j_{k-1}} \cdot \ldots \cdot D_{j_1}} \widehat{R}_{j_k}$$

$$+ (1-c) \sum_{j_2 \to j_1} \cdots \sum_{j_{k-1} \to j_{k-2}} \frac{c^{k-1}}{D_{j_{k-1}} \cdot \ldots \cdot D_{j_1}} + \ldots + (1-c) \sum_{j_1 \to i} \frac{c}{D_{j_1}} + (1-c).$$

Hereby, it is the intuition of this rule to exclude those nodes from the tree whose influence on PR of the root is weaker in the sense of (12). Then the extremal index is estimated by (9) using only generations of nodes of the truncated tree.

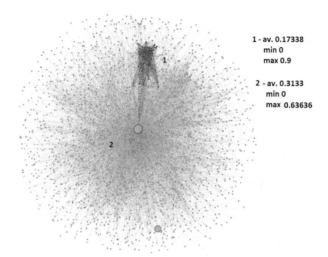

Fig. 7. Average, minimum and maximum of extremal indices of PR of nodes from two classes with MLM equal to 0.22 and 52.15 and colored by black - (1) and green - (2). (Color figure online)

To detect clusters we fix the values of u corresponding to the stability intervals of the plots in Fig. 5, i.e. $u_{PR} \approx 600$ and $u_{ML} \approx 75$ for PR and MLM, respectively. The scatter plot of the extremal indices of PR versus MLM is built for $\varepsilon = 0.01$. It shows diagonal trends which mean a similarity of the extremal indices of PR and MLM, see Fig. 2(b).

Figure 7 shows θ of PR for two classes with equal values of the MLM. Branching trees of depth equal to 7 associated with a node used to estimate θ may contain nodes lying outside these classes. The minimal index equal to zero is caused by the lack of exceedances over u w.r.t. PR of some nodes. The most valuable class with $MLM \approx 52.15$ has on average a mean cluster size approximately equal to $1/\theta = 1/0.313 \approx 3.195$, i.e. it includes at least 3 nodes with PR exceeding $u = 600$, and the class with $MLM \approx 0.22$ has a mean cluster size equal to 5.78.

In order to investigate the impact of ε we estimate the extremal index of PR of a triple of individual nodes for different values of ε by the blocks estimator (9) in Table 1. The threshold u corresponding to each estimate $\widehat{\theta}(u)$ is calculated by

Table 1. Blocks estimates of the extremal index of PR regarding three nodes in a Web graph with corresponding bootstrap estimates of u for different values of ε.

ε	"Black" node $PR = 6.48$			"Green" node $PR = 201.72$			"Grey" node $PR = 5031.31$		
	N	$\widehat{\theta}(u)$	u	N	$\widehat{\theta}$	u	N	$\widehat{\theta}$	u
0.01	20931	0.5	45	296146	0.86	1150	154449	0.77	1460
0.05	589	1	6.5	84372	0.65	1460	105386	0.68	1460

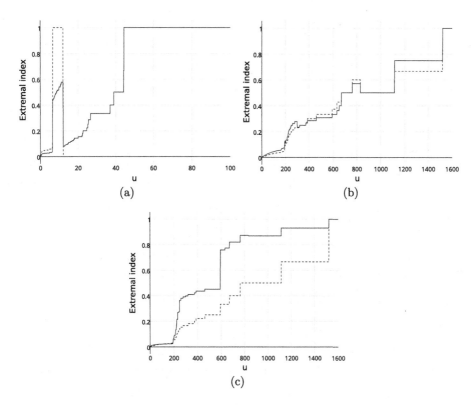

Fig. 8. Extremal index estimates by (9) for PR of the node from the black class (8(a)), of the node from the grey class (8(b)) and of the green node (8(c)) for $\varepsilon = 0.01$ (solid line) and $\varepsilon = 0.05$ (dashed line). (Color figure online)

the bootstrap algorithm described in Sect. 2.3. We select three nodes in Fig. 7: one is taken from the "black" class, one from the "green" class and one is a "grey" node located in the middle of the "green" class that has a large PR. The latter node does not belong to the "green" class but it has links to almost all nodes from the underlying network. The corresponding blocks estimates of the PRs of these nodes against the threshold u are shown in Fig. 8 for different values of ε. One may observe that the smaller the value of ε is the larger is the number N of the selected nodes in the truncated graph. This strongly impacts on the estimation of the extremal index. In order to calculate the blocks estimate well enough, we select a value u corresponding to the minimum of the bootstrap MSE (11), see Fig. 9. Then one can see in Fig. 9 the following tendency: the smaller ε corresponds to the larger optimal value of u. This outcome is achieved because the truncated branching tree contains a larger number of nodes in this case. Moreover, the "grey" node with the largest PR among all three nodes has the highest optimal u. One can select the following $u \approx 45, 1460, 1150$ from Fig. 9(a), (b) and (c) for $\varepsilon = 0.01$, respectively, for "black", "grey" and "green" nodes.

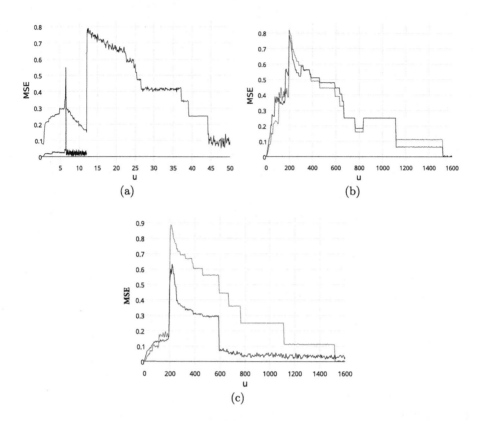

Fig. 9. The bootstrap estimate of the MSE against threshold u for three selected nodes and ε equal to 0.01 and 0.05, respectively: "black" node (9(a)), "grey" node (9(b)) and "green" node (9(c)) for $\varepsilon = 0.01$ (solid line) and $\varepsilon = 0.05$ (dashed line). (Color figure online)

This value u is the smallest threshold corresponding to the stability interval of smaller MSE. Hence, we obtain $\theta \approx 0.5, 0.86, 0.77$ for these nodes from Fig. 8 for $\varepsilon = 0.01$. Despite the PR of the "grey" node is the largest among all considered three nodes, its extremal index is the closest to one. This result implies that connections of this node are all arbitrary and independent. In other words, this node does not belong to a stable community with highly dependent links.

4   Conclusions

The paper is devoted to the stochastic analysis of a Web graph. Two characteristics of the node influence are considered, namely PageRank and a Max-Linear model. They are compared with regard to features of their underlying distributions and dependence structure. The latter dependence measure is represented by the extremal index of samples of the page rank variable.

Considering a Web graph, we propose a new clustering procedure of nodes by means of their extremal index values. Such clustering reflects the changes with regard to the extremal dependence structure of the Web graph. It may be an alternative to the clustering of nodes by the most distinct communities of nodes with a small number of edges between them. In our stochastic analysis approach we partition a real Web graph into clusters according to the extremal index values of the PageRank for equal MLM classes of nodes.

From the statistical point of view, the well-known nonparametric blocks estimator of the extremal index is modified in our paper with respect to random graphs. Considering the PageRank process corresponding to each node as an individual Thorny Branching Tree, we propose to utilize the new generations in such a tree as data blocks that are used by the blocks estimator. Due to loops in the graph such blocks may have common nodes. As the critical parameter of the blocks estimator is a threshold level, our next theoretical achievement is given by the proposal and the empirical study of a new bootstrap method to estimate this level. Due to the complexity of Web graphs several proposals to simplify the calculations of the extremal indices have been made. They include the truncation of the individual branching tree to calculate the extremal index of an individual node and replicating the same sample to select non-overlapping first generations as blocks to calculate the extremal index by the whole dataset of the node characteristics like PageRank or the Max-Linear model.

Our study of real Web graph data shows that PR and the MLM have similar tail and extremal indices. This result is in the agreement with our theoretical results [14]. It demonstrates the negligible impact of the dependence among generations of the branching trees associated with the nodes. The PR and MLM distributions are shown to be heavy tailed with an infinite variance.

Our future investigations will concern a theoretical study of the bootstrap procedure regarding the extremal index and a further study on the clustering of random graphs.

References

1. Beirlant, J., Goegebeur, Y., Teugels, J., Segers, J.: Statistics of Extremes: Theory and Applications. Wiley, Chichester (2004)
2. Brin, S., Page, L.: The anatomy of a large-scale hypertextual Web search engine. Comput. Netw. ISDN Syst. **30**(1), 107–117 (1998)
3. Chen, N., Litvak, N., Olvera-Cravioto, M.: PageRank in scale-free random graphs. In: Bonato, A., Graham, F.C., Prałat, P. (eds.) WAW 2014. LNCS, vol. 8882, pp. 120–131. Springer, Cham (2014). doi:10.1007/978-3-319-13123-8_10
4. Fortunato, S.: Community detection in graphs. Phys. Rep. **486**(3–5), 75–174 (2010)
5. Gissibl, N., Klüppelberg, C.: Max-Linear models on directed acyclic graphs. arXiv:1512.07522v1, pp. 1–33 (2015)
6. Hall, P.: Using the bootstrap to estimate mean squared error and select smoothing parameter in nonparametric problems. J. Multivar. Anal. **32**, 177–203 (1990)
7. Hill, B.M.: A simple general approach to inference about the tail of a distribution. Ann. Stat. **3**, 1163–1174 (1975)

8. Jelenkovic, P.R., Olvera-Cravioto, M.: Information ranking and power laws on trees. Adv. Appl. Probab. **42**(4), 1057–1093 (2010)
9. Leadbetter, M.R.: Probability theory and related fields. Zeitschrift für Wahrscheinlichkeitstheorie und Verwandte Gebiete **65**(2), 291–306 (1983)
10. Leskovec, J., Lang, K.J., Dasgupta, A., Mahoney, M.W.: Community structure in large networks: natural cluster sizes and the absence of large well-defined clusters. eprint arxiv:0810.1355 (2008)
11. Leskovec, J., Krevl, A.: SNAP Datasets: Stanford Large Network Dataset Collection (2014). http://snap.stanford.edu/data
12. Markovich, N.M.: Nonparametric Analysis of Univariate Heavy-Tailed Data. Wiley, Chichester (2007)
13. Markovich, N.M.: Clustering and hitting times of threshold exceedances and applications. Int. J. Data Anal. Tech. Strat. 1–18 (2017, to appear)
14. Markovich, N.M.: Extremes in random graphs models of complex networks. arXiv:1704.01302v1, 5 April 2017 (2017)
15. Markovich, N.M.: Analysis of clusters in network graphs for personalized web search. In: IFAC 2017 World Congress, Toulouse, France, 7–14 July 2017 (2017, to appear)
16. McElroy, T., Politis, D.N.: Moment-based tail index estimation. J. Statist. Plan. Infer. **137**(4), 1389–1406 (2007)
17. Volkovich, Y., Litvak, N.: On the exceedance point process for a stationary sequence. Adv. Appl. Probab. **42**(2), 577–604 (2010)

Prototyping Minimal Footprint NFC-Based User Access Control System for IoT Applications

Martin Stusek[1]([✉]), Jiri Pokorny[1], Krystof Zeman[1,2], Jaroslav Hajek[1], Pavel Masek[1], and Jiri Hosek[1]

[1] Department of Telecommunications,
Brno University of Technology, Brno, Czech Republic
`xstuse01@vutbr.cz`
[2] Peoples' Friendship University of Russia (RUDN University),
6 Miklukho-Maklaya St, Moscow 117198, Russian Federation

Abstract. User access control systems have become a standard part of security systems in many consumer as well as industrial applications. The majority of these systems utilize tokens to gain an access into restricted areas such as buildings, garages, and workplaces. They are available in different shapes and sizes, but their communication interface often utilizes NFC technology ensuring compatibility throughout a variety of distinct tokens. Main goal of this paper is to share most important hands-on experience acquired during the development of NFC-based user access control system with minimal deployment footprint. Common handheld/wearable devices like smartphones or smartwatches have been used to eliminate the need for another item to be carried by a user. In our Android-based implementation, users are authenticated via server application and their accounts can be managed by the user interface called Locker. Further, MySQL database acts as a storage for user data and for authentication purposes as well.

Keywords: Access control systems · Android · Authentication · IoT · NFC · Secure communication · Wireless communication

1 Introduction

The controlled access to restricted areas is all around us in modern everyday life. These areas cover personal life (home buildings) as well as professional life (workplaces, garages and others). Access cards, key tags or other items are used to unlock these locations. Mentioned items are designated as tokens, stationary devices that read these tokens are called card readers or terminals. Authentication of a token is usually performed via wireless connection. When a person needs to access certain area, at least one token is required. The problem arises when a person needs to access multiple areas with different tokens. Carrying multiple access items might not be a suitable due to several reasons i.e., security or comfort. RFID (Radio Frequency Identification) or lately NFC (Near

V.M. Vishnevskiy et al. (Eds.): DCCN 2017, CCIS 700, pp. 27–40, 2017.
DOI: 10.1007/978-3-319-66836-9_3

field communication) are the most common technologies for the communication between tokens and card readers. NFC technology has been growing more and more popular since it was developed in 2002 by Philips and Sony. It is a wireless technology working in short ranges up to 10 cm and using frequency of 13.56 MHz [7].

Since everyone owns a smartphone nowadays, new possibilities emerge to address this issue. Smartphones themselves contain various types of wireless communication interfaces such as BT (Bluetooth) or NFC which can be used to communicate with a card reader [13]. This would eliminate the necessity for any token and only the smartphone would be required. Of course, as with any smartphone application, if the smartphone owner loses the phone or someone steals it from him, theoretically, they could access the restricted areas of the smartphone owner. However, this security risk exists also with tokens. Smartphones have an advantage of other security protections such as PIN screen lock or a fingerprint reader. On the other hand, the passive tokens cannot be improved by adding such extra level of security.

This issue has been discussed in some recent research works and several solutions have been developed [13,14]. One of them is based on the BT technology. There is a BT module connected into Arduino Uno board on the terminal site. Smartphone with Android OS is used as a token emulating device. The application establishes a connection and sends out a unique identification code (UIC) and information separated by a comma.

Another approach can be using NFC instead of BT. Up to Android version 4.4, an SE (Secure element) was required for card emulation function. From this version, it was possible to emulate cards with no SE, but with a drawback of lower security. Authors of [3] realized the security limitations and introduced an alternative called Trusted Host-based Card Emulation (THCE) – a secure way for storing data on Android enabled devices. Our solution brings new security approach based on the implementation of AccountManager class [8], that encrypts data (user name, user date of birth and emulated card serial number - UID) on Android, thus disallowing an attacker to acquire any important data.

We also took into consideration the benefits of NFC over BT. NFC communicates in very short distances and there are only two devices that can communicate at the same time. This makes clear, who is communicating with whom. BT, on the other hand, can be connected to a network with up to 8 devices, which opens a possible door for new security breaches. Also, the setup time of the BT communication protocol is about 6 times higher than the one used in NFC and it makes it less convenient for everyday use [7]. More on benefits of NFC can be found in Sect. 2.

This paper is organized as follows. Section 2 compares considered technologies for short-range communication. Further in Sect. 3 we describe the NFC and its crucial features. Our practical implementation consists of HW (Hardware) and SW (Software) prototypes as well as our measurement results are introduced in Sect. 4. Finally, in Sect. 5 we summarize our work.

2   Comparison of BT and NFC Key Attributes

In this section, we describe the differences between considered short-range communication technologies and key factors for our decision-making process.

NFC and BT have been developed as wireless technologies targeting different applications (e.g., BT for handsfree, NFC for electronic payments). Important factors for our selection of suitable technology for access control system were (i) security, (ii) communication protocol complexity, and (iii) power consumption. Security of a technology is very important, especially in our case, because a security hole could allow access to people with malicious intentions into restricted areas. However, the protocol complexity plays the key role as well, especially in access response, meaning the longer time the less convenient the application is. If the application is not energy efficient, it might also mean less convenience, mostly when battery-powered mobile devices are used, as in our case, the smartphone is acting as the "token-device". These and other key differences in the technologies are described in Table 1. Following bullet points rationalize our decision for selecting NFC over BT:

- **Man in the middle attack** – An important feature of the NFC and BT technologies is the communication range. The short range can be inconvenient when accessing the card reader, but it also enables higher security. BT communication can be intercepted and possibly modified since there might be several meters of free space which can be abused by the attacker [6,10]. NFC communicates with distances up to only 10 cm – this physically prevents the attacker to intercept the communication.
- **Protocol complexity and power demands** – Since passive NFC tokens do not require any internal power supply, it is easy to deduce the approximate power consumption [9]. As mentioned in Table 1, communication setup time for BT is more than 6 times higher than for NFC, so this suggests that NFC uses less complex communication protocol than BT, but not necessarily with lower security [7].

3   Theoretical Background of NFC Technology

NFC technology provides set of standards for short range wireless communication among electronic devices i.e., smartphones, smartwatches and credit cards, that were originally designed to contactless payment transactions [18]. Massive expansion of NFC technology has begun with Android 2.3 Gingerbread, which brought native support of NFC to the system environment [15].

Communication chain, in case of NFC, consists of two basic elements; (i) NFC reader and (ii) access token. Further, the tokens could be divided into two groups: **passive** and **active** elements. NFC elements equipped with built-in power source, so-called active tokens, allow continuous data transition independently of NFC reader. Passive tokens do not have its own battery but contain built-in capacitor accumulating energy acquired from NFC reader through the

Table 1. Comparison of BT and NFC key attributes regarding requirements for access control systems utilizing wireless communication interfaces [5, 7, 16].

Parameter	Bluetooth 4.2	NFC
Communication range	100 m	10–20 cm
Data Rate	0.8–2.1 Mbps	0.02–0.4 Mbps
Cost	20 $	1 $
Frequency	2.4–2.4835 GHz	13.56 MHz
Security	128 bit SAFER+, E0, AES-CCM	AES 128
Network Topology	Piconets	Point-to-point
Devices per Network	8	2
Power Consumption	10–300 mW	1–300 mW
Setup Time	6 s	Less than 1 s

token's antenna. The capacitor then serves as a power source supplying the rest of the circuit. This implicates the need for NFC reader to initialize the connection. Next limitation resulting from restricted power source is maximum communication distance, which is in order of centimeters (10–20 cm) [4].

Based on these facts, the NFC forum defines three communication modes:

- **Read and Write** – This mode is selected when passive element i.e., NFC token is attached to the NFC reader. During the communication, it is possible to read and write data to the token. Transmitted data is encapsulated to the NDEF (NFC Data Exchange Format) messages defined by NFC forum [2].
- **Peer-to-Peer** – In case of two active devices (both equipped with independent power source), selected mode allows two-way communication in half-duplex mode with transmission speed up to 424 kbps. Data messages are transmitted over LLCP (Logical Link Control Protocol) protocol [17], that equals to the data link layer from the perspective of OSI (Open Systems Interconnection) model. LLCP ensures management of connection i.e., activation, monitoring and termination of both connection-oriented and connectionless communications [2].
- **Card emulation** – This mode allows communication of two active devices, where one of them emulates passive NFC token. Communication is initialized from NFC reader side and attached device is in role of passive NFC element. From this perspective host card emulation mode enables utilization of smartphone equipped with NFC technology as standard access token, therefore the emulation mode is used in created prototype on side of mobile device. This mode is implemented in Android operating system (since version 2.3 Gingerbread) and enables emulation of various NFC token types. However, up to version 4.4 (Kitkat), the utilization of additional secure element located in SIM (Subscriber Identity Module) or SD (Secure Digital) card, that controls the whole communication, was required. From Android version 4.4 it is possible to emulate NFC tokens without need of the secure

element. In this case, communication is fully managed from assigned application, that directly cooperates with NFC interface. Comparison of these two card-emulation approaches is depicted in Fig. 1 [2].

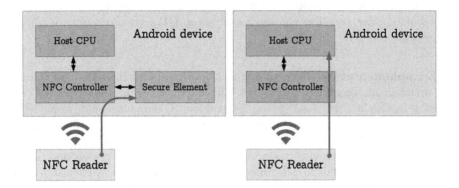

Fig. 1. Comparison of host card emulation methods.

3.1 Host Card Emulation in Android Environment

As it was mentioned in previous section, Android operating system supports several types of NFC tokens, which are predominantly used within contactless payments. Namely, Android supports emulation of cards based on ISO-DEP standard derived from ISO/IEC 14443-4 [12]. This norm defines four-layer communication stack that specifies parameters from the lowest layer i.e., physical characteristics, transmission power and communication frequency. Further, parameters of higher layers i.e., communication initialization, anti-collision loop and transmission protocol are defined. Tokens utilizing APDU messages based on ISO/IEC 7816-4 (Interindustry Commands for Interchange) are supported from the perspective of the application layer. However, backing of standard ISO/IEC 14443-4 type B, that differs in used modulation, is optional and fully in competence of device manufacturer [1].

The software part of NFC tokens emulation in Android environment is based on a component called Host Card Emulation Services, that is continuously running in background and does not require any user interaction. When the Android-powered device is attached to the NFC reader, an appropriate service is found and launched.

The Android HCE architecture allows emulation of several NFC tokens in one device, thus proper selection procedure of associated service is needed. The way of service selection is described in ISO/IEC 7816-4 standard that distinguishes each service according to the application ID (AID) containing up to 16-byte identifier. AID has to be unique identifier for each service launched on emulation device to prevent system from service collisions. The aforementioned standard defines several groups of identifiers that are based on the first four bits of AID. The most relevant groups are **A** for internationally and **D** for nationally registered services, respectively. For groups, it is also recommended to register new infrastructure in accordance to ISO/IEC 7816-5 standard that prevents infrastructure from AID collision. However, in case of created application, group **F** for unregistered application was used. But within this group, the AID collisions could appear, therefore thorough analysis of AID services running in mobile device is needed [11].

HCE Implementation. Created application comes with the host card emulation architecture based on Android 4.4 `HostApduService`, that declares two abstract methods that need to be overridden. Commands sent from NFC reader are processed in assigned HCE service, namely in `processCommandApdu()` method overriding the parent HostApduService class. Data between reader and mobile device is exchanged through the APDU messages that correspond to the application layer in OSI model. Since the communication on application layer is in half-duplex mode, the NFC reader always waits for response from mobile device.

Assigned service is selected according to the AID value that is part of the `SELECT AID` message sent from reader to mobile device. When the appropriate service is found, system aims all APDU messages to the selected service. This routing is active until next `SELECT AID` message is sent or until the connection is suspended.

In both of these cases, `onDeactivated()` method is called with an argument indicating which of the two happened.

4   Prototype Implementation and Measurements

This section contains description of developed prototype implementation of proposed access system that supports conventional NFC tokens and also HCE system.

The main benefit of created solution is its universality that allows utilization of classic approach with NFC cards together with modern emulated tokens. The system is designed to be lightweight, which allows its application on embedded devices with low performance. Other benefits come from utilization of NFC technology itself in comparison with BT. The communication range of NFC is in order of centimeters, therefore its power consumption is significantly lower and impact on battery life is negligible. Due to low range of NFC, it is not possible to eavesdrop the communication from more than a few centimeters. This also increases the security of this system in comparison with its competitors like BT.

From the perspective of communication chain, developed solution could be divided into (i) server side and (ii) client side as it is depicted in Fig. 2.

Server side provides users authentication methods and also provides graphical user interface for management of user accounts and locks. Client module contains NFC reader with wireless communication module that enables the device to communicate with authentication server.

Fig. 2. Architecture of prototype access system.

4.1   Client Side

The key element of client side is NFC reader based on chip PN532, that ensures reading of data from attached tokens. The most important parameter of the reader is wide support of NFC tokens formats, especially ISO/IEC 14443-4 standard with ISO/IEC 7816-4 data frames that could be emulated in mobile devices with Android [1].

Data obtained from NFC reader is processed and sent to the authentication server. Prime data processing on the client side is conducted on Arduino UNO microcontroller that communicates with NFC reader through I^2C interface. Utilization of universal I^2C bus allows connection of up to 127 devices. Therefore, access system could use several authentication methods i.e., fingerprints or pin code from keyboard in combination with NFC tokens that rapidly increase reliability of authentication.

Communication with the authentication server is provided by wireless module based on chip ESP 8266, which supports communication in 2.4 GHz frequency band fulfilling IEEE 802.11b/g/n standards. Data transmission between the authentication server and communication module is realized via TCP (Transmission Control Protocol) connection. Security and integrity of data transmission is guaranteed using TLS (Transport Layer Security) standard that encrypts communication and provides authenticity of data based on certificates.

Last element on the client side is the electromagnetic relay in role of controlled door lock. If the attached token is recognized and also has rights to enter into the restricted area, electromagnetic relay is switched.

4.2   Authentication Server

The centerpiece of created access system is the authentication server providing NFC tokens authentication and management of users accounts and locks. Further, database platform and graphical user interface that allows platform management are provided.

Database. Server provides MySQL relation database of all controlled locks, users accounts and its access tokens as depicted in Fig. 3. As shows entity-relation diagram it is possible to create groups of user accounts and locks which simplify the administration.

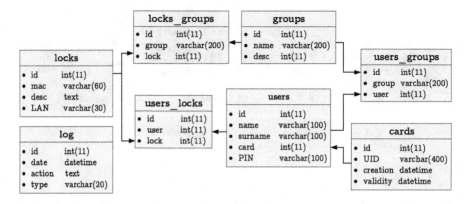

Fig. 3. Database entity-relationship diagram.

Database tables could be divided into four sections:

- **Users** – This table contains information about individual users i.e., name, surname and PIN (Personal Identification Number) that serves as user unique identifier. The last column of table carries information about a card assigned to the user. Association between these two tables is in ratio 1:1, thus each user could have just one card and the card has only one user.
- **Locks** – Information about all available access locks is contained in this table. As unique identifier mac column, that contains MAC (Media Access Control) address of wireless module deployed with lock, is used. For better user identification, the table contains `desc` column with text description of lock. Last table column contains IP address of lock, that can be used for unique identification in case of static address association.
- **Cards** – This table is tightly coupled with user records and table contains information about cards (tokens) assigned to the users. Each NFC token has a serial number (UID) serving as unique identifier that cannot be changed, therefore this value is used for card identification in created database. Last two columns of the table contain date of record creation and card expiration date.
- **Log** – Every user activity in access system i.e., attempt to access, enter to system and denial of access is recorded and the data is stored in a table called Log. Description of event is stored in column `action`, further, each record contains a column `type` that is used for event filtration in the administration interface.

User Interface. Management of user's accounts, cards and locks can be done through the graphical user interface called Locker; screenshot from homepage dashboard is depicted in Fig. 4. The interface is based on web technologies PHP (Hypertext Preprocessor), JavaScript and CSS (Cascading Style Sheets), thus is accessible as webpage in a web browser.

Besides basic management, the user interface allows creation of users and lock groups, coupled with database structure. This functionality may help in case of large areas with a high number of sections or users that can be easily divided into smaller groups.

Further, the user interface is used for management of devices' registration requests from Android application that is described in Sect. 4.2.

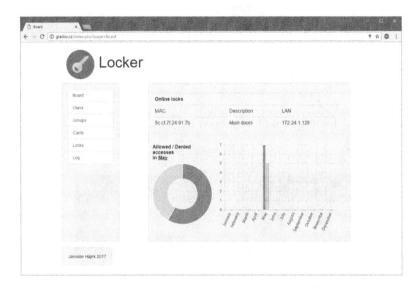

Fig. 4. Homepage dashboard of created server application.

User Authentication. A client side communicates with authentication server through the API (Application Programming Interface) of created application called Authentizer written in Java. Such approach with API provides higher level of security than direct access to the database. Therefore, created API limits the direct access to the data in database as much as possible.

As it is depicted in Fig. 5, server with Authentizer app listens on port 5631 – expecting TCP message containing keyword **verify**. The message has to contain an identifier (UID) of attached card and MAC address of wireless module assembled with the lock. At the server, the check if the card with received UID has rights to open the lock with MAC address, is performed through the received message analysis. In case of a match, server responds with message **open** and electromagnetic relay on the client side is opened. Otherwise, the response contains **end** and relay stays closed.

Android Application. Developed application requires operating system Android 4.4 or higher due to ability of card emulation without need of external

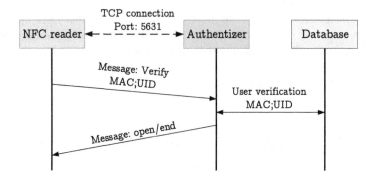

Fig. 5. NFC token authentication.

security element. Home screen of developed application with card cloning dialog is depicted in Fig. 6.

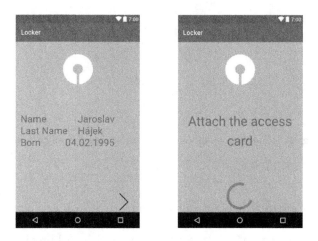

Fig. 6. Created Android application home screen.

Registration process starts when the NFC token is attached to the mobile phone and UID is read. Obtained card identifier together with user name and surname are sent to the authentication server, where existence of card with received UID is verified. In case of success, a response message to mobile device contains **yes** and user is requested to insert his date of birth. Server compares user's date of birth with PIN from database and creates registration notification. When system administrator accepts the request, then mobile device could be used as standard NFC token. Whole registration process is depicted in Fig. 7.

To ensure high level of security, user credentials and UID of emulated token are encrypted and stored using the **AccountManager**. This service provides

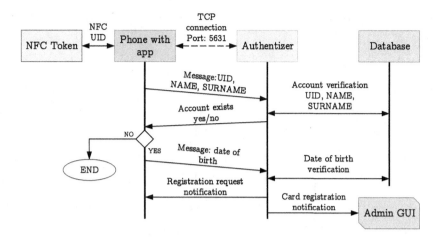

Fig. 7. Mobile device registration process.

encrypted storage for sensitive data i.e., password or access tokens that supports modern encryption standards AES-256 and 224–256 bits elliptic curves [8].

If the mobile device is already registered, the authentication process is the same as in case of normal NFC token, although the communication with NFC reader is managed by MyHostApduService that extends parent system class called HostApduService mentioned in Sect. 3.1. Before the connection is established, application verifies AID value in registered HCE services.

AID serves as unique identifier of registered services – in case of Locker its value is F0050607080910, where the letter **F** stands for unregistered AID value and the rest represents random series of number. When proper services are selected then APDU message, which contains command sent from NFC reader, is forwarded to the selected service. NFC reader sends the auth message and waits for response from mobile device. Response is created from combination of user name, surname and card UID that are hashed together. When the message is sent, user is informed of this event by notification. This indication serves a as simple protection against attackers, who want to stole data from mobile phone with intruder NFC reader.

4.3 Conducted Measurements

We conducted delay time measurements of our system – it was measured within the developed application running on Arduino Uno board. Starting point was the moment when a user attached a token or a smartphone, ending point the time when the command to open a lock was executed. These times were further subtracted and the resulting delay was obtained. Five sets of measurements were carried out with ten measurements on both NFC token and smartphone. The results are shown in box plot in Fig. 8.

Every measurement set is represented by the one group, where the bottom value is 5th percentile and the top is 95th percentile of all values. Whereas a

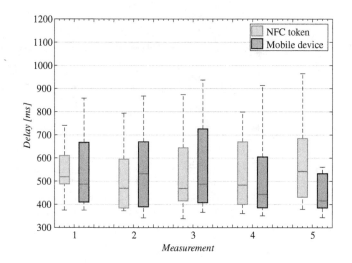

Fig. 8. Authentication delay measurements. (Color figure online)

coloured box represents 25th percentile on the bottom edge respective 75th per-centile on the top edge, the red line is the median. The difference between five sets of measurements was the time at which they took place. First measurement started at midnight with one hour pause between every other set. Measurement scenario with one hour period was selected to subdue differences in network load during the testing scenario.

We also measured the RTT (Round Trip Time) between client and server, the average time was 110 ms. This value tells us how much time of the total delay is the transmission time – the rest is the time consumed by authentication process at the server side.

5   Conclusion

In this paper, we discussed the possibility of the access token emulation on Android based systems using NFC wireless technology. We compared the capabilities of BT and NFC technologies and emphasized the benefits and drawbacks of both.

In the practical implementation, a prototype with Arduino Uno board and NFC reader was used on the client side, Raspberry PI served as the server side with an authentication algorithm and database running on it. GUI called Locker was implemented on the server for user account management. Finally, the Android application was developed to enable the access token emulation on a mobile device.

Overall, we designed a universal access system with NFC technology that works either with NFC cards or NFC emulated tokens. System is also very low power consuming, thus allowing to be utilized in embedded devices.

In our scenario, which was created to control door locks with attached NFC tokens, we measured authentication delays of NFC token and NFC emulating device. Maximum delay of emulated NFC was higher in four out of five sets of measurements. However, median values in both cases fluctuated in proximity of 500 ms. This proves that emulating an NFC token does not delay the communication significantly and it is possible to be used in our proposed configuration.

Currently, our system offers security on level of authentication server – the communication between client and server side is encrypted on session layer. The only insecure part is between client side and token. This could be further improved by implementing other levels of security we mentioned in the paper as the fingerprint reader, iris scanner or a keypad for pin number. Also, the computation power of the mobile phone can be utilized for data (command) encryption on application level. Asymmetric cryptography could be used for public key exchange and then the communication can be encrypted via AES cipher.

Acknowledgments. Research described in this paper was financed by the National Sustainability Program under grant LO1401. For the research, infrastructure of the SIX Center was used. This research was funded by the Technology Agency of Czech Republic project No. TF02000036. The publication was financially supported by the Ministry of Education and Science of the Russian Federation (the Agreement number 02.a03.21.0008).

References

1. Alattar, M., Achemlal, M.: Host-based card emulation: development, security, and ecosystem impact analysis. In: 2014 IEEE International Conference on High Performance Computing and Communications, 2014 IEEE 6th International Symposium on Cyberspace Safety and Security, 2014 IEEE 11th Intl Conf on Embedded Software and Systems (HPCC, CSS, ICESS), pp. 506–509, August 2014
2. Alliance, S.C.: Host card emulation (HCE) 101. A Smart Card Alliance Mobile and NFC Council White Paper (2014)
3. Armando, A., Merlo, A., Verderame, L.: Trusted host-based card emulation. In: 2015 International Conference on High Performance Computing and Simulation (HPCS), pp. 221–228. IEEE (2015)
4. Basyari, R.S., Nasution, S.M., Dirgantara, B.: Implementation of host card emulation mode over android smartphone as alternative ISO 14443A for Arduino NFC shield. In: 2015 International Conference on Control, Electronics, Renewable Energy and Communications (ICCEREC), pp. 160–165. IEEE (2015)
5. Coenen, M.: NFC in Android (2014). http://nfc-forum.org/wp-content/uploads/2014/03/Android-NFC-Forum-Developer-Spotlight-2.pdf
6. Cope, P., Campbell, J., Hayajneh, T.: An investigation of bluetooth security vulnerabilities. In: 2017 IEEE 7th Annual Computing and Communication Workshop and Conference (CCWC), pp. 1–7, January 2017
7. Coskun, V., Ozdenizci, B., Ok, K.: A survey on near field communication (NFC) technology. Wireless Pers. Commun. **71**(3), 2259–2294 (2013)
8. Developers, A.: Accountmanager (2017). https://developer.android.com/reference/android/accounts/AccountManager.html

9. Ensworth, J.F., Reynolds, M.S.: Ble-backscatter: ultralow-power IoT nodes compatible with bluetooth 4.0 low energy (BLE) smartphones and tablets. IEEE Trans. Microw. Theory Tech. (2017)
10. Haataja, K., Toivanen, P.: Two practical man-in-the-middle attacks on bluetooth secure simple pairing and countermeasures. IEEE Trans. Wireless Commun. **9**(1), 384–392 (2010)
11. ISO/IEC: Identification cards - integrated circuit cards - part 4: Organization, security and commands for interchange. Standard, International Organization for Standardization, Switzerland, August 2014
12. ISO/IEC: Identification cards - contactless integrated circuit cards - proximity cards - part 4: Transmission protocol. Standard, International Organization for Standardization, Switzerland, June 2016
13. Misra, S.: A very simple user access control technique through smart device authentication using bluetooth communication. In: 2014 International Conference on Electronics, Communication and Instrumentation (ICECI), pp. 1–4. IEEE (2014)
14. Morsalin, S., Islam, A.M.J., Rahat, G.R., Pidim, S.R.H., Rahman, A., Siddiqe, M.A.B.: Machine-to-machine communication based smart home security system by NFC, fingerprint, and PIR sensor with mobile android application. In: 2016 3rd International Conference on Electrical Engineering and Information Communication Technology (ICEEICT), pp. 1–6, September 2016
15. Prudanov, A., et al.: A trial of yoking-proof protocol in RFID-based smart-home environment. In: Vishnevskiy, V.M., Samouylov, K.E., Kozyrev, D.V. (eds.) DCCN 2016. CCIS, vol. 678, pp. 25–34. Springer, Cham (2016). doi:10.1007/978-3-319-51917-3_3
16. Scarfone, K., Padgette, J.: Guide to bluetooth security. NIST Spec. Publ. **800**(2008), 121 (2008)
17. Shobha, N.S.S., Aruna, K.S.P., Bhagyashree, M.D.P., Sarita, K.S.J.: NFC and NFC payments: a review. In: 2016 International Conference on ICT in Business Industry Government (ICTBIG), pp. 1–7, November 2016
18. Tabet, N.E., Ayu, M.A.: Analysing the security of NFC based payment systems. In: 2016 International Conference on Informatics and Computing (ICIC), pp. 169–174, October 2016

On-Line Traffic Management in New Generation Computer Networks

Helen Zaychenko and Yuriy Zaychenko$^{(\boxtimes)}$

Institute for Applied System Analysis, Igor Sikorsky Kiev Polytechnic Institute,
Peremogy Avenue, 37, Kiev 03057, Ukraine
zaychenkoyuri@ukr.net

Abstract. The problem of traffic management of different classes of service in NGN computer networks is considered. This problem is formulated as a rerouting problem of different classes flows under failures of channels and nodes while preserving the quality of service (QoS). The mathematical model of this problem is constructed and the algorithm of its solution is suggested. Experimental investigations of the suggested traffic management algorithm were carried out and its efficiency estimated.

Keywords: New generation networks · Traffic management · Flows rerouting · QoS

1 Introduction

Last years the problem of models and methods development for optimal traffic engineering(TE) in new generation networks (NGN) became very important due to wide practical implementation of these networks. On-line traffic control of different classes of service in NGN networks is executed in network routers so-called LSR (Label switching routers), in which additional functions TE are implemented [1,2]. LSR with functions of traffic engineering have two separate data bases: traditional Link-state Database (LSD) and special Traffic Engineering Database (TED) for traffic control. On-line traffic control is performed by rerouting of flows transmitted via virtual paths- so-called LSP (Label switching paths) [1–3] in case of congestion due to failures of communication equipment or channels. For practical implementation of on-line control the development of efficient rerouting algorithms is needed.

The goal of this paper is the development model and algorithm of optimal flows rerouting in NGN networks.

2 Problem Statement and Mathematical Model of Optimal Flows Rerouting in NGN

As it's well known an important peculiarity of NGN networks is the presence of different classes of service (CoS), each of which is served in routers (LSR)

© Springer International Publishing AG 2017
V.M. Vishnevskiy et al. (Eds.): DCCN 2017, CCIS 700, pp. 41–48, 2017.
DOI: 10.1007/978-3-319-66836-9_4

with corresponding priorities. Additionally, in NGN networks indices of Quality of Service (QoS) are introduced: Packets Transfer Delay (PTD), Packets Delay Variance (PDV) and Packets Loss Ratio.(PLR) [1–3]. Therefore the corresponding math model should take into account these properties of MPLS technology.

Let NGN network is given with structure $G = \{X, E\}, X = \{x_j\}j = \overline{1, n}$, where X is a set of network nodes (NN), $E = \{r, s\}$ is a set of channels, also are given capacities of channels μ_{rs}, $rs \in E$, so-called demand matrix $H(k) = \|H_{ij}(k)\|$, $i, j = \overline{1, n}$, where $h_{ij}(k)$ - is intensity of flow the k-th CoS, which is to be transmitted from node i to node j (Kbits/), flows distribution of all the classes $F(k) = [f_{rs}(k)]$, where $f_{rs}(k)$-is the intensity of flow of k-th class transmitted via channel (r, s), corresponding to matrix $H_{\sum}(k)$

Then the total value of flow for all the classes in the channel (r, s) will be equal to

$$f_{rs} = \sum_{k=1}^{K} f_{rs}(k) \tag{1}$$

In case if the service of different classes flows is executed with relative priorities ρ_k, decreasing with number of class, i.e. $\rho_1 > \rho_2 > \ldots > \rho_k$, $k = \overline{1, K}$, then the mean packets delay of k-th class is given by the following expression [5]

$$T_{av,k} = \frac{1}{H_{\sum}^{(k)}} \sum_{(r,s) \in E} \frac{f_{rs}^{(k)} \sum_{i=1}^{k} f_{rs}^{(i)}}{\left(\mu_{rs} - \sum_{i=1}^{k-1} f_{rs}^{(i)}\right) \cdot \left(\mu_{rs} - \sum_{i=1}^{k} f_{rs}^{(i)}\right)} \tag{2}$$

Assume that constraints on the value of PTD are introduced for all CoS in the form

$$T_{av,k} \leqslant T_{giv,k}, \tag{3}$$

where $T_{av,k}$ is a mean PTD of the k-th flow (s.), $T_{giv,k}$ is a constraint on this value.

Assume that virtual paths LSP $\prod_{ij}(k)$ for each connection (pair) (i, j), which are installed by protocol RSVP or SNMP are known [1,2].

Assume that channel (r_i, s_i) failed, denote such failure state z_i. It's demanded to reconfigure all the virtual connections of the failed channel (r_i, s_i) so that maximal satisfy the corresponding demands which were rejected (cancelled) while preserving the service of all other demands in full volume with given QoS - T_{av}.

Let call this problem the optimal *rerouting task in NGN network* under failures.

The mathematical model of this task takes the following form [5].

It's demanded to find such flows distribution $[f_{rs}(k)]$ which ensure maximal total flow value

$$H_{\sum} = \sum_{(i,j):(r_i,s_j) \in I_{ij}} h_{ij}^{(cor)} \to max \tag{4}$$

under constraints

$$T_{av}(F_{cor}^{(k)}) \leqslant T_{giv}, \quad k = \overline{1, K} \tag{5}$$

where $F_{cor}^{(k)}$ is the corrected flow of the k-th class (CoS) after rerouting.

3 The Algorithm of Optimal Rerouting in MPLS Network

The suggested method is based on the properties of maximal flow proved in [6].

Statement 1. *Maximal flow is the l flow by shortest paths in the conditional metric*

$$k_{rs} = \frac{\partial T_{av}}{\partial f_{rs}^{(k)}} \tag{6}$$

Statement 2. *Demand (i,j) dominates a demand (r,t) while transmission if*

$$l(\prod_{ij}^{min}) < l(\prod_{r,t}^{min}), \tag{7}$$

where $l(\prod_{ij}^{min}), l(\prod_{r,t}^{min})$ is the shortest path length for demands (i,j) and (r,t) in metrics (7) correspondingly. The suggested algorithm consist of two stages.

At the first stage all demands that used prior to failure the channel (r_i, s_j), are determined, timely rejected from service and the flows values are recalculated $F^{(k)} = \lfloor f_{rs}^{(k)} \rfloor, (r,s) \in E$.

At the second stage the capacities reserves are determined for all the channels and the flows of previously cancelled demands (connections) are optimally redistributed so that to achieve the maximal total flows value for cancelled demands $H_\Sigma \to max$.

1 Stage.

1. Find all demands (virtual connections) which used the failed (r_i, s_i). Denote them as $P_{r_i, s_i} = (i,j) : (r_i, s_i) \in \prod_{ij}$.
2. Timely cancel information transmission for rejected demands P_{r_i, s_i} and calculate new flows values - $F^{new}(k) = [f_{rs}^{new}(k)]$:

$$f_{rs}^{new} = \begin{cases} f_{rs} - \sum_{(i,j):(r_i,s_j)\in I_{ij}} h_{ij}^{(cor)} \to max, & if \quad (i,j) \in P_{r_i,s_i} \\ f_{rs}, & otherwise \end{cases} \tag{8}$$

In result obtain new flows distribution for all the classes $F^{new}(k) = [f_{rs}^{new}(k)]$, which include only demands which were not cancelled $(i,j) \setminus P_{r_i,s_i}$.
3. Determine the capacity reserves for all the channels

$$Q_{res} = \mu_{rs} - \sum_{k=1}^{K} f_{rs}^{new}(k). \tag{9}$$

Go to the second stage.

2 Stage. For the cancelled demands find the new routes (LSP) so that to ensure the following condition

$$\sum_{(i,j)\in P_{r_i,s_j}} h_{ij}^{(cor)} \to max \tag{10}$$

under constraints

$$T_{av}(F_{cor}^{(k)}) \leq T_{giv}, k = \overline{1, K} \qquad (11)$$

The second stage consists of k sub/stages at each of them execute rerouting and flow distribution of the k-th class. By this taking into account optimal flow property rerouting of flows for cancelled demands is performed in order of decreasing priorities until the capacities reserves will be exhausted and constant monitoring the fulfillment of constraints (11).

    1 sub/stage

    1 iteration

1. $k = 1$. At first consider the distribution of the cancelled demands of the first class.
2. Find the so-called conditional metrics $\frac{\partial T_{av}}{\partial f_{rs}^{(1)}} | f_{rs}^{new}(1)$
3. Find the shortest paths $\prod_{ij}^{min}(1)$ for all cancelled demands of the class $k = 1$.
4. Choose the demand $(i_1, j_1) \in P_{r_i,s_i}$ such that $l(\prod_{i_1,j_1}^{min}) = \min_{(i,j)} l(\prod_{i,j}^{min})$
5. Check the possibility of its transmission in full value by path $\prod_{i_1,j_1}^{min}$:

$$h_{i_1 j_1} < Q_{res}(\prod_{i_1,j_1}^{min}), \qquad (12)$$

where $Q_{res}(\prod_{i_1,j_1}^{min})$ is a reserve capacity of the path $\prod_{i_1,j_1}^{min}$:

$$Q_{res}(\prod_{i_1,j_1}^{min}) = \min_{(r,s)\in\prod_{i_1 j_1}} \{\mu_{rs} - f_{rs}\} \qquad (13)$$

If the condition (12) holds then distribute flow of the demand $h_{i_1 j_1}$ along the path $\prod_{i_1,j_1}^{min}$ and find the corrected flow distribution (FD)

$$f_{rs}^{cor}(1) = \begin{cases} f_{rs}^{new}(1) + h_{i_1,j_1} & \text{if } (i,j) \in P_{i_1,j_1}^{min} \\ f_{rs}^{new}(1), & \text{otherwise} \end{cases} \qquad (14)$$

Otherwise go to step 6.

6. Denote $h_{i_1,j_1}^{(a)} = Q_{res}(\prod_{i_1,j_1}^{min}) - \Delta$, where $h_{i_1,j_1}^{(a)}$ is a portion of demand value h_{i_1,j_1}, which may be transmitted over the path $\prod_{i_1,j_1}^{min}$, Δ is some threshold value.
7. Find the corrected flow distribution

$$f_{rs}^{cor}(k) = \begin{cases} f_{rs}^{new}(1) + h_{i_1,j_1}^{(a)} & \text{if } (r,s) \in P_{i_1,j_1}^{min} \\ f_{rs}^{new}(1), & \text{otherwise} \end{cases} \qquad (15)$$

Check the fulfillment of the constraint for $T_{av,1}$:

$$T_{av}(F_{cor}^{(1)}) \leq T_{giv,1} \qquad (16)$$

If the condition (16) is true, then $P_{r_i,s_i}^{new} = P_{r_i,s_i} \setminus (i_1, j_1)$ and go to step 8, otherwise go to step 9.

8. Check the condition $P_{r_i,s_i}^{new} \neq$ If it's true then go to next iteration and distribute the next demand of the first class $k = 1$. Otherwise go to step 9.
9. End of the sub/stage 1. Go to the sub/stage 2.

At this sub/stage redistribute cancelled demands of the class $k = 2$ so that not to violate the condition:

$$T_{av}(F_{cor}^{(2)}) \leq T_{giv,2} \qquad (17)$$

k sub/stage $(k > 1)$

At this sub/stage find reconfigured paths for cancelled demands of the k-th class.

Let $F^0(k) = [f_{rs}^0(k)]$-be a vector of multi-commodity flow of the k-th class and assume the following condition $T_{av}(F(k)|F(1), \ldots, F(k-1)) < T_{giv,k}$ holds.

1 iteration

1. Determine new conditional metrics $l_{rs}(k) = \frac{\partial T_{av}}{\partial f_{rs}(k)}|f_{rs}^0(k)$.
2. Find the shortest paths in metrics $l_{rs}(k)$ for all cancelled demands of the k-th class - $\prod_{ij}^{min}(k)$.
3. Search such demand $(i_k, j_k) \in P_{r_i,s_i}^{(k)}$ for which $l(\prod_{i_k,j_k}^{min}(k)) = \min\limits_{(i,j)\in P_{r_i s_i}^{(k)}} l(\prod_{i,j}^{min}(k))$
4. Determine the capacity reserve of the path $\prod_{i_k,j_k}^{min}$:

$$Q_{res}(\prod_{i_k,j_k 1}^{min}(k)) = \min\limits_{(r,s)\in\prod_{i_k,j_k}^{min}} ((\mu_{rs} - f_{rs}) - \varepsilon), \qquad (18)$$

where

$$f_{rs} = \sum_{i=1}^{k} f_{rs}^{(i)}. \qquad (19)$$

5. Check the condition:

$$h_{i_k j_k} < Q_{res}(\prod_{i_k,j_k}^{min}(k)). \qquad (20)$$

If it holds, then go to the step 6, otherwise to step 7.
6. Distribute the flow of the demand $h_{i_k j_k}$ in full volume $\prod_{i_k,j_k}^{min}$ and find new flows distribution:

$$f_{rs}^{new} = \begin{cases} f_{rs} - \sum\limits_{(i,j):(r_i,s_j)\in I_{ij}} h_{ij}^{(cor)} \to max, & \text{if } (i,j) \in P_{r_i,s_i} \\ f_{rs}, & \text{otherwise} \end{cases} \qquad (21)$$

Go to step 9.
7. Calculate $h_{i_k,j_k}^{(a)} = Q_{res}(\prod_{i_k,j_k}^{min}) - \Delta$.

8. Distribute the flow of the demand (i_k, j_k) by value $h^{(a)}_{i_k,j_k}$ over the path $\prod^{min}_{i_k,j_k}$ and find new flow distribution.

$$f^{new}_{rs} = \left\{ f_{rs} - f_{rs}(k) + h^{(a)}_{i_k,j_k}, \quad \text{if} \quad (r,s) \in \prod^{min}_{i_k,j_k} f_{rs}(k), \quad \text{otherwise} \right. \tag{22}$$

9. Check condition:

$$T_{av}(F^i(k)|F(1),\ldots,F(k-1)) < T_{giv,k} \tag{23}$$

If yes, then go to the step 10, otherwise decrease the value of flow to be transmitted of the demand (i_k, j_k) to such value $h^*_{i_k,j_k}$ that ensure the condition $T_{av}(F^i(k)|F(1),\ldots,F(k-1)) = T_{giv,k}$ and end of the sub/stage k.

10. Determine $P^{(k)}_{r_i,s_i} = P^k_{r_i,s_i} \setminus (i_k, j_k)$. Check the condition: $P^k_{r_i,s_i} =?$ If yes, then end of the sub/stage k, otherwise go to step1of the next iteration.

The sequence of iterations repeat until one of the next conditions holds:
(a) $T_{av}(F(k)|F(1),\ldots,F(k-1)) = T_{giv,k}$;
(b) $T_{av}(F(k+s)|F(1),\ldots,F(k-1)) = T_{giv,k+s}, a \leq s \leq K - k$;
(c) $P_{r_i,s_i} =$ and $T_{av}(F(k)|F(1),\ldots,F(k-1)) \leq T_{giv,k}$

In case (a) go to the next sub/ stage and find the flows distribution of the $k+1$ Class.

In case (b) when the condition for the less priority demand holds strictly (reached the bound) and $k + s < K$, then go to flow distribution (FD) for demands of the $k + s + 1$, as the flow distribution $F(k + s + 1)$ of the class $k+s+1$ doesn't influence the mean delay for higher priority flows $r < k+s+1$.

In case (c) go to the next stage $k + 1$ as in the case a).

The sequence of sub/stages ends when several conditions will hold strictly (reaches the bound) including the flow of the least priority, that is:

$$\exists k, T_{av}(F^{(k)}_{cor}) \geq T_{giv,k} \tag{24}$$

$$\exists k, T_{av}(F^{(K)}_{cor}) \geq T_{giv,K} \tag{25}$$

This means that free capacity resources are exhausted.

In result of the algorithm run the reconfigured paths for connections (i, j) are determined which were cancelled due to failure of channel or LSR. Naturally some of the less priority demands will be cancelled but the value of total flow transmitted in NG network after rerouting will be maximal.

Sub/stage 2 is similar to sub/stage 1. The cancelled demands CoS 2 redistribute until they will be distributed or the condition (8) will be violated, Then the end of sub/ stage 2 and go to sub/stage 3.

4   Experimental Investigations

For the estimation of the suggested algorithm of on-line traffic control and flows rerouting in NG network the experimental investigations were carried out. All

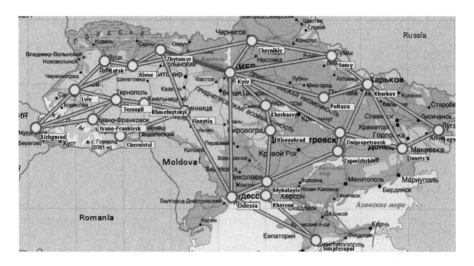

Fig. 1. Network topology

the experiments were performed at the global MPLS network whose structure is presented at the Fig. 1. The network consists of 25 nodes and 39 channels. 3 classes of flows are transmitted. For each class demand matrix H (k) was given.

The following constraints on the mean packets transfer delay were introduced for different classes $T_{giv,1}=0.1$s., $T_{giv,2}=0.5$s., $T_{giv,3}=1$s.

By these constraints the problem of optimal capacities choice and flows distribution was solved (CC FD) [5,6], in result capacities of all channels were determined and optimal flows distribution for all classes were found $F^{(k)} = \lfloor f_{rs}^{(k)} \rfloor$ as well as the total flow value for each class $H_{\sum}(k) = \sum\limits_{(i,j):(r_i,s_j)\in I_{ij}} h_{ij}(k), k = \overline{1,3}$.

Further in experiments different failure states were simulated: failure of one channel, two channels and three channels. In these states new flows distribution and actual flow value of the class k were determined $H_{\sum}^{\Phi}$.

Further the optimal rerouting using suggested algorithm was performed and new flows distribution $F_{cor}^{(k)}$ and total value of corrected flows for each class of service $H_{\sum,cor}(k)$ were found under constraints $T_{av}(F_{av}^{(k)}) \leq T_{giv}$.

Average total value of the transmitted flows before and after rerouting $H_{\sum}^{\Phi}$, $H_{\sum,cor}(k)$ correspondingly in % of nominal flows value in faultless state $H_{\sum}^{(0)}, H_{\sum}^{(1)}, \ldots, H_{\sum}^{(K)}$ are presented in the table 1 for different failure states.

The analysis of presented results shows the application of the suggested rerouting algorithm of on-line traffic control enables to increase the total value of transmitted flows in NGN network in case of failures. Note that such improvement refers to the high priority flows (k = 1, 2), that is natural as they are rerouted first of all and utilize the reserve capacity.

Table 1. Dependence total transmitted flows value for different failure states in % of $H_{\Sigma}^{(0)}(K)$

Type of failure states	Class of service, k=1		Class of service, k=2		Class of service, k=3	
	$H_{\Sigma}^{\Phi}(k)$	$H_{\Sigma,cor}(k)$	$H_{\Sigma}^{\Phi}(k)$	$H_{\Sigma,cor}(k)$	$H_{\Sigma}^{\Phi}(k)$	$H_{\Sigma,cor}(k)$
Failure of 1 channel	94.7	97.2	93	94.5	90	90.5
Failure of 2 channel	90.5	94.3	88	89.5	85	85.5
Failure of 3 channel	84.3	88.2	82	83	74	74

5   Conclusions

1. The problem of flows rerouting of different classes of service after failures in NGN networks is considered.
2. The mathematical model of this problem was constructed and flows rerouting algorithm based on the properties of maximal flow was suggested.
3. The experimental investigations were carried out and the efficiency of the suggested rerouting algorithm was estimated.

References

1. Nadeau, T.D.: MPLS Network Management: MIBs, Tools and Techniques. Morgan Kaufmann, San Francisco (2003). 529 pages
2. Farrel, A., Ayangar, A., Vasseur, J.P.: Inter-Domain MPLS and GMPLS Traffic Engineering. Resource Reservation Protocol-Traffic Engineering (RSVP-TE) Extensions, Cisco Systems Inc., February 2008
3. Alwain, V.: Advanced MPLS Design and Implementation. Cisco Press, Indianapolis (2002)
4. Goldstein, F.B., Goldstein, B.S.: MPLS Technologies and Protocols
5. Alwain, V.: Structure and implementation of modern MPLS technology. TRansl from English. - Publ. House Wiiliams, 480 p. (2004). (rus)
6. Helen, Z., Yuriy, Z.: MPLS networks:modeling, analysis and optimization.- Kiev.: NTUU KPI, 240 p. (2008). (rus)
7. Zaychenko, Y., Zaychenko, H.: New generation computer networks survivability analysis and optimization. In: Vishnevsky, V., Kozyrev, D., Larionov, A. (eds.) DCCN 2013. CCIS, vol. 279, pp. 73–81. Springer, Cham (2014). doi:10.1007/978-3-319-05209-0_6

Performance Modelling of Transmissions in Very Large Network Topologies

Monika Nycz[1], Tomasz Nycz[1], and Tadeusz Czachórski[2(✉)]

[1] Silesian University of Technology, Akademicka 16, 44–100 Gliwice, Poland
{Monika.Nycz,Tomasz.Nycz}@polsl.pl
[2] Institute of Theoretical and Applied Informatics, Polish Academy of Sciences,
Baltycka 5, 44–100 Gliwice, Poland
tadek@iitis.gliwice.pl

Abstract. Transient state queueing models help us to understand better the dynamics of internet transmissions and the performance of traffic control algorithms. Fluid flow approximation, due its simplicity, as it is based on first-order differential equations, is popular and frequently used – but even it, if applied to large topologies, is time and space consuming. Its algorithm is based on iterative calculations on large mutually interdependent structures. In consequence, the bottleneck of the method lies not in numerical computations but in storing and selection of data. This is why we investigate an approach in which a database (SAP HANA) and its language are used to implement the method. The model logic is represented by ETL (Extract, Transform and Load) customizable and user-friendly processes. The numerical examples are based on a real topology having over 100 000 nodes.

Keywords: Fluid flow approximation · Transient state analysis · Large network topologies · TCP networks · SAP HANA · SAP Data Services · ETL · Big data · Data analysis

1 Introduction

Transient state queueing models help us to understand better the dynamics of internet transmissions, the role of traffic control algorithms, and the influence of their parameters on the transmission quality of service. These models may be based on discrete event simulation or analytical approaches including Markov chains, diffusion approximation [3], and fluid-flow approximation, [7,8]. Markov chain models reflect usually isolated events as sending or receiving a packet and soon attain a complexity forbidding their reasonable solution. Diffusion approximation and fluid flow approximation operate on flows and are more tractable. Especially the latter method, due its simplicity, as it is based on first-order differential equations, is popular and frequently used. The results include a model of a network with general topology [7], analysis of RED queue in the bottleneck router [4,8], as well as its comparison with proportional-integral control [5]. In [6,18] a mixture of TCP and UDP flows was introduced and in [16] the method

© Springer International Publishing AG 2017
V.M. Vishnevskiy et al. (Eds.): DCCN 2017, CCIS 700, pp. 49–62, 2017.
DOI: 10.1007/978-3-319-66836-9_5

is adapted to the analysis of satellite networks. In [17] an adaptation of this approach to several flavors of TCP is presented. However, even this simple method,
if applied to large topologies, is time and space consuming. It turns out that
the bottleneck of modelling is not in numerical computations but in storing and
treatment of numerical results. A relevant software should be able to generate,
process and store large amounts of data generated by numerical calculations.
However, the analysis of the changes in vast computer networks, like the Internet, assumes iterative, step-based calculations on large structures that depend
on each other, so parallelization possibilities are limited.

This is why we investigate here an approach in which a large database and
its language are used to implement the method. The model logic is represented
by a user-friendly ETL (Extract, Transform and Load) process characteristic for
data warehousing.

2   Fluid Flow Approximation

In the last decade, the fluid-flow approximation, [7,8], is the most popular approach for modelling transient states in wide area networks, especially the Internet. It can take into account the role of TCP congestion window and active
queue management in IP routers. The method uses first-order ordinary linear
differential equations that are solved numerically. Let us first consider a single
connection model including, sender, receiver and one intermediate router; it is
congestion router, others may be omitted. The basic model equations focus on

- the changes of the queue in the congestion node, which are defined as the
  input stream reduced by output stream

$$\frac{dq(t)}{dt} = \frac{W(t)}{R(q(t))} - \mathbf{1}(q(t) > 0) \cdot C$$

where q is the queue length at congestion node, W is the congestion window
(CWND) size of the flow, R is the round trip time, that is the time needed for
information about the current network state to propagate through network,
C is constant transmission speed of the congestion node. It may also take
into account the unresponsive (UDP) flows, treated as a noise $u(t)$

$$\frac{dq(t)}{dt} = \frac{W(t)}{R(q(t))} - \mathbf{1}(q(t) > 0) \cdot (C - u(t)).$$

It is supposed in [6] that $u(t)$ is a stationary process with mean u_0 and its
variations are on a finer time-scale, hence it may be regarded as constant and
it results in diminishing the bandwidth seen by responsive flows;
- the changes in transmission rates of the TCP flow, the window increases in
  the absence of loss on the path, and decreases otherwise

$$\frac{dW(t)}{dt} = \frac{1}{R(q(t))} - \frac{W(t)}{2} \cdot \frac{W(t - \tau)}{R(q(t - \tau))} p(t - \tau)$$

where $p(t - \tau)$ is packet loss probability and τ is time elapsed between the loss of a packet and decision of a sender to change the congestion window size (the model assumes that it is equal to the round trip time). The equation may be extended to take into account the slow start procedure and time-out losses [8];

– the value of the round trip time is determined as the queue delay in the congestion node and the total link propagation delay Lp

$$R_i(q(t)) = \frac{q(t)}{C} + Lp.$$

– the IP routers have mechanisms preventing overloading their buffers, such as RED [1] which proactively drop packets when queues exceed certain established thresholds on moving average queue

$$x_n(t) = \alpha \cdot q_n(t) + (1 - \alpha) \cdot x_{n-1}(t)$$

with a probability $p(t)$:

$$p(x_n(t)) = \begin{cases} 0, & 0 \leqslant x_n(t) < th_{min} \\ \dfrac{x_n(t) - th_{min_n}}{th_{max} - th_{min}} p_{max}, & th_{min} \leqslant x_v(t) \leqslant th_{max} \\ 1, & th_{max} < x(t) \leqslant B \end{cases},$$

where x_n is the weighted moving average queue length and th_{min}, th_{max} are the thresholds values, p_{max} is a parameter, B is maximum queue size.

We may use also another control mechanism, e.g. classical PID controllers reacting on the difference between current and desired queue size; we are experimenting with such controllers having non-integer integration and differentiation parts.

The linearization around working point (W_0, p_0, q_0, u_0) gives transfer functions, expressed here in terms of their Laplace transforms, [5]

$$W(s) = -P_{win}(s)e^{-sR_0}p(s), \qquad l(s) = \frac{K}{R_0}W(s),$$

$$q(s) = P_{que}(s)\left[l(s) + u(s)\right], \qquad p(s) = q(s)C_{aqm}(s)$$

where

$$P_{win}(s) = \frac{\dfrac{R_0 C_{eff}^2}{2K^2}}{s + \dfrac{2K}{R_0^2 C_{eff}}}, \qquad P_{que}(s) = \frac{1}{s + \dfrac{C_{eff}}{C}\dfrac{1}{R_0}}, \qquad C_{eff} = C - u_0.$$

and $C_{aqm}(s)$ is the Laplace transform of transfer function of the AQM controller. To determine it in case of classical RED algorithm, we made proceed as in [8]: the process of taking the moving average (denoted below as a continuous variable x) is modelled as

$$\frac{dx}{dt} = \log_e(1 - w)/\Delta - \log_e(1 - w)/\Delta q(t)$$

where Δ is the interarrival time of packets and is taken as $\Delta = 1/C$. Hence, the transfer function of RED mechanism having changes of current queue δq at the entrance and changes of packet loss δp as the output has the form

$$C_{aqm}(s) = L_{red}\frac{k}{k + s}$$

where

$$k = -\ln(1 - w)/\Delta = -C\ln(1 - w) \qquad \text{and} \qquad L_{red} = \frac{p_{max}}{th_{max} - th_{min}}.$$

The above equations may be replaced by others if the AQM algorithm changes.

This way a control loop schema is formulated and analyzed. Below, we present some notions on the stability analysis following [13]. The stability of the system may be investigated with the use of Nyquist criterion. In general, the closer the transfer function $G(j\omega)$ representing the whole control loop transfer functions, comes to encircling the $(-1 + j0)$ point, the more oscillatory is the system response. The closeness of $G(j\omega)$ locus to the $(-1 + j0)$ point can be used as a measure of the margin of stability.

Another example of flexibility of the method is its adaptation to the energy saving routers [12] where we introduce two independent thresholds put on current queue length. They determine whether the router should go into energy-saving mode or not. In energy-saving state the node is not servicing packets - it collects packets in the buffer. The algorithm is as follows:

- when the current queue length q decreases and reaches the first threshold T_A, the router's service is switched off. That results in the router's queue growth,
- when the current queue length q increases and reaches the second threshold T_B, the router's service is switched on. It is the moment, when we assume that the router has its queue long enough to start transmitting,
- In the other cases, the state of the router is not changed.

Assuming the T_B threshold a bit higher than T_A, (hysteresis) we prevent the situation, that reaching the threshold value by the router causes constant switching between normal and energy-saving modes.

$$\frac{dq(t)}{dt} = \begin{cases} \dfrac{W(t)}{R(q(t))} - \mathbf{1}(q(t) > 0) \cdot C, & q \geqslant T_B \\[2ex] \dfrac{W(t)}{R(q(t))} & q \leqslant T_A \\[2ex] \dfrac{W(t)}{R(q(t))} - \mathbf{1}(q(t) > 0) \cdot C, & q > T_A, \ N \text{ mode} \\[2ex] \dfrac{W(t)}{R(q(t))} & q < T_B, \ ES \text{ mode} \end{cases}$$

where: N – normal, ES – energy-saving.

The approach may be also adapted to other than Reno control mechanisms. The TCP Vegas tries to estimate the available bandwidth on the basis of changes in RTT and may increase or decrease CWND by 1 packet. It calculates R_{Base}, the minimum value of the RTT (no queueing time) and *expected rate*

$$expected = \frac{W(t)}{R_{Base}}.$$

The *actual rate* depends on the current value $R(t)$ of the RTT:

$$actual = \frac{W(t)}{R(t)}$$

The Vegas mechanism is based on three thresholds: α, β and γ, where α and β refer to the Additive-Increase/Additive-Decrease (AI/AD) paradigm, while γ is related to the modified slow-start phase (SS), [2]

$$\frac{dW_i(t)}{dt} = \underbrace{\frac{W(t - R(t)) * W(t - R(t))}{R(t)} p_0(t - R(t))}_{SS}$$

$$+ \underbrace{\frac{1}{R(t - R(t))} p_1(t - R(t))}_{AI} - \underbrace{\frac{1}{W(t - R(t))R(t - R(t))} p_2(t - R(t))}_{AD}$$

where

$$p_0 = \begin{cases} 1 & \text{for } \dfrac{W(R - R_{Base})}{R} \leq \gamma \\ 0 & \text{otherwise} \end{cases}$$

$$p_1 = \begin{cases} 1 & \text{for } \gamma \leq \dfrac{W(R - R_{Base})}{R} \leq \alpha \\ 0 & \text{otherwise} \end{cases} \qquad p_2 = \begin{cases} 1 & \text{for } \dfrac{W(R - R_{Base})}{R} \geq \beta \\ 0 & \text{otherwise} \end{cases}$$

The approach presented above for one connection may be easily extended to a network having any number of nodes, any topology and any number of connections. Coming back to TCP Reno, in a general case we have

$$\frac{dq_j(t)}{dt} = \sum_{i=1}^{K_j} \frac{W_i(t)}{R_i(\boldsymbol{q}_i(t))} - 1(q_j(t) > 0) \cdot C_j$$

$$\frac{dW_i(t)}{dt} = \frac{1}{R_i(\boldsymbol{q}_i(t))} - \frac{W_i(t)}{2} \cdot \frac{W_i(t - \tau_i)}{R_i(\boldsymbol{q}_i(t - \tau_i))}$$

$$\cdot \left(1 - \prod_{j \in V_i}(1 - p_{ij}(t - \tau_i))\right),$$

where

q_j is the queue at node v,

W_i is the congestion window size of a flow i,

R_i is the round trip time for this flow,

$\boldsymbol{q}_i$ is the set of queues in connection i,

C_j is the transmission speed in node j,

K_j is the number of flows crossing node j,

τ_i is the round trip time at the flow i,

p_{ij} is loss probability at node j for flow i,

V_i is the set of nodes a connection i is traversing. The total link i propagation delay is

$$R_i(\boldsymbol{q}_i(t)) = \sum_{j=1}^{K_i} \frac{q_j(t)}{C_j} + \sum_{j=1}^{K_i-1} Lp_j \ .$$

K_i is the number of nodes in a flow i.

The fluid-flow differential equations are solved numerically. However, if we consider a thousand- or million-node topologies, the calculations generate a large amount of data. In our example we have 134 023 nodes, 50 000 flows and 1000 modelling steps, and we obtained more than 50 million of generated rows for losses, 50 million for flows and 134 millions for routers data. The standard solution is to create a dedicated software structure for storing and analyzing these data. The use of the dedicated structure leads to the necessity of development of a new code, each time a new demand comes. Moreover, considering large structures, the computing power for the analysis of obtained results on standard PC is insufficient. Thus, the more universal solution appears to model with the use of a database, in particular the ETL processes that feed the database with generated data.

3   Implementation

The studies consisted of a few steps. The first one involved the selection of a Internet topology with thousands of nodes and the preparation of structures to store the input data and the generated numerical data. In the second phase, the model of fluid-flow approximation was implemented using the ELT/ETL tool. In the final stage we carried out the modelling for loaded input data, analyzed the results and visualized them using e.g. Gephi [9] tool.

The selected real topology contained only links between network points, hence we needed to generate the initial parameters of the model, such as the initial queue length, the maximum size of the buffer, the initial congestion window size, the links delay, etc. By using the algorithm described in [10,11], we had selected pairs of edge nodes in the network and chose 50 thousands of routes between these points. The whole model configuration data was stored in tabular form in a .CSV file and then loaded into prepared structures in the database. The schema comprised:

- tables with the values for nodes and flows (Routers, Flows),
- table with routes (Paths),
- table with losses for each connection (Losses),
- tables with historical data (HistRouters, HistFlows).

As a storage for data we used SAP HANA Platform – in-memory database directed to accelerate real-time responsiveness, analyze data on the fly, perform advanced analytic of large volumes of data – developed by SAP SE, [14].

In this phase, the main focus was placed on checking the possibility to implement the model numerical logic. The implementation consisted in writing the fluid-flow approximation algorithm in the language of the ELT/ETL tool, which in this case was SAP Data Services, [15] – a dedicated data integration and transformation software for SAP HANA Platform. Within the tool capabilities, among others, there are data discovering, cleansing, validating, enhancing, integrating, querying. The capabilities are implemented as special building blocks, which the end user can apply to create data processing jobs using SAP Data Services Designer. Jobs, the only objects that can be executed, may consist of work-flows (the possible processing paths using data-flows and the definition of the order of their execution) and data-flows (the smallest unit representing the flow of data through data transformation steps). In our case we only needed data-flow level in the algorithm, because there were no alternative paths or complex dependencies. Our solution was mostly based on Query blocks of SAP Data Services. In this article we discuss one of few tested implementations.

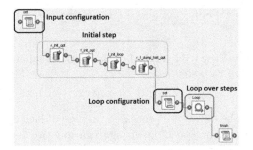

Fig. 1. ETL Job View - the main components of the algorithm

The implementation phase focused on the possibility to implement numerical logic. The fluid-flow algorithm, Fig. 1, was divided into two parts: initialization (initial step) and calculations (loop over steps, Fig. 2). Within each part the data were processed. In initialization, the values (such as queue length, congestion window size, drop probability, etc.) were computed in time $t = 0$. In calculations, in turn, the values were computed within the time range [$step\_size; total\_time$].

Few methods were tested in order to obtain full pushdown of the logic into the SAP HANA database. In this paper we select exemplary data flow, within which the flows parameters per single step were computed, Figs. 3 and 4.

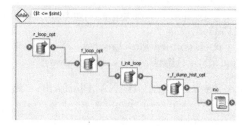

Fig. 2. ETL Loop View - the elements of the calculations part

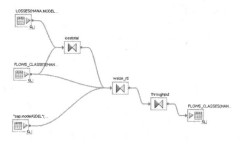

Fig. 3. ETL Data Flow View - one of few tested methods of computation of parameters within single step for all flows

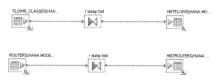

Fig. 4. ETL Data Flow View - the method of saving the computed routers and flows parameters at the end of each step

Within each part the data were processed. In case of lack of tool functionality (like exponent function) or speeding up aggregations we use the SAP HANA models as a data processing element.

To calculate model parameters within Query blocks (which are similar to SQL SELECT statements) we translated all fluid-flow equations into SQL-like scripting language of SAP Data Services and set them as the expressions defining the new values in particular columns.

The analysis performed over the data can be very fast (up to few second, depending on the complexity). The possibility to extract knowledge from the data is only limited by the ability to write the query in SQL language. Here we demonstrate the answers to few exemplary questions about network changes and the results.

– The 15 highest percentage queue loads in particular moment in time ($t = 13$ s)

```
SELECT hr.idrouter, hr.queue/r.buffer*100 AS load
FROM HistRouters hr
INNER JOIN Routers r
ON r.idrouter = hr.idrouter
WHERE time = 13
ORDER BY load DESC, idrouter
```

Input set records without filters: 134 157 023.
Records processed: 134 023.
Query executed in: ≈320 ms.

– The highest value of RTT time in secs ($t \in [0; 100]$)

```
SELECT MAX(rtt)
FROM HistFlows
```

Input set records: 50 050 000.
Records processed: 1.
Query executed in: ≈1.7 ms.

– Loss rates in flows in particular time interval ($t \in [50.01; 60]$)

```
SELECT idflow, time, loss
FROM Losses
WHERE time BETWEEN 50.01 AND 60 ORDER BY time, idflow
```

Input set records without filters: 44 559 219.
Records processed: 4 594 719.
Query executed in: ≈2.64 s.

– The most frequently congested router (above 50%, $t \in [0; 100]$)

```
SELECT c.idrouter, c.cnt
FROM
(
SELECT hr.idrouter, COUNT(hr.time) AS cnt
FROM HistRouters hr
INNER JOIN Routers r
ON hr.idrouter = r.idrouter
WHERE hr.queue/r.buffer*100>=50
GROUP BY hr.idrouter
) c
INNER JOIN
(
SELECT MAX(cnt) AS maxcnt
FROM
(SELECT hr.idrouter, COUNT(hr.time) AS cnt
```

```
FROM HistRouters hr
INNER JOIN Routers r
ON hr.idrouter = r.idrouter
WHERE hr.queue/r.buffer*100>=50
GROUP BY hr.idrouter)
) m
ON m.maxcnt = c.cnt
```

Input set records without filters: 134 157 023.
Records processed: 54 457 002.
Query executed in: ≈3.66 s.

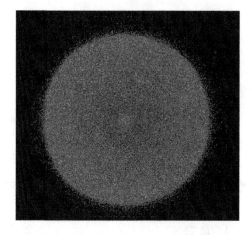

Fig. 5. The loads of nodes in the modelled network for time = 1 s. (Color figure online)

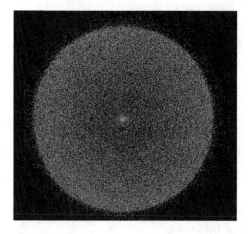

Fig. 6. The loads of nodes in the modelled network for time = 3 s. (Color figure online)

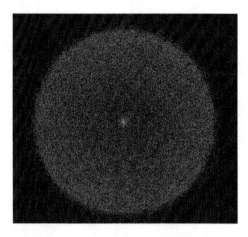

Fig. 7. The loads of nodes in the modelled network for time = 4 s. (Color figure online)

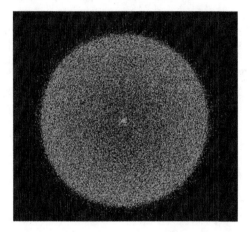

Fig. 8. The loads of nodes in the modelled network for time = 8 s. (Color figure online)

The main benefit of the presented approach is the flexibility of modifying and testing the solutions. Moreover, during the load (calculations), new data successively appeared in the database, thus the detection of changes in the network could be performed in the meantime.

4   Exemplary Numerical Results

The results are retrieved by querying the data stored in SAP HANA database, extracted into flat files and imported into the visualization tool. In Figs. 5, 6, 7, and 8 we present the snapshots of the "network life" – four images, in specially selected moments in time, representing the loads of each queue in the whole network, presented using Gephi. The loads (percentage) are defined by the colours: from 0% (green), through 50% (yellow), up to 100% (red).

The average queue load [%] in particular moments of observation

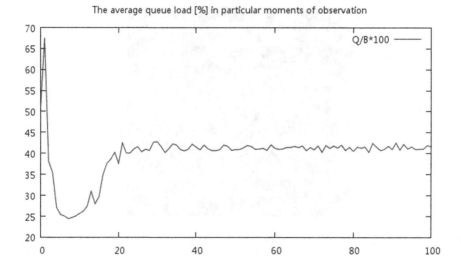

Fig. 9. The average over all nodes of queue filling (actual size compared to maximum queue volume) in the modelled network.

The average throughput of the connections in particular moments of observation

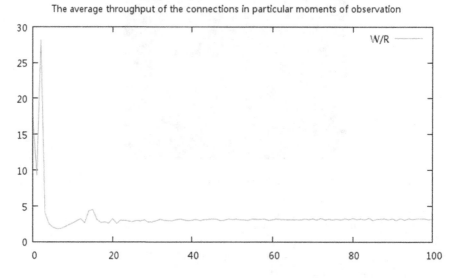

Fig. 10. The average over all nodes of flow intensities in the modelled network.

The modeling started with queues half filled. The initial window size was randomly set on the values in range [1, 150]. As a result, the observation started in the situation of a critical moment, where most of the nodes were congested. With further time growth the network overload diminished and stabilized at the same level.

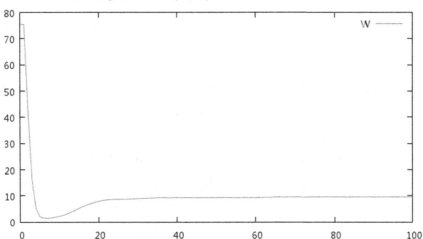

Fig. 11. The average over all nodes of window size in the modelled network.

Then in Figs. 9, 10, and 11 we present aggregated values, representing the global character of the changes in the network. Thus, we could say what the average queue lengths, flow intensities and window sizes were and how they changed as a function of time. We note that the network tends to a steady state.

5   Conclusions

The use of the conjunction of ETL process and SAP HANA database resulted in flexibility of the solution, especially in logic modifications, and the capabilities of fast data analysis. We see that the fluid-flow equations can be implemented as an ELT/ETL job, which is much more customizable than standard native code implementation. Moreover, we obtain a powerful tool to discover the knowledge on network states, changes, trends, etc. As a next step we will focus on further time-based optimization of the algorithms.

Acknowledgments. This work was supported by a grant *Fluid-Flow Approximation using ETL Process and SAP HANA Platform* of Hasso-Plattner-Institute (HPI) in Potsdam, Germany and makes contribution to the researches of the COST ACROSS Autonomous Control for a Reliable Internet of Services project.

References

1. Braden, B., et al.: Recommendations on queue management and congestion avoidance in the internet. RFC 2309, IETF (1998)
2. Domański, A., Domańska, J., Pagano, M., Czachórski, T.: The fluid flow approximation of the TCP vegas and reno congestion control mechanism. In: Czachórski, T., Gelenbe, E., Grochla, K., Lent, R. (eds.) ISCIS 2016. CCIS, vol. 659, pp. 193–200. Springer, Cham (2016). doi:10.1007/978-3-319-47217-1_21

3. Gelenbe, E.: On approximate computer systems models. J. ACM **22**(2), 261–269 (1975)
4. Hollot, C.V., Misra, V., Towsley, D., et al.: A control theoretic analysis of RED. In: Proceedings of IEEE/INFOCOM (2001)
5. Hollot, C.V., Misra, V., Towsley, D., Dong, W.B.: Analysis and design of controllers for AQM routers supporting TCP flows. IEEE Trans. Autom. Control **47**(6), 945–959 (2002). Special issue on Systems and Control Methods for Communication Networks
6. Hollot, C.V., Liu, Y., Misra, V., Towsley, D., et al.: Unresponsive flows and AQM performance. In: Proceedings of IEEE INFOCOM (2003)
7. Liu, Y., Presti, F.L., Misra, V., Towsley, D., Gu, Y.: Fluid models and solutions for large-scale IP networks. ACM/SigMetrics **31**, 91–101 (2003)
8. Misra, V., Gong, W.B., Towsley, D.: A fluid-based analysis of a network of AQM routers supporting TCP flows with an application to RED. In: Conference on Applications, Technologies, Architectures and Protocols for Computer Communication (SIGCOMM 2000), pp. 151–160 (2000)
9. Gephi: The Open Graph Viz Platform. http://gephi.github.io/
10. Nycz, M., Nycz, T., Czachórski, T.: An analysis of the extracted parts of opte internet topology. In: Gaj, P., Kwiecień, A., Stera, P. (eds.) CN 2015. CCIS, vol. 522, pp. 371–381. Springer, Cham (2015). doi:10.1007/978-3-319-19419-6_35
11. Nycz, M., Nycz, T., Czachórski, T.: Modelling dynamics of TCP flows in very large network topologies. In: Abdelrahman, O.H., Gelenbe, E., Gorbil, G., Lent, R. (eds.) ISCIS 2015. LNEE, vol. 363, pp. 251–259. Springer, Cham (2016). doi:10.1007/978-3-319-22635-4_23
12. Nycz, M., Nycz, T., Czachórski, T.: Fluid-flow approximation in the analysis of very large energy-aware networks, submitted
13. Ogata, K.: Modern Control Engineering. Prentice Hall of India, New Dehli (1977)
14. SAP HANA Platform - Capabilites. http://hana.sap.com/capabilities.html
15. SAP Data Services. http://hana.sap.com/capabilities.html
16. Sridharan, M., et al.: Tuning RED parameters in satellite networks using control theory. In: Proceedings of Performance and Control of Next Generation Communication Networks, SPIE, Orlando Florida, 7–11 September 2003, vol. 5244, pp. 145–153 (2003)
17. Srikant, R.: The Mathematics of Internet Congestion Control. Springer Series: Systems and Control: Foundations and Applications. Birkhäuser, Boston (2004)
18. Li, W., Zeng-zi, L., Yan-ping, C., Ke, X.: Fluid-based stability analysis of mixed TCP and UDP traffic under RED. In: Proceedings of the 10th International Conference on Engineering of Complex Computer Systems (ICECCS) (2005)

Multiservice Queueing System with MAP Arrivals for Modelling LTE Cell with H2H and M2M Communications and M2M Aggregation

V.M. Vishnevsky[1]([✉]), K.E. Samouylov[2], V.A. Naumov[3], A. Krishnamoorthy[4], and N. Yarkina[2]

[1] V.A. Trapeznikov Institute of Control Sciences of Russian Academy of Sciences,
Moscow, Russia
vishn@inbox.ru
[2] Department of Applied Probability and Informatics,
RUDN University, Moscow, Russia
ksam@sci.pfu.edu.ru, nat.yarkina@mail.ru
[3] Service Innovation Research Institute (PIKE), Helsinki, Finland
valeriy.naumov@pfu.fi
[4] Department of Mathematics, CMS College, Kottayam, India
achyuthacusat@gmail.com

Abstract. The paper address resource allocation in an LTE call with both human-to-human and machine-to-machine (M2M) communications. M2M aggregation and access barring are considered as a means to reduce congestion due to the nature of M2M traffic. The cell is modelled as a multiservice queueing system with streaming and elastic customers. Resources for M2M communications are allocated in batches of fixed size; requests for them arrive according to a Markovian arrival process. We obtain the stationary probability distribution of the system and formulas for its performance measures and propose an algorithm for their computation.

Keywords: Queueing system · Markovian Arrival Process · Map · Teletraffic theory · LTE · Internet of Things IoT · Machine-to-machine communication · M2M · Machine-type communication · MTC · Aggregation · Access barring

1 Introduction

Until recently, the evolution of mobile communication networks was primarily aimed at boosting data transmission rates and assuring the quality of service, due, for the most part, to the increasing use of multimedia and interactive

This work has been financially supported by the Russian Science Foundation and the Department of Science and Technology (India) via grant #16-49-02021 and INT/RUS/RFS/16 for the joint research project by the V.A. Trapeznikov Institute of control Sciences and the CMS College Kottayam.

V.M. Vishnevskiy et al. (Eds.): DCCN 2017, CCIS 700, pp. 63–74, 2017.
DOI: 10.1007/978-3-319-66836-9_6

applications. However, the expansion of machine-to-machine (M2M) communications(also known as Machine-Type Communications, MTC) and the imminent rise of the Internet of Things (IoT) impose on networks substantially different requirements,as a result of the differing characteristics of the traffic generated by IoT applications and the limitations often placed on M2M devices (low complexity and cost, long battery lifetime, etc.).

Cellular networks have several major advantages making them a strong contender for a place in the emerging IoT infrastructure, including ubiquitous network coverage, full mobility and roaming support, operators having a strong position on the market and benefiting from a solid supplier base, the possibility to provide a wide range of IoT applications far beyond the simplest, low-end transmitting devices (e.g., smart metering), extensive access control, QoS and security management capabilities. However, cellular networks in their present state (especially post-2G) are not fully suitable for a large-scale massive IoT deployment, which implies the capacity to handle efficiently infrequent transmissions of small volumes of data from an extremely large number of inexpensive and energy-efficient devices. Among the main challenges of providing massive IoT connectivity in LTE networks, researchers point out congestion and system overload in the access and core network, but also the need for extra coverage for difficult propagation environments (such as basements), simplified signalling procedures to accommodate low-end transmitting devices and extend their battery lifetime, M2M devices identification, etc. [1,5,7,10].

A considerable effort has been put in to address these issues and to adapt the existing cellular network infrastructure to massive IoT. In 3GPP Release 13 of LTE, which appeared in 2016 and marked the beginning of LTE Advanced Pro, M2M communications and IoT received major focus. A dedicated standard for massive IoT - Narrowband Internet of Things (NB-IoT) - was prepared by 3GPP in a very short space of time to become part of Release 13. It allows an LTE network operator to use the existing infrastructure to deploy an access network for simple stationary IoT devices (with data rates up to 200 kbps), thus aiming at competing with fast-growing LPWAN (Low-Power Wide-Area Network) technologies, such as SigFox and LoRa.

In the coming years, cellular networks are expected to provide connectivity to a wide range of M2M applications and 5G has the potential to become the main enabler for a full-scale IoT [10], however radio resource sharing among conventional mobile service subscribers and a large number of M2M devices remains a topical issue. One general approach to reduce signalling overload and to increase spectrum efficiency in cellular IoT is M2M aggregation, which could be implemented, for instance,by means of a relay-based data aggregation scheme [8] or the use of virtual carriers [4]. Also, in order for LTE networks to become a viable means for providing IoT connectivity, network resources must be allocated in such a way that M2M traffic does not affect adversely human subscribers, who still provide the main income to the network operator. For delay-tolerant M2M, this can be achieved through the use of overload control mechanisms, such as Enhanced Access Barring, introduced in LTE Rel-11 [1,7].

In [6], a Markov model of dynamic radio resource scheduling in an LTE cell withM2M aggregation and overload control giving priority to H2H traffic is proposed. In the present study, we model a similar network cell with M2M and H2H connectivity in terms of the mathematical teletraffic theory,however, unlike in [6], requests for M2M transmissions arrive according to a more general Markovian Arrival Process (MAP). The explicit expressions for the probability measures along with the computational algorithm proposed in the study can be used for evaluating the efficiency of a resource allocation scheme in meeting QoS requirements and adjust its parameters. The rest of the paper is organised as follows. In Sect. 2, simplifying assumptions regarding the operation of the LTE cell with H2H and M2M communications are made. In Sect. 3, we model the cell as a multiservice queueing system and in Sect. 4 derive its steady-state distribution and performance measures. In Sect. 5, a computational algorithm is proposed, and in Sect. 6 numerical results are presented. Finally, Sect. 7 concludes the paper.

2   Assumptions About Cell Operation

We consider uplink transmission in an LTE cell that provides connectivity to M2M devices and a streaming service (e.g., voice or video calls) to human subscribers using conventional user equipment (UE). For simplicity, we assume that all H2H UEs and all M2M devices in the cell have the same signal/noise ratio and do not change their positions relative to the base station. Thus, all the radio links have similar characteristics and the transmission rate depends only on the number of allocated radio resource units.Here, resource unit is an abstract unit of the cell capacity corresponding to the minimum applicable data rate (for instance, in bps) and comprising signalling and control overhead. We assume that the maximum through put of the cell equals Cresource units. As we mentioned previously, M2M data packets are likely to be of a very small size and arrive from an extremely large number of devices. By consequence, allocating resources to M2M devices in the same manner as to conventional UEscan be inefficient, in particular, due to signaling and control procedures and overhead. A possible solution to this is group resource allocation to M2M devices, or M2M aggregation. Thus, we shall assume that resources to M2M devices in the LTE cell under consideration are allocated in batches of a fixed size that are shared equally among a certain number of M2M devices. Finally, as the traffic load on the cell increases, the base station scheduler has to adjust the amount of resources allocated to users in accordance with the QoS measures predefined by the operator, for example, the blocking probability for telephony requests or the average transmission time of M2M data packets. In view of this, we assume that access barring is in place and a part of the cell capacity (R resource units) is available to H2H traffic only and cannot be used for M2M transmission, thus giving priority to H2H over M2M.

3   System Description

The operation of the LTE cell described above can be expressed in terms of teletraffic theory and modelled by means of a multiservice queueing system with streaming customers corresponding to H2H connections and elastic customers representing data transmission from M2M devices. We consider a queueing system with two types of customers, streaming and elastic, and C servers, Rof which are available to streaming customers only, whereas $C_E = C - R$ are available to both types of customers (see Fig. 1). Elastic customers are served in accordance with Egalitarian Processor Sharing discipline (EPS, [11]), but each such customer in the system requires at least d_E servers at all times. Also, servers are allocated to elastic customers in batches of $c \geq d_E$.

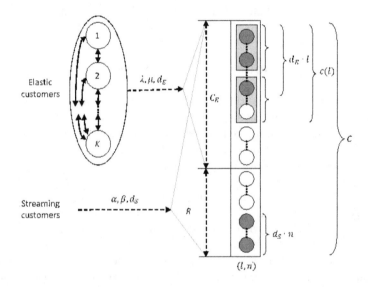

Fig. 1. Queueing system with two types of customers and MAP arrivals

We assume that elastic customers arrive according to a Markovian arrival process (MAP). More specifically, they are generated by a source, which can be in one of K states with transitions defined by two $K \times K$ matrices $\mathbf{Q_1}$ and $\mathbf{Q_0}$. Elements of $\mathbf{Q_1}$ are non-negative and represent transitions accompanied by customer arrivals, whereas elements of $\mathbf{Q_0}$ represent transitions without arrivals. Matrix $\mathbf{Q} = \mathbf{Q_0} + \mathbf{Q_1}$ is the infinitesimal generator of the Markov chain describing transitions of the source. We assume that $\mathbf{Q_1}$ is a non-zero matrix, and $\mathbf{Q}$ is irreducible [3,9]. Let $\mathbf{q}$ denote the row vector of the stationary probabilities of the states of the source, and let $\mathbf{1}$ be a column vector of ones. Now, the arrival rate of elastic customers is $\lambda = \mathbf{q}\mathbf{Q_1}\mathbf{1}$.

Let the required service times of elastic customers (assuming that the customer is served by just one server) be exponentially distributed with parameter μ, and let $M = \lfloor c/d_E \rfloor = max\{y \in \mathbb{N}y \le \frac{c}{d_E}\}$ define the maximum number of elastic customers that can be served simultaneously by one batch of servers. Since the number of server batches allocated to elastic customers cannot exceed $S = C_E/c = maxy \in \mathbb{N} \le \frac{C_E}{c}$, the maximum number of elastic customers in the system equals $L = MS$. Suppose that there are l elastic customers in the system. The total number of servers allocated to these customers equals $c(l) = c\lceil l/M \rceil = cminy \in \mathbb{N} \ge l/M$; the capacity of these servers is equally shared among all the customers present, i.e. each customer is served by $\frac{c(l)}{l}$ servers. The arrival of an $(l+1)$-th elastic customer results in one of the following:

- If at the time of arrival $c(l+1) = c(l)$, then the arriving customer is admitted to the system, but no additional servers are allocated. The capacity of the servers occupied by elastic customers is redistributed among $(l+1)$ customers (each elastic customer is served by $\frac{c(l+1)}{l+1} = \frac{c(l)}{l+1}$ servers instead of $\frac{c(l)}{l}$; the service rate decreases).
- If at the time of arrival $c(l+1) > c(l), c(l+1)?C_E$, and among servers available to elastic customers no less than c servers are free, then the arriving customer is admitted to the system and an additional batch of c servers is allocated. Here, the service rate of all elastic customers increases as the capacity of the new batch is shared among them.
- Otherwise (if an additional batch is required, but less than c servers among those available to elastic customers are free) the arriving customer is blocked without affecting the future arrivals.

An admitted elastic customer is in service until its remainder service time equals zero and departs the system thereafter. At the time of any elastic departure, the servers are redistributed among the remaining $l-1$ elastic customers as follows:

- If $c(l-1) = c(l)$ (i.e., the remaining elastic customers cannot be served by a smaller number of batches), then no servers are released; the capacity of servers occupied by elastic customers prior to the departure is redistributed equally among the remaining elastic customers.
- If $c(l-1) < c(l)$, then a batch of c servers is released and the remaining elastic customers are served by $c(l-1)$ servers, which capacity is equally shared among them.

As for the streaming customers, we assume that their inter arrival and service times are independent and exponentially distributed with parameters α and β respectively, and each streaming customer occupies exactly d_S servers at any time while in the system. Suppose that there are l elastic and n streaming customers in service. Then $C-c(l)$ servers are available to streaming customers and the maximum of $N(l) = \lfloor (C - c(l))/d_S \rfloor$ streaming customers can be served simultaneously. At arrival, a streaming customer is admitted if he finds at least d_S servers free. An admitted streaming customer occupies servers for the time of service and releases the servers and departs the system once the service is completed. If at streaming arrival $n + 1 > N(l)$, i.e. there are less than d_E servers free, then the customer is blocked and does not return. In addition, with this notation, an elastic arrival is blocked if it finds $n > N(l + 1)$.

4   Steady-State Distribution and Stationary Characteristics

Define l(t) and n(t) to be, respectively, the number of elastic and streaming customers in service, and let k(t) denote the state of the source at time $t \geq 0$. The composite stochastic process $X(t) = (l(t), n(t), k(t)), t \geq 0$, is a Markov process with state space

$$\chi = \{(l, n, k) | 0 \leq l \leq L, n \geq 0, c(l) + nd_S \leq C, 1 \leq k \leq K\}.$$

The transition rate matrix of process X(t) is block tridiagonal :

$$A = \begin{bmatrix} D_0 & \Lambda_0 & & & \\ M_1 & D_1 & \Lambda_1 & & \\ & M_2 & \ddots & \ddots & \\ & & \ddots & D_{L-l} & \Lambda_{L-l} \\ & & & M_L & D_L \end{bmatrix} \quad (1)$$

In **A**, the blocks located in the i-th block row and the j-th block column are $K(N(i)+1 \times K(N(j)+1)$ block matrices composed of square matrices of size K. Let **I** denote the identity matrix of order K, and extend the definition of N(l) by setting $N(l) = N(L)$ for $l > L$. Now, the diagonal blocks of **A** are block matrices of the form

$$D_i = \begin{cases} D_{i,1}, if N(l) = N(l + 1), \\ [D_{i,1} D_{i,2}], otherwise, \end{cases} \quad (2)$$

$$D_{i,1} = \begin{bmatrix} F_{i,0} & \alpha I & & & & \\ \beta I & F_{i,1} & \alpha I & & & \\ & 2\beta I & \ddots & & \ddots & \\ & & \ddots & F_{i,N(l+1)} - l & \alpha I & \\ & & & N(l+1)\beta I & F_{i,N(l+1)} & \\ & & & & (N(l+1)+1)\beta I & \end{bmatrix}, 0 \leq l \leq L, \quad (3)$$

$$D_{i,2} = \begin{bmatrix} \alpha I & & & & & \\ G_{i,N(l+1)+1} & \alpha I & & & & \\ (N(l+1)+2)\beta I & G_{i,N(l+1)+2} & \alpha I & & & \\ & (N(l+1)+3)\beta I & \ddots & \ddots & & \\ & & & \ddots & G_{i,N(i)-1} & \alpha I \\ & & & & N(l)\beta I & G_{i,N(l)} \end{bmatrix}, \quad (4)$$

$$0 \le l \le L, \quad (5)$$

$$D_{L,l} = \begin{bmatrix} G_{L,0} & \alpha I & & & \\ \beta I & G_{L,1} & \alpha I & & \\ & 2\beta I & \ddots & \ddots & \\ & & \ddots & G_{L,N(L)} - l & \alpha I \\ & & & (N(L)\beta I & G_{L,N(L)} \end{bmatrix}, \quad (6)$$

where

$$F_{ij} = Q_0 - (\alpha + j\beta + c(l)\mu)I, G_{ij} = Q - (\alpha + j\beta + c(l)\mu)I, j \neq N(l),$$

$$F_{l,N(l)} = Q_0 - (N(l)\beta + c(l)\mu)I, G_{l,N(l)} = Q - (N(l)\beta + c(l)\mu)I.$$

The off-diagonal blocks of **A** are rectangular block diagonal matrices of the form

$$\Lambda_i = \begin{cases} \Lambda_{i,1}, if N(l+1) = N(l+2), \\ [\Lambda_{i,1}\Lambda_{i,2}], otherwise, \end{cases}$$

$$\Lambda_{i,l} = \begin{bmatrix} Q_1 & & & & \\ & Q_1 & & & \\ & & \ddots & & \\ & & & Q-1 & \\ & & & & Q_1 \end{bmatrix}, \quad (7)$$

$$\Lambda_{i,2} = \begin{bmatrix} & & \\ Q_1 & & \\ & \ddots & \\ & & Q_1 \end{bmatrix}, \tag{8}$$

$$M_i = \begin{cases} M_{i,1}, if N(l-1) = N(l), \\ [M_{l,1}O], otherwise, \end{cases}$$

$$\Lambda_{i,2} = \begin{bmatrix} c(l)\mu I & & & \\ & c(l)\mu I & & \\ & & \ddots & \\ & & & c(l)\mu I \end{bmatrix}. \tag{9}$$

The steady-state probability distribution of X(t) can be expressed in vector form in accordance with partitioning (1) of generator $\mathbf{A}$ as $\mathbf{p} = (\mathbf{p_0}, \mathbf{p_1}, \ldots, \mathbf{p_L})$, where $\mathbf{p_l}$ is a vector of length $K(N(l)+1)$ and of the form $\mathbf{p_l} = (\mathbf{p_{l,0}}, \mathbf{p_{l,1}}, \ldots, \mathbf{p_{l,N(L)}})$, $0 \leq l \leq L$. An algorithm for its computation is proposed in Sect. 5. Knowing the stationary distribution if is easy to calculate the blocking probabilities for streaming and elastic customers:

$$B_S = \sum_{l=0}^{L} P_{l,N(l)} 1, \tag{10}$$

$$B_E = \frac{1}{\lambda} \left(\sum_{l=0}^{L-1} \sum_{n=N(l+1)+1)}^{N(l)} p_{l,n} Q_1 1 + \sum_{n=1}^{N(l)} p_{L,n} Q_1 1 \right) \tag{11}$$

Elastic service times have a phase-type distribution defined by the stationary distribution, $\widetilde{\mathbf{p}}$, of the Markov chain embedded at the arrival instants of elastic customers, and the infinitesimal generator, $\widetilde{\mathbf{A}}$, of the Markov process that stops at the departure instant of the elastic customer under consideration. Therefore, knowing distribution $\widetilde{\mathbf{p}}$ and matrix $\widetilde{\mathbf{A}}$, one can easily obtain the probability moments of elastic service times. Stationary distribution $\widetilde{\mathbf{p}}$ can be obtained from distribution $\mathbf{p}$ of process X(t) using the well-knows formulas [3], whereas matrix $\widetilde{\mathbf{A}}$ is obtained from matrix $\mathbf{A}$ by removing the first block column and the first block row, and by multiplying coefficient c(l) by $(1 - \frac{1}{l})$ in formula (3).

5   Computational Algorithm

With the partitioning of distribution $\mathbf{p}$ defined above, we can express the balance equation $\mathbf{pA} = \mathbf{0}$ in the form

$$p_0 D_{0,1} + c(1)\mu p_1 = 0, p_0 D_{0,2} = 0, \tag{12}$$

$$p_{l-1}\Lambda_{l-1,1} + p_l D_{l,1} + c(l+1)\mu p_l + 1 = 0, 0 < l < L, \tag{13}$$

$$p_{l-1}\Lambda_{l-1,2} + p_l D_{l,2} = 0, 0 < l < L, \tag{14}$$

$$p_{L-1}\Lambda_{L-1,2} + p_L D_{L,1} = 0, \tag{15}$$

To solve this system efficiently, we will apply an algorithm, similar to the one proposed in [2]. This algorithm is a variant of Gaussian block elimination and contains three phases: *Phase 1*. Compute auxiliary matrices $H_l, 0 \leq l \leq L$, of the size $K(N(0)+1) \times K(N(l)+1)$ using the formulas

$$H_0 = I,$$

$$H_1 = -\frac{1}{c(1)\mu} D_{0,1},$$

$$H_{l+1} = -\frac{1}{c(l+1)\mu}(H_{l-1}\Lambda_{l-1,1} + H_l D_{l,1}, l = 1, 2, \dots, L - 1. \tag{16}$$

Phase 2. Obtain p_0 by solving the system of linear equations $p_0 B = b$. Here, vector b is obtained by replacing the last K elements of a zero vector of length $K(N(0)+1)$ with vector q. Matrix B is a square of order $K(N(0)+1)$ non-singular matrix of the form $B = [B_0, B_1, \dots, B_{L-l}, B_L]$, where $B_0 = D_{02}, B_l = H_{l-1}\Lambda_{l-1,2} + H_l D_{l,2}, l = 1, 2, \dots, L - 1$, and matrix B_L is obtained by replacing the last K columns of $H_{L-1}\Lambda_{L-1}+H_L D_L$ with the matrix $\sum_{l=0}^{L} H_l(u_l \otimes I)$, where u_l is a column vector of ones of length $N(l) + 1$ and $\otimes$ denotes the Kronecker product. *Phase 3*. Compute the stationary distribution using the expressions

$$p_l = p_0 H_i, l = 1, \dots, L, \tag{17}$$

or the performance measures of the system, in particular, the blocking probability of streaming customers

$$B_S = \sum_{l=0}^{L} p_0 \widetilde{H}_l 1, \tag{18}$$

where matrix $\widetilde{H}_l$ is composed of the last Kcolumns of matrix H_l, and the blocking probability of elastic customers

$$B_E = \frac{1}{\lambda}(\sum_{l=0}^{L-1} p_0 \widetilde{H}_l + p_0 H_L)Q_1 1, \tag{19}$$

where matrix $\widetilde{\widetilde{H}}_l$ is obtained from matrix H_l by removing the first $K(N(l+1)+1$ columns.

6   Numerical Results

We use the algorithm presented in the previous section to evaluate the blocking probabilities for M2M and H2H connections in at LTE cell with peak throughput of 50 Mbps. We assume that 10 Mbps of the total capacity are reserved for H2H traffic only, the minimum required data rate for M2M connections is set to 100 kbps, and H2H subscribers receive a streaming video service requiring 3 Mbps. Let the resource unit equal 100 kbps. Now, the structural parameters to have the following values: $K = 3$,

$$Q_1 = \begin{bmatrix} 0,1 & 0,1 & 0,1 \\ 1 & 1 & 1 \\ 5 & 5 & 5, \end{bmatrix}, Q_0 = \begin{bmatrix} -0,5 & 0,1 & 0,1 \\ 1 & -5 & 1 \\ 5 & 5 & -25 \end{bmatrix}, \tag{20}$$

resulting in $\lambda = 0.8035714$; $\mu = 50$, which corresponds to the mean M2M packet length of 250 bytes, and $\beta = 0.0055$, corresponding to the mean H2H service time of about 3 min.

Figures 2 and 3 show the blocking probabilities as functions of the arrival rate α of H2H requests, which grows from 0.0075 to 0.15 requests per second. In Fig. 2, the blocking probabilities are computed for c = 60, c = 120 and c = 200 (i.e., resources for M2M traffic are allocated in batches of 6, 12 and 20 Mbps respectively). It can be seen from Fig. 2, that, for the values under consideration, an increase of the batch size results in a higher blocking probability of M2M requests without affecting the blocking probability of H2H requests. In Fig. 3, the blocking probabilities are plotted for c = 60, c = 45, c = 30 and c = 15 (i.e., the size of the M2M batch is 6, 4.5, 3 and 1.5 Mbps respectively). Here, the curves coincide for all H2H blocking probabilities and M2M blocking probabilities for c = 45 and c = 30, whereas the M2M blocking probability for c = 15 remains zero. This is due to the choice of structural parameters, as 20 resource units of the total capacity cannot be occupied by H2H connections.

Figures 4 and 5 depict the blocking probabilities as functions of M2M requests arrival rate λ. Here, we increase element $(1,1)$ of matrix Q_1, which leads to the

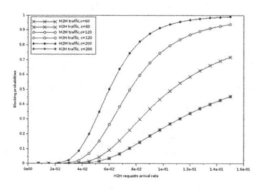

Fig. 2. Blocking probabilities as functions of the H2H arrival rate, c = 60, 120, 200

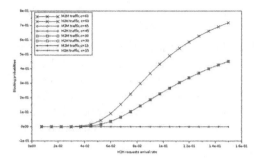

Fig. 3. Blocking probabilities as functions of the H2H arrival rate, c = 60, 45, 30, 15

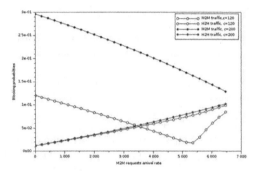

Fig. 4. Blocking probabilities as functions of the M2M arrival rate, c = 120, 200

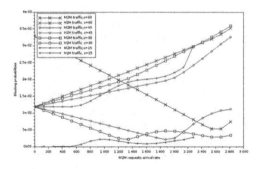

Fig. 5. Blocking probabilities as functions of the M2M arrival rate, c = 60, 45, 30, 15

growth of λ plotted on the X-axis. The arrival rate of H2H requests is constant and equals $\alpha = 0.05$. The figures show that the model allows to evaluate the influence of different structural parameters, such as the size of the batch, but also the total and the reserved capacity and the H2H required throughput on the performance measures. It should be added, that the algorithm performs well numerically within the load ranges shown on the figures. However, as the arrival rates grow further, matrix becomes ill-conditioned, which is subject to further research.

7   Conclusion

The proposed model allows to study the combination of streaming H2H and data M2M traffic in an LTE network cell with M2M aggregation and access barring mechanisms. It can be used for evaluating resource allocation schemes in LTE to accommodate massive M2M and to adjust their parameters so that QoS requirements are met. The algorithms proposed performs well numerically in a wide range of parameters values. Its extension to higher arrival rates is subject to further research.

References

1. Andreev, S., Galinina, O., Pyattaev, A., Gerasimenko, M., Tirronen, T., Torsner, J., Sachs, J., Dohler, M., Koucheryavy, Y.: Understanding the IoT connectivity landscape - a contemporary M2M radio technology roadmap. IEEE Commun. Mag. **53**(9), 32–40 (2015)
2. Basharin, G.P., Naumov, V.A.: Lösungsmethoden für lineare algebraishe Gleichungssysteme stationarer charakteristiken. In: Handbuch der Bedienungstheorie I, pp. 387–430. Akademie-Verlag, Berlin (1983)
3. Basharin, G.P., Naumov, V.A.: Simple matrix description of peaked, smooth traffic, its applications. In: Proceedings of the 3rd International Seminar on Teletraffic Theory "Fundamentals of Teletraffic Theory", pp. 38–44. VINITI, Moscow (1984)
4. Beale, M.: Future challenges in efficiently supporting M2M in the LTE standards. In: Proceedings of the 10th Wireless Communications and Networking Conference WCNCW 2012, Paris, pp. 186–190 (2012)
5. Bhat, P., Dohler, M.: Overview of 3GPP machine-type communication (MTC) standardization. In: Anton, C., Dohler, M. (eds.) Machine-to-Machine (M2M) Communications - Architecture, Performance and Applications. Elsevier, Amsterdam (2015)
6. Buturlin, I., Gudkova, I.A., Chukarin, A.V.: On radio resource allocation scheme model with fixed capacities for machine type communications in ITE network. T-Comm - Telecommun. Transp. **8**, 14–18 (2014). In Russian
7. Ghavimi, F., Chen, H.H.: M2M communications in 3GPP LTE/LTE-A networks: architectures, service requirements, challenges, and applications. IEEE Commun. Surv. Tutor. **17**(2), 525–549 (2015)
8. Mehmood, Y., Görg, C., Timm-Giel, A.: A radio resource sharing scheme for IoT/M2M communication in LTE-A Downlink. In: IEEE ICC2016-Workshops: W07-Workshop on Convergent Internet of Things, pp. 296–301 (2016)
9. Naumov, V.A.: Markovskie modeli potokov trebovaniy [Markovian Models of Arrival Processes]. pp. 67–73. UDN Publisher, Moscow (1987). In Russian
10. Palattella, M.R., Dohler, M., Grieco, A., Rizzo, G., Torsner, J., Engel, T., Ladid, L.: Internet of things in the 5G era: enablers, architecture, and business models. IEEE J. Sel. Areas Commun. **34**(3), 510–527 (2016). doi:10.1109/JSAC.2016.2525418. [7397856]
11. Yashkov, S.F., Yashkova, A.S.: Processor sharing: a survey of the mathematical theory. Autom. Remote Control **68**(9), 1662–1731 (2007). http://dx.doi.org/10.1134/S0005117907090202

Hidden Markov Models in Long Range Dependence Traffic Modelling

Joanna Domańska[2]([✉]), Adam Domański[1], and Tadeusz Czachórski[2]

[1] Institute of Informatics, Silesian Technical University,
Akademicka 16, 44-100 Gliwice, Poland
adamd@polsl.pl

[2] Institute of Theoretical and Applied Informatics, Polish Academy of Sciences,
Baltycka 5, 44–100 Gliwice, Poland
{joanna,tadek}@iitis.gliwice.pl

Abstract. Hidden Markov Models (HMM) have been widely used in several areas of computer science. Conventional HMMs are well-known for their efficiency in modeling short-term dependencies between adjacent elements, but some researchers concluded that they cannot grasp long-range interactions between distant elements. Long-range dependence (LRD) of data refers to temporal similarity present in the data. Various studies demonstrated the presence of LRD at network traffic on several levels of communications protocols. This paper concerns the HMM-traffic source capability to capture the LRD appeared in real network traffic. We used several estimators of Hurst parameter to evaluate the LRD. Not all LRD processes mandatorily have a definable Hurst parameter, but the value of H between 0.5 and 1 is usually considered the standard measure of LRD.

Keywords: Hidden Markov Models · Long-range dependence · Performance evaluation · Network traffic

1 Introduction

Computer networks modeling helps developers to predict the behaviour of proposed networks, to characterize network load, to locate overloaded nodes and to predict the network behaviour with the increasing load. The necessity of computer modeling appears in many areas of computer networks design and exploitation, in the initial design phase of network mechanisms, allowing a realistic assessment of the quality and comparison the proposed mechanism with the existing solutions as well as to adapt the network configuration and the network protocols parameters to the specific purposes during the use phase.

For the proper evaluation of computer network performance it is necessary to create not only appropriate models of network mechanisms but also the realistic packet traffic models. The combination of these two elements allows us to obtain correct results of network modeling that fit measurements performed in real, existing objects.

© Springer International Publishing AG 2017
V.M. Vishnevskiy et al. (Eds.): DCCN 2017, CCIS 700, pp. 75–86, 2017.
DOI: 10.1007/978-3-319-66836-9_7

Research related to the Internet traffic aims to provide a better understanding of the modern Internet, inter alia, by presenting the current characteristics of Internet traffic based on a large number of experimental data and introducing issues related to the internet traffic modeling. The understanding of the traffic nature of the modern Internet is important for the Internet community. It supports optimization and development of protocols and network devices, and improves the network applications security and the protection of network users.

Measurements and statistical analysis of packet network traffic, performed already in the 90s, show that this traffic displays a complex statistical nature [1]. There are in this traffic statistic phenomena such as: self-similarity, long-range dependence and burstiness. During the last two decades, self-similarity and long-range dependence (LRD) became an important research domain. Extensive measurements demonstrated the self-similarity and LRD of network traffic on several levels of communication protocols.

Self-similarity of a process means that the change of time scales does not influence the statistical characteristics of the process. It results in long-distance autocorrelation and makes possible the occurrence of very long periods of high (or low) traffic intensity. These features have a great impact on a network performance. They enlarge mean queue lengths at buffers and increase the probability of packet losses, reducing this way the quality of services provided by a network.

In consequence, it is needed to propose new or to adapt known types of stochastic processes while modeling these negative phenomena in network traffic. Traditionally, the traffic intensity has been regarded as a stochastic process and represented in queueing models by short term dependencies. Several models have been introduced for the purposes of modeling self-similar processes in the network traffic area. These models of traffic sources use: fractional Brownian Motion [2], chaotic maps [3], fractional Autoregressive Integrated Moving Average (fARIMA) [4], wavelets and multifractals and processes based on Markov chains: SSMP (Special Semi-Markov Process) [6], MMPP (Markov-Modulated Poisson Process) [5,17], BMAP (Batch Markovian Arrival Process) - for modeling network traffic.

All above mentioned traffic models have their advantages and drawbacks, but many arguments weigh in favour of processes based on Markov chains. The analytical and numerical methods associated with Markov chains are relatively well developed. These models are easy to simulate. Additionally, their advantage consists in the fact that they allow the use of traditional and well known queueing models and modeling techniques for computer networks performance analysis. The correct traffic model should not only replicate the required statistical characteristics of a real traffic but should also allow to use them as a generator in models of network mechanisms.

The aim of this article is the use of Hidden Markov Model (HMM) for LRD traffic modeling. To the best of authors' knowledge, no work has been done on applying HMM models to LRD network traffic modeling.

The rest of this paper is organized as follows. Section 2 briefly describes the issue of the Hidden Markov Models (HMMs). Section 3 presents few methods of

Hurst parameter estimation. The Hurst parameter H expresses the degree of the self-similarity. Section 4 presents how to obtain the LRD traffic source based on HMM modelling. This section also shows the analysis of LRD in the obtained HMM traces. Some conclusions are given in Sect. 5.

2   Hidden Markov Models

Hidden Markov Models (HMMs) [7] is a statistical modelling tool for systems with hidden internal states that can be observed and measured only indirectly. Recently, the interest in HMM-based models has grown and HMM models have been proposed as a tool for several network traffic related research problems. HMM may be seen as a probabilistic function of a (hidden) Markov chain. This Markov chain is composed of two variables:

- the hidden-state variable, the temporal evolution of which follows a Markov-chain behavior, $x_n \in \{s_1, \ldots, s_N\}$ represent the (hidden) state at discrete time n with N being the number of states,
- the observed variable which stochastically depends on the hidden state ($y_n \in \{o_1, \ldots, o_M\}$ and represents the observable at discrete time n with M being the number of observables).

An HMM is characterized by the set of parameters:

$$\lambda = \{\mathbf{u}, \mathbf{A}, \mathbf{B}\}$$

where:

- $\mathbf{u}$ is the initial state distribution, where $u_i = Pr(x_1 = s_i)$
- $\mathbf{A}$ is the $N \times N$ state transition matrix, where $A_{i,j} = Pr(x_n = s_j \mid x_{n-1} = s_i)$
- $\mathbf{B}$ is the $N \times M$ observable generation matrix, where $B_{i,j} = Pr(y_n = o_j \mid x_n = s_i)$

HMM models have been used to model the states of packet channels via corresponding loss probabilities and end-to-end delay distributions. Similar works have been proposed to model wired and wireless packet channels. A few modeling works using HMMs to model traffic sources at packet level are present in literature. The article [37] proposed a Hidden Markov Model for Internet traffic sources at packet level, jointly analyzing Inter Packet Time and Packet Size. The cited paper aims at modeling the average behavior of a single session. The authors underlined that the study of the superposition of several sessions generated by multiple sources may indeed lead to the generation of an aggregate traffic showing long range dependence and self-similarity characteristics, but such investigation falls beyond the scope of their work.

Many researchers concluded that conventional HMMs are well-known for their efficiency in modeling short-term dependencies between adjacent elements, but they cannot capture long-range interactions between distant elements [14]. Our paper confirms the HMM-traffic source capability to capture the LRD appeared in real network traffic.

3   Hurst Parameter Estimators

Self-similarity is an often-observed natural phenomenon. The term was introduced by Mandelbrot [10] for explaining water level pattern of river Nile observed
by Hurst. Let $X(t)$ be a stochastic process representing increment process (e.g.
in bytes/second). In this case X takes a form of a discrete time series $\{X_t\}$,
where $t = 0, 1, \ldots, N$. The sequence $X^{(m)}(k)$ is obtained by averaging $X(t)$ over
non-overlapping blocks of length m:

$$X^{(m)}(k) = \frac{1}{m} \sum_{i=1}^{m} X((k-1)m + i), \qquad k = 1, 2, \ldots. \tag{1}$$

Let $Y(t)$ be a continuous-time process representing the traffic volume, i.e. $X(t) = Y(t) - Y(t-1)$. $Y(t)$ is exactly self-similar when it is equivalent, in the sense of
finite dimensional distributions, to $a^{-H} Y(at)$, where $t > 0$, $a > 0$, and $0 < H < 1$
is the Hurst parameter. The process $Y(t)$ may be nonstationary. The Hurst
parameter H expresses the degree of the self-similarity.

Long-range dependence of data refers to temporal similarity present in the
data. LRD is associated with stationary processes. If a process $X(k)$ is second-
order stationary with variance σ^2 and autocorrelation function $r(k)$, then it has
LRD only if its autocorrelation function is non-summable, $\sum_n r(n) = \infty$. That
means that the process exhibits similar fluctuations over a wide range of time
scales.

Not all LRD processes mandatorily have a definable Hurst parameter but
the value of H between 0.5 and 1 is usually considered the standard measure
of LRD. The parameter can be estimated in a number of ways [18]. The *R/S
statistic*, *aggregated variance* and *periodogram* are well known methods with a
significant history of use, the *local Whittle's estimator* and *wavelet based methods*
are newer techniques and they perform relatively well.

The aggregate variance method [9] uses the plot of $\log[Var(X^{(m)})]$ as defined
in Eq. (1) versus $\log m$. The estimated value of Hurst parameter is obtained by
fitting a simple least squares line through the resulting points in the plane. The
the value of asymptotic slope β between -1 and 0 suggests LRD and estimated
Hurst parameter is given by $H = 1 - \beta/2$.

Another time-domain based technique of Hurst parameter estimation is called
R-S Plot [10]. The *R-S* method, one of the oldest techniques, is based on Central
Limit Theorem. Let $R(n)$ be the range of the data aggregated over blocks of
length n and $S^2(n)$ be the sample variance of data aggregated at the same scale.
The rescaled range of X over a time interval n is defined as the ratio R/S:

$$\frac{R}{S}(n) = S^{-1}(n) \left[\max_{0 \leq t \leq n} (X(t) - t\overline{X}(n)) - \min_{0 \leq t \leq n} (X(t) - t\overline{X}(n)) \right] \tag{2}$$

where $\overline{X}(n)$ is the sample mean over the time interval n, and $S(n)$ is standard deviation. For LRD processes, the ratio has the following characteristic for large n:

$$\frac{R}{S} \sim \left(\frac{n}{2}\right)^H .$$

(3)

A log-log plot of $\frac{R}{S}(n)$ versus n should have a constant slope H as n becomes large.

The method using *Periodogram* in log-log scale [11] is frequency domain method, the periodogram is defined by:

$$I_X(\omega) = \frac{1}{2\pi n} \left| \sum_{j=1}^n X_j e^{ij\omega} \right|^2$$

(4)

A log-log plot $I_X(\omega_{n,k})$ versus $\omega_{n,k} = \frac{2\pi k}{n}$ should have a slope of $1 - 2H$ around $\omega = 0$.

Whittle's estimator is a semiparametric maxminimum likelihood estimator which assumes a functional form for estimate the spectral density at frequencies near zero, [12,13]. To estimate Hurst parameter one should minimize the function:

$$Q(H) = \sum_j \left[\log f_j(\omega_j) + \frac{\log I_X(\omega_j)}{f_j(\omega_j)} \right],$$

(5)

where $f(\omega) = c\omega^{2H-1}$.

The wavelet-based Hurst parameter estimators are based on the shape of the power spectral density function of the LRD process. Wavelets can be thought of as akin to Fourier series but using waveforms other than sine waves. Wavelet analysis has been applied in Hurst parameter estimation due to its powerful properties.

4 LRD Traffic Source Based on HMM

To learn the HMM model parameters we used as a training sequence the traces generated using Fractional Gaussian noise. Fractional Gaussian noise (fGn) has been proposed in [10] as a model for the long-range dependence postulated to occur in a variety of hydrological and geophysical time series. Nowadays, fGn is one of the most commonly used self-similar processes in network performance evaluation and the only stationary Gaussian process being exactly self-similar. In this paper we use a fast algorithm, first introduced in [15], for generating approximate sample paths for a fGn process, . We have generated the sample traces with the Hurst parameter with the range of 0.50 to 0.90.

We consider an HMM in which the state and the observable variables are discrete. Given a sequence of observable variables $y = (y_1, y_2, \ldots, y_L)$ referred to as the *training sequence*, we want to find the set of parameters such that the likelihood of the model $L(\mathbf{y}; \lambda) = Pr(\mathbf{y} \mid \lambda)$ is maximum. We solved it via the Baum-Welch algorithm, a special case of the Expectation-Maximization algorithm [16],

Table 1. HMM transition probabilities

From/To	A	B	C	D
A	0.47939038	0.02110681	0.02110681	0.47939038
B	0.02060962	0.47889319	0.47889319	0.0206096
C	0.02060962	0.47889319	0.47889319	0.02060962
D	0.47939038	0.02110681	0.02110681	0.47939038

Table 2. HMM emission probabilities

States/Symbols	"0"	"1"
A	0.2376875	0.7787210
B	0.7623125	0.2212790
C	0.7787210	0.2376875
D	0.2212790	0.7623125

that iteratively updates the parameters in order to find a local maximum point of the parameter set.

Next we used the HMM trained with the fGn data as the traffic source model. The estimators described in the Sect. 3 were used to evaluate the long-range dependence of the generated traces.

Tables 1 and 2 present the example set of obtained HMM parameters. Table 3 shows Hurst parameters calculated with the use of five estimators described in Sect. 3. This LRD analysis is based on the data generated by the HMM traffic source described above (trace 2) and by the two others obtained HMM traffic sources. Trace 1 is an example of generated traffic with lower Hurst parameter and trace 3 - with higher H parameter. For all cases the traffic intensity λ is the same and closed to 0.5.

The Figs. 1, 2 and 3 show respectively the example of traffic fluctuations over different time scales which are obtained from HMM traces described above.

Table 3. Hurst parameter estimates for traces generated by HMM - 1 000 000 samples

	Trace 1	Trace 2	Trace 3
Estimator	Hurst parameter		
R/S Method	0.88	0.83	0.56
Aggregate variance method	0.66	0.64	0.59
Periodogram method	0.82	0.76	0.65
Whittle method	0.92	0.80	0.65
Wavelet-based method	0.84	0.70	0.58

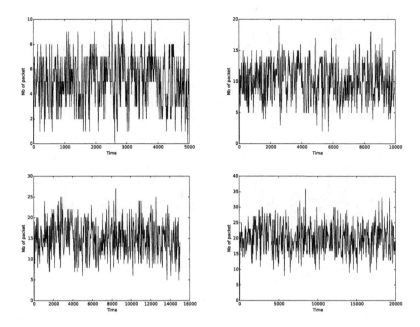

Fig. 1. Traffic fluctuations over different time scales (trace 1)

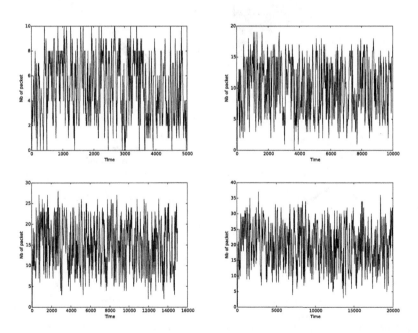

Fig. 2. Traffic fluctuations over different time scales (trace 2)

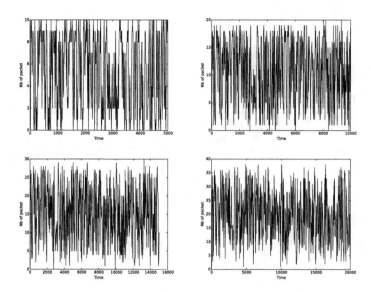

Fig. 3. Traffic fluctuations over different time scales (trace 3)

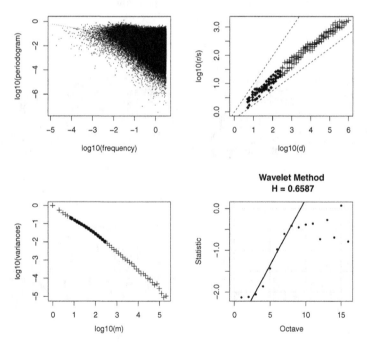

Fig. 4. The periodogram, the R-S plot, the variance-time plot and the wavelet based analysis (trace 1)

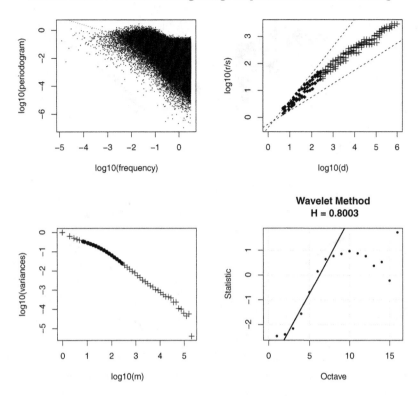

Fig. 5. The periodogram, the R-S plot, the variance-time plot and the wavelet based analysis (trace 2)

Figures 4, 5 and 6 show a graphical presentation of the degree of self-similarity of these three HMM processes calculated using four methods:

– the periodogram
– the R-S plot
– the variance-time plot
– the wavelet analysis; it displays the variance of wavelet coefficients within each scale of the discrete wavelet transformation.

The analysis presented in this section confirm the presence of long-term dependencies in the traffic generated by HMM traffic sources developed by the authors.

In most cases obtained HMM models generated non LRD traffic. Positive results were obtained only in a few described above cases and they do not depend on the initial configuration of the HMM nor on the degree of LRD.

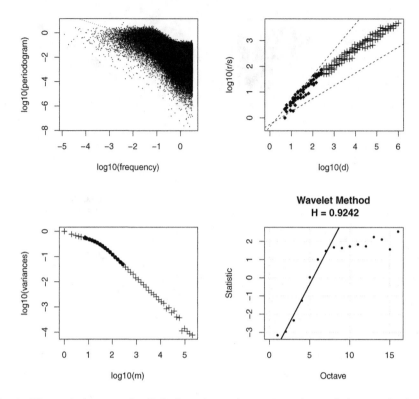

Fig. 6. The periodogram, the R-S plot, the variance-time plot and the wavelet based analysis (trace 3)

5   Conclusions

Our article [8] presented the very first step in modeling LRD traffic using Hidden Markov Models. The well-known Bellcore traces are used as a training sequence to learn HMM model parameters.

In this paper, we use as training sequences synthetic traffic traces generated by the Fraction Gaussian Noise. This traffic model allows us to obtain the traces with different degree of LRD. We used several estimators of Hurst parameter to confirm the LRD. Our results confirm also that although the Hurst parameter is well defined mathematically, it is problematic to measure it properly [19]. There are several methods to estimate the Hurst parameter but they often produce conflicting results [20]. Many researches conclude that wavelet technique is faring well in comparative studies [19,21].

Our experiments show that it is possible to generate LRD traffic (with different Hurst parameters) with the use of classical HMM model.

References

1. Willinger, W., Leland, W.E., Taqqu, M.S.: On the self-similar nature of ethernet traffic. IEEE/ACM Trans. Netw. **2**, 1–15 (1994)
2. Norros, I.: On the use of fractional Brownian motion in the theory of connectionless networks. IEEE J. Selected Areas Commun. **13**(6), 953–962 (1995)
3. Erramilli, A., Singh, R.P., Pruthi, P.: An application of determinic chaotic maps to model packet traffic. Queueing Syst. **20**(1–2), 171–206 (1995)
4. Harmantzis, F., Hatzinakos, D.: Heavy network traffic modeling and simulation using stable FARIMA processes. In: 19th International Teletraffic Congress, Beijing, China (2005)
5. Andersen, A.T., Nielsen, B.F.: A markovian approach for modeling packet traffic with long-range dependence. IEEE J. Selected Areas Commun. **16**(5), 719–732 (1998)
6. Robert, S., Boudec, J.Y.L.: New models for pseudo self-similar traffic. Perform. Eval. **30**(1–2), 57–68 (1997)
7. Rabiner, L.R.: A tutorial on Hidden Markov Models and selected applications in speech recognition. Proc. IEEE **77**(2), 257–286 (1989)
8. Domańska, J., Domański, A., Czachórski, T.: Internet traffic source based on hidden markov model. In: Balandin, S., Koucheryavy, Y., Hu, H. (eds.) NEW2AN/ruSMART -2011. LNCS, vol. 6869, pp. 395–404. Springer, Heidelberg (2011). doi:10.1007/978-3-642-22875-9_36
9. Beran, J.: Statistics for long-Memory Processes. Chapman & Hall, New York (1994)
10. Mandelbrot, B.B., Wallis, J.: Computer experiments with fractional gaussian noises. Water Resour. Res. **5**(1), 228–241 (1969)
11. Geweke, J., Porter-Hudak, S.: The estimation and application of long memory time series models. J. Time Ser. Anal. **4**(4), 221–238 (1983)
12. Robinson, P.M.: Gaussian semiparametric estimation of long range dependence. Ann. Stat. **23**, 1630–1661 (1995)
13. Park, C., Hernandez-Campos, F., Long, L., Marron, J., Park, J., Pipiras, V., Smith, F., Smith, R., Trovero, M., Zhu, Z.: Long range dependence analysis of internet traffic. J. Appl. Stat. **38**(7), 1407–1433 (2011)
14. Yoon, B.J., Vaidyanathan, P.P.: Context-sensitive hidden Markov models for modeling long-range dependencies in symbol sequences. IEEE Trans. Sig. Process. **54**, 4169–4184 (2006)
15. Paxson, V.: Fast, approximate synthesis of fractional Gaussian noise for generating self-similar network traffic. ACM SIGCOMM Comput. Commun. Rev. **27**(5), 5–18 (1997)
16. Bilmes, J.A.: A Gentle Tutorial od the EM Algorithm and its Application to Parameter Estimation for Gaussian Mixture and Hidden Markov Models, University of Berkeley (1998)
17. Domańska, J., Domański, A., Czachórski, T.: Modeling packet traffic with the use of superpositions of two-state MMPPs. In: Kwiecień, A., Gaj, P., Stera, P. (eds.) CN 2014. CCIS, vol. 431, pp. 24–36. Springer, Cham (2014). doi:10.1007/978-3-319-07941-7_3
18. Domańska, J., Domański, A., Czachórski, T.: Estimating the intensity of long-range dependence in real and synthetic traffic traces. In: Gaj, P., Kwiecień, A., Stera, P. (eds.) CN 2015. CCIS, vol. 522, pp. 11–22. Springer, Cham (2015). doi:10.1007/978-3-319-19419-6_2

19. Clegg, R.G.: A practical guide to measuring the Hurst parameter. Int. J. Simul. **7**(2), 3–14 (2006)
20. Karagiannis, T., Molle, M., Faloutsos, M.: Long-range dependence: ten years of internet traffic modeling. IEEE Internet Comput. **8**(5), 57–64 (2004)
21. Stolojescu, C., Isar, A.A.: Comparison of Some Hurst Parameter Estimators. In: 13th International Conference on Optimization of Electrical and Electronic Equipment, Brasov, Romania, pp. 1152–1157 (2012)

Integration Data Model for Continuous Service Delivery in Cloud Computing System

V.V. Efimov[1](✉), S.V. Mescheryakov[1], and D.A. Shchemelinin[2]

[1] St. Petersburg Polytechnic University,
Polytekhnicheskaya 29, St. Petersburg 195251, Russia
2vadim@inbox.ru, serg-phd@mail.ru
[2] RingCentral Inc., San Mateo, CA 94404, USA
dshchmel@gmail.com

Abstract. New data model and approach are proposed to integrate monitoring events, incident management, problem management, change management, and other ITIL processes, which are usually managed by separate operations departments and tools in a big International IT Company. New integration architecture and monitoring applications are introduced, allowing provisioning a full-cycle continuous delivery of telecommunication services in a globally distributed cloud computing system.

Keywords: Integration data model · Continuous service delivery · Internet telecommunications · Cloud computing

1 Introduction

In big International IT Companies, providing multiple cloud services in globally distributed regions, a big amount of computing resources in data centers, including physical servers, virtual machines, custom software applications, is being operated [1]. This paper is based on particular research of real data from RingCentral Company [2], which is provisioning 24/7 Internet telecommunications services in 27 countries of North America, West Europe, Pacific and South-East Asia.

RingCentral globally distributed infrastructure consists of more than 10 K servers located in 6 data centers all over the world. All the events from remote servers are collected into monitoring system, having overall big data traffic of 13 K values per second. Each measurement is analyzed for possible anomalies, such as critically high CPU usage, lack of free disk space, out of memory, etc. More than 3 K exceptions are detected and resolved daily. Each anomaly may result in either computing redundancy loss or performance degradation or even service outage. In average, the 3 incidents with customer impact are detected daily.

Repetitive alarm is a problem, which needs a root cause to analyze and a hot fix to apply to prevent such incidents in future. Average amount of changes in RingCentral Company, including emergency patches and planned updates, is about 30 per day and has grown twice given 2016 statistics (see Fig. 1).

© Springer International Publishing AG 2017
V.M. Vishnevskiy et al. (Eds.): DCCN 2017, CCIS 700, pp. 87–97, 2017.
DOI: 10.1007/978-3-319-66836-9_8

Given the global scalability and big amount of the changes to the production system as well as its fast growing trend, the integration of all the management processes is needed for the purpose of the improvement of the operations efficiency.

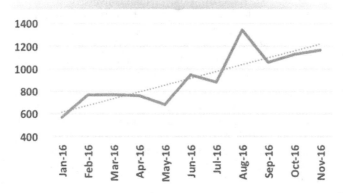

Fig. 1. Number of changes in RingCentral cloud infrastructure in 2016 per month

2   Cloud Operations

To operate effectively a cloud infrastructure and restore the custom services in case of the incident, the following management processes are traditionally handled in IT companies [3, 4]:

1. Events monitoring — consolidation and analysis of health data from remote servers.
2. Incident management — the process to monitor and restore the cloud services as soon as possible with minimal customer impact.
3. Problem management — the process to identify the root cause of repetitive incidents and apply a permanent fix.
4. Change management — planned system upgrade and maintenance with no downtime, providing 24/7 service availability.

The problem of big globally distributed IT companies is that each ITIL process is managed by a separate operations department, which may be geographically located in different regions and use different automation tools (Table 1).

The results of the analysis of the ITIL processes are as follows:

1. ITIL processes are logically correlated, meaning that each process may initiate another one (any change to the system may raise new monitoring events, which may cause an incident and a subject for root cause analysis of the problem, which, in its turn, results in a change to apply a fix to the system).

2. ITIL processes have data relationships (the result of any process is an input for another one as shown in Fig. 2).

Based on the analysis of the ITIL processes, the following requirements for new integration data model are formulated:

1. Ability to fast and reliable identification of proper time window to apply a fix to the system, taking into consideration the other planned changes, the current state of the system, ongoing and the recent incidents and known problems.
2. The problem management requires tracking of repetitive incidents and the status of root cause fixes.

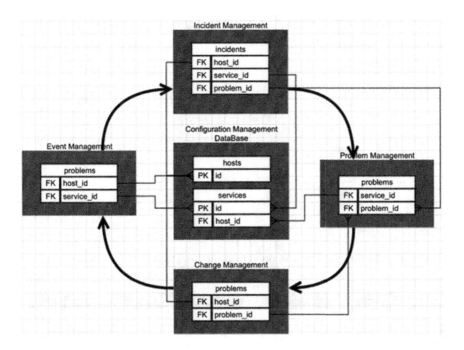

Fig. 2. Integration data model for continuous service delivery

Table 1. Operations processes and automation tools

Process	Department	Automation tool
Event monitoring	Monitoring team	Zabbix monitoring system
Incident management	Network operation center	Incident management portal
Problem management	System analysts and problem managers	JIRA tracking system
Change management	Change managers and deployment team	Auto-deployment System

3. The incident management requires visibility of relationships between the recent changes and current anomalies to be able to identify the faulty change and roll it back.

3   Integrated Cloud Operations Architecture

The proposed integration approach is implemented in the globally distributed infrastructure of the International RingCentral Company (see Fig. 3). For better visibility of the ongoing monitoring status and possible anomalies of all the

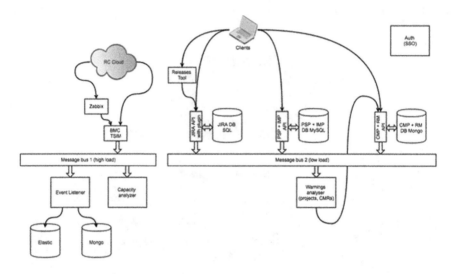

Fig. 3. Integrated cloud operations architecture at RingCentral

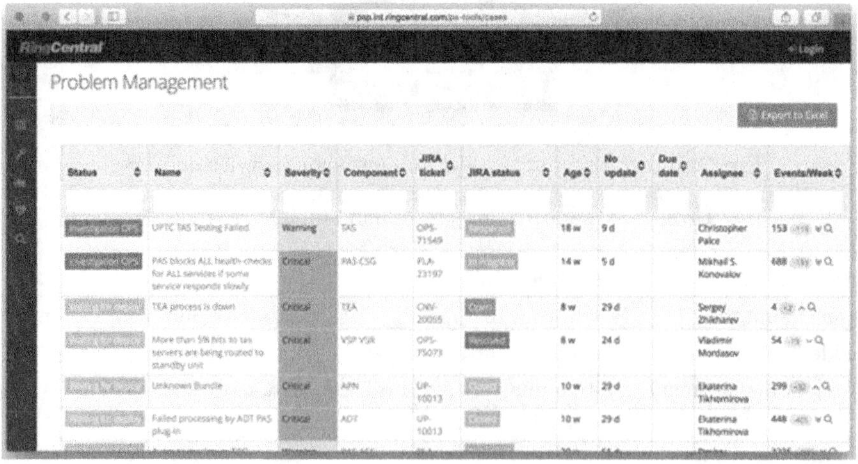

Fig. 4. Integrated interface, showing the problem name (Name), related JIRA ticket, amount of correlated alarms last week (Events/Week)

remote hosts in the globally distributed regions, the integrated user interface (see Fig. 4) is specially designed, allowing automation of the correlation between the monitoring events, recent hot changes and planned upgrades, known chronic problems and corresponding workarounds to apply a hot fix to restore cloud services with no customer impact.

4   KPI Evaluation and Integration

According to ITIL, the following standard metrics are used to evaluate the Quality of Service (QoS) and Key Performance Indicators (KPIs) in IT companies [6]:

1. Time to detect (TTD), mean time to detect (MTTD).
2. Time to communicate (TTC), mean time to communicate (MTTC).
3. Time to restore (TTR), mean time to restore (MTTR).
4. Turnaround time (TAT), mean turnaround time (MTAT).

TTD is a control reaction to an incident or an alert in the monitoring system, which cannot be immediate due to objective delay in data extraction and delivery. Basically, TTD is a term of ITIL Incident Management, which specifies the time between the actual event on remote device and the time it is detected in the monitoring system [7]. TTD is usually limited by the company policy and is determined by the time of the acknowledgment of an alert via integrated interface (see Fig. 4).

According to ITIL, TTC is the time of nonverbal notification and communication between involved people [8]. At RingCentral Company, TTC is computed as the time of a case created in JIRA incident tracking system [9] and escalated to appropriate executor, depending on the severity of an incident.

TTR in ITIL is a basic measure of the service downtime needed to restore or replace a failed component [10]. Sometimes TTR is interpreted as "time to response" that is not correct. TTR should include post-restore verification and is finished only after a QoS control is passed.

Mathematically, MTTR is estimated using the expression:

$$MTTR = \sum (SoU - SoD)/NoF, \tag{1}$$

where SoU — Start of Uptime; SoD — Start of Downtime; NoF — Number of Failures.

MTTR is usually a part of Service Level Agreement (SLA) [11]. At RingCentral Company with 24/7 service availability, the declared SLA is 99.99% as minimum, which is KPI of IT industry level.

TAT is standard ITIL category of performance engineering for overall evaluation of incident management effectiveness [12]. For Change Management in ITIL, TAT stands for the entire time needed to complete a change request [13]. According to particular company policy, TAT is computed as total time duration between the start of a task and the finish of its execution, meaning that the customer impact is ended, QoS verification test is OK and JIRA incident is closed. Therefore, TAT is a sum of TTD and TTC and TTR, or the average values of those KPIs:

$$TAT = TTD + TTC + TTR \tag{2}$$

$$MTAT = MTTD + MTTC + MTTR \tag{3}$$

Most of the TAT time is needed for TTR to repair the anomaly. To speed up the Incident Management process in many cases, a predefined remediation procedure is applied, either manually or automatically initiated from the monitoring system. Customer impact is also taken into account and is estimated depending on the number of affected users, dropped connections, capacity degradation, loss of redundancy, type of the service, geographic location, working weekday, business hours, duration of the service outage, etc.

At RingCentral Company, Zabbix enterprise-class solution [14] is implemented to collect all the monitoring data from remote hosts into a centralized SQL database with a specified polling interval, recalculate and verify the trigger expressions after each of the latest values received, and notify about the events when a certain value exceeded the specified threshold. The historical data and the trends are visualized as charts and graphs, custom screens and dashboards of specially designed web application (see Fig. 4).

To apply the proposed approach to the Incident Management and integrate with JIRA incident tracking system, the custom SQL queries are introduced and are being executed from Zabbix server against JIRA and other databases on daily, weekly and monthly schedule. Special aggregated metrics are configured in Zabbix to automatically evaluate the specified ITIL KPIs.

The type of the alert, the name of the affected host and the timestamp of the alarm registration are stored in Zabbix monitoring database. The actual date/time of the event on remote host can be extracted from JIRA ticket description once it is properly documented. A good example of JIRA case is shown in Fig. 5.

Fig. 5. JIRA sample case

Another option to automatically define the exact date/time of the event independently of the data transfer delay is to send it along with the polling value from remote host to Zabbix server as shown in Fig. 6. Estimation of the time difference between a real event happened on remote host and a timestamp registered in Zabbix monitoring system (reflected in Fig. 4) would give TTD value.

TTC is usually a manual process but is possible to automate if register the date/time of JIRA document creation and ignore a relatively fast action to assign a case to the predefined escalation person. Support engineer, who is on duty or on call, is involved into the Root Cause Analysis (RCA) after an incident is fixed and, therefore, is a subject of KPI evaluation as well.

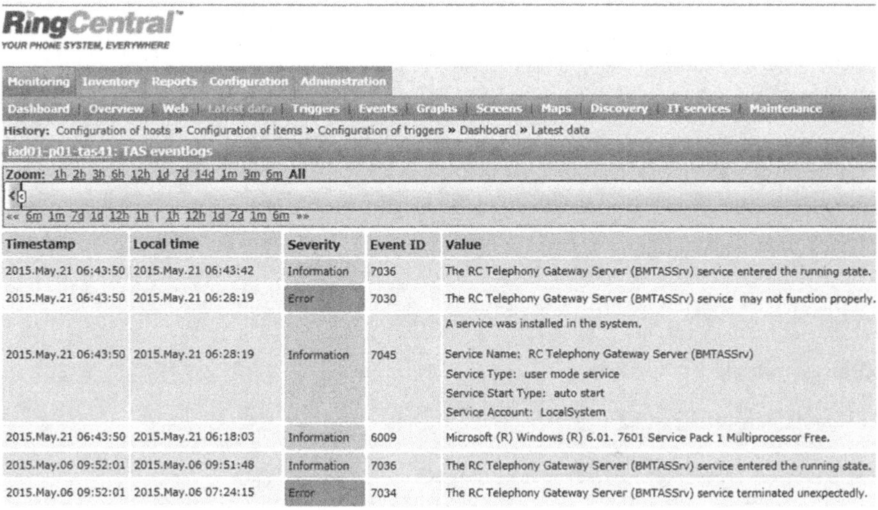

Fig. 6. Zabbix real time events on remote host

```
select ifnull(sum(c)-count(*),0) from
(select i.short_desc,
 (select group_concat(t.tag_name)
  from inc_has_tags h
  left outer join service_tags t
  on h.service_tags_id_tag=t.id_tag
  where h.incident_inc_id=i.inc_id
  group by h.incident_inc_id
 ) tags,
 count(*) c
 from incident i
 where create_start_date<curdate() and
 create_start_date>=adddate(curdate(),-1)
 group by i.short_desc, tags
 having count(*)>1
) as dupes;
```

Fig. 7. Sample SQL code to query out the repetitive alerts from the monitoring database

TTR is easy to calculate as time duration between JIRA incident created and closed, or moved to resolved status, meaning that the custom service is restored and QoS is tested.

Besides the standard ITIL KPIs, the additional incident related metrics are collected in the database and are being calculated by Zabbix monitoring system on a regular daily, weekly and monthly basis for the purpose of the Incident Management process improvement. For example, the following SQL code shown in Fig. 7 is implemented to figure out and analyze the repetitive incidents.

Items	KPI_IMP
% incidents detected by Zabbix	74
% incidents escalated	0
% incidents received during NOC manual hourly testing	11
% incidents received from customer support	0
% incidents received from other sources	5
Avg incident create to resolve duration, minutes	65
Changes related to incidents	CMR-16346,CMR-16362
Incidents resolved/closed	19
Incidents resolved/closed with level-0 impact	15
Incidents resolved/closed with level-1 impact	4
Incidents resolved/closed with level-2+ impact	0
Max customer impact, minutes	240
Max customers affected	0
Max number of dropped calls	5000
Total incidents	19
Total incidents with level-0 impact	15
Total incidents with level-1 impact	4
Total incidents with level-2+ impact	0

Fig. 8. Example of the Incident Management daily report

Items	KPI_CMP
% changes caused by incidents	4
% changes implemented within target time	84
% changes implemented without rollback plan	20
% changes implemented without testing	24
% changes requiring scheduled outages	0
% changes with post implementation feedback	40
% emergency changes	12
% no risk changes	64
% notification changes	56
Avg change implementation time, minutes	1h 7m
Changes implemented	25
Changes implemented - emergency	3
Changes implemented - major	1
Changes implemented - notification	14
Changes implemented - stage	0
Changes implemented - standard	7
Total changes approved	26
Total changes canceled/rejected	1
Total changes created	24
Total changes scheduled	31

Fig. 9. Example of the Change Management daily report

Figure 8 shows an example of the Incident Management report generated automatically on a daily and monthly basis in Zabbix monitoring system at RingCentral Company. Historical data is also available for reporting and analysis as both the tables and charts and graphs.

Change Management is also an imortant part of ITIL to track all the changes to production infrastructure, either planned upgrade to a new release or hot fix of an incident. Additional KPI metrics and corresponding SQL queries against the database are created in Zabbix similar to Incident Management KPIs described above. An example of the daily report of Change Management KPIs at Ring-Central Company is shown in Fig. 9.

Daily KPI reports are extended to a week, month and other custom time periods. For regular weekly, monthly and quarterly reports, the aggregated KPI calculated metrics are introduced and Zabbix triggers with certain thresholds are specified for the Incident and Change Management processes as well as for the analytics. Example in Fig. 10 shows the 2-months statistics of the change implementation time. One case is exceeded the 4 h limit, which is not a problem and is allowed on the weekend, while the average time 1 h 49 min is less than 50% of the overall maintenance window usage, which means that the Change Management may have a better planning and more efficient workload of the operations resources of IT Company.

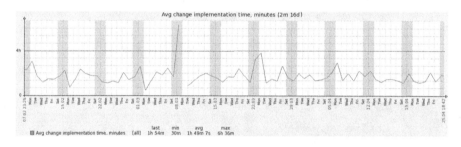

Fig. 10. Example of the change implementation time history

5   Conclusions

The results of the analysis of ITIL processes in big International IT companies with globally distributed services are presented. The correlation between repetitive incidents, chronic problems, and planned/unplanned changes is figured out.

Standard ITIL KPIs and additional metrics are implemented in Zabbix monitoring system to automatically evaluate and analyze the efficiency of the Incident and Change Management processes in IT Company. Not only Zabbix but optionally any other monitoring solution with a centralized database can be used for integration purposes.

Integration data model and the integrated applications are implemented for better visualization and efficiency of cloud operations. The benefit of new integration approach is automatic establishment of the relationship between the repetitive anomaly, statistics, known problem, and the progress of its resolution.

Additional study of the other automation tools, such as software testing and auto-deployment of applications into the cloud infrastructure, which are currently not integrated, is a subject of future research.

Acknowledgments. This work is implemented at and is partially supported by the International RingCentral Company [2].

References

1. Bernstein, D.: Containers and cloud: from LXC to Docker to Kubernetes. IEEE Cloud Comput. **1**(3) (2014). http://ieeexplore.ieee.org/document/7036275/
2. RingCentral Inc. http://ringcentral.com
3. ITIL 2011 Edn. https://en.wikipedia.org/wiki/ITIL
4. Ardulov, Y., Shchemelinin, D., Mescheryakov, S.: Monitoring and remediation of cloud services based on 4R approach. In: The 41st International IT Capacity and Performance Conference by CMG, San Antonio, TX, USA (2015). http://www.cmg.org/publications/conference-proceedings/conference-proceedings2015/
5. Lanubile, F., Ebert, C., Prikladnicki, R., Vizcaino, A.: Collaboration tools for global software engineering. IEEE Softw. **27**(2) (2010). http://ieeexplore.ieee.org/abstract/document/5420797/
6. Mescheryakov, S., Shchemelinin, D., Efimov, V.: Adaptive control of cloud computing resources in the internet telecommunication multiservice system. In: The 6th International Congress on Ultra Modern Telecommunications and Control Systems, St. Petersburg, Russia (2014). http://ieeexplore.ieee.org/xpl/articleDetails.jsp?arnum-ber=7002117
7. Fault Detection and Isolation. https://en.wikipedia.org/wiki/Fault_de-tection_and_isolation
8. Nonverbal Communication. http://en.wikipedia.org/wiki/Nonverbal_communication
9. Atlassian JIRA. https://www.atlassian.com/software/jira
10. Mean Time to Repair. http://en.wikipedia.org/wiki/Mean_time_to_repair
11. Service Level Agreement. http://en.wikipedia.org/wiki/Service-level_agreement
12. Performance Engineering. http://en.wikipedia.org/wiki/Performance_engineering
13. Turnaround Time. Performance Management. http://en.wikipedia.org/wiki/Turnaround_time
14. Zabbix Enterprise-class Monitoring Solutions. http://www.zabbix.com/solution

A Retrial Queueing System with Abandonment and Search for Priority Customers

A. Krishnamoorthy, V.C. Joshua$^{(\boxtimes)}$, and Ambily P. Mathew

Department of Mathematics, CMS College, Kottayam, India
achyuthacusat@gmail.com, vcjcms@gmail.com, ambilycms@gmail.com

Abstract. We consider a single server retrial queueing model with two types of customers, abandonment of customers and search for priority customers. Two types of customers namely type I and type II arrive according to two independent Poisson processes and upon arrival, if the server is busy, join orbits I and II respectively. Orbit I is of finite capacity while orbit II is of Infinite capacity. Retrial attempts are made by both types of customers at constant rates and each unsuccessful retrial attempt is followed by abandonment of customers from the respective orbits with some probability. Type I customer has higher priority over type II customers. Search for customers in orbit 1 is introduced to reduce their waiting time and also to minimize the loss of such customers due to abandonment. Condition for stability is established and the system state distribution is computed. Several performance measures of the system that influences the efficiency are derived and are illustrated graphically/numerically.

Keywords: Retrial queues · Orbit · Priority customers · Search · Abandonment

1 Introduction

The theory of retrial queues is recognized as an important part of queueing theory and mathematical modelling of telecommunication networks. The revolutionary changes happening in the field of information and communication theory boosted the applicability of queueing theory and other mathematical modelling techniques. A significant contribution has been and is being made by Russian scientiests and its wide range of applicability made it an important area within which active research has been continuously going on. Falin [10] and Falin and Templeton [11] gives a brief account of retrial queues and points out the inability of standard queueing models cannot be applied in solving a number of practically important problems. Artalejo and Gomez Corral [3] focuses on the application of algorithmic methods in studying retrial queues. Artalejo [4,5] gives a classified and accessible bibliography of research on retrial queus.

In the case of classical queue, Neuts and Ramalhoto [17] the concept of search by the server at a service completion epoch. In the case of $M/G/1$ queues

© Springer International Publishing AG 2017
V.M. Vishnevskiy et al. (Eds.): DCCN 2017, CCIS 700, pp. 98–107, 2017.
DOI: 10.1007/978-3-319-66836-9_9

with retrials search for orbital customers was introduced by Artalejo et al. [2]. Chakravarthy et al. [6] analysed multiserver queues with search of customers from the orbit. Krishnamoorthy et al. [15] incorporated non persistency of customers in $M/G/1$ Retrial queues with orbital search. Gomez Corral [8] done the stochastic analysis of a retrial queue with general retrial times. Choi and Chang [7] investigated retrial queue with finite capacity and geometric loss. Kazuki Kajiwara and Phung-Duc [13] analysed multiserver queues with Guard channel for Priority and retrial customers. Angelika [1] derived Stability Condition of a Retrial Queueing System with Abandoned and Feedback Customers.

In the present model, we consider two types of customers namely type I and type II of which type I customers are treated as valuable customers and the loss of such customers due to abandonment and the finiteness of the orbit will affect the cost effectiveness of the model. So we introduce search in the orbit of type I customers and as a result they get some sort of higher priority over type II customers in orbit II and primary customers of both the types. The search for orbital customers in $M/G/1$ retrial queues was introduced by Artalejo et al. [2]. Dudin et al. [9] extended the search mechanism to structurally complex single server model in which customers leave the system for ever without getting service.

In this paper we cosider retrial queues with two types of calls, abandonment of customers and search for valuable customers. Steady state probabilities are computed using Neuts' [18] Matrix Geometric methods. The rate matrix is computed using Logarithmic reduction Algorithm [16].

Simulation is an important numerical tool for the study of complex queueing models. Simulation methodology can effiectively be utilised for exploiting the stochastic elements associated with queueing models under consideration. The simulation outputs associated with the estimation of various performance measures can be analysed in detail and the analysis helps us to study cost effectiveness of the model. Simulation methods for queues are described by Glynn et al. in [12]. Konovalov and Razumchik gives a description of algorithmic simulation framework of the selection of efficicient Decision rules in Bank's Manual underwriting process [14].

2   Model Description

In this section we consider a single server retrial queueing system with two types of customers, namely type I and Type II arrive according to two independent Poisson processes with rates λ_1 and λ_2 respectively. Type I customers have higher priority over type II customers. Any type I customer, who upon arrival finds the server busy leaves the service area and joins orbit I which is of finite capacity, say K. Any type II customer, who upon arrival, finds the server busy, joins orbit II which is of infinite capacity. Retrial attempts are made by both types of cusomers from their respective orbits at constant retrial rates. Suppose that retrials from orbit I and orbit II are at rates μ_1 and μ_2 respectively. Any retrial attempt made by an orbital customer will result in a success if the server is idle at the time of the retrial and the customer will immediately enters in to service. If the retrials are not successful, then customers in orbit I and orbit II, abandon the system with probabilities γ_1 and γ_2 and return to their respective orbits with probabilities $1 - \gamma_1$ and $1 - \gamma_2$ respectively. Search for customers is introduced in retrial queueing models to reduce the idle time of the server and to ensure the maximum utitisation of the service facility. In this model search is for customers in orbit I since the customers in orbit I are of higher priority than those in orbit II and the search mechanism will reduce the waiting time of such customers and also improve the efficiency of the model under consideration. Since the loss of type I customers from the system will cost more, search for customers in orbit I helps to minimize the loss of such customers due to abondonment. Search time is assumed to be negligible. Immediately after each service completion epoch, the server either goes in for search of type I customers from orbit I with probability p or remains idle with probability $1-p$. Service time distribution H(.) is assumed to be of Phase type with irreducible representation $PH(\beta, S)$ and is of dimension m. The vector S^0 is given by $S^0 = -Se$.

Let $N_1(t)$ be the number of customers in orbit I, $N_2(t)$ the number of customers in orbit II, $C(t)$ the server status and $S(t)$ the phase of the service in progress. $C(t) = 0, 1,$ or 2 depending on whether the server is idle, busy with a typeI customer or busy with a type II customer respectively.

Let $\{N_2(t), N_1(t), C(t), S(t)\}$ the Markov process describing the process under consideration. This model can be considered as a Level Independent Quasi-Birth-Death $LIQBD$ process and a solution is obtained by Matrix Analytic Method. Refer Neuts [18] or Latouche and Ramaswamy [16] for a detailed study of Matrix Analytic Methods. We define the state space of the QBD under consideration and analyse the structure of its infinitesimal generator.

The state space consists of all quadraples of the form (n_2, n_1, i, j) where $n_2 \geq 0, 0 \leq n_1 \leq K, C(t) = 1, 2$ and $S(t) = 1, 2, ..., m$ when the server is busy and triplets of the form $n_2, n_1, 0$ where $n_2 \geq 0, 0 \leq n_1 \leq K$ when the server is idle.

The infinitesimal generator Q of the LIQBD describing the $M_2, M_1/PH/1$ retrial queue with search for priority customers and abandonment of customers is of the form

$$
\begin{pmatrix}
B & A_0 & O & \cdots\cdots\cdots \\
A_2 & A_1 & A_0 & O \cdots\cdots\cdots \\
O & A_2 & A_1 & A_0 & O \cdots\cdots \\
O & O & A_2 & A_1 & A_0 & O \cdots \\
 & \ddots & \ddots & \ddots & \ddots & \ddots & \ddots \\
 & & \ddots & \ddots & \ddots & \ddots & \ddots & \ddots
\end{pmatrix}
$$

where B, A_0, A_1, A_2 are all square matrices of order $3(K+1)$ whose entries are block matrices. A_0 represents the arrival of a customer to the system; that is transition from level $n \to n+1$. A_2 represents departure of a customer after service completion: $n \to n-1$, A_1 describes all transitions in which the level does not change (transitions within levels). The structure of the matrices are as follows:

$$\mathbf{B} = \begin{pmatrix}
-\lambda_1-\lambda_2 & \lambda_1\beta & \lambda_2\beta & 0 & \cdots & \cdots & \cdots \\
S^0 & S-(\lambda_1+\lambda_2)I & 0 & 0 & \lambda_1 I & \cdots & \\
S^0 & 0 & S-(\lambda_1+\lambda_2)I & 0 & 0 & \lambda_1 I & \\
0 & \mu_1\beta & 0 & -(\lambda_1+\lambda_2+\mu_1) & \lambda_1\beta & \lambda_2\beta & \cdots \\
0 & pS^0\otimes\beta+\gamma_1\mu_1 I & 0 & (1-p)S^0 & S-(\lambda_1+\lambda_2+\gamma_1\mu_1)I & \cdots & \lambda_1 I \\
0 & pS^0\otimes\beta & \gamma_1\mu_1 I & (1-p)S^0 & 0 & S-(\lambda_1+\lambda_2+\gamma_1\mu_1)I & \ddots \\
\ddots & \ddots & \ddots & \ddots & \ddots & \ddots & \ddots \\
\cdots & \cdots & \cdots & & & & \\
0 & \mu_1\beta & 0 & -(\lambda_1+\lambda_2+\mu_1) & \lambda_1\beta & \lambda_2\beta & \cdots \\
0 & pS^0\otimes\beta+\gamma_1\mu_1 I & 0 & (1-p)S^0 & S-(\lambda_1+\lambda_2+\gamma_1\mu_1)I & 0 & \lambda_1 I \\
0 & pS^0\otimes\beta & \gamma_1\mu_1 I & (1-p)S^0 & 0 & S-(\lambda_1+\lambda_2+\gamma_1\mu_1)I & \cdots \\
\ddots & \ddots & \ddots & \ddots & \ddots & \ddots & \ddots \\
\cdots & \cdots & \cdots & \cdots & \cdots & \cdots & \cdots \\
\cdots & \cdots & \mu_1\beta & 0 & -(\lambda_1+\lambda_2+\mu_1) & \lambda_1\beta & \lambda_2\beta \\
\cdots & \cdots & pS^0\otimes\beta+\gamma_1\mu_1 I & 0 & (1-p)S^0 & S-(\lambda_2+\gamma_1\mu_1)I & \\
\cdots & \cdots & pS^0\otimes\beta & \gamma_1\mu_1 I & (1-p)S^0 & 0 & S-(\lambda_2+\gamma_1\mu_1)I
\end{pmatrix}$$

$$\mathbf{A}_1 = \begin{pmatrix}
-(\lambda_1+\lambda_2+\mu_2) & \lambda_1\beta & \lambda_2\beta & 0 & \cdots & \\
S^0 & S-(\lambda_1+\lambda_2+\gamma_2\mu_2)I & 0 & 0 & \lambda_1 I & \cdots \\
S^0 & 0 & S-(\lambda_1+\lambda_2+\gamma_2\mu_2)I & 0 & 0 & \lambda_1 I \\
0 & \mu_1\beta & 0 & -(\lambda_1+\lambda_2+\mu_1+\mu_2) & \lambda_1\beta & \lambda_2\beta \\
0 & pS^0\otimes\beta+\gamma_1\mu_1 I & 0 & (1-p)S^0 & S-(\lambda_1+\lambda_2+\gamma_1\mu_1+\gamma_2\mu_2)I & \\
0 & pS^0\otimes\beta & \gamma_1\mu_1 I & (1-p)S^0 & 0 & \ddots \\
\ddots & \ddots & \ddots & \ddots & \ddots & \ddots \\
\cdots & \cdots & \cdots & \cdots & \cdots & \cdots \\
\cdots & \cdots & \cdots & \cdots & \cdots & \\
\cdots & \cdots & \cdots & \cdots & \cdots & \cdots \\
\ddots & \ddots & \ddots & \ddots & \ddots & \\
\cdots & \cdots & \cdots & \cdots & & \\
\cdots & \cdots & \mu_1\beta & 0 & -(\lambda_1+\lambda_2+\mu_1+\mu_2) & \lambda_1\beta & \lambda_2\beta \\
\cdots & \cdots & pS^0\otimes\beta+\gamma_1\mu_1 I & 0 & (1-p)S^0 & S-(\lambda_2+\gamma_1\mu_1+\gamma_2\mu_2)I \\
\cdots & pS^0\otimes\beta & \gamma_1\mu_1 I & (1-p)S^0 & 0 & \ddots
\end{pmatrix}$$

$$\mathbf{A}_0 = \begin{bmatrix} U & & \\ & \ddots & \\ & & U \end{bmatrix}$$

where

$$U = \begin{pmatrix} 0 & 0 & 0 \\ 0 & \lambda_2 I & 0 \\ 0 & 0 & \lambda_2 I \end{pmatrix}$$

$$\mathbf{A}_2 = \begin{bmatrix} V & & \\ & \ddots & \\ & & V \end{bmatrix}$$

where

$$V = \begin{pmatrix} 0 & 0 & \mu_2\beta \\ 0 & \gamma_2\mu_2 I & 0 \\ 0 & 0 & \gamma_2\mu_2 I \end{pmatrix}$$

The matrix $\mathbf{A} = \mathbf{A_0} + \mathbf{A_1} + \mathbf{A_2}$ can be written as

$$A = \begin{pmatrix} B_1^0 & B_0 & O & \cdots & \cdots \\ B_2 & B_1 & B_0 & \ddots & \cdots \\ O & \ddots & \ddots & \ddots & O \\ \cdots & \cdots & B_2 & B_1 & B_0 \\ \cdots & \cdots & \cdots & B_2 & B_1^K \end{pmatrix}$$

where

$$B_1^0 = \begin{pmatrix} -(\lambda_1 + \lambda_2 + \mu_2) & \lambda_1\beta & (\lambda_2 + \mu_2)\beta \\ S^0 & S - \lambda_1 I & 0 \\ S^0 & 0 & S - \lambda_1 I \end{pmatrix}$$

$$B_0 = \begin{pmatrix} 0 & 0 & 0 \\ 0 & \lambda_1 I & 0 \\ 0 & 0 & \lambda_1 I \end{pmatrix}$$

$$B_1 = \begin{pmatrix} -(\lambda_1 + \lambda_2 + \mu_1 + \mu_2) & \lambda_1\beta & (\lambda_2 + \mu_2)\beta \\ (1-p)S^0 & S - (\lambda_1 + \gamma_1\mu_1)I & 0 \\ (1-p)S^0 & 0 & S - (\lambda_1 + \gamma_1\mu_1)I \end{pmatrix}$$

$$B_2 = \begin{pmatrix} 0 & \mu_1\beta & 0 \\ 0 & pS^0 \otimes \beta + \gamma_1\mu_1 I & 0 \\ 0 & pS^0 \otimes \beta & \gamma_1\mu_1 I \end{pmatrix}$$

$$B_1^N = \begin{pmatrix} -(\lambda_1 + \lambda_2 + \mu_1 + \mu_2) & \lambda_1\beta & (\lambda_2 + \mu_2)\beta \\ (1-p)S^0 & S - \gamma_1\mu_1 I & 0 \\ (1-p)S^0 & 0 & S - \gamma_1\mu_1 I \end{pmatrix}$$

We see that $\mathbf{A}$ is an irreducible infinitesimal generator matrix(ref) and so there exists the stationary vector π of $\mathbf{A}$ such that

$$\pi\mathbf{A} = \mathbf{0}, \pi\mathbf{e} = \mathbf{1} \tag{1}$$

where $\pi = (\pi_1, \pi_2, \ldots, \pi_K)$ whose components are $\pi_i = (\pi_{i0}, \pi_{i1}, \pi_{i2})$, for $0 \le i \le K$. and $\pi_{i1} = (\pi_{i11}, \pi_{i12} \ldots, \pi_{i1m})$ and $\pi_{i2} = (\pi_{i21}, \pi_{i22} \ldots, \pi_{i2m})$. Solving these equations we get

$$\pi_i = \pi_K \prod_{j=0}^{K-1-i} H_{K-i-j} \tag{2}$$

for $i = 0, 1, \ldots K - 1$ where the sequence of matrices H_i are defined as

$$H_i = -B_2[H_{i-1}B_0 + B_1]^{-1}$$

for $i = 1, 2, \ldots\ldots, K - 1$ and

$$H_0 = -B_2[B_1^0]^{-1}.$$

The normalising condition is given by the equation

$$\pi_K \left[\sum_{i=0}^{K-1} \prod_{j=0}^{K-1-i} H_{K-1-j} + I \right] \mathbf{e} = 1 \qquad (3)$$

The stability condition is given by

$$\pi \mathbf{A_0 e} < \pi \mathbf{A_2 e}.$$

i.e.

$$\pi_K \left[\sum_{i=0}^{K-1} \prod_{j=0}^{K-1-i} H_{K-1-j} + I \right] \mathbf{Ue} < \pi_\mathbf{K} \left[\sum_{i=0}^{K-1} \prod_{j=0}^{K-1-i} \mathbf{H_{K-1-j}} + \mathbf{I} \right] \mathbf{Ve}. \qquad (4)$$

In terms of the The stability conditionin terms of the component subvectors is given by

$$[\lambda_2 - \gamma_2\mu_2] \sum_{i=0}^{K} \sum_{j=1}^{2} \sum_{k=1}^{m} \pi_{ijk} < \mu_2 \sum_{i=0}^{K} \pi_{i0}.$$

3   Matrix Analytic Solution

The Quasi-birth-death processes can be conveniently and efficiently solved using the *Matrix Analytic Method*.

The stationary distribution of the Markov process under consideration is obtained by solving the set of equations $\mathbf{xQ} = \mathbf{0}$, $\mathbf{xe} = \mathbf{1}$. Let $\mathbf{x}$ be decomposed conformally with $\mathbf{Q}$. Then
$\mathbf{x} = (\mathbf{x_0}, \mathbf{x_1}, \mathbf{x_2}, \ldots\ldots)$ where $\mathbf{x_i} = (\mathbf{x_{i0}}, \mathbf{x_{i1}}, \ldots\ldots \mathbf{x_K})$

$$x_{ij} = (x_{ij0}, \mathbf{x_{ij1}}, \mathbf{x_{ij2}})$$

for

$$j = 1, 2, \ldots\ldots, K$$

whereas for $k = 1, 2$, the vectors

$$\mathbf{x_{ijk}} = (\mathbf{x_{ijk1}}, \mathbf{x_{ijk2}} \ldots\ldots\ldots, \mathbf{x_{ijkm}}).$$

x_{ijkl} is the probability of being in state $(ijkl)$ for $k = 1, 2, i \geq 0, j = 1, 2 \ldots\ldots K, l = 1, 2, \ldots\ldots m$ and $\mathbf{x_{ij0}}$ is the probability of being in state$(i, j, 0)$. From $\mathbf{xQ} = \mathbf{0}.$, it may be shown that there exists a constant matrix $\mathbf{R}$ such that $\mathbf{x_i} = \mathbf{x_{i-1}R}$. The subvectors $\mathbf{x_i}$ are geometrically related by the equation $\mathbf{x_i} = \mathbf{x_0R^i}$ R can be obtained from the matrix quadratic equation

$$R^2 A_2 + R A_1 + A_0 = O \qquad (5)$$

R can be obtained by successive substitution procedure $R(0) = 0$ and

$$R_{k+1} = -V - R_k^2 W$$

where $V = A_2 A_1^{-1}, W = A_0 A_1^{-1}$ or x_0 can be evaluated using $\mathbf{xe} = \mathbf{1}$

3.1   Some Performance Measures of the System

In this section we evaluate some performance measures of the system.

1. Expected Number of customers in the system

$$E[N] = \sum_{i=0}^{\infty} i \sum_{j=0}^{K} j \mathbf{x_{ij0}} + \sum_{i=0}^{\infty} i \sum_{j=0}^{K} (j+1) \sum_{s=1}^{m} \mathbf{x_{ij1s}} + \sum_{i=0}^{\infty} (i+1) \sum_{j=0}^{K} j \sum_{s=1}^{m} \mathbf{x_{ij2s}}$$

2. Expected Number of customers in orbit I

$$E[N_1] = \sum_{i=0}^{\infty} \sum_{j=0}^{K} j \mathbf{x_{ij0}} + \sum_{j=0}^{K} j \sum_{i=0}^{\infty} \sum_{r=1}^{2} \sum_{s=1}^{m} \mathbf{x_{ijrs}}$$

3. Expected Number of customers in orbit II

$$E[N_2] = \sum_{i=0}^{\infty} i \sum_{j=0}^{K} \mathbf{x_{ij0}} + \sum_{i=0}^{\infty} i \sum_{j=0}^{K} \sum_{r=1}^{2} \sum_{s=1}^{m} \mathbf{x_{ijrs}}$$

4. Expected Number of customers abandoned from orbit I

$$E[N_{aban1}] = \sum_{i=0}^{\infty} \sum_{j=0}^{K} \sum_{r=1}^{2} \sum_{s=1}^{m} j \gamma_1 \mu_1 \mathbf{x_{ijrs}}$$

5. Expected Number of customers abandoned from orbit II

$$E[N_{aban2}] = \sum_{i=0}^{\infty} \sum_{j=0}^{K} \sum_{r=1}^{2} \sum_{s=1}^{m} i \gamma_2 \mu_2 \mathbf{x_{ijrs}}$$

6. Probability that the server is idle

$$a_0 = \sum_{i=0}^{\infty} \sum_{j=0}^{K} \mathbf{x_{ij0}}$$

7. Probability that the server is busy with type I customer

$$a_1 = \sum_{i=0}^{\infty} \sum_{j=0}^{K} \sum_{s=1}^{m} \mathbf{x_{ij1s}}$$

8. Probability that the server is busy with type II customer

$$a_2 = \sum_{i=0}^{\infty} \sum_{j=0}^{K} \sum_{s=1}^{m} \mathbf{x_{ij2s}}$$

9. The probability that a customer of type I is blocked from entering the system upon arrival

$$P_{block(typeI)} = \sum_{i=0}^{\infty} \sum_{r=1}^{2} \sum_{s=1}^{m} \mathbf{x_{iKrs}}$$

4   Numerical Examples

We consider a single server system in which customers of type I and type II arrive according to a poisson process with rates λ_1 and λ_2 respectively. Suppose that the service is exponential with mean service rate ν. In the present model we first illustrate the effect of search probability on the number of customers in the system and the probability that the server is idle. Numerical results are done by using simulation techniques. Let K, the maximum number of customers in the Orbit I be 100.

From Tables 1 and 2, it can be infered that as expected, the probability that the server is idle and the expected number of customers in the system decreases with increasing values of the probability p with which the server searches for customers from orbit I. For Table 1 we have

$$\lambda_1 = 2, \lambda_2 = 3, \nu = 5, \mu_1 = 2, \mu_2 = 3, \gamma_1 = 0.6, \gamma_2 = 0.7$$

Table 1. Effect of search success probability p.

p	N	a_0
0.1	1.877007764	0.420448008
0.2	1.819894917	0.414916092
0.3	1.775582399	0.408386011
0.4	1.727738793	0.404507702
0.5	1.689207994	0.397936296
0.6	1.662106678	0.39348927
0.7	1.621625934	0.390806184
0.8	1.599075927	0.387882619
0.9	1.57728109	0.384563212

For Table 2, we take $\lambda_1 = 1$, $\lambda_2 = 2$, $\nu = 5$, $\mu_1 = 4$, $\mu_2 = 5$, $\gamma_1 = 0.7$, $\gamma_2 = 0.7$

We have introduced search in orbit I to minimize the expected number of abandoned customers from orbit I and also to minimise the idle time of the server. The numerical results in Table 3 shows that the expeted number of abandoned customers from orbit 1 decreases with increasing value of search success probability, p. The number of customers abandoned from orbit II increases with increasing values of p since as p increases the idle time of the server is reduced and as a result the number of sucessful retrials from orbit II decreases which results in the abandonment of customers from orbit II. For the results in Table 3, we take

$$\lambda_1 = 10, \lambda_2 = 15, \nu = 5, \mu_1 = 10, \mu_2 = 12, \gamma_1 = 0.6, \gamma_2 = 0.7, p = .5$$

Table 2. Effect of search success probability p.

p	N	a_0
0.1	0.721804481	0.570844189
0.2	0.713102888	0.57007455
0.3	0.71116057	0.569944736
0.4	0.705437642	0.569419931
0.5	0.697892615	0.567688565
0.6	0.694974951	0.566081789
0.7	0.689376435	0.566066395
0.8	0.685385248	0.56457097
0.9	0.680184129	0.564330148

Table 3. Expected number of abandonments.

Search success probability	Expected number of abandonments	
p	Orbit I	Orbit2
0	4.937433764	8.700333274
0.1	4.870248393	8.762917016
0.2	4.781446625	8.819218354
0.3	4.738514516	8.887289707
0.4	4.681411964	8.935210822
0.5	4.594032819	9.007115559
0.6	4.541464845	9.073393536
0.7	4.477382067	9.130821335
0.8	4.400131124	9.175793742
0.9	4.333457057	9.234141287
1	4.283148845	9.288529334

Acknowledgement. A. Krishnamoorthy and V.C Joshua thanks the Department of Science and Technology, Goverment of India for the support given under the Indo-Russian Project $INT/RUS/RSF/P - 15$.

References

1. Angelika, A.B., Rabhi, A., Yahiaoui, L.: Stability condition of a retrial queueing system with abandoned and feedback customers. Appl. Appl. Math. **10**(2), 667–677 (2015)
2. Artalejo, J.R., Joshua, V.C., Krishnamoorthy, A.: An M/G/1 retrial queue with orbital search by server. In: Artalejo, J.R., Krishnamoorthy, A. (eds.) Advances in Stochastic Modelling, pp. 41–54. Notable Publications, New Jersy (2003)

3. Artalejo, J.R., Gomez-Corral, A.: Retrial Queueing Systems: A Computational Approach. Springer-Verlag, Berlin (2008)
4. Artalejo, J.R.: Accessible bibliography of research on retrial queues. Math. Comput. Modell. **30**, 1–6 (1999)
5. Artalejo, J.R.: A classified bibliography of research on retrial queues Progress in 1990–1999. Top **7**, 187–211 (1999)
6. Chakravarthy, S.R., Krishnamoorthy, A., Joshua, V.C.: Analysis of a multi-server retrial queue with search of customers from the orbit. Perform. Eval. **63**(8), 776–798 (2006)
7. Choi, B.D., Chang, Y.: $MAP_1, MAP_2/M/C$Retrial Queue with the Retrial Group of Finite Capacity and Geometric Loss. Math. Comput. Modell. **30**, 99–113 (1999)
8. Corral Gomez, A.: Stochastic Analysis of a single server retrial queue with general retrial times. Nav. Res. Log. **46**, 561–581 (1999). John-Wiley And Sons
9. Dudin, A.N., Krishnamoorthy, A., Joshua, V.C., Tsarenkov, G.: Analysis of BMAP/G/1 retrial system with search of customers from the orbit. Eur. J. Oper. Res. **157**, 169–179 (2004)
10. Falin, G.I.: A survey of retrial queues. Queueing Syst. **7**(2), 127–167 (1990)
11. Falin, G.I., Templeton, J.G.C.: Retrial Queues. Chapman and Hall, London (1997)
12. Glynn, P.W., Iglehart, D.L.: Simulation method for queues:an overview. Queueing Syst. **3**, 221–256 (1988)
13. Kajiwara, K., Phung-Duc, T.: Multiserver Queue with Guard Channel for Priority and Retrial Customers. Int. J. Stoch. Anal. **2016**, 23 (2016)
14. Konovalov, M., Razumchik, R.: Simulation and selection of efficient Decision rules in Bank's manual underwriting process. In: Procedings of 30th European Conference on Modeling and Simulation. ISBN: 978-0-9932440-2-4
15. Krishnamoorthy, A., Deepak, T.G., Joshua, V.C.: An M/G/1 retrial queue with nonpersistent customers and orbital search. Stoch. Anal. Appl. **23**, 975–997 (2005)
16. Latouche, G., Ramaswami, V.: Introduction to Matrix Analytic Methods in Stochastic Modeling. SIAM, Philadelphia (1999). Mathematics - 334 pages
17. Neuts, M.F., Ramalhoto, A.: Service model in which the server is required to search for customers. J. Appl. Prob. **21**, 157–166 (1984)
18. Neuts, M.F.: Matrix-Geometric Solutions in Stochastic models-An Algorithmic Approach. The Johns Hopkins University Press, Baltimore and London (1981)

Usage of Video Codec Based on Multichannel Wavelet Decomposition in Video Streaming Telecommunication Systems

Kirill Bystrov[✉], Alexander Dvorkovich, Viktor Dvorkovich, and Gennady Gryzov

Moscow Institute of Physics and Technology, Institutsky lane 9, 141700 Dolgoprudny, Moscow region, Russia
kirill.bystrov@phystech.edu, dvork.alex@gmail.com, v.dvorkovich@mail.ru, gryzov@gmail.com
https://mipt.ru/en/

Abstract. The amount of transmitted video content is rising sharply in modern telecommunication systems. This is especially true for streaming video. In this regard, the requirements are increased not only in terms of the compression ratio and the quality of the transmitted image, but also in terms of video codec performance.

The usage of multichannel wavelet transform may be a way to increase the compression ratio while maintaining the same quality of reconstructed images.

Therefore, the aim of the study is to implement multichannel wavelet decomposition based video codec and to make use of it in real-time mode.

Multichannel wavelet video codec based on new filter banks specially designed for wavelet decomposition of images into a certain number of channels has been implemented within the research.

Video compression results for various number of decomposition channels are presented in the efficiency evaluation of multichannel wavelet video codec.

Keywords: Discrete Wavelet Transform · Video codecs · Streaming video · Telecommunication systems · Video coding performance

1 Introduction

Block orthogonal transforms are widely used in intra-frame video compression algorithms and thereby enabling to reduce the size of transmitted video content. Historically, discrete cosine transform (DCT) and integer approximation of DCT are generally used in video codecs. Discrete wavelet transform (DWT) may be considered as an alternative to DCT. DWT enables to avoid "blocking" artifacts which occur while using DCT at low bitrates and consequently provides a theoretical opportunity for achieving better quality of reconstructed images.

Two-channel DWT scheme based on low-pass and high-pass FIR filters is usually used (for example, in JPEG2000 standard [1]). The increase in the number

© Springer International Publishing AG 2017
V.M. Vishnevskiy et al. (Eds.): DCCN 2017, CCIS 700, pp. 108–119, 2017.
DOI: 10.1007/978-3-319-66836-9_10

of channels should enable to improve the compression ratio while maintaining the same quality of the reconstructed images due to a more compact representation of the signal energy along the frequency subbands [2]. These considerations are confirmed by experience of implementing three-channel DWT in Dirac video codec [3].

Therefore, the aim of the study is to implement multichannel DWT video codec and to evaluate its efficiency. Estimation of compression ratio at a certain level of distortion is used as the criteria of effectiveness. Three- and five-channel banks of FIR filters are selected for testing the video codec based on multichannel DWT.

The novelty of the study is the implementation of the video codec which supports video coding with various number of channels, including wavelet decomposition of images into five channels, what is practically implemented in video codecs for the first time. The practical value of these innovations lies both in the ability to verify the practical applicability of various filter banks in video coding and in the potential of using this codec for streaming video after implementation of necessary modifications and optimization.

2   Image Processing Using DWT

If filter bank is correctly selected, then an increase in the number of channels (i.e. in the number of used filters) should lead to greater compactness of the image energy in the low-frequency domain. Consider multichannel discrete wavelet transform in the case of the three-channel scheme. Transform coefficients are obtained at the analysis stage: the convolution of the input signal and filters' impulse characteristics (low-pass filter H, middle-pass filter B, high-pass filter G) (see Fig. 1) is performed and then its result is decimated according to selected decimation scheme (see Fig. 2). The initial signal is reconstructed at the synthesis stage: the inverse wavelet transform is performed by the convolution of subbands transform coefficients, which are supplemented by zero taps according to decimation scheme, and those of corresponding reconstruction filters (low-pass filter K_h, middle-pass filter K_b, high-pass filter K_g).

Filters frequency responses are determined from (1), considering that H and G filters are symmetric, i.e. $h_n = h_{-n}$, $g_m = g_{-m}$, and B filter is antisymmetric, i.e. $b_0 = 0$, $b_l = b_{-l}$. Reconstruction filters characteristics K_h, K_b and K_g are determined from (2). Initial signal reconstruction condition is determined from (3), where $H_\triangle, H_\triangledown, H_\bigcirc, B_\triangle, B_\triangledown, B_\bigcirc, G_\triangle, G_\triangledown, G_\bigcirc$ - are filters characteristics after corresponding decimation (see Fig. 2).

$$
\begin{cases}
H(x) = h_0 + 2 \sum_{n=1}^{N} h_n cos(\pi n x), \\
B(x) = 2j \sum_{l=1}^{L} b_l sin(\pi l x), \\
G(x) = g_0 + 2 \sum_{m=1}^{M} g_m cos(\pi m x).
\end{cases}
\tag{1}
$$

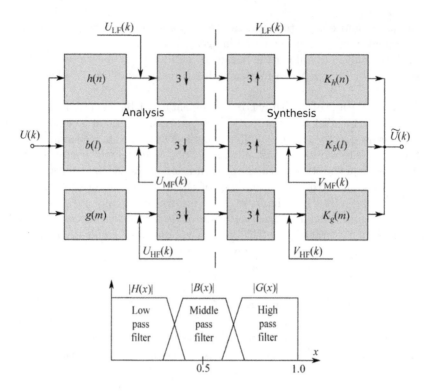

Fig. 1. The structural scheme of three-channel subband transform system

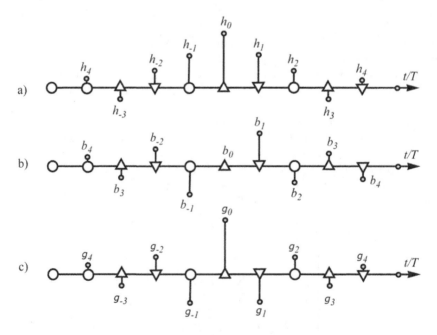

Fig. 2. LF(a), MF(b), HF(c) reactions in response to the unit impulse

$$\begin{cases} K_h(x) = kh_0 + 2 \sum_{n=1}^{N_1} kh_n cos(\pi nx), \\ K_b(x) = 2j \sum_{l=1}^{L_1} kb_l sin(\pi lx), \\ K_g(x) = kg_0 + 2 \sum_{m=1}^{M_1} kg_m cos(\pi mx). \end{cases} \qquad (2)$$

$$\begin{cases} H_\triangle(x)K_h(x) + B_\triangle(x)K_b(x) + G_\triangle(x)K_g(x) = 1, \\ H_\triangledown(x)K_h(x) + B_\triangledown(x)K_b(x) + G_\triangledown(x)K_g(x) = 1, \\ H_\bigcirc(x)K_h(x) + B_\bigcirc(x)K_b(x) + G_\bigcirc(x)K_g(x) = 1. \end{cases} \qquad (3)$$

$N + L + M + 2$ equations are required for computing all unknown values h_n, b_l, g_m. The necessary conditions: the determinant of the system must be a constant, $H(0) = \text{const}$, $H(1) = 0$ (i.e. H is low-pass filter), $B(0) = 0$, $B(1) = 0$ (i.e. B is middle-pass filter), $G(0) = 0$, $G(1) = \text{const}$ (i.e. G is high-pass filter). Equations system must be supplemented by additional conditions, for example by the equality to zero of filter characteristics derivatives at the boundaries of the range and etc. [4], in order that it would be determined.

All expressions for five-channel scheme to (4), (5) and (6) are similar to (1), (2) and (3), but with more variables and more complex equations system [5]. It should also be noted that H and G filters are symmetric, i.e. $h_n = h_{-n}$, $g_k = g_{-k}$; D and C filters are antisymmetric, i.e. $d_0 = 0$, $d_l = \text{-}d_{-l}$, $c_0 = 0$, $c_r = \text{-}c_{-r}$ (see Figs. 3 and 4).

$$\begin{cases} H(x) = h_0 + 2 \sum_{n=1}^{N} h_n cos(\pi nx), \\ D(x) = 2j \sum_{l=1}^{L} d_l sin(\pi lx), \\ B(x) = 2j \sum_{s=1}^{S} b_s sin(\pi sx), \quad b_0 \equiv 0, \quad or \\ B(x) = b_0 + 2 \sum_{s=1}^{S} b_s cos(\pi sx), \\ C(x) = 2j \sum_{r=1}^{R} d_r sin(\pi rx), \\ G(x) = g_0 + 2 \sum_{k=1}^{K} g_k cos(\pi kx). \end{cases} \qquad (4)$$

$$\begin{cases} K_h(x) = kh_0 + 2\sum_{n=1}^{N_1} kh_n cos(\pi nx), \\[2mm] K_d(x) = 2j\sum_{l=1}^{L_1} kd_l sin(\pi lx), \\[2mm] K_b(x) = 2j\sum_{s=1}^{S_1} kb_s sin(\pi sx), \quad kb_0 \equiv 0, \quad or \\[2mm] K_b(x) = kb_0 + 2\sum_{s=1}^{S_1} kb_s cos(\pi sx), \\[2mm] K_c(x) = 2j\sum_{r=1}^{R_1} kd_r sin(\pi rx), \\[2mm] K_g(x) = kg_0 + 2\sum_{k=1}^{K_1} kg_k cos(\pi kx). \end{cases} \tag{5}$$

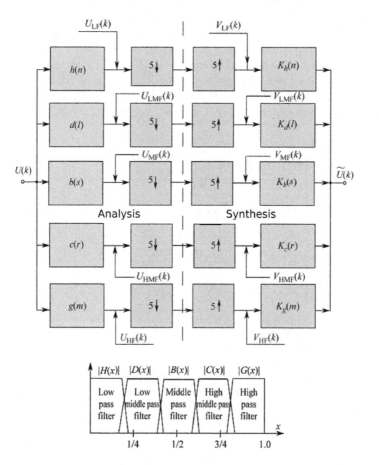

Fig. 3. The structural scheme of five-channel subband transform system

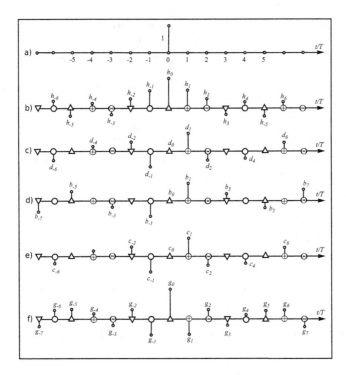

Fig. 4. LF(b), LMF(c), MF(d), HMF(e), HF(f) reactions in response to the unit impulse(a)

$$
\begin{cases}
H_\triangle(x)K_h(x) + D_\triangle(x)K_d(x) + B_\triangle(x)K_b(x) + C_\triangle(x)K_c(x) + G_\triangle(x)K_g(x) = 1, \\
H_\triangledown(x)K_h(x) + D_\triangledown(x)K_d(x) + B_\triangledown(x)K_b(x) + C_\triangledown(x)K_c(x) + G_\triangledown(x)K_g(x) = 1, \\
H_\bigcirc(x)K_h(x) + D_\bigcirc(x)K_d(x) + B_\bigcirc(x)K_b(x) + C_\bigcirc(x)K_c(x) + G_\bigcirc(x)K_g(x) = 1, \\
H_\ominus(x)K_h(x) + D_\ominus(x)K_d(x) + B_\ominus(x)K_b(x) + C_\ominus(x)K_c(x) + G_\ominus(x)K_g(x) = 1, \\
H_\oplus(x)K_h(x) + D_\oplus(x)K_d(x) + B_\oplus(x)K_b(x) + C_\oplus(x)K_c(x) + G_\oplus(x)K_g(x) = 1.
\end{cases} \quad (6)
$$

3   Practical Implementation

It is required either to develop new video codec or to implement multichannel wavelet transform in an existing one for the effectiveness evaluation of the practical usage of multichannel wavelet transform in video coding. The second way was chosen and Schrodinger video codec [6] was selected for the given task. The choice of this codec is determined by the following factors: open and free source code; the usage of wavelet transform for video compression.

The process of multichannel wavelet video codec development includes:

1. analysis of Schrodinger's structure;
2. identification of "implicit two-channeling" (such implementation of functions and data structures that enables only two-channel transform) and required changes;

3. "implicit two-channeling" removal and source code modification for working with various number of wavelet transform channels;
4. computation of filter banks for multichannel wavelet transform;
5. multichannel wavelet transform implementation;
6. testing the codec with specially designed filter banks and evaluation of its effectiveness.

Three- and five-channels banks of filters were used for testing efficiency of developed video codec. Analysis filter bank 23/23/23 (see Table 1) and synthesis filter bank 13/13/13 (see Table 2) were used for three-channel wavelet transform; 5/5/5/5/5 analysis and synthesis filter banks (see Tables 3 and 4) were used for five-channel transform. Mallat's pyramid [7] with 2 levels was used for efficiency evaluation of these filter banks.

A direct comparison of the given wavelet video codec with x264 and x265 codecs is not correct enough because the last ones are significantly optimized in terms of quality (motion estimation and compensation, entropy coder, etc.). Meanwhile, their test results are also presented to demonstrate the potential of wavelet codec under development, keeping in mind appropriate further improvements.

The following video sequences of standard (704×576 - 4cif), enhanced (1280×720 - 720p) and high (1920×1080 - 1080p) definition were selected for testing:

4cif – city, crew, harbour, ice, soccer; [8]
720p – mobcal, parkrun, shields, stockholm; [9]
1080p – blue_sky, pedestrian_area, rush_hour, station2, sunflower, tractor. [9]

Table 1. Analysis filter bank 23/23/23 for three-channel wavelet transform

n	$h(n)$	$b(n)$	$g(n)$
0	0.717215	0	-0.685303
1	0.482119	-0.687905	-0.499947
2	0.100785	0.052845	0.146139
3	-0.0572617	0.192827	0.0322635
4	-0.0434293	-0.0239495	-0.0249847
5	-0.00257065	0.0337432	0.00924183
6	0.00762182	-0.00319313	-0.00460337
7	0.0016123	-0.00829397	-0.000219346
8	0.00091449	0.000815896	-0.000671062
9	0.000107057	-0.0000712195	-0.0000620485
10	-0.000411828	0.000273966	0.000238688
11	0.0000812236	-0.0000540336	-0.0000470757

Table 2. Synthesis filter bank 13/13/13 for three-channel wavelet transform

n	$kh(n)$	$kb(n)$	$kg(n)$
0	0.663679	0	0.662462
1	0.499683	0.688063	−0.509985
2	0.155318	0.0473184	0.0954987
3	−0.0494516	−0.153701	0.0783513
4	−0.0573942	−0.0260899	−0.069081
5	−0.00810589	−0.00475089	−0.00853267
6	0.0123629	0.00151849	0.0195878

Table 3. Analysis filter bank 5/5/5/5/5 for five-channel wavelet transform

n	$h(n)$	$d(n)$	$b(n)$	$c(n)$	$g(n)$
0	0.4472135955	0	0.59628479442586	0	0.666666666
1	0.4472135955	0.36	0.22360679743054	0.608604962188	−0.5
2	0.4472135955	0.608604962188	−0.52174919464347	−0.36	0.166666667

Table 4. Synthesis filter bank 5/5/5/5/5 for five-channel wavelet transform

n	$kh(n)$	$kd(n)$	$kb(n)$	$kc(n)$	$kg(n)$
0	0.4472135955	0	0.59628479442586	0	0.666666666
1	0.4472135955	−0.36	0.22360679743054	−0.608604962188	−0.5
2	0.4472135955	−0.608604962188	−0.52174919464347	0.36	0.166666667

4   The Results

The dependence of distortion level on bitrate is one of the video codec effectiveness criteria. The PSNR metric of reconstructed images was used as the distortion level estimate in test video sequences. Video coding was performed in Intra mode because Schrodinger's motion estimation and compensation algorithms are significantly inferior to x264/x265 analogs and they have not been modified yet. Graphs for some of the tested video sequences are presented in the paper.

Three-channel filter bank results are comparable to those of x264 for most video sequences (see Figs. 5 and 6) and, moreover, they are better than x264 results and close to those obtained by x265 for some 1080p test videos (see Figs. 7 and 8).

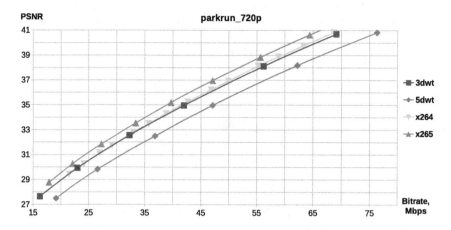

Fig. 5. Parkrun

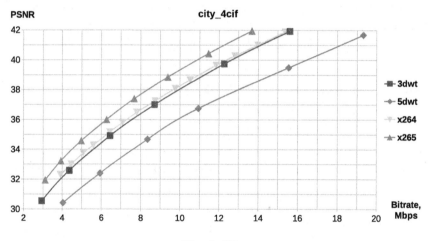

Fig. 6. City

Nevertheless, three-channel wavelet transform results are significantly inferior to those of x264/x265 for some other test videos (see. Fig. 9). It is attributed not to the transform efficiency but to the difference in entropy coding between x264/x265 and Schrodinger - the last uses simpler realization of entropy coder. Results comparable to x264/x265 are expected for such video sequences (and perhaps even better for the rest of tested videos) due to entropy coding improvement in wavelet codec.

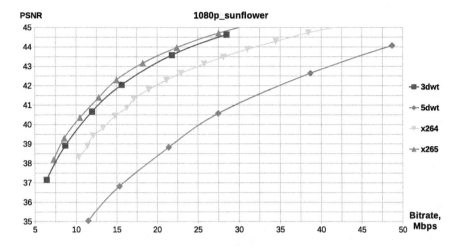

Fig. 7. Sunflower

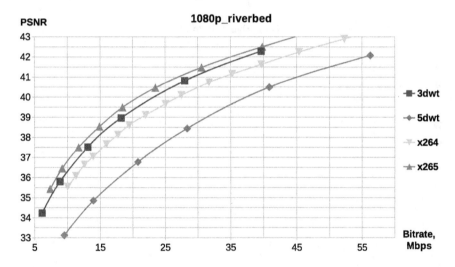

Fig. 8. Riverbed

It should be noted that results close to those of x264 are obtained by wavelet codec also in motion prediction mode for individual video sequences (see Fig. 10) for which motion estimation is inefficient (for example, video "riverbed"). This shows that motion prediction algorithm improvement in wavelet video codec should increase its effectiveness to x264/x265 level also in inter-frame coding mode.

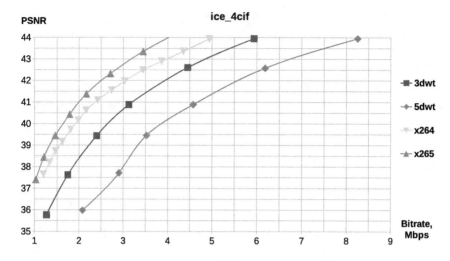

Fig. 9. Ice

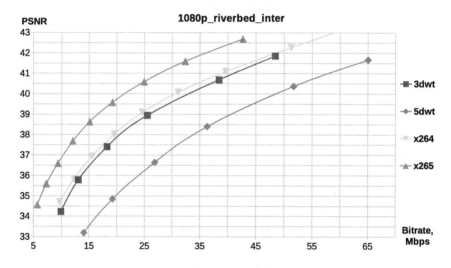

Fig. 10. "Riverbed" with motion estimation

5   Conclusions

Video codec based on multichannel wavelet transform was developed within the study. Its compression effectiveness in Intra coding mode was close to that of x264 in the same mode for most test video sequences; furthermore, it was close to x265 compression effectiveness for some other tested videos. This result is important to note especially against the background of the fact that x264 and x265 use intra-frame prediction algorithms, which considerably increase compression ratio, whereas implemented wavelet codec does not contain any comparable analog of those. Wavelet video codec is inferior to x264/x265 in motion prediction mode what may be attributed to the lack of comparable in efficiency motion estimation and compensation algorithms and entropy coder. Results comparable to those of x264/x265 should be provided by specified modules improvement in wavelet codec. Therefore, it is concluded that practical applicability of wavelet decomposition based video codec is proved.

It is easy to note that computationally simple five-channel filter banks used in the research (see. Tables 3 and 4) are much inferior to three-channel wavelet transform realization that makes use of well-designed filter banks (see Tables 1 and 2). It may be accounted for the poor ability of this filter bank to split the signal energy by frequencies. More complex and long filters usage for five-channel wavelet transform should give better results than for three-channel filter bank.

It should also be noted that the parallelization of computations in multi-channel wavelet transform module was not used in this video codec version - the simplest implementation was used. Computations transfer to OpenCL and wavelet video codec optimization should significantly improve its performance what will enable to use optimized wavelet codec for streaming video.

Acknowledgments. This work was supported by Russian Ministry of Education and Science under Grant ID RFMEFI58115X0015.

References

1. Taubman, D.S., Marcellin, M.W.: JPEG2000: standard for interactive imaging. Proc. IEEE **90**, 1336–1357 (2002)
2. Dvorkovich, V.P., Dvorkovich, A.V.: Digital Video Information Systems (Theory and Practise). Technosphere, Moscow (2012)
3. Prokhorov, I.B., Gryzov, G.Y.: Implementation of 3-band wavelet decomposition in Dirac video codec. In: Digital Signal Processing and its Applications: 17th International Conference Proceedings, Moscow, pp. 507–509 (2015)
4. Dvorkovich, A.V., Dvorkovich, V.P.: The methodology of multichannel wavelet filter banks computation for image decomposition within video compression. In: 3rd International Conference "Engineering & Telecommunication - En&T 2016", pp. 19–21. Books of Abstracts, Moscow/Dolgoprudny (2016). (In Russian)
5. Dvorkovich, V.P., Dvorkovich, A.V.: Window Functions for Harmonic Analysis of Signals. Technosphere, Moscow (2016)
6. Schrodinger video codec. http://schrodinger.sourceforge.net/schrodinger_faq.php
7. Mallat, S.G.: A theory for multiresolution signal decomposition : the wavelet representation. IEEE Trans. Pattern Anal. Mach. Intell. **11**, 674–693 (1989)
8. 4cif test video sequences. ftp://ftp.tnt.uni-hannover.de/pub/svc/testsequences/
9. 720p/1080p test video sequences. http://media.xiph.org/video/derf/

Automated Classification of a Calf's Feeding State Based on Data Collected by Active Sensors with 3D-Accelerometer

Valentin Sturm[1], Dmitry Efrosinin[1,2]([✉]), Natalia Efrosinina[1], Leonie Roland[3],
Michael Iwersen[3], Marc Drillich[3], and Wolfgang Auer[4]

[1] Johannes Kepler University, Altenbergerstrasse, 69, 4030 Linz, Austria
{valentin.sturm,dmitry.efrosinin}@jku.at
[2] Peoples' Friendship University of Russia (RUDN University),
Miklukho-Maklaya St 6, Moscow 117198, Russia
[3] Clinical Unit for Herd Health Management in Ruminants,
University Clinic for Ruminants, Department for Farm Animals and Veterinary
Public Health, University of Veterinary Medicine Vienna, 1210 Vienna, Austria
{michael.iwersen,marc.drillich}@vetmeduni.ac.at
[4] Smartbow GmbH, Jutogasse 3, 4675 Weibern, Austria
wolfgang.auer@smartbow.at
http://www.jku.at/stochastik
http://www.vetmeduni.ac.at/
http://www.smartbow.at/

Abstract. The paper deals with the problem of time series classification for the feeding state of calves by means of features evaluated for acceleration real-time data sets. The eartags equipped with an active sensor were developed for location and animal activity identification. Video records synchronized with a sensor data were collected from three calves. After the data preprocessing including the reconstruction of lost information, filtering and frequency stabilization, new time series were used to develop a machine-learning algorithm with equidistant and non-equidistant time series segmentation method based on a modified Kolmogorov-Smirnov statistic. The proposed classification method has achieved a good recognition quality for the feeding state with a best overall accuracy of approximately 94%. Thus this methodology is useful in identifying the feeding state and we may expect the possibility to generalize it to the multi-state case as well. The further improvement of the algorithm is a subject of our future research.

Keywords: Eartag · Time series classification · Kolmogorov-Smirnov statistic · Machine-learning algorithm

1 Introduction

A system with automated identification and classification of farm animal behaviour has the aim to improve livestock management and the efficiency of production by integration of this system into daily farm routnes. Such a system can

© Springer International Publishing AG 2017
V.M. Vishnevskiy et al. (Eds.): DCCN 2017, CCIS 700, pp. 120–134, 2017.
DOI: 10.1007/978-3-319-66836-9_11

serve as an element of the future smart farm. Some devices and state recognition methods have already been tested with different accuracy to identify farm animal behaviour. In Alvarenga et al. [1] a tri-axial accelerometer was attached to a halter on the under-jaw of each sheep. Among the forty-four features the authors have extracted the five most important features, including X-axis and movement variation in form of a sum of increments for the acceleration vector's coordinates. As it was shown, the used features had the largest contribution to the quality of recognition of five common behaviours such as lying, standing, walking, running and grazing. A decision-tree algorithm was used to classify the features and has exhibited the highest accuracy of 85.5%. Sheep behaviour was analysed as well in Marais et al. [4] by means of another device attached around the sheep's neck. It turned out that the most important feature is the maximum and minimum value for each axis in a frame. For classification the authors have used linear and quadratic discriminant analysis. The classification accuracy has reached a value of 89.7% with the usage of ten different features. The problem of the cow behaviour pattern recognition using a three-dimensional accelerometer placed over the neck was studied by Martiskainen et al. [6] and Kuankid et al. [3]. In first case the multi-class support vector machine classification models were constructed based on nine most important features to identify eight recorded behaviours. The average accuracy of the proposed classification model was equal to 94%. In the latter case the authors have implemented a simple behavioural technique based on the mean and variance reference values with an average accuracy of about 92.6%. Another device with equipped accelerometer was used by Robert et al. [5] to classify three activity states of the calves. It was attached to the lateral side of the right rear leg. Seven main features were used to train the classification model. While the accuracy of identification of lying and standing activities was high, the walking accuracy was significantly lower. The average classification accuracy for the proposed data sets was approximately 88.3%.

Hence we conclude that accelerometers can be used successfully to generate data sets for accurate description of farm animal behaviour. To make such devices feasible for farmers they must satisfy certain constraints including cost, reliability, usability and power consumption. The Austrian company Smartbow GmbH has developed a new identification system based on active low-energy high-frequency sensors with 3D-accelerometer integrated into eartags which record measurements at 10 Hz. Comparing to the alternative systems, the eartags have a number of advantages, i.e. they are very comfortable for animals and the energy supply of the sensor is sufficient for several years. But the eartags can not be located with specified orientation as in previous studies where a fixed coordinate axis for some state was perpendicular to the ground. The separate usage of axis accelerations for evaluation of the features has in this case a very restrictive effect to the quality of classification. Hence we are interested in orientation-independent features, which should consist of all components of the corresponding vectors. Another difficult moments are related to the incompleteness of the data samples due to signal collisions, noisiness, strong non-stationarity and frequency instability of the data sets evaluated by the eartags. Moreover, according to the data material, many feeding states have hardly distinguishable boundaries. Therefore appropriate data transformation methods as well as a suitable feature vector and a new classifier must be found out for the recognition problem under study.

In this paper we suggest a novel method to identify the milk feeding state from the acceleration data sets and determine the accuracy, sensitivity and precision. The video records synchronized with a sensor data are used as gold standard for the state recognition. After the data preprocessing including the reconstruction of the lost information through a stochastic simulation, solution of the problem of the unbalanced data sets for different states, filtering and frequency stabilization, new time series were used to develop a machine-learning algorithm with equidistant and non-equidistant time series segmentation method based on a modified Kolmogorov-Smirnov statistic. The described classification methods have achieved a quite good recognition quality for the feeding state with an overall accuracy of up to 94%, a sensitivity of 66% and a precision of 51%. Thus the proposed methodology is useful in identifying the feeding state and we may expect the possibility to generalize it to the multi-state case as well. The further improvement of the algorithm is a subject of our future research.

The rest of the paper is organized as follows: The Materials and Methods are given in Sect. 2. Section 3 presents the main results including an experiment methodology.

2   Material and Methods

For training and testing our methods we consider the data of 3 calves on 5 consecutive days. These datasets consist of two different types, namely acceleration data provided by the Smartbow eartags, see Fig. 1, and video evaluation of the same 5 days, based on which we assume to know the true behaviour in every moment.

Fig. 1. Smartbow Eartag

The 3 calves were kept in small boxes and had ad libitum access to milk, offered by a feeding bucket with an artificial teat. Four cameras were positioned at different locations in the barn to monitor the calves during the observation period. The resulting videos were carefully examined to classify the current behaviour on a granularity level of one second. In total we have 15 different datasets for 24 h each. The acceleration data is returned and recorded in three different axes, with a frequency of roughly 10 Hz, in $10^{-3}g$ steps and cut-off boundaries of $[-2048, 2048]$,

$$\mathbf{a} = \{(a_x(t), a_y(t), a_z(t)) : t \in T)\}. \tag{1}$$

Due to the nature of the sensor fixation on the ear, the possible change of relative coordinates over time because of rotating or other movement of the RFID-chip and the lack of gyroscope, the individual coordinates of (1) cannot be interpreted. Thus a orientation-independent signal has to be evaluated, e.g. the signal vector magnitude (SVM) of the acceleration data,

$$\mathbf{x} = \{x_t : t \in T\}, \text{ where } x_t := \sqrt{a_x^2(t) + a_y^2(t) + a_z^2(t)}. \tag{2}$$

These data sets are used as a main data material for the feature extraction and subsequent classification.

2.1   Terminology of Binary Classification and Model Quality

As our goal is to classify the behaviour of calves with the help of acceleration data, in particular to detect time periods, where calves are drinking, we are distinguishing between two relevant states:

- Calf is drinking (1)
- Calf is not drinking (0),

and we deal with the problem of so called binary classification. As our granularity is one second, we can split up the day in parts of length $1\,\mathrm{s}$, which brings us to a total of 864000 time points to consider a day, and for every part we want to decide, whether the calf is currently in the state 0 or 1.

Suppose we have given a series of data (x_t) with $x_t \in \{0, 1\}, 1 \le t \le N$, which represent the real states. As a result of a prognosis method we get a second time series $(\tilde{x}_t)$ with $\tilde{x}_t \in \{0, 1\}, 1 \le t \le N$, which represents the assumed states at every moment. If less than $5\,\mathrm{min}$ passed between two consecutive drinking events, they were considers as a single one. Based on this we define the following four values, which are conveniently arranged into a so called confusion matrix:

	Prognosis=1	Prognosis=0
Actual = 1	Tp	Fn
Actual = 0	Fp	Tn

Based on these four entries of the confusion matrix, we can build many different statistical quantities, which are used to represent some aspect of quality. We present a small selection:

Name	Formula
Sensitivity (Sens.)	$\dfrac{Tp}{Tp+Fn}$
Specifity (Spec.)	$\dfrac{Tn}{Tn+Fp}$
Precision (Prec.)	$\dfrac{Tp}{Tp+Fp}$
Accuracy (Acc.)	$\dfrac{Tp+Tn}{Tp+Fn+Fp+Tn}$
Youden's Index $(\mathcal{J})$	$\dfrac{Tp}{Tp+Fn}+\dfrac{Tn}{Tn+Fp}-1$
F_1-Score	$2\dfrac{\text{Sensitivity}\cdot\text{Precision}}{\text{Sensitivity}+\text{Precision}}$
Cohens κ	[8]

2.2 Algorithm of Activity Recognition

We want to give a small overview of the designed algorithm. It consists roughly of the following steps:

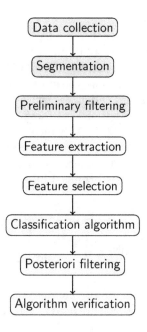

2.3   Segmentation

We define two different procedures, based on which we distinguish two different schemes we want to compare: Equidistant Segmentation (ES) and change-point detection based on modified Kolmogorov-Smirnov statistics (KSS). The first method is more straight forward, we just split up the day in equidistant intervals, which are partly overlapping

$$\{x_t \ldots x_{t+l-1}\}, \{x_{t+l/a} \ldots x_{t+l+l/a-1}\}, \{x_{t+2l/a} \ldots x_{t+l+2l/a-1}\}, \ldots$$

where a corresponds to the number of times a certain interval, except for the outermost, is overlapped by other intervals. In our examples l was set to 600 and a was set to 1, so no overlapping occurred.

Next we describe the alternative segmentation method based on a modified Kolmogorov-Smirnov statistic, see e.g. in [2, pp. 37–80]. Suppose we have given a time series $\mathbf{x} = \{x_1, \ldots, x_N\}$ with k change-points at times $t_1, \ldots, t_k$:

$$1 < \alpha \le t_1 \le \cdots \le t_l \le \beta < N,$$

where $\alpha < \dfrac{N}{2} < \beta$. The modified Kolmogorov-Smirnov-Statistic is of the form:

$$Y_N(n, \delta; \mathbf{y}) := \left[\left(1 - \frac{n}{N} \right) \frac{n}{N} \right]^\delta \left(\frac{1}{n} \sum_{k=1}^{n} y_k - \frac{1}{N-n} \sum_{k=n+1}^{N} y_k \right), \qquad (3)$$

where $n \in [1, N-1] \cap \mathbb{N}, \delta \in [0,1]$ and $\mathbf{y} = \{y_k : k \in [1, N] \cap \mathbb{N}\}$ is the realization of the diagnostic sequence. As it was shown in [2], the choice $\delta = 1$ in (3) provides the minimum false alarm probability, that is the probability of a decision making about the presence of change points when this is not true, the choice $\delta = 0$ provides minimum probability of false tranquillity, that is the probability of a decision making about the lack of change-points when this is not true and the choice $\delta = 1/2$ leads to the minimum of estimation error probability for a change-point. The statistic $Y_N(n, 1; \mathbf{y})$ for a sample of a normal distributed diagnostic sequence $\mathbf{y}$ with a change-point in mean value is illustrated in Fig. 2. According to [2], the statistic

$$\tilde{Y}_N(\delta; \mathbf{y}) := \sqrt{N} \max_{1 \le n \le N} |Y_N(n, \delta; \mathbf{y})|,$$

has an asymptotic distribution in form

$$\lim_{N \to \infty} \mathbb{P}[\tilde{Y}_N(1; \mathbf{y}) > c] = 2 \sum_{k=1}^{\infty} (-1)^{k+1} e^{-2\left(\frac{kc}{\sigma} \right)^2} = F(c).$$

We can calculate this statistic for every point n in the considered interval. Moreover, we define a point $\hat{n}$ is a (preliminary) change-point, if $Y_N(\hat{n}, \delta) > c$, where

$$\hat{n} := \underset{1 \le n \le N}{\mathrm{argmax}} \left| \sqrt{N} Y_N(n, \delta; \mathbf{y}) \right|$$

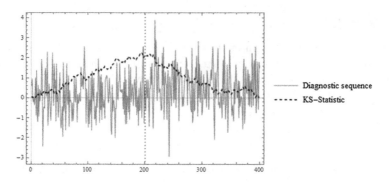

Fig. 2. Depiction of the statistic for a realisation of the stochastic process $X_1 \dots X_{500} \sim \mathcal{N}(0,1), X_{501} \dots X_{1000} \sim \mathcal{N}(0.5,1)$

As a time series y_k one can choose a transformation of the original series x_k in form $y_k = f(x_k)$, where the choice of this specific function can be influenced by the type of change-point one wants to find. Given a time series $\mathbf{x} := \{x_1, x_2, \dots, x_N\}$, we try to find some change-points in the first step and therefore we use the diagnostic sequence

$$y(x_k) := \alpha_1 f_1(x_k) + \alpha_2 f_2(x_k) + \alpha_3 f_3(x_k), \tag{4}$$

with

$$f_1(x_k) := \frac{\operatorname{argmax}_\omega \hat{f}(\omega, \{x_{k-d}, \dots, x_{k+d}\}) - \mu_1}{\sigma_1},$$

$$f_2(x_k) := \frac{\hat{\sigma}^2(\{x_{k-d}, \dots, x_{k+d}\}) - \mu_2}{\sigma_2},$$

$$f_3(x_k) := \frac{\hat{\gamma}(1, \{x_{k-d}, \dots, x_{k+d}\}) - \mu_3}{\sigma_3},$$

where $\alpha_1, \alpha_2, \alpha_3 \in [-1,1]$, $|\alpha_1| + |\alpha_2| + |\alpha_3| = 1$, $f_1(x_k)$ describes the argument where the periodogram is maximized, $f_2(x_k)$ calculates the sample variance and $f_3(x_k)$ the sample auto-covariance of order 1. The functions are normalized by subtracting the mean and dividing by the empirical standard deviation of a whole day. So for every point, we calculate the linear combination of three different diagnostic sequences in a sliding window of length $2d + 1$. Our goal is to find optimal values $\alpha_1^*, \alpha_2^*, \alpha_3^*$, which minimize the following cost function $\hat{g}$:

$$\hat{g}(\alpha_1, \alpha_2, \alpha_3) := \frac{1}{l} \sum_{i=1}^{l} \left(C_0 \sum_{t \in S_{i,0}} \sigma(a_t - c) + C_1 \sum_{t \in S_{i,1}} \sigma(c - a_t) \right), \tag{5}$$

with

$$\sigma(x) := \frac{1}{1 + e^{-x}}, \ a_t := \max_{\hat{t} \in \{t-k, \ldots, t+k\}} \sqrt{2k+1} \, |Y_{2k+1}(n, 1)| \, ,$$

$$S_{i,0} := \{t \in I | x_t = 0\}, \ T_{i,0} = |S_{i,0}|,$$

$$S_{i,1} := \{t \in I | x_t = 1\}, \ T_{i,1} = |S_{i,1}|,$$

$$C_0 := \begin{cases} \dfrac{c_0}{T_{i,0}} & \text{if } T_{i,0} > 0 \\ 0 & \text{if } T_{i,0} = 0 \end{cases}, \quad C_1 := \begin{cases} \dfrac{c_1}{T_{i,1}} & \text{if } T_{i,1} > 0 \\ 0 & \text{if } T_{i,1} = 0 \end{cases},$$

$$c_0, c_1 \in (0, 1), \ c_0 + c_1 = 1.$$

The function $\hat{g}$ builds an average value of the defined sums over all considered intervals. To solve the optimization problem, we decide to use the data of one calf and discretize the space of possible arguments

$$(\alpha_1, \alpha_2, \alpha_3) \in \{-1, -0.75, \ldots, 1\}^3, |\alpha_1| + |\alpha_2| + |\alpha_3| = 1.$$

Due to symmetry we can further restrict $\alpha_1 \geq 0$. The two sums in (5) describe a measure for the amount of points inside a drinking interval for which the KS-statistic is above some threshold c respectively the amount of points outside drinking events, for which the statistic is below the same threshold. To calculate the value of $\hat{g}$, we need the actual change-points t_i. We extract c out of the given training data, e.g. as a empiric quantile of values of the KS-statistic. The choices

$$k = d = 50, \ C_0 = 0.1, \ C_1 = 0.9$$

lead to

$$(\alpha_1^*, \alpha_2^*, \alpha_3^*) = (0, -0.5, 0.5)$$

as a solution of the discretized minimization problem, which can also be seen in Fig. 3.

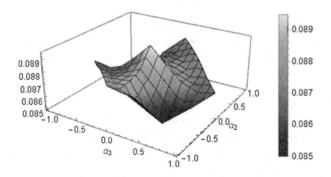

Fig. 3. Interpolated graph of $\hat{g}$ for various values of α_2 and α_3

The next step, after deciding for a diagnostic sequence, is to find the corresponding change-points:

Algorithm 1 (Kolmogorov-Smirnov segmentation)

1. *Given the datasets* $s_i = \{x_{i1}, \ldots, x_{iN}\}$, *split up* s_i *in intervals of equal length* l:

$$\{x_{i1}, \ldots, x_{il}\}, \{x_{i,l+1}, \ldots, x_{i,2l}\}, \ldots,$$
$$\{x_{i,(\lfloor N/l \rfloor -1)l+1}, \ldots, x_{i,\lfloor N/l \rfloor l}\}, \{x_{i,\lfloor N/l \rfloor)l+1}, \ldots, x_{iN}\}.$$

2. *Calculate the KS-statistic for every interval* $\{x_a, \ldots, x_b\}$ *with the diagnostic sequence*

$$y(x_k) = \alpha_1 f_1(x_k) + \alpha_2 f_2(x_k) + \alpha_3 f_3(x_k), \ k \in [a+d, b-d] \cap \mathbb{N}.$$

3. *If the inequality*

$$\tilde{Y}_N(\delta, \mathbf{y}) > c, \ \mathbf{y} = \{y(x_k) : k \in [a+d, b-d] \cap \mathbb{N}\}$$

holds, append the point x_t *where the statistic is maximized to a list as a candidate change-point and bisect the corresponding interval in the following way,*

$$\{x_a, \ldots, x_t\}, \{x_t, \ldots, x_b\}.$$

4. *Repeat step 3 and 4 for every newly formed interval until either a certain minimal interval length* $l_{\min}$ *is reached or no further change-point candidate is found.*

5. *Merge the resulting list of candidates with the points* $1, N$ *and* $kl, 1 \leq k \leq \lfloor N/l \rfloor l$ *to build an ascending list* $\{\tilde{t}_1, \tilde{t}_2, \ldots, \tilde{t}_m\}$ *and form new intervals*

$$\{x_{\tilde{t}_k+1}, \ldots x_{\tilde{t}_{k+1}}\}.$$

6. *Calculate KS-statistic for all newly formed intervals with two diagnostic sequences*

$$y_1(x_k) := f_2(x_k),$$
$$y_2(x_k) := f_3(x_k).$$

7. *For every new interval determine the maxima of* $\tilde{Y}_N(\delta; \mathbf{y}_1)$ *and* $\tilde{Y}_N(\delta; \mathbf{y}_2)$ *and compare them with the corresponding thresholds* c_1, c_2. *If either none, one or both values lie above the threshold we have found 0, 1 or 2 change-points respectively.*

8. *Divide the dataset* s_i *into equidistant parts and put these parts in one of two groups depending on the minimal length of the interval they are lying in, according to the change-points found before.*

In summary, we split up the day according to the discovered change-points and afterwards make a equidistant segmentation and divide this segments in two groups, based on the length of the interval they are lying in. We decided for a critical minimal length to split the data in 2 approximately equal-sized groups, this choice can also be subject to parameter tuning (Fig. 4).

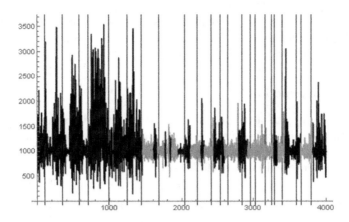

Fig. 4. Depiction of the KSS method applied to a selected excerpt of our data

2.4   Preliminary Filtering

To reduce noise in the acceleration data, we apply a Gaussian denoising filter as illustrated in Fig. 5, which seems to be appropriate for our given datasets. Of course, based on the nature of the raw data, one can use a different filter such as median filter, mean filter, moving-average-Filter, bandpass filter, low-pass/highpass filter or wavelet transform to stabilize the frequencies and reduce the noise.

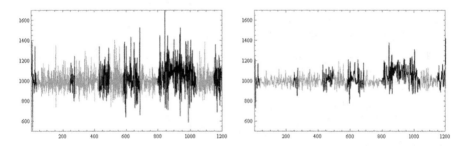

Fig. 5. Section of our unfiltered data (left), and the same section after application of a Gaussian filter (right)

We calculate the vector $(\theta_1, \ldots, \theta_n)$ of a fixed number n of statistics, specified below, for all intervals. Afterwards we define boundaries $(\theta_{i,\min}, \theta_{i,\max})$, $i = 1, \ldots, n$, for these parameters and exclude an interval from further consideration as a possible drinking interval if

$$(\theta_1, \ldots, \theta_n) \notin [\theta_{1,\min}, \theta_{1,\max}] \times \cdots \times [\theta_{n,\min}, \theta_{n,\max}].$$

We based the decision of these boundaries on an minimum amount of sensitivity ($Sens > 0.99$) of the remaining intervals, while we empirically tried to maximize the precision. Of course a more sophisticated method could be implemented to find a even better filtering. In our example we chose variance, skewness and kurtosis as the features.

2.5   Feature Extraction

Let $\mathbf{x} = \{x_1, \ldots, x_n\}$ be a part of our time series. Based on this data we want to extract the orientation-insensitive features which describe the nature of the time series and hopefully are going to help us in our following classification.
A short selection of used Features follows:

Feature	Equation
Mean square deviation	$\frac{1}{n} \sum\limits_{i=1}^{n} x_i^2$
Empirical variance	$\frac{1}{n-1} \sum\limits_{i=1}^{n} (x_i - \bar{x})^2$
Empirical skewness	$\dfrac{\frac{1}{n} \sum\limits_{i=1}^{n} (x_i - \bar{x})^3}{\left(\frac{1}{n} \sum\limits_{i=1}^{n} (x_i - \bar{x})^2 \right)^{3/2}}$
Empirical kurtosis	$\dfrac{\frac{1}{n} \sum\limits_{i=1}^{n} (x_i - \bar{x})^4}{\left(\frac{1}{n} \sum\limits_{i=1}^{n} (x_i - \bar{x})^2 \right)^{2}}$
Empirical autocorrelations with lag h, $h = 1, \ldots, 11$	$\frac{1}{(n-h)\sigma^2} \sum\limits_{i=1}^{n-h} (x_i - \bar{x})(x_{i+h} - \bar{x})$
p-quantiles x_p, $p = 0.01, 0.1, 0.5, 0.9, 0.99$	$x_p := x_{\lfloor np+1 \rfloor}$
Periodogram-LSE with $f(\omega) := a + b\log(\omega)$	a, b
Parameter of fitted $\mathcal{AR}$- /$\mathcal{MA}$-process	$\phi_1, \ldots \phi_6, \theta_1, \ldots \theta_6$

In the Case of KSS, we split up the feature vectors in two groups according to the grouping in Algorithm 1.

2.6   Feature Selection Method

To select a fitting feature subspace we used the work of [7], which considers correlation based feature selection. We implemented some of the algorithmic approaches and decided for a 1-step and 2-step backward elimination algorithm to reduce the amount of considered features. As a measure for correlation we used the absolute value of the classical Neyman-Correlation coefficient

$$K(x, y) := \frac{|\sum_{i=1}^{n}(x_i - \overline{x})(y_i - \overline{y})|}{\sqrt{\sum_{i=1}^{n}(x_i - \overline{x})^2 \cdot \sum_{i=1}^{n}(y_i - \overline{y})^2}}. \quad (6)$$

All features were normalized $\hat{f}_i = \dfrac{f_i - \mu_i}{\sigma_i}$ on a daily and animal-individually basis before calculating (6) and searching for the feature subspace. In the case of using ES we calculate the features subspace for every training set, where in the case of KSS we calculate the corresponding subspace for both groups of features for every training set.

2.7 Classification Algorithm

To eliminate the bias of using the data of one calf for finding the parameters of KSS, we exclude these datasets from further classification. We assign the class label 1 to a feature vector if at least 5 s of drinking occur in the corresponding interval, otherwise we assign class label 0. To estimate the quality of our two different procedures with the remaining 10 days of records we use the following cross-validation approach. First we construct 10 different splits of the data into training and validation data. Therefore we choose every combination of 2 days out of 5 and say in every split that 6 days (3 days of 2 calves) serve as our training data and the remaining 4 days as our validation data. In every split we further choose all combinations of 4 days out of the 6 training days. In the case of ES we build a classifier for every chosen subset, whereas in the case of KSS we build two classifier for every chosen subset, one for every group of intervals. That procedure yields a total of $\binom{6}{4} = 15$ respectively 30 classifiers in every split. In every classifying procedure we randomly oversampled the rarer class in the training set to obtain a balanced amount of class examples. Next we build a ensemble learner, which connects the 15 beforehand calculated algorithms and decides for class 1 or 0 according to the number of different algorithms which decide for 1 or 0 respectively. In the case of KSS this procedure is independently used on both groups. We decided for a certain minimum amount $c_{\min}$ to vote for class 1 as we maximized the function

$$f(c_{min}; Sens., Prec.) = (Sens. + Prec.) \cdot \text{sgn}(\max\{Sens. - 0.8, 0\}) \quad (7)$$

on the training data. In words, we tried to maximize the sum of precision and sensitivity while keeping a minimum sensitivity. Suppose we have n different classifiers $C_1 \dots C_n$ with

$$C_i : \mathbb{R}^k \to \{0, 1\}$$

$$x \to c_i,$$

then we define a new classifier with the following function mapping from the feature space to class label

$$F : \mathbb{R}^{n+k} \to \{0, 1\}$$

$$F(x, c_1, c_2, \ldots c_n, c_{\min}) := \text{sgn} \left(\max\{\sum_{i=1}^{n} c_i - c_{\min}, 0\} \right) \qquad (8)$$

After classifying, the segments which are chosen as belonging to class Drinking are further merged into bigger intervals if at most 5 min are between subsequent segments. An additional posteriori filtering step can be implemented which may further improve the results, this approach is not presented here.

3   Experimental Results

We used 5 different built-in classification algorithms in Mathematica ©11 with the following method descriptions (Tables 1 and 2):

- Neural Network (NN): This classifier is composed of layers of artificial neuron units. Each unit computes its value as a function of the unit values in the previous layer. Information is processed layer by layer from the feature layer to the output layer which gives the class probabilities. It is also called a feed-forward neural network or a multi-layer perceptron.
- Nearest Neighbors (KNN): The nearest neighbours classifier infers the class of a new example by analyzing its nearest neighbors in the feature space. In its simplest form, it picks the commonest class amongst the "k"-nearest neighbors.
- Naive Bayes (NB): The naive Bayes classifier assumes that features are generated independently given the class and uses Bayes' theorem to predict the class.

Table 1. Quality comparison of different used methods with 1-step backward elimination

Classifier	Measure	ES	KSS	Classifier	Measure	ES	KSS
	Accuracy	0.940	0.938		Accuracy	0.918	0.919
NN	Cohens κ	0.542	0.522	KNN	Cohens κ	0.381	0.410
	F1-Score	0.573	0.554		F1-Score	0.423	0.452

Classifier	Measure	ES	KSS	Classifier	Measure	ES	KSS
	Accuracy	0.824	0.853		Accuracy	0.884	0.895
NB	Cohens κ	0.315	0.349	SVM	Cohens κ	0.421	0.437
	F1-Score	0.379	0.407		F1-Score	0.470	0.484

Classifier	Measure	ES	KSS
	Accuracy	0.862	0.882
LR	Cohens κ	0.379	0.413
	F1-Score	0.434	0.464

Table 2. Quality comparison of different used methods with 2-step backward elimination

Classifier	Measure	ES	KSS	Classifier	Measure	ES	KSS
	Accuracy	0.943	0.941		Accuracy	0.923	0.926
NN	Cohens κ	0.545	0.536	KNN	Cohens κ	0.407	0.418
	F1-Score	0.575	0.566		F1-Score	0.447	0.456

Classifier	Measure	ES	KSS	Classifier	Measure	ES	KSS
	Accuracy	0.833	0.853		Accuracy	0.881	0.896
NB	Cohens κ	0.325	0.345	SVM	Cohens κ	0.418	0.439
	F1-Score	0.387	0.404		F1-Score	0.468	0.485

Classifier	Measure	ES	KSS
	Accuracy	0.836	0.865
LR	Cohens κ	0.340	0.389
	F1-Score	0.400	0.444

- Support vector machine (SVM): The support vector machine classifier separates the training data into two classes using a maximum-margin hyperplane. The original feature space can be mapped into a higher dimensional space to improve linear separability.
- Logistic regression (LR): This classifier models the class probabilities with logistic functions of linear combination of features.

In the following tables we summarized the average accuracy, Cohens κ and F1-Score of the different used classification methods for our 10-fold cross validation, where in each comparison higher values are highlighted. Moreover we present the confusion matrix (Table 3) for the best result.

Table 3. Confusion matrix with total amount in $10^{-1}s$ (left) and percentages (right) of the neural network method combined with 2-step backward elimination and equidistant segmentation and table with quality measures

	Prognosis=1	Prognosis=0		Prognosis=1	Prognosis=0
Actual = 1	$1.333 \cdot 10^6$	$6.896 \cdot 10^5$	Actual = 1	3.86%	2.00%
Actual = 0	$1.280 \cdot 10^6$	$3.126 \cdot 10^7$	Actual = 0	3.70%	90.44%

Measure	Sens.	Spec.	Prec.	Acc.	$\mathcal{J}$	F_1	κ
Value	0.659	0.961	0.510	0.943	0.620	0.575	0.545

4   Conclusion

The results show that both the ES and KSS method are feasible for achieving good results in detecting drinking events in calves. The results are astounding when keeping in mind, that in KSS we effectively halve the amount of samples in our classification algorithms and still get comparable or even better results than with ES. With defining additional or other diagnostic sequences, this procedure can be generalized to other problems. Some problems we tried to overcome were the noisiness of the data, the usage of only the magnitude of acceleration, and the imbalanced class frequency. The classification results can be further used to estimate the drinking amount with regression, which is subject to further research.

Acknowledgements. This work was funded by the Austrian Research Promotion Agency (FFG), Project No. 848610 and Smartbow GmbH. The publication was financially supported by the Ministry of Education and Science of the Russian Federation (the Agreement number 02.a03.21.0008).

References

1. Alvarenga, F.A.P., Borges, I., Palkovic, L., Rodina, J., Oddy, V.H., Dobos, R.C.: Using a three-axis accelerometer to identify and classify sheep behaviour at pasture. Appl. Anim. Behav. Sci. **181**, 91–99 (2016)
2. Brodsky, B.E., Darkhovsky, B.S.: Nonparametric Methods in Change-Point Problems. Kluwer Academic Publishers, The Netherlands (1993)
3. Kuankid, S., Rattanawong, T., Aurasopon, A.: Classification of the cattle's behaviours by using accelerometer data with simple behavioural technique. In: Proceedings of 2014 APSIPA Annual Summit and Conference, Siem Reap, Cambodia, pp. 1368–1372 (2014)
4. Marais, J., Le Roux, S.P., Wolhuter, R., Niesler, T.: Automatic classification of sheep behaviour using 3-axis accelerometer data. Livestock Sci. **196**, 42–48 (2017)
5. Robert, B., White, B.J., Renter, D.G., Larson, R.L.: Evaluation of three-dimensional accelerometers to monitor and classify behaviour patterns in cattle. Comput. Electron. Agric. **67**, 80–84 (2009)
6. Martiskainen, P., Järvinen, M., Skön, J.O., Tiirikainen, J., Kolehmainen, M., Mononen, J.: Cow behaviour pattern recognition using a three-dimensional accelerometer and support vector machines. Appl. Anim. Behav. Sci. **119**, 32–38 (2009)
7. Hall, M.A.: Correlation-based feature selection for machine learning. Ph.D. thesis. The University of Waikato (1999)
8. Cohen, J.: A coefficient of agreement for nominal scales. Educ. Psychol. Measur. **20**(1), 37–46 (1960)

Bounding the Risk Probability

Igor Nikiforov$^{(\boxtimes)}$

Université de Technologie de Troyes, UTT/ICD/LM2S, UMR 6281, CNRS,
12, rue Marie Curie, CS 42060, 10004 Troyes Cedex, France
Igor.Nikiforov@utt.fr
http://www.utt.fr

Abstract. For some safety–critical applications, it is important to calculate the probability that a discrete time autoregressive (AR) process
leaves a given interval at least once during a certain period of time. For
example, such AR process can be interpreted as a temporally correlated
safety indicator and the interval as a target zone of the process. It is
assumed that the safety of the system under surveillance is compromised
if the above-mentioned probability becomes too important. This problem has been previously studied in the case of known distributions of
the innovation process. Let us assume now that the distributions of the
innovation and initial state are unknown but some special bounds for
the cumulative distribution functions and/or for the probability density
functions are available. Numerical methods to calculate the bounds for
the above-mentioned probability are considered in the paper.

Keywords: Risk probability · Distribution bounding · Autoregressive
process · Numerical integration

1   Introduction and Motivation

For some safety–critical applications, it is important to calculate the probability
that a discrete time autoregressive (AR) process $\{Q_n\}_{n\geq 1}$ leaves a given open
interval $]-h,h[$ (or an open ball in the multidimensional case) during a certain period of time T. For example, such AR process can be interpreted as a
temporally correlated safety indicator and the interval $]-h,h[$ as a target zone
of the process. It is assumed that the safety of a system is compromised if the
above-mentioned probability becomes too important.

It is assumed that the AR process is represented by the following equation:

$$Q_n = (1-\lambda)Q_{n-1} + \lambda y_n, \quad n = 1,2,3,\ldots, \tag{1}$$

where $0 < \lambda < 1$ is the autoregressive coefficient and the independent random
variables (vectors) y_n obey a certain distribution F_y, i.e., $y_n \sim F_y$.

I. Nikiforov—The author gratefully acknowledges the research and financial support
of this work from the Thales Alenia Space, France.

© Springer International Publishing AG 2017
V.M. Vishnevskiy et al. (Eds.): DCCN 2017, CCIS 700, pp. 135–145, 2017.
DOI: 10.1007/978-3-319-66836-9_12

In safety–critical applications, the safety is defined by the probability that the AR process reaches one of the barriers $-h$ or h during the time period T. First of all, let us define the stopping time N:

$$N = \inf\left\{n \geq 1 : |Q_n| \geq h\right\} \tag{2}$$

and, next, the probability of the event $\{N \leq T\}$ provided that the starting point is defined as follows $Q_0 = u$, i.e., $\mathbb{P}\left(N \leq T | Q_0 = u\right)$. Under assumption that the starting point u obeys a certain distribution, i.e., $u \sim F_0$, we are usually interested in the calculation of the following probability:

$$\mathbb{P}\left(N \leq T | u \in \right] - h, h\left[\right). \tag{3}$$

In the case of the vector process AR, the stopping time is defined as follows:

$$N = \inf\left\{n \geq 1 : \|Q_n\|_2 \geq h\right\}. \tag{4}$$

From the practical point of view, this probability is crucially important in the following scenarios.

Scenario 1. Equations (1), (2) and (4) define the Geometric Moving Average (GMA) chart, which is used for detecting the abrupt changes in the properties of random sequences, see papers [1–4] and books [5,6]. Traditionally, the Average Run Length (ARL) to false alarm and the worst-case mean detection delay are used as statistical performance measures of the GMA charts. The disadvantage of the conventional criterion for safety-critical applications consists in the existence of the right "tail" of the conditional distribution of the detection delay (see discussion in [7]). Strictly speaking, a small worst-case *mean* detection delay need not imply that the probability of the transient change detection is negligible. Moreover, for safety-critical applications, it is more convenient to use the probability of false alarm instead of the conventional ARL to false alarm. For this reason, recently, another statistical measure has been proposed for the transient change detection in safety-critical applications: the probability of missed detection and the probability of false alarm during a certain period [7–9]. It is easy to see that both the above mentioned probabilities for GMA chart are defined in Eq. (3), where the probabilities calculated under two probabilistic measures before and after change-point.

Scenario 2. Equations (1), (2) and (4) define the first-passage-problem for the AR process. For some safety–critical applications, the state Q_n can be viewed as a risk indicator. For instance, Q_n is the estimation error. This estimation is calculated by using strongly correlated observations. In such situation, the probability that the AR process leaves an interval $] - h, h[$ (or an open ball of radius h, $\|X\|_2 < h$) at least once during a certain period of time is interpreted as a risk for the system under surveillance. The interval $] - h, h[$ (or the open ball $\|X\|_2 < h$) is interpreted as a target zone (confidence interval) of the process. It is assumed that the safety of the process is compromised if the above-mentioned probability becomes too important.

2   Original Contribution

This problem has been previously studied in the case of known distributions of the innovation process $\{y_n\}_{n\geq 1}$. The behaviour of the random walk between absorbing boundaries in the continuous and discrete time and the related first-passage-problem have been studied in many books and papers starting from [10] (see, for example, [11,12]). The first-passage-problem for the AR process (or Ornstein-Uhlenbeck process in continuous time) has been studied in [1–4,11] by using analytical and numerical methods, mainly for calculating the expectation of first passage time.

Let us assume now that the distributions of the innovation $\{y_n\}_{n\geq 1}$ and initial state Q_0 are non-Gaussian and moreover unknown but some special bounds for the Cumulative Distribution Functions (CDF) and/or Probability Density Functions (PDF) are available. In the rest of the paper we discuss numerical methods to calculate the bounds for the probability defined in Eq. (3) when such bounds are available. Moreover, let us consider the vector AR process. In this case, the stopping time is defined by Eq. (4). In the rest of the paper we extend the numerical methods to calculate the bounds for the probability defined in Eq. (3) to the vector AR process.

3   Bounding the Risk for the Scalar AR Process

The very first idea to calculate the probability that the random walk is absorbed by boundaries $-h$ and h by solving integral equations is due to [13,14]. A very pedagogical introduction to the first-passage-problem can be found in [11, Chap. 2]. The adaptation of general equations to the case of AR process can be found in [2]. There are two different manners of representing the recursive equation: (i) by using the probability that the absorption occurs at one of the barriers $-h$ or h at or before the n-th step, i.e. $\mathbb{P}(N \leq n|Q_0 = u)$, see [11,13,14]; (ii) by using the probability that the absorption occurs at one of the barriers $-h$ or h at the n-th step, i.e. $\mathbb{P}(N = n|Q_0 = u)$, see [2].

First Manner. Let us denote the probability that the absorption occurs at one of the barriers $-h$ or h at or before the n-th step by $\widetilde{p}_n(u) = \mathbb{P}(N \leq n|Q_0 = u)$. The recursive equation is given by

$$\widetilde{p}_n(u) = \mathbb{P}\left(|(1-\lambda)u - \lambda y_1| > h\right) + \frac{1}{\lambda}\int_{-h}^{h}\widetilde{p}_{n-1}(z)f_y\left(\frac{z-(1-\lambda)u}{\lambda}\right)dz,$$

where $n = 1, 2, 3, \ldots, T$ and

$$\widetilde{p}_0(u) = \begin{cases} 0 & \text{if } -h < x < h \\ 1 & \text{if } x = -h \\ 1 & \text{if } x = h \end{cases}.$$

Second Manner. We will use the second manner in the rest of the paper. The recursive equation is given by:

$$p_n(u) = \frac{1}{\lambda} \int_{-h}^{h} p_{n-1}(z) f_y \left(\frac{z - (1-\lambda)u}{\lambda} \right) dz, \quad n = 2, 3, \dots, T, \qquad (5)$$

where $p_n(u) = \mathbb{P}(N = n | Q_0 = u)$, $f_y(x)$ is the PDF of y_n. The initial condition $p_1(u)$ is calculated in the following manner

$$p_1(u) = \mathbb{P}\left(|(1-\lambda)u - \lambda y_1| > h \right)$$
$$= 1 - F_y \left(\frac{h - (1-\lambda)u}{\lambda} \right) + F_y \left(\frac{-h - (1-\lambda)u}{\lambda} \right). \qquad (6)$$

Finally, the probability of the event $\{1 \leq N \leq T\}$ provided that $Q_0 = u$ is given by

$$\mathbb{P}(1 \leq N \leq T | Q_0 = u) = \sum_{n=1}^{T} p_n(u). \qquad (7)$$

If the initial condition $Q_0 = u$ is a random variable, we have to randomize the result in the following manner (under assumption that the distribution of initial state is known):

$$\mathbb{P}\left(1 \leq N \leq T | u \in [-h, h]\right) = \frac{\int_{-h}^{h} f_{Q_0}(x) \sum_{n=1}^{T} p_n(x) dx}{\int_{-h}^{h} f_{Q_0}(x) dx}, \quad u \sim F_{Q_0}. \qquad (8)$$

where $f_{Q_0}(x)$ is the PDF of F_{Q_0}.

Assumption 1. *Let us assume that the CDF $F_y(x)$ of the innovation process $\{y_n\}_{n \geq 1}$ and the CDF $F_{Q_0}(x)$ of the initial state Q_0 obey the following inequality*

$$\underline{F}_y(x) \leq F_y(x) \leq \overline{F}_y(x) \quad and \quad \underline{F}_{Q_0}(x) \leq F_{Q_0}(x) \leq \overline{F}_{Q_0}(x) \quad for \ x \in \mathbb{R}.$$

Lemma 1. *Let us consider that Assumption 1 is satisfied. Then the upper bound for the probability $p_n(u)$ is given by*

$$p_n(u) \leq \overline{p}_n(u) = \overline{p}_{n-1}(h)\overline{F}_y \left(\frac{h - (1-\lambda)u}{\lambda} \right)$$

$$- \overline{p}_{n-1}(-h)\underline{F}_y \left(\frac{-h - (1-\lambda)u}{\lambda} \right)$$

$$- \int_{-h}^{h} \underline{F}_y \left(\frac{z - (1-\lambda)u}{\lambda} \right) \mathbb{I}_{\{\overline{p}'_{n-1}(z) \geq 0\}} \overline{p}'_{n-1}(z) dz$$

$$- \int_{-h}^{h} \overline{F}_y \left(\frac{z - (1-\lambda)u}{\lambda} \right) \mathbb{I}_{\{\overline{p}'_{n-1}(z) < 0\}} \overline{p}'_{n-1}(z) dz, \qquad (9)$$

where $n = 2, 3, \dots, T$, $\mathbb{I}_{\{A\}} = \begin{cases} 1 \ if \ A \ is \ true \\ 0 \ if \ A \ is \ false \end{cases}$ is the indicator function of the event A, $\overline{p}'_{n-1}(z) = \dfrac{d\overline{p}_{n-1}(z)}{dz}$ and the upper bound for the probability $p_1(u)$ is given by

$$p_1(u) \leq \overline{p}_1(u) = 1 - \underline{F}_y\left(\frac{h - (1 - \lambda)u}{\lambda}\right) + \overline{F}_y\left(\frac{-h - (1 - \lambda)u}{\lambda}\right). \quad (10)$$

Proposition 1. *Let us consider that Assumption 1 is satisfied. Then the upper bound for the probability* $\mathbb{P}\left(1 \leq N \leq T | u \in [-h, h]\right)$ *is given by*

$$\mathbb{P}\left(1 \leq N \leq T | u \in [-h, h]\right) \leq \frac{1}{a}\left[\overline{p}_T(h)\overline{F}_{Q_0}(h) - \overline{p}_T(-h)\underline{F}_{Q_0}(-h)\right.$$
$$- \int_{-h}^{h} \underline{F}_{Q_0}(x)\mathbb{I}_{\{\overline{p}_T'(x) \geq 0\}}\overline{p}_T'(x)dx$$
$$\left. - \int_{-h}^{h} \overline{F}_{Q_0}(x)\mathbb{I}_{\{\overline{p}_T'(x) < 0\}}\overline{p}_T'(x)dx\right], \quad (11)$$

where $a = \underline{F}_{Q_0}(h) - \overline{F}_{Q_0}(-h)$, $\overline{p}_T(x) = \sum_{n=1}^{T}\overline{p}_n(x)$ *and* $\overline{p}_T'(x) = \dfrac{d\overline{p}_T(x)}{dx}$.

4 Bounding the Risk for the Vector AR Process

Let us assume now that $Q_n \in \mathbb{R}^m$, $m \geq 2$. The extension of the scalar first-passage-problem to the vector case defined by Eqs. (1) and (4) leads to the following recursive equation

$$p_n(U) = \frac{1}{\lambda^m}\underset{\|Z\|_2 \leq h}{\int \cdots \int} p_{n-1}(Z)f_y\left(\frac{Z - (1 - \lambda)U}{\lambda}\right) dZ, \quad n = 2, 3, \ldots, T, \quad (12)$$

where $p_n(U) = \mathbb{P}(N = n | Q_0 = U)$, $f(X)$ is the PDF of y_n, $X, U, Z \in \mathbb{R}^m$, $dZ = dz_1 \cdots dz_m$. The initial condition $p_1(U)$ is calculated in the following manner

$$p_1(U) = \mathbb{P}\left(\|(1 - \lambda)U - \lambda y_1\|_2 > h\right)$$
$$= \frac{1}{\lambda^m}\underset{\|Z\|_2 \geq h}{\int \cdots \int} f_y\left(\frac{Z - (1 - \lambda)U}{\lambda}\right) dZ. \quad (13)$$

The probability of the event $\{1 \leq N \leq T\}$ provided that $Q_0 = U$ is given by previously defined Eq. (7) and the probability $\mathbb{P}\left(1 \leq N \leq T | u \in [-h, h]\right)$ by previously defined Eq. (8).

Assumption 2. *Let us assume that the CDF* $F_y(X) = \prod_{i=1}^{m} F_{y,i}(x_i)$ *of the innovation process* $\{y_n\}_{n \geq 1}$ *and the CDF* $F_{Q_0}(X) = \prod_{i=1}^{m} F_{Q_0,i}(x_i)$ *of the initial state* Q_0 *obey the following inequality*

$$\underline{F}_{y,i}(x) \leq F_{y,i}(x) \leq \overline{F}_{y,i}(x) \text{ and } \underline{F}_{Q_0,i}(x) \leq F_{Q_0,i}(x) \leq \overline{F}_{Q_0,i}(x) \text{ for } x \in \mathbb{R},$$

where $i = 1, \ldots, m$.

For the sake of simplicity, we consider the case of $m = 2$ in the rest of the paper.

Lemma 2. *Let us consider that Assumption 2 is satisfied. Then the upper bound for the probability $p_n(U)$ is given by Lemma 1, where z is replaced with z_1, u with u_2, and the function $\overline{p}_{n-1}(z)$ is replaced with the function $\overline{I}(z_1, u_2)$. This latter function is given by the following equation*

$$
\overline{I}(z_1, u_2) = \overline{p}_{n-1}\left(z_1, \sqrt{h^2 - z_1^2}\right) \overline{F}_{y,2}\left(\frac{\sqrt{h^2 - z_1^2} - (1 - \lambda)u_2}{\lambda}\right)
$$

$$
- \overline{p}_{n-1}\left(z_1, -\sqrt{h^2 - z_1^2}\right) \underline{F}_{y,2}\left(\frac{-\sqrt{h^2 - z_1^2} - (1 - \lambda)u_2}{\lambda}\right)
$$

$$
- \int_{-\sqrt{h^2 - z_1^2}}^{\sqrt{h^2 - z_1^2}} \underline{F}_{y,2}\left(\frac{z_2 - (1 - \lambda)u_2}{\lambda}\right) \mathbb{I}_{\{\overline{p}'_{n-1}(z_1, z_2) \geq 0\}} \overline{p}'_{n-1}(z_1, z_2) dz_2
$$

$$
- \int_{-\sqrt{h^2 - z_1^2}}^{\sqrt{h^2 - z_1^2}} \overline{F}_{y,2}\left(\frac{z_2 - (1 - \lambda)u_2}{\lambda}\right) \mathbb{I}_{\{\overline{p}'_{n-1}(z_1, z_2) < 0\}} \overline{p}'_{n-1}(z_1, z_2) dz_2, \quad (14)
$$

where $n = 2, 3, \ldots, T$, $\overline{p}'_{n-1}(z_1, z_2) = \dfrac{\partial \overline{p}_{n-1}(z_1, z_2)}{\partial z_2}$. *The upper bound for the probability $p_1(U)$ is given by*

$$
1p_1(u_1, u_2) \leq \overline{p}_1(u_1, u_2) = 1 - \underline{I}_1(h, u_2) \underline{F}_{y,1}\left(\frac{h - (1 - \lambda)u_1}{\lambda}\right)
$$

$$
+ \underline{I}_1(-h, u_2) \overline{F}_{y,1}\left(\frac{-h - (1 - \lambda)u_1}{\lambda}\right)
$$

$$
+ \int_{-h}^{h} \overline{F}_{y,1}\left(\frac{z_1 - (1 - \lambda)u_1}{\lambda}\right) \mathbb{I}_{\{\underline{I}'_1(z_1, u_2) \geq 0\}} \underline{I}'_1(z_1, u_2) dz_1
$$

$$
+ \int_{-h}^{h} \underline{F}_{y,1}\left(\frac{z_1 - (1 - \lambda)u_1}{\lambda}\right) \mathbb{I}_{\{\underline{I}'_1(z_1, u_2) < 0\}} \underline{I}'_1(z_1, u_2) dz_1, \quad (15)
$$

where

$$
\underline{I}_1(z_1, u_2) = \underline{F}_{y,2}\left(\frac{\sqrt{h^2 - z_1^2} - (1 - \lambda)u_2}{\lambda}\right) - \overline{F}_{y,2}\left(\frac{-\sqrt{h^2 - z_1^2} - (1 - \lambda)u_2}{\lambda}\right)
$$

and $\underline{I}'_1(z_1, u_2) = \dfrac{\partial \underline{I}_1(z_1, u_2)}{\partial z_1}$.

Proposition 2. *Let us consider that Assumption 2 is satisfied. Then the upper bound for the probability $\mathbb{P}\left(1 \leq N \leq T | u_1^2 + u_2^2 < h^2\right)$ is given by*

$$
\mathbb{P}\left(1 \leq N \leq T | u_1^2 + u_2^2 < h^2\right) \leq \frac{1}{a}\left[I_0(h)\overline{F}_{Q_0,1}(h) - I_0(-h)\underline{F}_{Q_0,1}(-h)\right.
$$

$$
- \int_{-h}^{h} \underline{F}_{Q_0,1}(x_1) \mathbb{I}_{\{I'_0(x_1) \geq 0\}} I'_0(x_1) dx_1
$$

$$
\left. - \int_{-h}^{h} \overline{F}_{Q_0,1}(x_1) \mathbb{I}_{\{I'_0(x_1) < 0\}} I'_0(x_1) dx_1\right], \quad (16)
$$

where

$$I_0(x_1) = \overline{p}_T\left(x_1, \sqrt{h^2 - x_1^2}\right)\overline{F}_{Q_0,2}\left(\sqrt{h^2 - x_1^2}\right)$$

$$- \overline{p}_T\left(x_1, -\sqrt{h^2 - x_1^2}\right)\underline{F}_{Q_0,2}\left(-\sqrt{h^2 - x_1^2}\right)$$

$$- \int_{-\sqrt{h^2-x_1^2}}^{\sqrt{h^2-x_1^2}} \underline{F}_{Q_0,2}(x_2)\mathbb{I}_{\{\overline{p}'_T(x_1,x_2)\geq 0\}}\overline{p}'_T(x_1,x_2)dx_2$$

$$- \int_{-\sqrt{h^2-x_1^2}}^{\sqrt{h^2-x_1^2}} \overline{F}_{Q_0,2}(x_2)\mathbb{I}_{\{\overline{p}'_T(x_1,x_2)<0\}}\overline{p}'_T(x_1,x_2)dx_2,$$

$$I_0'(x_1) = \frac{dI_0'(x_1)}{dx_1}, \ \overline{p}_T(X) = \sum_{n=1}^T \overline{p}_n(X), \ \overline{p}'_T(x_1,x_2) = \frac{\partial \overline{p}_T(x_1,x_2)}{\partial x_2},$$

$$a = I(h)\underline{F}_{Q_0,1}(h) - I(-h)\overline{F}_{Q_0,1}(-h) + \int_{-h}^{h} \underline{F}_{Q_0,1}(x_1)\mathbb{I}_{\{I'(x_1)\geq 0\}}I'(x_1)dx_1$$

$$+ \int_{-h}^{h} \overline{F}_{Q_0,1}(x_1)\mathbb{I}_{\{I'(x_1)<0\}}I'(x_1)dx_1,$$

$$I'(x_1) = \frac{dI(x_1)}{dx_1}, \ and \ I(x_1) = \underline{F}_{Q_0,2}\left(\sqrt{h^2 - x_1^2}\right) - \overline{F}_{Q_0,2}\left(-\sqrt{h^2 - x_1^2}\right).$$

5 Using the Bounds for the PDF

Sometimes it is necessary to use the bounds for the PDF of Q_0 and/or for the PDF of y_n. Such kind of bounds are usually used to overbound the distributions with excess-mass functions (see, for instance, [15]).

Assumption 3. *Let us assume that the PDF $f_y(X)$ of the innovation process $\{y_n\}_{n\geq 1}$ and the PDF $f_{Q_0}(X)$ of the initial state Q_0 obey the following inequality*

$$f_y(X) \leq \overline{f}_y(X) \ and \ f_{Q_0}(X) \leq \overline{f}_{Q_0}(X) \ for \ X \in \mathbb{R}^m.$$

Let us assume that Assumption 3 is satisfied. Then the above-mentioned recursive equations have to be replaced with the following inequalities

$$p_n(U) \leq \frac{1}{\lambda^m} \underset{\|Z\|_2 \leq h}{\int \cdots \int} p_{n-1}(Z)\overline{f}_y\left(\frac{Z - (1-\lambda)U}{\lambda}\right) dZ, \ n = 2, 3, \ldots, T, \quad (17)$$

and the initial condition $p_1(U)$ is also upper bounded in the following manner

$$p_1(U) \leq \frac{1}{\lambda^m} \underset{\|Z\|_2 \geq h}{\int \cdots \int} \overline{f}_y\left(\frac{Z - (1-\lambda)U}{\lambda}\right) dZ. \quad (18)$$

Unfortunately, this method of the risk overbounding can lead to non-exploitable results due to the recursive character of the above mentioned equations. This problem can be especially important for large values of T. Hence, a special attention should be paid to the choice of the upper bounds $\overline{f}_y(X)$ and $\overline{f}_{Q_0}(X)$.

6   Numerical Examples

The first example is devoted to the usage of the bounds for the CDFs of y_n and Q_0. Let us consider a scalar AR process given by Eq. (1) with unknown distributions of the innovation process $\{y_n\}_{n \geq 1}$ and the initial state Q_0. It is assumed that Assumption 1 is satisfied with the following bounds for the CDF $F_y(x)$ of the innovation process $\{y_n\}_{n \geq 1}$ and the CDF $F_{Q_0}(x)$ of the initial state Q_0:

$$\underline{F}_y(x) = \mathcal{N}\left(\mu, \frac{1 - (1 - \lambda)^2}{\lambda^2}\right) \leq F_y(x) \leq \overline{F}_y(x) = \mathcal{N}\left(-\mu, \frac{1 - (1 - \lambda)^2}{\lambda^2}\right)$$

and

$$\underline{F}_{Q_0}(x) = \mathcal{N}(\mu, 1) \leq F_{Q_0}(x) \leq \overline{F}_{Q_0}(x) = \mathcal{N}(-\mu, 1),$$

where $\mu = 1$, $\lambda \in [0.05, 0.95]$. The boundary h of the stopping time (2) is chosen such that $h = 4$. Let us now compare the probability $\mathbb{P}(1 \leq N \leq T | u \in [-h, h])$ with $T = 50$ defined by Eq. (8) and calculated for the Gaussian innovation process and the initial state with the "worst case expectation", i.e., $\mu = 1$, with the upper bound given by Proposition 1 for unknown distributions of the innovation process and initial state. These probabilities as functions of λ are presented in Fig. 1. The probability that the absorption occurs at one of the barriers $-h = -4$ or $h = 4$ at or before the 50-th step for the known Gaussian AR process with $\mu = 1$ is shown in dashed line and the upper bound for this probability for an AR process with unknown distributions is shown in solid line in Fig. 1.

The second example is devoted to the usage of the excess-mass function for the PDF of y_n. Let us consider a scalar AR process given by Eq. (1) with $1 - \lambda = 0.99$, and the initial position $u \in [-6, 6]$. It is assumed that Assumption 3 is satisfied with the following excess-mass bound for the PDF $f_y(x)$ of the innovation process $\{y_n\}_{n \geq 1}$:

$$f_y(x) \leq \overline{f}_y(x) = c \cdot f(x), \tag{19}$$

where $f(x)$ is the PDF of the Gaussian distribution $\mathcal{N}\left(0, \frac{1-(1-\lambda)^2}{\lambda^2}\right)$ and $c = 1.1$. The boundary h of the stopping time (2) is chosen such that $h = 6$. The results of comparison between the Gaussian AR process and the AR process with the innovation bounded with the excess-mass function are presented in Figs. 2 and 3. The probability of absorption at barriers at or before the n-th step for the Gaussian AR process as a function of u calculated by using Eqs. (5) and (6) is shown in Fig. 2. The probability of absorption at barriers at or before the n-th step for the AR process with a PDF of the innovation process which respects the excess-mass bound given by (19) as a function of u calculated by using Eqs. (17) and (18) is shown in Fig. 3.

By randomizing the results with the initial CDF $F_{Q_0}(x) = \mathcal{N}(0, 1)$ (see Eq. (8)), we get the probability $\mathbb{P}(1 \leq N \leq T | u \in [-h, h]) = 2.202 \cdot 10^{-8}$ for the Gaussian AR process and $\mathbb{P}(1 \leq N \leq T | u \in [-h, h]) = 4.595 \cdot 10^{-7}$ for the

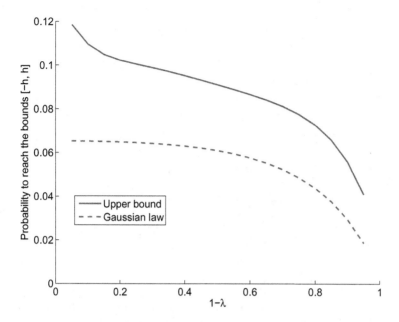

Fig. 1. The probability of absorption at barriers and its upper bound.

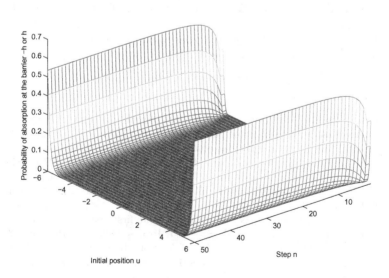

Fig. 2. The probability of absorption at barriers at or before the n-th step for the Gaussian AR process.

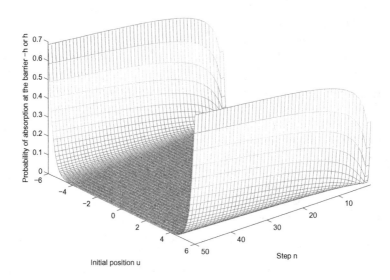

Fig. 3. The probability of absorption at barriers at or before the n-th step for the AR process with a PDF of the innovation process which respects the excess-mass bound given by (19).

AR process with a PDF of the innovation process which respects the above-mentioned excess-mass bound. It is obvious that the choice of the excess-mass bound given by (19) is not optimal and it is used here for illustration only.

Acknowledgments. The author gratefully acknowledges the research and financial support of this work from the Thales Alenia Space, France.

References

1. Robinson, P.B., Ho, T.Y.: Average run lengths of geometric moving average charts by numerical methods. Technometrics **20**(1), 85–93 (1978)
2. Crowder, S.V.: A simple method for studying run-length distributions of exponentially weighted moving average charts. Technometrics **29**(4), 401–407 (1987)
3. Novikov, A.A.: On the first exit time of an autoregressive process beyond a level and an application to the "Change-Point" problem. Teor. Veroyatnost. i Primenen. **35**(2), 282–292 (1990)
4. Novikov, A.A., Kordzakhia, N.: Martingales and first passage times of AR(1) sequences. Stochast. Int. J. Probab. Stochast. Process. **80**(2–3), 197–210 (2008)
5. Basseville, M., Nikiforov, I.V.: Detection of Abrupt Changes: Theory and Application. Information and System Sciences Series. Prentice Hall Inc, Englewood Cliffs (1993)
6. Tartakovsky, A., Nikiforov, I., Basseville, M.: Sequential Analysis: Hypothesis Testing and Changepoint Detection. Chapman & Hall/CRC Monographs on Statistics & Applied Probability. Taylor & Francis, New York (2014)

7. Bakhache, B., Nikiforov, I.: Reliable detection of faults in measurement systems. Int. J. Adapt. Control Signal Process. **14**(7), 683–700 (2000)
8. Guépié, B.K., Fillatre, L., Nikiforov, I.: Sequential detection of transient changes. Seq. Anal. **31**(4), 528–547 (2012)
9. Guépié, B.K., Fillatre, L., Nikiforov, I.: Detecting a suddenly arriving dynamic profile of finite duration. IEEE Trans. Inf. Theory **63**(5), 3039–3052 (2017)
10. Wald, A.: Sequential Analysis. Wiley, New York (1947)
11. Cox, D.R., Miller, H.D.: The Theory of Stochastic Processes. Wiley, New York (1965)
12. Shepp, L.A.: A first passage problem for the Wiener process. Ann. Math. Stat. **38**(6), 1912–1914 (1967)
13. Kemperman., J.: The general one-dimensional random walk with absorbing barriers with applications to sequential analysis. 's-Gravenhage: Excelsiors foto-offset, Dissertation, Amsterdam (1950)
14. Page, E.S.: An improvement to Wald's approximation for some properties of sequential tests. J. R. Stat. Soc. Ser. B Methodol. **16**(1), 136–139 (1954)
15. Rife, J., Walter, T., Blanch, J.: Overbounding SBAS and GBAS error distributions with excess-mass functions. In: Proceedings of the GNSS 2004 International Symposium on GNSS/GPS, Sydney, Australia, 6–8 December (2004)

Simulation of Finite-Source Retrial Queueing Systems with Collisions and Non-reliable Server

Ádám Tóth[1]([✉]), Tamás Bérczes[1], János Sztrik[1], and Anna Kvach[2]

[1] Faculty of Informatics, University of Debrecen,
Egyetem tér 1, Debrecen 4032, Hungary
adamtoth102@gmail.com, {berczes.tamas,sztrik.janos}@inf.unideb.hu
[2] Tomsk State University, Tomsk, Russia
kvach_as@mail.ru

Abstract. The aim of the present paper is to build a simulation program to investigate finite-source retrial queuing system with collision of the customers where the server is subject to random breakdowns and repairs depending on whether it is idle or busy. All the random variables involved in the model construction are assumed to be independent and generally distributed. The novelty of the investigation is to carry sensitivity analysis of the performance measures using various distributions. Several figures show the effect of different distributions on the performance measures such as mean and variance of number of customers in the system, mean and variance of response time, mean and variance of time a customer spent in service, mean and variance of sojourn time in the orbit.

Keywords: Simulation · Sensitivity analysis · Finite-source queuing system · Closed queuing system · Collision · Unreliable server · Retrial queue

1 Introduction

Retrial queues have been commonly used to depict many real situations emerging in telephone switching systems, telecommunication networks, computer networks and computer systems, call centers, wireless communication systems, etc. In many practical situations it is important to bear in mind that the rate of generation of new calls decreases as the number of customers in the system increases. This can be achieved with the use of finite-source, or quasi-random input models. Retrial queues with quasi-random input are recent interest in modeling cellular mobile networks, computer networks and local-area networks with random access protocols, and with multiple-access protocols, see, for example, [3,8].

In practice a few components of the system are prone to random breakdowns so it is important to study reliability of retrial queues with server breakdowns and repairs. Due to this it has a heavy influence on the performance measures of the system. Finite-source retrial queues with unreliable server have been investigated in several recent papers for example, [2,5,7,13,14].

© Springer International Publishing AG 2017
V.M. Vishnevskiy et al. (Eds.): DCCN 2017, CCIS 700, pp. 146–158, 2017.
DOI: 10.1007/978-3-319-66836-9_13

In many cases including data transmission from disparate sources there is a possibility to be conflict for a limited number of channels or other facilities. Several sources launching uncoordinated attempts can produce collisions leading to the loss of the transmission and consequently the necessity for retransmission. An essential matter is to develop workable procedures for allaying the conflict and corresponding message delay. There have been recent results on retrial queues with collision in [1, 4, 9–12].

The aim of the present paper is to investigate such systems with unreliable server which are finite source, and collisions can take place. In this paper we build simulation models using SimPack, a collection of C/C++ libraries and executable programs for computer simulation [6], to receive the desired performance measures. In this collection various algorithms are supported connected with simulation including discrete event simulation, continuous simulation and combined (multi-model) simulation. The novelty of this work is to provide sensitivity analysis using various distributions.

2 System Model

Let us consider a finite source retrial queueing system in which the number of sources is denoted by N and each of them can generate request with rate λ/N, that is the source time is exponentially distributed with parameter λ/N. If a customer finds the server idle it enters into service instantly. The service times are supposed to be gamma distributed with parameter α and β. When the server is engaged with a request, an arriving (from the orbit or the source) customer evokes a collision with a customer under service and both requests are directed towards the orbit. From the orbit it retries to be served after an exponentially distributed time with parameter σ/N. It is supposed that the service unit fails after some time which is an exponentially distributed random variable with parameter γ_0 when it is busy and with parameter γ_1 when it is idle.

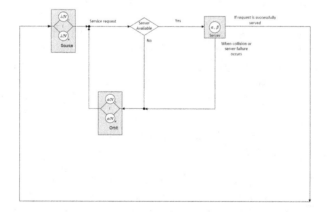

Fig. 1. System model

Immediately upon the breakdown it is forwarded for repair and the restoration time is also exponentially distributed random variable with parameter γ_2. We suppose that when the server is unavailable every source is eligible to generate customers and sends it to the unit, and the customers from the orbit may retry to the server. Moreover, in this model we suppose that the interrupted request gets into the orbit instantaneously and all of its services are independent of each other. When the submission is successful, the requests go back to the source. All the random variables involved in the model construction are assumed to be independent of each other (Fig. 1).

3   Simulation Results

3.1   Scenario A

The following table shows the input parameters of Scenario A (see Table 1).

Table 1. Numerical values of model parameters

Case	N	λ/N	γ_0	γ_1	γ_2	σ/N	α	β
1	100	0.01	0.1	0.1	1	0.01	0.5	0.5
2	100	0.01	0.1	0.1	1	0.01	1	1
3	100	0.01	0.1	0.1	1	0.01	2	2

Figure 2 shows the steady-state distribution of the three investigated cases. It is observed the mean number of customers increases as α and β are getting larger. *Case 2* is a special case because when $\alpha = 1$ it represents the exponential distribution. From the shape of the curves it is clearly visible that the steady-state distribution of the cases are normally distributed. The next table presents the considered performance measures in relation with the different cases (see Table 2).

In Table 2 the notations mean the followings: $E(NS)$ and $D^2(NS)$ - mean number and variance of customers, $E(T)$ and $D^2(T)$ - mean and variance of response time, $E(W)$ and $D^2(W)$ - mean and variance of waiting time, $E(S)$ and $D^2(S)$ - mean and variance of successful service time, $E(IS)$ - mean interrupted service time.

Table 2. Numerical results

Case	E(NS)	$D^2(NS)$	E(T)	$D^2(T)$	E(W)	$D^2(W)$	E(S)	$D^2(S)$	E(IS)
1	63.6842	27.9734	175.3073	65657.3454	174.5884	65434.6696	0.3147	0.1979	0.4041
2	70.5912	24.3012	239.9734	105273.4267	238.9734	104918.6389	0.4784	0.2289	0.5217
3	75.1825	21.2439	302.8106	151781.1411	301.5377	151277.6006	0.6472	0.2095	0.6257

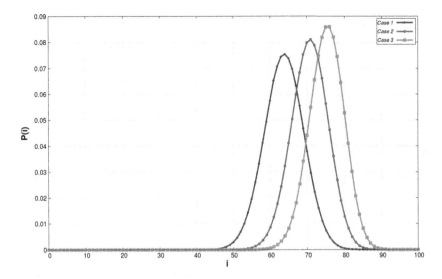

Fig. 2. Comparison of steady-state distributions

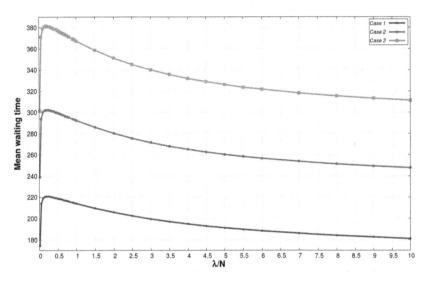

Fig. 3. Mean waiting time vs. intensity of incoming customers

Figure 3 represents the conformation of mean waiting time. The same parameters are (see Table 1) used as in case of Fig. 2 but here the running parameter is λ/N. As it is expected with the increment of λ/N mean waiting time increases as well but an interesting phenomenon is noticeable namely after λ/N is greater than 0.1 mean waiting time starts to decrease.

Figure 4 shows the effect of the inter-arrival time on the mean successful service time. Because of the opportunity of collisions it can easily happen that one

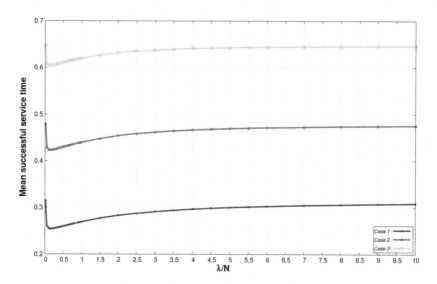

Fig. 4. Mean successful service time vs. intensity of incoming customers

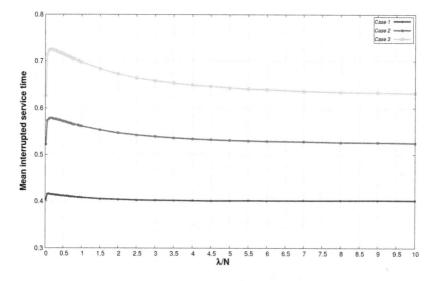

Fig. 5. Mean interrupted service time vs. intensity of incoming customers

job occurs multiple times in the service unit. Under mean successful service time we mean the average of those specific intervals when jobs are served appropriately. When λ/N is 0.01 the mean successful time is the highest then it starts to decrease as λ/N is increasing. This lasts till λ/N reaches 0.1, after it starts to increase.

On Fig. 5 the development of the mean interrupted service time can be seen. The opposite process happens compared to Fig. 4. It is worth mentioning when the service time is exponentially distributed then the sum of the mean inter-rupted service time and the mean successful service time is equal to 1/rate parameter.

3.2 Scenario B

Scenario B is very similar to Scenario A except that now the distribution of inter-arrival times of the customers is not exponential but gamma distributed. The next table (see Table 3) presents the input parameters of Scenario B.

Table 3. Numerical values of parameters of Scenario B

Case	N	α	β	γ_0	γ_1	γ_2	σ/N	α_1	β_1/N
1	100	1	1	0.1	0.1	1	0.01	0.5	0.01
2	100	1	1	0.1	0.1	1	0.01	1	0.01
3	100	1	1	0.1	0.1	1	0.01	2	0.01

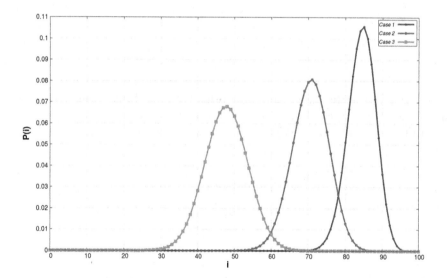

Fig. 6. Steady-state distributions of Scenario B

Figure 6 displays the steady-state distribution of Scenario B. Now the service time is exponentially distributed ($\alpha = 1$) and the inter-arrival time is gamma distributed. In these cases the steady-state distributions are still normally distributed and when α_1 is greater than 1 the mean number of requests in the

Table 4. Numerical results

Case	E(NS)	$D^2(NS)$	E(T)	$D^2(T)$	E(W)	$D^2(W)$	E(S)	$D^2(S)$	E(IS)
1	84.3609	14.1827	270.0351	128831.5059	269.0354	128420.4087	0.4502	0.2039	0.5495
2	70.5912	24.3012	239.9734	105273.4267	238.9734	104918.6389	0.4784	0.2289	0.5217
3	47.7859	34.2376	183.0164	69830.9728	182.0164	69573.992	0.5462	0.2982	0.4538

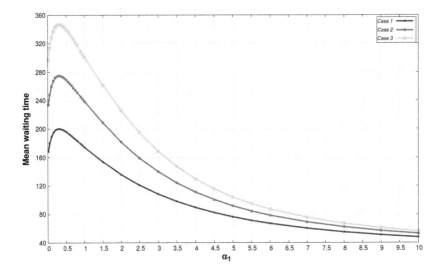

Fig. 7. Mean waiting time vs. shape parameter

system is significantly lower than in the other cases. In Table 7 the main performance measures can be found in connection with the cases (Table 4).

In Figs. 7, 8 and 9 the service time distribution of the *Cases* is the following (Table 5):

Table 5. Parameters of service time

Case	α	β
1	0.5	0.5
2	1	1
3	2	2

All the other parameters are according to Table 2. The running parameter is α_1 so in this way the impact of different distributions on the various performance measures can be discovered. First the mean waiting time (Fig. 7), after an initial jump mean waiting time starts to monotonically decrease resulting that as α_1 is getting bigger the less time the customers spend in the system. At the end the values of separate cases are almost equal.

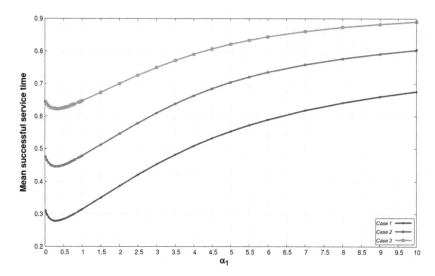

Fig. 8. Mean successful service time vs. shape parameter

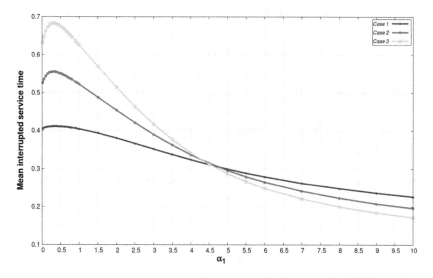

Fig. 9. Mean interrupted service time vs. shape parameter

As a consequence of Fig. 7 it is not surprising how the mean successful and interrupted service time change in function of α_1. In those cases when the mean waiting time reaches the highest values the mean successful service time are the lowest. The mean interrupted service time acts the same as in case of the mean waiting time. It is interesting that after α_1 is higher than 4.5 we observe that in *Case 3* the values of the mean interrupted service time is the lowest among the

Cases despite the fact that at the very start it possesses the highest values of mean interrupted service time.

3.3   Scenario C

In Scenario C not just inter-arrival but also the retrial time is general distributed with the same parameters. Below Table 6 shows the input parameters of Scenario C.

Table 6. Numerical values of parameters of Scenario C

Case	N	α	β	γ_0	γ_1	γ_2	α_1	β_1/N
1	100	1	1	0.1	0.1	1	0.5	0.01
2	100	1	1	0.1	0.1	1	1	0.01
3	100	1	1	0.1	0.1	1	2	0.01

Table 7 contains the main performance measures in connection with the cases.

Table 7. Numerical results

Case	E(NS)	$D^2(NS)$	E(T)	$D^2(T)$	E(W)	$D^2(W)$	E(S)	$D^2(S)$	E(IS)
1	81.5384	16.9599	220.8314	82735.764	219.8314	82377.8317	0.3398	0.1183	0.6602
2	70.5912	24.3012	239.9734	105273.4267	238.9734	104918.6389	0.4784	0.2289	0.5217
3	56.6635	30.1636	261.483	146264.4778	260.4827	145919.0907	0.626	0.3915	0.3743

This modification has no significant effect on the steady-state distribution (see Fig. 10). Of course the mean customers in the system is quite disparate but the distribution remains normal. Also when α_1 is less than 1 it results higher mean number of customers in the system compared to when it is more than one.

As in earlier in Scenario B in Figs. 7, 8 and 9 the service time distribution of the *Cases* is the following (Table 8):

Table 8. Parameters of service time

Case	α	β
1	0.5	0.5
2	1	1
3	2	2

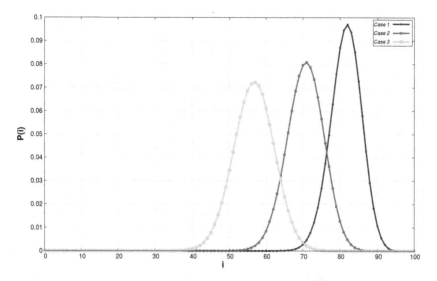

Fig. 10. Steady-state distributions of Scenario C

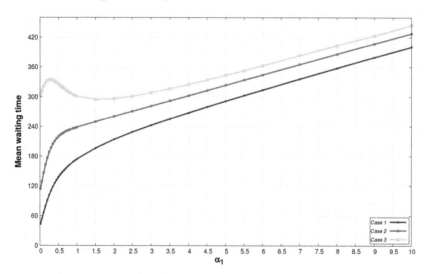

Fig. 11. Mean waiting time vs. shape parameter

Presently both inter-arrival and retrial distribution changes as α is increasing which results higher values of mean waiting time (Fig. 11). With the exception of *Case 3* the others are monotonically increasing. This is true for the mean successful service time (Fig. 12) but now *Case 3* rises from the beginning.

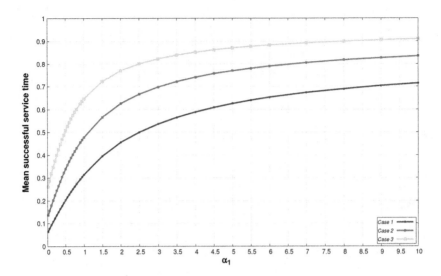

Fig. 12. Mean successful service time vs. shape parameter

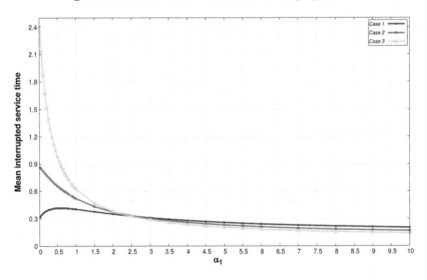

Fig. 13. Mean interrupted service time vs. shape parameter

The last figure shows the effect of the inter-arrival and inter-request time on the mean interrupted service time. At the beginning the difference is quite high among the *Cases* especially *Case 3* has high values compared to the others but it is valid till α_1 is smaller than 1 then the values of the *Cases* are nearly identical (Fig. 13).

4   Conclusions

In this paper a finite-source retrial queueing model was introduced with unreliable server and the possibility of collision. In all cases in all scenarios it turns out that the steady-state distribution of the number of customers in the service facility is normally distributed. We used SimPack to carry out stochastic simulation showing the effect of different distribution of service, inter-arrival and retrial times on several main performance measures.

Acknowledgment. The work of Tamás Bérczes was supported in part by the project EFOP-3.6.2-16-2017-00015 supported by the European Union, co-financed by the European Social Fund.

References

1. Ali, A.A., Wei, S.: Modeling of coupled collision and congestion in finite source wireless access systems. In: 2015 IEEE Wireless Communications and Networking Conference (WCNC), pp. 1113–1118. IEEE (2015)
2. Almási, B., Roszik, J., Sztrik, J.: Homogeneous finite-source retrial queues with server subject to breakdowns and repairs. Math. Comput. Modelling **42**(5–6), 673–682 (2005)
3. Artalejo, J.R., Gómez-Corral, A.: Retrial Queueing Systems. A Computational Approach. Springer, Berlin (2008)
4. Balsamo, S., Rossi, G.-L.D., Marin, A.: Modelling retrial-upon-conflict systems with product-form stochastic petri nets. In: Dudin, A., De Turck, K. (eds.) ASMTA 2013. LNCS, vol. 7984, pp. 52–66. Springer, Heidelberg (2013). doi:10.1007/978-3-642-39408-9_5
5. Dragieva, V.I.: Number of retrials in a finite source retrial queue with unreliable server. Asia-Pac. J. Oper. Res. **31**(2), 23 (2014)
6. Fishwick, P.A.: Simpack: getting started with simulation programming in C and C++. In: J.S. et al. (ed.) WSC 1992 Proceedings of the 24th Conference on Winter Simulation, pp. 154–162. ACM, New York (1992)
7. Gharbi, N., Dutheillet, C.: An algorithmic approach for analysis of finite-source retrial systems with unreliable servers. Comput. Math. Appl. **62**(6), 2535–2546 (2011)
8. Kim, J., Kim, B.: A survey of retrial queueing systems. Ann. Oper. Res. **247**(1), 3–36 (2016)
9. Kim, J.S.: Retrial queueing system with collision and impatience. Commun. Korean Math. Soc. **25**(4), 647–653 (2010)
10. Kvach, A., Nazarov, A.: Sojourn time analysis of finite source markov retrial queuing system with collision. In: Dudin, A., Nazarov, A., Yakupov, R. (eds.) ITMM 2015. CCIS, vol. 564, pp. 64–72. Springer, Cham (2015). doi:10.1007/978-3-319-25861-4_6
11. Nazarov, A., Kvach, A., Yampolsky, V.: Asymptotic analysis of closed markov retrial queuing system with collision. In: Dudin, A., Nazarov, A., Yakupov, R., Gortsev, A. (eds.) ITMM 2014. CCIS, vol. 487, pp. 334–341. Springer, Cham (2014). doi:10.1007/978-3-319-13671-4_38

158 Á. Tóth et al.

12. Peng, Y., Liu, Z., Wu, J.: An M/G/1 retrial G-queue with preemptive resume priority and collisions subject to the server breakdowns and delayed repairs. J. Appl. Math. Comput. **44**(1–2), 187–213 (2014). doi:10.1007/s12190-013-0688-7
13. Wang, J., Zhao, L., Zhang, F.: Analysis of the finite source retrial queues with server breakdowns and repairs. J. Ind. Manag. Optim. **7**(3), 655–676 (2011)
14. Zhang, F., Wang, J.: Performance analysis of the retrial queues with finite number of sources and service interruptions. J. Korean Stat. Soc. **42**(1), 117–131 (2013)

Retrial Tandem Queue with *BMAP*-Input and Semi-Markovian Service Process

Valentina Klimenok[1]($\boxtimes$), Olga Dudina[1], Vladimir Vishnevsky[2], and Konstantin Samouylov[3]

[1] Department of Applied Mathematics and Computer Science, Belarusian State University, 220030 Minsk, Belarus
klimenok@bsu.by, taramin@mail.ru
[2] Institute of Control Sciences of Russian Academy of Sciences, Moscow, Russia
vishn@inbox.ru
[3] Peoples' Friendship University of Russia, Moscow, Russia
ksam@sci.pfu.edu.ru

Abstract. We consider a tandem queueing system consisting of two stations. The input flow at the single-server first station is described by a *BMAP* (batch Markovian arrival process). If a customer from this flow meets the busy server, it goes to the orbit of infinite size and tries its luck later on in exponentially distributed random time. The service time distribution at the first station is assumed to be semi-Markovian. After service at the first station a customer proceeds to the second station which is described by a multi-server queue without a buffer. The service time by the server of the second station is exponentially distributed. We derive the condition for the stable operation of the system and determine the stationary distribution of the system states. Some key performance measures are calculated and illustrative numerical results are presented.

Keywords: Tandem retrial queue · Batch Markovian arrival process · Semi-Markovian service process · Asymptotically quasi-Toeplitz Markov chain

1 Introduction

Retrial queues are good mathematical models for many telecommunication networks such as telephone switching systems, cellular mobile networks, local area networks under the protocol of random multiple access, etc. So, despite their complexity, retrial queueing systems are popular object for investigations. They have been extensively studied under a variety of scenarios, for references see, e.g., survey [1] and books [2,3].

The analysis of current situation makes clear the great importance of tandem retrial queueing system with a batch Markovian arrival process (*BMAP*). Tandem queues can be used for modeling real-life networks of linear topology as well as for the validation of general decomposition algorithms in networks, see, e.g., [4–6]. Thus, tandem queueing systems have found much interest in the

© Springer International Publishing AG 2017
V.M. Vishnevskiy et al. (Eds.): DCCN 2017, CCIS 700, pp. 159–173, 2017.
DOI: 10.1007/978-3-319-66836-9_14

literature. An extensive survey of early papers on tandem queues can be seen in [7]. Most of these papers are devoted to exponential queueing models. Over the last three decades or so, the efforts of many investigators in tandem queues were in weakening the distribution assumptions on the service times as well as on the arrivals. In particular, the arrival process should be able to capture any correlation and burstiness that are commonly seen in the traffic of modern communication networks [6]. Such an arrival process was introduced in [8] and ever since this process is referred to as a batch Markovian arrival process ($BMAP$). $BMAP$ includes many input flows considered previously, such as stationary Poisson, Erlangian, Hyper-Markovian, Phase-Type (PH) renewal process, Markov Modulated Poisson Process ($MMPP$) and their superpositions. Tandem queues with the $BMAP$ input or its ordinary counterparts were considered in [9–16].

At the same time, to the best of our knowledge, only the papers [13, 16, 17] are devoted to investigation of tandem queues with $BMAP$ input where the effect of retrials is taken into account. The paper [13] considers the $MAP/PH/1 \rightarrow \cdot/PH/1/K+1$ tandem retrial queue. The paper [16] deals with a tandem retrial queue with two Markovian flows and reservation of servers on the second station is studied. The paper [17] is devoted to the tandem queue $BMAP/G/1 \rightarrow \cdot/M/N/0$ with retrials and group occupation of servers of the second station. Here the customers receive service individually, and upon completion of a service the customer's type is determined. This type identification is necessary to determine the nature of service, if any, offered at station 2. The customer's type is classified based on the number of servers (resources) required to process the request of the customer. The simultaneous initiation or occupation of several servers to a customer's request is typical for the so called non-elastic traffic in communication networks.

In this paper, we consider a generalization of the model [17] to the case of correlated semi-Markovian service process at the first station. Unlike the system [17], where service times at the first station are independent identically distributed random variables, in this paper we assume that these times can be significantly dependent and distributed according to different laws. This assumption makes the system under consideration more adequate model of real-life systems and processes but more difficult for analytical investigation.

The rest of the paper is organized as follows. In Sect. 2, the mathematical model is described. In Sect. 3, the embedded Markov chain is investigated, the condition for existence is derived, and the stationary distribution of the system states at the service completion epochs at the first station is calculated. The stationary distribution at an arbitrary time is derived in Sect. 4. The system performance measures are given in Sect. 5. In Sect. 6, the numerical examples illustrating the behavior of the system characteristics depending on its parameters are presented. Finally, the conclusions are given in Sect. 7.

2   Mathematical Model

We consider tandem queue consisting of two stations.

The first station has a single server. Customers arrive at the first station in batches. The input flow of batches is described by the $BMAP$. The $BMAP$ is defined by the underlying process ν_t, $t \geq 0$, which is an irreducible continuous time Markov chain with state space $\{0, ..., W\}$, and with the matrix generating function $D(z) = \sum_{k=0}^{\infty} D_k z^k$, $|z| \leq 1$. Arrivals occur only at epochs of the jumps in the underlying process ν_t, $t \geq 0$. The intensities of the transitions of the process ν_t accompanied by the arrival of a batch of size k are defined by the matrices D_k, $k \geq 0$. The matrix $D(1)$ is the infinitesimal generator of the process ν_t. The stationary distribution vector $\boldsymbol{\theta}$ of this process satisfies the equations $\boldsymbol{\theta}D(1) = \mathbf{0}, \boldsymbol{\theta}\mathbf{e} = 1$, where $\mathbf{e}$ is a column vector consisting of 1's, and $\mathbf{0}$ is a row vector of 0's. The average intensity λ (fundamental rate) of the $BMAP$ is given by $\lambda = \boldsymbol{\theta}D'(z)|_{z=1}\mathbf{e}$. The average intensity λ_b of group arrivals is defined by $\lambda_b = \boldsymbol{\theta}(-D_0)\mathbf{e}$. The coefficient of variation c_{var} of intervals between successive group arrivals is defined by $c_{var}^2 = 2\lambda_b\boldsymbol{\theta}(-D_0)^{-1}\mathbf{e} - 1$. The coefficient of correlation c_{cor} of the successive intervals between group arrivals is given by $c_{cor} = (\lambda_b\boldsymbol{\theta}(-D_0)^{-1}(D(1) - D_0)(-D_0)^{-1}\mathbf{e} - 1)/c_{var}^2$. For more information about the $BMAP$, its history and properties see, e.g., [8].

If the arriving batch of primary customers meets a free first station server upon arrival, one customer automatically starts a service and the rest of the batch go to so called orbit. If the server is busy at an arrival epoch, all customers of the batch go to the orbit. From the orbit, they try their luck later on after a random amount of time. We assume that the total flow of retrials is such as the probability of generating a retrial attempt in the interval $(t, t + \Delta t)$ is equal to $\alpha_i\Delta t + o(\Delta t)$ when the number of customers in the orbit is equal to i, $i > 0$, $\alpha_0 = 0$. The orbit capacity is supposed to be unlimited. We do not fix the explicit dependence of the intensity α_i on i assuming only that $\lim_{i\to\infty} \alpha_i = \infty$. Note that such dependence describes the classic retrial strategy ($\alpha_i = i\alpha$) and the linear strategy ($\alpha_i = i\alpha + \gamma$, $\alpha > 0$) as special cases.

The service time of primary and repeated customers at the first station is governed by the semi-Markovian process m_t, $t \geq 0$. It is characterized by the state space $\{1, ..., M\}$ and the semi-Markovian kernel $B(t) = (B_{m,m'}(t))_{m,m'=\overline{1,M}}$. The successive service times of customers are defined as the sojourn times of the process m_t, $t \geq 0$, in its states. The average service time is calculated as $b_1 = \boldsymbol{\delta}\int_0^\infty tdB(t)\mathbf{e}$ where $\boldsymbol{\delta}$ is the unique solution to the system $\boldsymbol{\delta}B(\infty) = \boldsymbol{\delta}, \boldsymbol{\delta}\mathbf{e} = 1$.

After receiving service at the first station a customer proceeds to the second station which is represented by N independent identical servers. The service time by a server is exponentially distributed with the parameter $\mu > 0$. The service of an arbitrary customer at the second station requires a random number η of servers. Here η is an integer-valued random variable with the distribution

$q_n = P\{\eta = n\}$, $q_n \geq 0$, $n = \overline{0, N}$, $\sum\limits_{n=0}^{N} q_n = 1$. No queue is allowed between the first and the second station.

In case a customer completes the service at the first station and does not see required number of free servers at the second station, it leaves the system forever with the probability p, $0 \leq p \leq 1$. With the probability $1-p$ the customer waits until the required number of servers of the second station becomes free and then occupies these servers immediately. The waiting period is accompanied by blocking the first station server operating.

For further use in the sequel, we introduce the following notation:

- I is an identity matrix of appropriate dimension. When needed the dimension of the matrix will be identified with a suffix;

- $\otimes$ and $\oplus$ are symbols of the Kronecker product and sum of matrices, see, e.g., [18];

- $\bar{W} = W + 1$;

- $P(n, t)$, $n \geq 0$, are coefficients of the matrix expansion $e^{D(z)t} = \sum\limits_{n=0}^{\infty} P(n, t) z^n$, $|z| \leq 1$. The (ν, ν')th entry of the matrix $P(n, t)$ defines the probability that n customers arrive in the $BMAP$ during the interval $(0, t]$ and the state of the underlying process of the $BMAP$ at the epoch t is ν' given $\nu_0 = \nu$, $\nu, \nu' = \overline{0, W}$;

- $\tilde{D}_k = D_k \otimes I_M$, $\hat{D}_k = I_{N+1} \otimes \tilde{D}_k$, $k \geq 0$, $\hat{D}(z) = \sum\limits_{k=0}^{\infty} \hat{D}_k z^k$, $|z| \leq 1$;

- $H(t) = (H_{r,r'}(t))_{r,r'=\overline{0,N}}$, where $H_{r,r'}(t) = 0$ for $r \leq r'$ and, for $r > r'$, $H_{r,r'}(t)$ is the distribution function with the Laplace-Stieltjes transform $h_{r,r'}(s) = \prod\limits_{l=r'+1}^{r} l\mu(l\mu + s)^{-1}$;

- Q_m, $m = \overline{1, 3}$, are square matrices:

$$Q_1 = \begin{pmatrix} q_0 & q_1 & \cdots & q_N \\ 0 & q_0 & \cdots & q_{N-1} \\ \vdots & \vdots & \ddots & \vdots \\ 0 & 0 & \cdots & q_0 \end{pmatrix}, \quad Q_3 = \begin{pmatrix} 0 & \cdots & 0 & q_N \\ 0 & \cdots & 0 & q_{N-1} \\ \vdots & \ddots & \vdots & \vdots \\ 0 & \cdots & 0 & q_0 \end{pmatrix},$$

$$Q_2 = diag\{ \sum\limits_{n=N-r+1}^{N} q_n, r = \overline{0, N}\};$$

- $\tilde{Q}_m = Q_m \otimes I_{\bar{W}}$, $\hat{Q}_m = \tilde{Q}_m \otimes I_M$, $m = \overline{1, 3}$; $\bar{Q} = \hat{Q}_1 + p\hat{Q}_2$;
- $Q = \tilde{Q}_1 + p\tilde{Q}_2 + (1-p)E\tilde{Q}_3$,
- $E = \bar{I}_{N+1} \otimes I_M$, where $\bar{I}_{\bar{W}}$ is a square matrix of size $N + 1$ whose below-diagonal entries are equal to 1 and the rest entries are zeroes.

3 The Stationary Distribution of the Embedded Markov Chain

Let t_n denote the time of the nth service completion at the first station. Consider the process

$$\xi_n = \{i_n, r_n, \nu_n, m_n\}, n \geq 1,$$

where

- i_n is the number of customers in the orbit at the epoch t_n, $i_n \geq 0$;
- r_n is the number of busy servers at the second station at the epoch $t_n - 0$, $r_n = \overline{0, N}$;
- ν_n is the state of the $BMAP$ underlying process at the epoch t_n, $\nu_n = \overline{0, W}$;
- m_n is the state of the service directing process m_t at the epoch $t_n + 0$, $m_n = \overline{1, M}$.

It is easy to see that the process ξ_n, $n \geq 1$, is a four-dimensional Markov chain which describes the process of the system operation at the service completion epochs.

Enumerate the states of the chain ξ_n, $n \geq 1$, in the lexicographic order and form the square matrices $P_{i,l}$, $i, l \geq 0$, of size $(N+1)\bar{W}M$ of transition probabilities of the chain from the states having the value i of the first component to the states having the value l of this component.

Lemma 1. The transition probability matrices $P_{i,l}$ are defined as follows:

$$P_{i,l} = 0, \, l < i - 1, \, i > 1,$$

$$P_{i,l} = \bar{Q} A_i [\alpha_i \Omega_{l-i+1} + \sum_{k=1}^{l-i+1} \hat{D}_k \Omega_{l-i-k+1}] + (1-p)[\sum_{n=0}^{l-i+1} \mathcal{H}_n \hat{Q}_3 A_{i+n} \alpha_{i+n} \Omega_{l-i-n+1}$$

$$+ \sum_{n=0}^{l-i} \mathcal{H}_n \hat{Q}_3 A_{i+n} \sum_{k=1}^{l-i-n+1} \hat{D}_k \Omega_{l-i-n-k+1}], \, l \geq max\{0, i-1\}, \, i \geq 0, \tag{1}$$

$$A_i = \int_0^\infty e^{-\alpha_i t} e^{(\Delta \oplus \tilde{D}_0)t} dt \otimes I_M = (\alpha_i I - \Delta \oplus \tilde{D}_0)^{-1} \otimes I_M, \, i \geq 0,$$

$$\Omega_n = \int_0^\infty e^{\Delta t} \otimes P(n, t) \otimes dB(t), \quad \mathcal{H}_n = \int_0^\infty dH(t) \otimes P(n, t) \otimes I_M, \, n \geq 0,$$

$$\Delta = \begin{pmatrix} 0 & 0 & 0 & \cdots & 0 & 0 \\ \mu & -\mu & 0 & \cdots & 0 & 0 \\ 0 & 2\mu & -2\mu & \cdots & 0 & 0 \\ \vdots & \vdots & \vdots & \ddots & \vdots & \vdots \\ 0 & 0 & 0 & \cdots & N\mu & -N\mu \end{pmatrix}.$$

Proof. Formula (1) becomes clear if we take into account the probabilistic interpretation of the matrices which appear in the right hand side of (1).

The matrix $\bar{Q}$ defines probabilities that a customer served at the first station finds the required number of idle servers at the second station and occupies them or does not find the required number of idle servers and leaves the system.

The matrix Δ is an infinitesimal generator of the death process which describes the evolution of the number of occupied servers at the second station between two consecutive service completion epochs at the first station.

The matrix $A_i \alpha_i$ defines probabilities that, given i customers stay in the orbit after the service completion epoch at the first station, the next service at this station will be initiated by a customer from the orbit. The matrix $A_i \hat{D}_k$ has the analogous probabilistic sense with the only difference that the next service at the first station is initiated by a primary customer arriving in k-size batch.

The matrices Ω_n and $\mathcal{H}_n$ defines probabilities that during the service time at the first station and during the blocking time, respectively, n customers arrive into the system.

The matrix $(1-p)\mathcal{H}_n \tilde{Q}_3$ defines probabilities that a customer served at the first station does not find the required number of idle servers at the second station, causes the blocking of the first station server and during the blocking time n customers arrive into the system.

Using the above probabilistic interpretations and the total probability formula, we get expression (1) for the transition probability matrices.

Lemma 1 is proved.

It is seen from (1) that transition probability matrices $P_{i,l}$ depend on i and l and this dependence can not be reduced to the dependence on the difference $l - i$ only. It means that the Markov chain ξ_n, $n \geq 1$, is a level dependent one. At the same time the dependence of i vanishes when i tends to ∞ and the matrices $P_{i,l}$ approach to matrices that depend on the values i and l only via the difference $l - i$. It implies that the chain under consideration belongs to the class of asymptotically quasi-Toeplitz Markov chains (AQTMC), see [19]. So, the further investigation of the ξ_n, $n \geq 1$, will be based on the results given in [19].

Let us denote

$$Y_k = \lim_{i \to \infty} P_{i,i+k-1}, \; k \geq 0, \tag{2}$$

and let $Y(z)$ be the generating function of the matrices Y_k, $k \geq 0$. The matrices $\tilde{Y}_k$, $k \geq 0$, can be considered as transition probability matrices of a quasi-Toeplitz Markov chain $\tilde{\xi}_n$, $n \geq 1$, with the same state space as the chain ξ_n, $n \geq 1$. The chain $\tilde{\xi}_n$, $n \geq 1$, is called as limiting chain relative to the chain ξ_n, $n \geq 1$.

Corollary 1. The Markov chain ξ_n, $n \geq 1$, belongs to the class of asymptotically quasi-Toeplitz Markov chain. The generating function of its limiting chain transition probability matrices has form

$$Y(z) = [\bar{Q} + (1-p)\mathcal{H}(z)\hat{Q}_3]\Omega(z), \tag{3}$$

where

$$\Omega(z) = \sum_{n=0}^{\infty} \Omega_n z^n = \int_0^{\infty} e^{\Delta t} \otimes e^{D(z)t} \otimes dB(t),$$

$$\mathcal{H}(z) = \sum_{n=0}^{\infty} \mathcal{H}_n z^n = \int_0^{\infty} dH(t) \otimes e^{D(z)t} \otimes I_M, \ |z| \leq 1.$$

Proof. It is seen from (1) that transition probability matrices $P_{i,l}$ depend on i and l and this dependence can not be reduced to the dependence on the difference $l - i$ only. It means that the Markov chain ξ_n, $n \geq 1$, is a level dependent one. At the same time the dependence of i vanishes when i tends to ∞ and the matrices $P_{i,l}$ approach to matrices that depend on the values i and l only via the difference $l - i$. It implies that the chain under consideration belongs to the class of asymptotically quasi-Toeplitz Markov chains (AQTMC), see [19].

Using (2) and Lemma 1 we get the expression (3) for the generating function $Y(z)$.

In what follows we will use the results for asymptotically quasi-Toeplitz Markov chains given in [19] to derive the ergodicity condition and calculate the stationary distribution.

Theorem 1. The sufficient condition for ergodicity of the Markov chain ξ_n, $n \geq 1$, is the fulfillment of the inequality

$$\rho = \lambda[b_1 + (1 - p)\boldsymbol{y} \int_0^{\infty} t dH(t) Q_3 \mathbf{e}] < 1, \tag{4}$$

where the vector $\boldsymbol{y}$ is the unique solution of the system

$$\boldsymbol{y} Q^- \int_0^{\infty} e^{\Delta t} \otimes d\tilde{B}(t) = \boldsymbol{y}, \ \boldsymbol{y}\mathbf{e} = 1. \tag{5}$$

$$Q^- = Q_1 + pQ_2 + (1 - p)EQ_3, \ \tilde{B}(t) = \boldsymbol{\delta} B(t)\mathbf{e}.$$

Proof. The matrix $Y(1)$ is an irreducible one. So, as follows from [19], the sufficient condition for ergodicity of the chain ξ_n, $n \geq 1$, is the fulfillment of the inequality

$$\boldsymbol{x} Y'(1)\mathbf{e} < 1, \tag{6}$$

where $\boldsymbol{x}$ is the unique solution of the system

$$\boldsymbol{x} Y(1) = \boldsymbol{x}, \ \boldsymbol{x}\mathbf{e} = 1. \tag{7}$$

Let the vector $\boldsymbol{x}$ be of the form

$$\boldsymbol{x} = \boldsymbol{y} \otimes \boldsymbol{\theta} \otimes \boldsymbol{v} \tag{8}$$

where θ is the vector of the stationary distribution of the $BMAP$, y and v some stochastic vectors of size $N+1$ and M, respectively, and the vector y is the unique solution of system (5).

It is verified by the direct substitution that the vector x of form (8) is the unique solution of system (7) and $v = \delta$. The last equation means that the vector v coincides with the stationary distribution vector of the service process $m_t, t \geq 0$.

We differentiate (3) at the point $z = 1$ and substitute the obtained expression for $Y'(1)$ and the vector x of form (8) into inequality (6). Then using that the vector $v = \delta$, we derive inequality (4).

Theorem 1 is proved.

Remark 1. Inequality (7) is intuitively clear on noting that the vector y gives the stationary distribution of the number of busy servers at the second station given the server of the first station works under overload conditions. Then $y \int_0^\infty t dH(t) Q_3 \mathbf{e}$ defines the average blocking time of the server of the first station under overload condition and ρ is the system load.

In what follows we suppose that inequality (4) is fulfilled.

Denote the stationary state probabilities of the Markov chain $\xi_n = \{i_n, r_n, \nu_n, m_n\}, n \geq 1$, by

$$\pi(i, r, \nu, m), \ i \geq 0, \ r = \overline{0, N}, \ \nu = \overline{0, W}, \ m = \overline{1, M}.$$

Form the row vectors of these probabilities

$$\boldsymbol{\pi}(i, r, \nu) = (\pi(i, r, \nu, 1), \ldots, \pi(i, r, \nu, M)),$$

$$\boldsymbol{\pi}(i, r) = (\boldsymbol{\pi}(i, r, 0), \boldsymbol{\pi}(i, r, 1), \ldots, \boldsymbol{\pi}(i, r, W)),$$

$$\boldsymbol{\pi}_i = (\boldsymbol{\pi}(i, 0), \boldsymbol{\pi}(i, 1), \ldots, \boldsymbol{\pi}(i, N)), i \geq 0.$$

To calculate the vectors $\boldsymbol{\pi}_i, i \geq 0$, we use the numerically stable algorithm (see [19]) which has been elaborated for calculating the stationary distribution of the multi-dimensional asymptotically quasi-Toeplitz Markov chain.

The algorithm is based on censoring technique and asymptotic properties of the chain under consideration. It consists of the next principal steps:

1. Calculate the matrix G as the minimal nonnegative solution of the matrix equation $G = Y(G)$;
2. For preassigned sufficiently large integer i_0 calculate the matrices G_{i_0-1}, $G_{i_0-2}, \ldots, G_0$ using the equation of the backward recursion

$$G_i = (I - \sum_{l=i+1}^\infty P_{i+1,l} G_{l-1} G_{l-2} \ldots G_{i+1})^{-1} P_{i+1,i}, \ i = i_0 - 1, i_0 - 2, \ldots, 0,$$

with the boundary condition $G_i = G, i \geq i_0$.

3. Calculate the matrices $\Phi_l, l \geq 1$, using recurrent formulas

$$\Phi_l = (\bar{P}_{0,l} + \sum_{i=1}^{l-1} \Phi_i \bar{P}_{i,l})(I - \bar{P}_{l,l})^{-1}, l \geq 1;$$

4. Calculate the vector π_0 as the unique solution to the system

$$\pi_0(I - \bar{P}_{0,0}) = \mathbf{0}, \ \pi_0(I + \sum_{l=1}^{\infty} \Phi_l)\mathbf{e} = 1.$$

5. Calculate the vectors π_l by $\pi_l = \pi_0 \Phi_l, l \geq 1$.

4 The Stationary Distribution at an Arbitrary Time

Consider now the process of the system states at an arbitrary time

$$\zeta_t = \{i_t, r_t, \nu_t, m_t, k_t\}, \ t \geq 0,$$

where

- i_t is the number of customers in the orbit;
- r_t is the number of busy servers at the second station;
- ν_t is the state of the arrival directing process;
- m_t is the state of the service directing process;
- k_t is a random which takes values 0, 1, 2 depending on whether the server of the first station is idle, serves a customer, or it is blocked at time t, $t \geq 0$.

The process ζ_t, $t \geq 0$, is non-Markovian. It can be classified as semi-regenerative processes, for definition see [20]. The stationary distribution of this process can be related to the stationary distribution of the embedded Markov chain ξ_n, $n \geq 1$, using the results [20] for Markov renewal and semi-regenerative processes.

Let

$$p(i, r, \nu, m, k) == \lim_{t \to \infty} P\{i_t = i, \ r_t = r, \ \nu_t = \nu, \ m_t = m, \ k_t = k\},$$

$$i \geq 0, \ r = \overline{0, N}, \ \nu = \overline{0, W}, m = \overline{1, M}, \ k = \overline{0, 2},$$

be the steady-state probabilities of the process ζ_t, $t \geq 0$.

Define the vectors of these probabilities

$$\boldsymbol{p}(i, r, \nu, k) = (p(i, r, \nu, 1, k), ..., p(i, r, \nu, M, k)),$$

$$\boldsymbol{p}(i, r, k) = (\boldsymbol{p}(i, r, 0, k), ..., \boldsymbol{p}(i, r, W, k)), \ \boldsymbol{p}_i(k) = (\boldsymbol{p}(i, 0, k), ..., \boldsymbol{p}(i, N, k)).$$

Theorem 2. The non-zero stationary probability vectors $\boldsymbol{p}_i(k)$, $i \geq 0$, $k = \overline{0,2}$, are related to the stationary probability vectors $\boldsymbol{\pi}_i$, $i \geq 0$, of the embedded Markov chain ξ_n, $n \geq 1$, as follows:

$$\boldsymbol{p}_0(0) = \lambda^{-1}\boldsymbol{\pi}_0[\bar{Q} + (1-p)\mathcal{H}_0\hat{Q}_3]A_0,$$

$$\boldsymbol{p}_i(0) = \lambda^{-1}[\boldsymbol{\pi}_i\bar{Q}A_i + (1-p)\sum_{l=0}^{i}\boldsymbol{\pi}_l\mathcal{H}_{i-l}\hat{Q}_3A_i], \ i \geq 1,$$

$$\boldsymbol{p}_i(1) = \lambda^{-1}\{\sum_{l=0}^{i}\boldsymbol{\pi}_l[\bar{Q}A_l \sum_{k=1}^{i-l+1}\hat{D}_k\tilde{\Omega}_{i-l-k+1} + (1-p)\sum_{k=0}^{i-l}\mathcal{H}_k\hat{Q}_3A_{l+k}$$

$$\times \sum_{m=1}^{i-l-k+1}\hat{D}_m\tilde{\Omega}_{i-l-k-m+1}] + \sum_{l=0}^{i+1}\boldsymbol{\pi}_l[\bar{Q}A_l\alpha_l\tilde{\Omega}_{i-l+1}$$

$$+ (1-p)\sum_{k=0}^{i-l+1}\mathcal{H}_k\hat{Q}_3A_{l+k}\alpha_{l+k}\tilde{\Omega}_{i-l-k+1}]\}, \ i \geq 0,$$

$$\boldsymbol{p}_i(2) = \lambda^{-1}(1-p)\sum_{l=0}^{i}\boldsymbol{\pi}_l\sum_{k=0}^{i-l}(\mathcal{H}_k + \delta_{0,k}I)\hat{Q}_2\int_0^{\infty}e^{-\mu Rt} \otimes P(i-l-k,t) \otimes I_M dt,$$

where $\delta_{i,j}$ is Kronecker's symbol, $\tilde{\Omega}_n = \int_0^{\infty}e^{\Delta t} \otimes P(n,t) \otimes (I_M - \nabla_B(t))dt, n \geq 0$, $\nabla_B(t)$ is the diagonal matrix with diagonal elements $(B(t)e)_j, j = \overline{1,M}$.

Corollary 2. The stationary probability vectors $\boldsymbol{p}_i$, $i \geq 0$, can be calculated by the following formula:

$$\boldsymbol{p}_0 = \boldsymbol{p}_0(0), \ \boldsymbol{p}_i = \boldsymbol{p}_i(0) + \sum_{m=1}^{2}\boldsymbol{p}_{i-1}(m), \ i \geq 1.$$

5   Performance Measures

Basing on the stationary distribution we can calculate different performance characteristics of the system. The most important performance measures are calculated as follows:

- Mean number of customers at the first station at the service completion epoch at this station and at an arbitrary time

$$\tilde{L} = \boldsymbol{\Pi}'(1)e, \qquad L = \boldsymbol{P}'(1)e,$$

where $\boldsymbol{\Pi}(z) = \sum_{i=0}^{\infty}\boldsymbol{\pi}_iz^i$, $\boldsymbol{P}(z) = \sum_{i=0}^{\infty}\boldsymbol{p}_iz^i$, $|z| \leq 1$.

- Variance of the mean number of customers at the first station at the service completion epoch at this station and at an arbitrary time

$$\tilde{D} = \boldsymbol{\Pi}''(1)\mathbf{e} + \tilde{L} - \tilde{L}^2, \qquad D = \boldsymbol{P}''(1)\mathbf{e} + L - L^2.$$

- The vector of the stationary distribution of the number of busy servers at the second station at the service completion epoch at the first station and at an arbitrary time

$$\tilde{\boldsymbol{r}} = \boldsymbol{\Pi}(1)(I_{N+1} \otimes \mathbf{e}_{\bar{W}M}), \qquad \boldsymbol{r} = \boldsymbol{P}(1)(I_{N+1} \otimes \mathbf{e}_{\bar{W}M}).$$

- Mean number of busy servers at the second station at the service completion epoch at the first station and at an arbitrary time

$$\tilde{N}_{busy} = \tilde{\boldsymbol{r}}diag\{r, r = \overline{0, N}\}\mathbf{e}, \qquad N_{busy} = \boldsymbol{r}diag\{r, r = \overline{0, N}\}\mathbf{e}.$$

- The probability that an arbitrary customer leaves the system or causes the blocking of the server at the first station

$$P_{loss} = p\boldsymbol{\Pi}(1)\hat{Q}_2\mathbf{e}, \qquad P_{block} = (1 - p)\boldsymbol{\Pi}(1)\hat{Q}_2\mathbf{e}.$$

- The probability of immediate access to the first station server

$$P_{imm} = -\lambda^{-1} \sum_{i=0}^{\infty} \boldsymbol{p}_i(0)(\mathbf{e}_{(N+1)M} \otimes D_0\mathbf{e}).$$

6 Numerical Examples

The proposed algorithms for calculating the stationary distributions were realized as computer programm using tools of software "Sirius++" [21]. In this section we present numerical examples demonstrating the behavior of the performance measures as function of the system load and intensity of retrials.

We consider $BMAP$ having the fundamental rate $\lambda = 10$, the correlation coefficient $c_{cor} = 0.1$ and characterized by the matrices

$$D_0 = \begin{pmatrix} -5.39233 & 4.33008 \times 10^{-5} \\ 1.74403 \times 10^{-5} & -0.70174 \end{pmatrix}, \ D_1 = D_3 = \begin{pmatrix} 1.60698 & 0.01072 \\ 0.11744 & 0.09307 \end{pmatrix},$$

$$D_2 = \begin{pmatrix} 2.14264 & 0.01429 \\ 0.15659 & 0.12410 \end{pmatrix}.$$

Semi-Markovian kernel $B(t)$ is defined as $B(t) = diag\{B_1(t), B_2(t)\}P$, where the transition matrix P has the form $P = \begin{pmatrix} 0.6 & 0.4 \\ 0.35 & 0.65 \end{pmatrix}$,

$B_1(t)$ and $B_2(t)$ define the Erlangian distributions of order 3 with the parameters 20 and 50, respectively. The average service time b_1 is equal to 0.102.

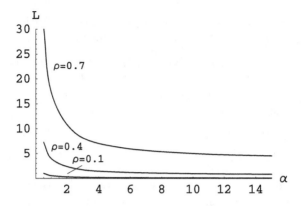

Fig. 1. The mean number of customers at the first station as a function of retrial rate α for different values of system load ρ.

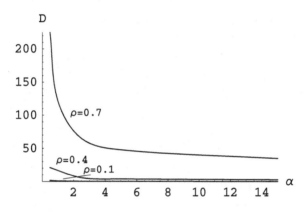

Fig. 2. The variance of the mean number of customers at the first station as a function of retrial rate α for different values of system load ρ.

We assume that the number of servers at the second station $N = 3$, the probability p is equal to 0.6. The parameters of service at the second station are as follows: $q_0 = 0$, $q_1 = 0.9$, $q_2 = q_3 = 0.05$, and the service rate $\mu = 3$.

We consider the classical retrial strategy: $\alpha_i = i\alpha$, $i \geq 0$, and vary the retrial rate α in the interval $[0.5, 15]$.

Figures 1, 2 illustrate the dependence of mean number L of customers and the variance D of the number of customers at the first station on the value α for three different values of system load ρ.

The dependence of the loss probability P_{loss} and the probability P_{imm} of immediate access to the first station server on the value α under the different values of system load is presented in Figs. 3 and 4.

The system load ($\rho = 0.1$, 0.4, 0.7) in this experiment varies by means of scaling the fundamental rate ($\lambda = 0.8$, 3.12, 5.45, respectively). In turn, the

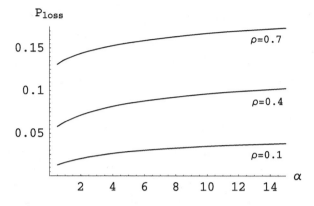

Fig. 3. The loss probability as a function of retrial rate α for different values of system load ρ.

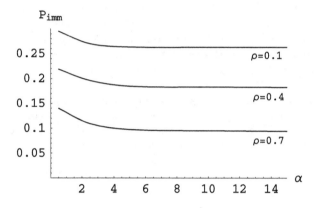

Fig. 4. The probability of immediate access to the first station server as a function of retrial rate α for different values of system load ρ.

fundamental rate λ of the $BMAP$ varies by multiplying the matrices D_k, $k = \overline{0,3}$, by some positive constant.

Figures show that the mean number of customers, the variance of the number of customers at the first station and the loss probability increase, while the probability of immediate access to the first station server without visiting the orbit decrease when the input intensity λ (and the system load ρ) increases. It confirms the fact that the system load has a great impact on the performance measures: the increase of the load makes the quality of service in the system worse.

It is also seen from the figures that all characteristics are sensitive with respect to the value of retrial rate α. Under the fixed value of system load ρ, the probability P_{imm} becomes worse, while characteristics L and D become better when α increases, but when α becomes greater than 4 the change of these characteristics becomes unessential. The increase of retrial rate α has a negative influence on the loss probability P_{loss}. This probability increases when α increases.

The obtained results illustrate the importance of computational investigation of the tandem queue under consideration and can be used for correct prediction of system operation.

7   Conclusion

In this paper, the tandem queue with repeated attempts and semi-Markovian service process is investigated. The processes of the system states at embedded epochs and at arbitrary time are studied. The condition for the stationary distribution existence is derived and the algorithms for calculating the stationary distribution are presented. The key performance measures are calculated. The numerical examples demonstrating the dependence of the performance measures on the system load and intensity of retrials are given. The results of this paper can be exploited for capacity planning, performance evaluations, and optimization of real-life tandem queues and two-node networks with the random multiple access to the first station in the case of correlated bursty traffic as well as for validation of general networks decomposition algorithms.

Acknowledgments. This work has been financially supported by the Russian Science Foundation and the Department of Science and Technology (India) via grant No 16-49-02021 (INT/RUS/RSF/16) for the joint research project by the V.A. Trapeznikov Institute of Control Sciences and the CMS College Kottayam.

References

1. Gomez-Corral, A.: A bibliographical guide to the analysis of retrial queues through matrix analytic techniques. Ann. Oper. Res. **141**, 163–191 (2006)
2. Artalejo, J.R., Gomez-Corral, A.: Retrial Queueing Systems: A Computational Approach. Springer, Berlin (2008)
3. Falin, G., Templeton, J.: Retrial Queues. Chapman and Hall, London (1997)
4. Balsamo, S., Persone, V.D.N., Inverardi, P.: A review on queueing network models with finite capacity queues for software architectures performance prediction. Perform. Eval. **51**, 269–288 (2003)
5. Ferng, H.W., Chao, C.C., Peng, C.C.: Path-wise performance in a tree-type network: per-stream loss probability, delay, and delay variance analysis. Perform. Eval. **64**, 55–75 (2007)
6. Heindl, A.: Decomposition of general tandem networks with $MMPP$ input. Perform. Eval. **44**, 5–23 (2001)
7. Gnedenko, B.W., Konig, D.: Handbuch der Bedienungstheorie. Akademie Verlag, Berlin (1983)
8. Lucantoni, D.M.: New results on the single server queue with a batch Markovian arrival process. Commun. Stat.-Stoch. Models **7**, 1–46 (1991)
9. Breuer, L., Dudin, A.N., Klimenok, V.I., Tsarenkov, G.V.: A two-phase $BMAP/G/1/N \rightarrow PH/1/M - 1$ system with blocking. Autom. Rem. Control **65**, 117–130 (2004)

10. Gomez-Corral, A.: A tandem queue with blocking and Markovian arrival process. Queueing Syst. **41**, 343–370 (2002)
11. Gomez-Corral, A.: On a tandem G-network with blocking. Adv. Appl. Probab. **34**, 626–661 (2002)
12. Gomez-Corral, A., Martos, M.E.: Performance of two-station tandem queues with blocking: the impact of several flows of signals. Perform. Eval. **63**, 910–938 (2006)
13. Gomez-Corral, A., Martos, M.E.: A matrix-geometric approximations for tandem queues with blocking and repeated attempt. Oper. Res. Lett. **30**, 360–374 (2002)
14. Klimenok, V.I., Breuer, L., Tsarenkov, G.V., Dudin, A.N.: The $BMAP/G/1/N \rightarrow PH/1/M - 1$ tandem queue with losses. Perform. Eval. **61**, 17–40 (2005)
15. Klimenok, V., Kim, C.S., Tsarenkov, G.V., Breuer, L., Dudin, A.N.: The $BMAP/G/1 \rightarrow \cdot/PH/1/M$ tandem queue with feedback and losses. Perform. Eval. **64**, 802–818 (2007)
16. Kim, C.S., Klimenok, V., Taramin, O.: A tandem retrial queueing system with two Markovian flows and reservation of channels. Comput. Oper. Res. **37**, 1238–1246 (2010)
17. Klimenok, V.I., Taramin, O.S.: Tandem service system with batch Markov flow and repeated calls. Autom. Rem. Control **71**, 1–13 (2010)
18. Graham, A.: Kronecker Products and Matrix Calculus with Applications. Ellis Horwood, Chichester (1981)
19. Klimenok, V.I., Dudin, A.N.: Multi-dimensional asymptotically quasi-Toeplitz Markov chains and their application in queueing theory. Queueing Syst. **54**, 245–259 (2006)
20. Cinlar, E.: Introduction to Stochastic Process. Prentice-Hall, N.J. (1975)
21. Dudin, A.N., Klimenok, V.I., Tsarenkov, G.V.: Software "Sirius++" for performance evaluations of modern communication networks. In: Amborski, K., Meuth, H. (eds.) Proceedings of the 16th European Simulation Multiconference, Darmstadt, 3–5 June 2002, pp. 489–493. SCS, Netherlands (2002)

On Search Services for Internet of Things

Dmitry Namiot[1(✉)] and Manfred Sneps-Sneppe[2]

[1] Faculty of Computational Mathematics and Cybernetics,
Lomonosov Moscow State University,
GSP-1, 1–52, Leninskiye Gory, Moscow 119991, Russia
`dnamiot@gmail.com`
[2] Ventspils International Radio Astronomy Centre,
Ventspils University College,
Inzenieru 101a, Ventspils 3601, Latvia
`manfreds.sneps@gmail.com`

Abstract. The Internet of Things (IoT) data services are designed to be available to other services and users by demand. In other words, in practice, the services should be available at any time and at any location. But the big question here is how the other parties become aware of existing services. Sure, for some use cases we could have a static map of available services. But of course, it does not cover all the possible IoT deployment scenarios. So, the search (discovery) should be one of the keys for Internet of Things. And we can see here the whole analogue with the Internet itself. But for Internet of things, we will face a lot of challenges for search services. For example, we will see a lot of resources (billions of devices), heterogeneity, dynamic nature of services, streaming data, etc. In this paper, we discuss existing approaches for implementing IoT search, as well as approaches for setting standards in this area.

Keywords: Internet of Things · Search · Services

1 Introduction

This paper presents an extended version of our presentation for DCCN 2017 conference. IoT data services are designed to be available to end-users and other services (e.g., in Machine to Machine communications) by demand. So, they should be available, in practice, at any time and location. But besides that, the consumers (other users and other processes) should be aware of existing services. In other words, IoT system should not provide sensing data only, but also inform users (other services) about their availability. And this information could be provided not only by requests but proactively also.

Traditionally, Internet search is more oriented to static data (web sites). But for IoT streaming (dynamic flow) is one of the main features. Also, with IoT search, we can face other challenges. For example, the amount of IoT devices (data sources) is much bigger than the whole set of web sites. Another challenge is heterogeneity for IoT systems. Many authors note that existing solutions for

© Springer International Publishing AG 2017
V.M. Vishnevskiy et al. (Eds.): DCCN 2017, CCIS 700, pp. 174–185, 2017.
DOI: 10.1007/978-3-319-66836-9_15

searching, accessing, and using the information on the Internet and the Web will not be applicable or will remain far from adequate for practical and large-scale dynamic IoT applications [1]. As it is stated in [2], for example, Internet of Things has become a common buzzword nowadays on the Web. However, there is no search tool currently in place for discovering and learning about the different types of IoT elements. As per [3], web search engines for IoT are the new frontier.

In this paper, we would like to discuss the challenges and requirements for IoT search. The authors also are involved in the local standardization group that works on localization standards for Smart Cities, Internet of Thins and cyber-physical systems in Russia. So, we would like to discuss the proposals for standards in IoT search area.

Classically, web search engines rely on automated programs (search bots, web crawlers). They are responsible for discovering (recognizing) new sites and visiting them. For the each visited site such bot makes a copy of the content and provides this copy for search engine processing. A search engine can analyze accumulated content (it could be done offline) and create a so-called search index. This constantly updated index is a base for creating answers for search queries [4]. The response for the search process query will be presented as s list of URLs (website addresses). And what is also important here, the whole data transfer between the user (e.g., web browser), website and the search engine server will be performed according to HTTP standards.

It is obviously, that for IoT devices many of them will be "sleeping" for some time (or even almost always). For example, some sensor is in low power mode and will be activated (automatically switched on) only due to some special conditions. Below this temperature, the sensor will be inactive in a sleep mode. So, in general, this temperature sensor will not be discoverable by web crawlers sent out by a traditional search engine as it will be sleeping most of the time and will not respond.

As a second problem, we must highlight the fact that security requirements for IoT networks very often discard external requests at all. Especially, it is true for so-called Industrial Internet of Things [5].

No doubt, that a search is one for basic Internet services. The search service plays a significant role in Internet success [6]. It is why search services for Internet of Things are so important. The rest of the paper is organized as follows. In Sect. 2, we describe related works. In Sect. 3, we discuss the proposals for IoT search standards.

2   Related Works

In this section, we would like to discuss several approaches for IoT search. One of the early papers [7] proposes web pages with micro-formats [8] for sensors. They denote an HTML page that represents a sensor as a sensor page, and an HTML page that represents a real-world entity as an entity page. In order to be able to display sensor-specific information to the user and at the same time provide this data accordingly structured for the indexer, authors follow the concept of micro-formats. The micro-formats provided semantic data for indexing "bots". The whole architecture is illustrated in Fig. 1.

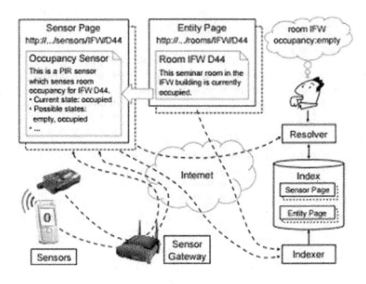

Fig. 1. Web of things search [7]

Authors present this as a "real-time" search engine. As a similar conception we could mention Microsearch [9], MAX [10], and Snoogle [11]. As per this model, all sensor nodes attached to real-world objects should provide a textual description in terms of keywords. But of course, such approach is not applicable for the dynamic content [12].

Another approach is presented in SPITFIRE [13]. It belongs to so-called Semantic Sensor Web [14]. The Semantic Sensor Web proposes annotating sensor data with semantic meta-data. The provided meta-data should be machine-understandable through vocabulary definitions. This vocabulary definition could be some ontology, for example. And of course, any formally-defined annotations lets automate some task (e.g., integration, maintenance).

In this connection, we should mention Sensor Markup Language (SenML). It is an emerging solution for representing device parameters and measurements. SenML is replacing proprietary data formats and is being accepted by more and more vendors [15]. In Distributed Image Search [16], authors are trying to find cameras that captured similar images. The similarity is defined via most relevant matching sensors and images.

In paper [17], authors propose an infrastructure for shared sensing. It is illustrated in Fig. 2.

This system (SenseWeb) let discover sensors based on their static meta-data. And the geographical location is also part of meta-data. The architecture's key

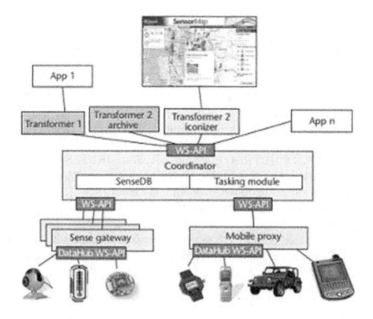

Fig. 2. The infrastructure for shared sensing [17]

components are the coordinator, sensors, sensor gateways, mobile proxy, data transformers, and applications. On this schema, a transformer is responsible for data semantics conversion. For example, a transformer might extract the people count from a video stream, converts units, fuse data, etc. The transformers are expendable, so end-users might use various transformers for different sensor data using suitable domain–specific algorithms.

An interesting approach is presented in paper [18]. XGSN is an Open source semantic sensing middleware for Web of Things. Using semantically rich models we can solve heterogeneity issues for sensing (IoT) systems. XGSN plays the role of a fully distributed data acquisition middleware with semantic annotation capabilities. But as per XGSN manual, virtual sensors in XGSN are set up using a configuration descriptor (an XML document). In other words, it is a static definition.

Sensors in IoT systems can produce various data streams. So, search for data streams is an important part of IoT platforms. In this connection, we could mention Stream Analytics Query Language (SAQL). SAQL queries could be run over the sensor data to find interesting patterns from the incoming stream of data [19].

In the same time, the real-time utilization of IoT data streams suggests a paradigm shift to new horizontal and distributed architecture, because existing cloud-based centralized architecture will cause large delays for providing service and waste many resources on the cloud and on networks [20]. This shift leads to such conceptions as Edge Computing [21], Fog Computing [22], and Cloudlets

[23]. For such conceptions, the data processing is executed on those components in the edge of networks to mitigate server load and to decrease the response time. Such solutions are presented for many IoT platforms [24].

3   On IoT Search Standards

In general, we could mention the following initiatives in IoT interoperability: Hypercat (PAS-212) from Hypercat and BSI, WoT from W3C, Semantic web standards from W3C, CoRE Resource Directory from IETF, Service Layer Core Protocol Specification from oneM2M, Web-Enabled Data Distribution Service from OMG, and several not finished documents from ISO/IEC (e.g., ISO/IEC AWI 21823-1, ISO/IEC CD 20924, ISO/IEC CD 30141).

As the standardization forces behind IoT search, we can mention Internet Engineering Task Force (IETF) [25] and Hypercat consortium [26]. A key solution for the IoT search problem from IETF is a new type of search engine called a Resource Directory (RD) [27].

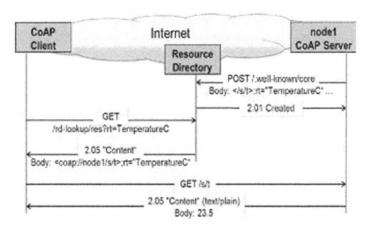

Fig. 3. Resource directory lookup and registration [28]

A Resource Directory is a repository for Web Links. Web Links describe resources hosted on other web servers, which are called endpoints (EP). Web Links are also called resource directory entries. Any endpoint in this approach is a web server associated with a scheme, IP address, and port. So, a physical node (a physical computer or computing device) may host one or more endpoints. Any RD implements a set of HTTP interfaces (REST model) for endpoints. REST interfaces let register and maintain Web Links as well as lookup resources. As per specification, an RD can be logically segmented by the use of domains.

As per Core Resource Directory draft specification, in many M2M applications, direct discovery of resources is not practical due to sleeping nodes, disperse networks, or networks where multicast traffic (polling resources) is inefficient. And RD hosts descriptions of IoT (M2M) resources held on other servers, allowing lookups to be performed for those resources. It is illustrated in Fig. 3.

In general, it is a distributed search engine, with multiple RDs assigned for a given geographical area. The sensors (IoT devices) should register their web addresses (URIs) to their local RD. It is very important, that sensor will do that on its own initiative (a push model). For example, it could be performed during the installation of a new sensor (during its first working session). Of course, before a sensor (an endpoint in IETF manual) can make use of an RD, it must first know the RD's address and port, and the base URI information for endpoint's REST API. Discovery of the RD base URI is performed by sending either a multicast or unicast GET request to some well-known core RD.

After discovering the location of an RD, an endpoint may register its resources (sensing meta-data) using the registration interface. This interface accepts a HTTP POST requests from an endpoint containing the list of resources to be added to the directory. And of course, any registration could be updated or even completely removed.

Then when a search request is sent to the RD, the RD will first perform access control and other security checks. As per this specification, only authorized parties are allowed to discover the relevant information. RD Lookup allows lookups for domains, groups, endpoints, and resources. And lookup request is simple HTTP GET command. The following example shows a client performing a resource lookup with the example look-up location /rd-lookup/ [27]:

GET /rd-lookup/res?rt=temperature

As per its declaration, Hypercat is a Global Alliance and standard (PAS 212 [30]) driving secure and interoperable Internet of Things for Industry and cities. The Hypercat specification allows IoT clients to discover information about IoT assets over the web [26]. This specification (it is being wrapped as PAS-212 standard) will allow inter-exchange of data between data hubs in different domains.

The idea of Hypercat is based on the IoT model from British Telecom (Fig. 4).

The middle level here is so-called Data Hub. This hub obtains data from sensors and presents them for the application. Hypercat specification describes data in this hub. The whole idea behind Hypercat is to use for data discovery and interoperability the same model as used by the Internet. So, data access should be performed via standard web protocols (HTTPS), with the well-known REST model and data should be presented in the JSON format. All the resources should be presented as URI. Shortly, Hypercat is JSON-based hypermedia catalogue format for exposing collections of URIs.

Note, that most (if not any) of IoT search approaches are based on REST model. And for REST (dislike web services) common model for meta-data still does not exist [31]. The core element in this model is a catalogue. Each Hyper-Cat catalogue may expose any number of URIs. A Hypercat catalogue is a file representing an unordered collection of resources on the web. Each item in a

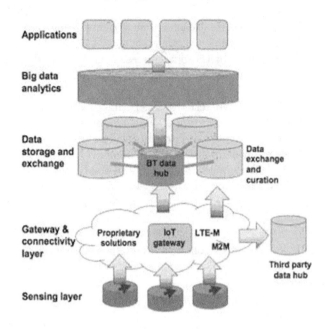

Fig. 4. British Telecom IoT model [29]

catalogue refers to a single resource by its URI, which may itself be a further catalogue.

Any catalogue contains items. For example, a catalogue is a group of sensors, and an item is an individual sensor. And any item is a JSON object too. Any catalogue could have own meta-data, and any item could have meta-data too.

The simplest Hypercat server is a traditional HTTP server. So, all operations (create a catalogue, add an item, create meta-data, etc.) are HTTP commands. Of course, catalogues could be updated dynamically.

And search functionality is an ordinary HTTP GET command, where search parameters are HTTP query parameters. On the top level, the model is very simple. We have so-called well-known URI for the catalogue on the given server (/cat). It lets us request a list of URIs via simple HTTP GET. And each URI in obtained list is either some resource or a new catalogue (a new endpoint).

Let us describe it with more details. All URIs in a catalogue must be unique within this catalogue. It lets us address them individually. A catalogue may provide metadata for itself and metadata for each own item. Any element in the catalogue (and catalogue itself) is a JSON object. Some of the properties for any catalogue object are mandatory. For example: *items* is a property with a list (array) of items and *catalogue-metadata* if an array of metadata. And all arrays (elements) are JSON arrays (JSON elements). So, for end-user application we have a well-known URI for a catalogue and well-known elements for its data. It is illustrated in Fig. 5.

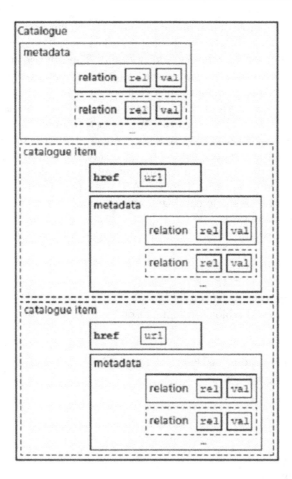

Fig. 5. Hypercat catalogue

Any item is (in any array) is a JSON object and it also could have some mandatory properties. A property *href* describes an identifier for the resource item (it is an URI as a JSON string). And *item-metadata* is a JSON array of metadata objects.

A metadata object is again a JSON object. All metadata objects must own two mandatory properties: *rel* and *val*. The first one describes a predicate (verb) in the for if URI, and the second one describes an entity (noun). The following example for Hypercat specification illustrates the property (it suppose to be a color) and its value

{ "rel" : "urn:X-hypercat:rels:isColour", "val" : "blue" }

Of course, it is up to an application to decide that some property named *isColor* should be interpreted as a color.

Further, this principle of mandatory elements (with known naming) and additional elements is consistently maintained for all concepts. So, for example, relationships could be also mandatory and optional. The are two mandatory relationships: human readable description in English and content-type. An optional relationship, for example, includes such element as *urn:X-hypercat:rels: supportsSearch*. Its value describes an URI of a defined search mechanism as a JSON string. In other words, a catalogue describes its own end-point for search queries. A simple search is performed by providing a query string (as per RFC 1738) to a catalogue. If multiple search parameters are supplied, the server must return the intersection of items where all search parameters match in a single item, combining the parameters with boolean AND. As per specification, a "simple search" searches only a single catalogue resource and it does not include other linked or nested catalogues.

The following example describes some catalogue with weather data:

```
catalogue-metadata:{
0: {
"rel":"urn:X-hypercat:rels:isContentType",
"val":"application/vnd.tsbiot.catalogue+json"
},
1: {
"rel":"urn:X-hypercat:rels:hasDescription:en",
"val":"Weather description"
}
}
items:
0: {
    href: /weather
    item-metadata: {
        0: { "val": "weather", "rel": urn:X-hypercat:rels:hasDescription:en}
        1: { "val": "+15", "rel":"Temperature Celsius"} } }
```

In this example, URI /weather should return the temperature. As it could be seen, it is almost an RDF-like approach, because each *relationship* and *value* can be a URI, and values could also be primitives. So, on practice a Hypercat document could be converted to an appropriate RDF document. It allows an integration of a Hypercat approach with the W3C suite of standards and ongoing development.

Hypercat specification is a basic for British Standards Institute PAS-212 standard. This standard (Automatic resource discovery for the Internet of Things - Specification) will make it much easier to discover Internet of Things data [32]. It has been developed in conjunction with the Hypercat Alliance. This specifications covers:

– A mandatory file format for representing a catalogue of linked-data resources, annotated with metadata.

– Recommendations for catalogue access in file transport; security mechanisms to protect access and to prove provenance; search functions; subscription mechanisms; well-known entry-points and machine-readable hints to ease usability

To support the implementation of HyperCat, a HyperSpace portal [33] presents some HyperCat-compatible solutions for vertical industries (Smart House, Logistics, etc.). The Fig. 6 illustrates a real catalogue from British Telecom. Another interesting example is Smart City project CityVerve (Manchester) [34]. CityVerve Manchester is creating a blueprint for smarter cities worldwide. And it uses Hypercat-catalogue as a core solution.

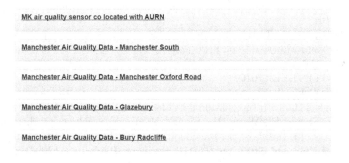

Fig. 6. IoT catalogue

As per IEEE, HyperCat represents a pragmatic starting point to solving the issues of managing multiple data sources, aggregated into multiple data hubs, through linked data and web approaches. Hypercat incorporates a lightweight, JSON-based approach based on a technology stack already used by a large population of web developers. So, it supports a low barrier to entry into IoT search, thereby encouraging the growth of IoT ecosystems [35].

W3C has agreed to collaborate with Hypercat to pursue a "shared vision" for IoT [36]. Hypercat (PAS-212 from BSI) corresponds to the W3C's promoted idea of web of things. As per their memorandum [35], W3C, Hypercat, and BSI are seeking to collaborate on promoting a shared vision for countering fragmentation of the IoT, and to encourage more companies to take an active role in driving work on standards as the Web evolves from a Web of pages into a Web of things. Actually, Hypercat and WoT Metadata from W3C are very similar. So, we think that future iterations of the W3C approach are likely to be harmonized with the Hypercat (PAS-212) approach.

In our opinion, adopting Hypercat in Russia could be a low-risk solution and will bring immediate results in IoT development and adoption. Hypercat presents a very good opportunity to accelerate IoT adoption in Russia. We should mention also, that Hypercat-based solution could be very easily integrated with other software.

References

1. Barnaghi, P., Sheth, A.: On searching the internet of things: requirements and challenges. IEEE Intell. Syst. **31**(6), 71–75 (2016)
2. Mosab, F., et al.: Topical search engine for Internet of Things. In: 2014 IEEE/ACS 11th International Conference on Computer Systems and Applications (AICCSA). IEEE (2014)
3. Carlton, A.: Web search engines for IoT: The new frontier. http://www.networ kworld.com/article/3111984/internet-of-things/web-search-engines-for-iot-the-new-frontier.html. Accessed Apr 2017
4. Buttcher, S., Clarke, C.L.A., Cormack, G.V.: Information Retrieval: Implementing and Evaluating Search Engines. Mit Press, Cambridge (2016)
5. Wurm, J., et al.: Security analysis on consumer and industrial IoT devices. In: 2016 21st Asia and South Pacific Design Automation Conference (ASP-DAC). IEEE (2016)
6. Ward, M.R., Lee, M.J.: Internet shopping, consumer search and product branding. J. Prod. Brand Manage. **9**(1), 6–20 (2000)
7. Ostermaier, B., et al.: A real-time search engine for the web of things. Internet of Things (IOT). IEEE (2010)
8. Khare, R., Celik, T.: Microformats: a pragmatic path to the semantic web. In: Proceedings of the 15th International Conference on World Wide Web. ACM (2006)
9. Tan, C.C., Sheng, B., Wang, H., Li, Q.: Microsearch: when search engines meet small devices. In: Indulska, J., Patterson, D.J., Rodden, T., Ott, M. (eds.) Pervasive 2008. LNCS, vol. 5013, pp. 93–110. Springer, Heidelberg (2008). doi:10.1007/978-3-540-79576-6_6
10. Yap, K.-K., Srinivasan, V., Motani, M.: MAX: human-centric search of the physical world. In: Proceedings of the 3rd International Conference on Embedded Networked Sensor Systems. ACM (2005)
11. Wang, H., Tan, C.C., Li, Q.: Snoogle: a search engine for pervasive environments. IEEE Trans. Parallel Distrib. Syst. **21**(8), 1188–1202 (2010)
12. Miorandi, D., et al.: Internet of things: vision, applications and research challenges. Ad Hoc Netw. **10**(7), 1497–1516 (2012)
13. Pfisterer, D., et al.: Spitfire: toward a semantic web of things. IEEE Commun. Mag. **49**(11), 40–48 (2011)
14. Sheth, A., Henson, C., Sahoo, S.S.: Semantic sensor web. IEEE Internet Comput. **12**(4), 78–83 (2008)
15. Su, X., et al.: Connecting IoT sensors to knowledge-based systems by transforming SenML to RDF. Procedia Comput. Sci. **32**, 215–222 (2014)
16. Yan, T., Ganesan, D., Manmatha, R.: Distributed image search in camera sensor networks. In: Proceedings of the 6th ACM Conference on Embedded Network Sensor Systems. ACM (2008)
17. Grosky, W.I., et al.: Senseweb: an infrastructure for shared sensing. IEEE Multimedia **14**(4), 8–13 (2007)
18. Calbimonte, J.-P., et al.: XGSN: an Open-source Semantic Sensing Middleware for the Web of Things. TC/SSN@ ISWC (2014)
19. Get started with Azure Stream Analytics to process data from IoT devices https://docs.microsoft.com/en-us/azure/stream-analytics/stream-analytics-get-started-with-azure-stream-analytics-to-process-data-from-iot-devices. Accessed Apr 2017

20. Yasumoto, K., Yamaguchi, H., Shigeno, H.: Survey of real-time processing technologies of iot data streams. J. Inform. Process. **24**(2), 195–202 (2016)
21. Hu, Y.C., et al.: Mobile edge computing-A key technology towards 5G. ETSI White Paper 11 (2015)
22. Bonomi, F., Milito, R., Natarajan, P., Zhu, J.: Fog computing: a platform for internet of things and analytics. In: Bessis, N., Dobre, C. (eds.) Big Data and Internet of Things: A Roadmap for Smart Environments. SCI, vol. 546, pp. 169–186. Springer, Cham (2014). doi:10.1007/978-3-319-05029-4_7
23. Mineraud, J., et al.: A gap analysis of Internet-of-Things platforms. Comput. Commun. **89**, 5–16 (2016)
24. Namiot, D., Sneps-Sneppe, M.: On open source mobile sensing. In: Balandin, S., Andreev, S., Koucheryavy, Y. (eds.) NEW2AN 2014. LNCS, vol. 8638, pp. 82–94. Springer, Cham (2014). doi:10.1007/978-3-319-10353-2_8
25. Internet Engineering Task Force. https://www.ietf.org/. Accessed Apr 2017
26. Hypercat consortium. http://www.hypercat.io/. Accessed Apr 2017
27. Resource Directory. https://tools.ietf.org/html/draft-ietf-core-resource-directory-10. Accessed Apr 2017
28. Ishaq, I., et al.: Flexible unicast-based group communication for CoAP-enabled devices. Sensors **14**(6), 9833–9877 (2014)
29. Fisher, M., Davies, J.: Creating internet of things ecosystems. J. Inst. Telecommun. Professionals **9**, 10–14 (2015)
30. PAS 212:2016 Automatic resource discovery for the Internet of Things. Specification http://shop.bsigroup.com/forms/PASs/PAS-212-2016-download/. Accessed Apr 2017
31. Sneps-Sneppe, M., Namiot, D.: Metadata in SDN API for WSN. In: 2015 7th International Conference on New Technologies, Mobility and Security (NTMS). IEEE (2015)
32. Internet of Things interoperability specification is published https://www.bsigroup.com/en-GB/about-bsi/media-centre/press-releases/2016/july/Internet-of-Things-interoperability-specification-is-published/. Accessed Jun 2017
33. Hyperspace https://hyperspace.center/. Accessed Jun 2017
34. CityVerve http://www.cityverve.org.uk/. Accessed Jun 2017
35. Davies, J.: Hypercat: resource Discovery on the Internet of Things http://iot.ieee.org/newsletter/january-2016/hypercat-resource-discovery-on-the-internet-of-things.html. Accessed Jun 2017
36. Web of things at the Hypercat Showcase https://www.w3.org/blog/wotig/2016/10/05/web-of-things-at-the-hypercat-showcase/. Accessed Jun 2017

Some Features of a Finite-Source M/GI/1 Retrial Queuing System with Collisions of Customers

Anatoly Nazarov[1,2], János Sztrik[3], and Anna Kvach[1(✉)]

[1] National Research Tomsk State University,
36 Lenina ave., Tomsk 634050, Russia
nazarov.tsu@gmail.com, kvach_as@mail.ru
[2] Department of Applied Probability and Informatics,
Peoples' Friendship University of Russia,
Miklukho-Maklaya str. 6, Moscow 117198, Russia
[3] Faculty of Informatics, University of Debrecen,
Egyetem tér 1, Debrecen 4032, Hungary
sztrik.janos@inf.unideb.hu

Abstract. In this paper a finite-source M/GI/1 retrial queuing system with collisions of customers is considered. The definition of throughput of the system as average number of customers, which are successfully served per unit time is introduced. It is shown that at some combinations of system parameter values and probability distribution of service time of customers the throughput can be arbitrarily small, and at another values of parameters throughput can be greater than the service intensity. Applying method of asymptotic analysis under the condition of unlimited growing number of sources it is proofed that limiting distribution of the number of retrials/transitions of the customer into the orbit is geometric and the sojourn/waiting time of the customer in the orbit follows a generalized exponential distribution. In addition, the mean sojourn time of the customer under service is obtained.

Keywords: Closed queuing system · Finite-source queuing system · Retrial queue · Collisions · Asymptotic analysis · Throughput · Sojourn time · Limiting distribution · Number of transitions into an orbit

1 Introduction

Retrial queues have been widely used to model many problems arising in telephone switching systems, telecommunication networks, computer networks and systems, call centers, etc. In many practical situations it is important to take into account the fact that the rate of generation of new primary calls decreases as the number of customers in the system increases. This can be done with the help of finite-source, or quasi-random input models, see, for example [1,2,7].

Another very important component of queuing models is the collisions of the customers. In the main model it is assumed that if an arriving customer finds

© Springer International Publishing AG 2017
V.M. Vishnevskiy et al. (Eds.): DCCN 2017, CCIS 700, pp. 186–200, 2017.
DOI: 10.1007/978-3-319-66836-9_16

the server busy, it involves into collision with customer under service and they both moves into the orbit. See, for example [3,4,6].

The aim of the present paper is to investigate such systems which has the above mention properties, that is finite-source, retrial and collisions of customer in the case of non-Markov service. This article paper is a continuation of the paper [5] in which the asymptotic distribution of the number of customers in the system was investigated.

2   Model Description and Notations

Let us consider a closed retrial queuing system of type M/GI/1//N with collision of the customers. The number of sources is N and each of them can generate a primary request during an exponentially distributed time with rate λ/N. A source cannot generate a new call until end of the successful service of this customer. If a primary customer finds the server idle, he enters into service immediately, in which the required service time has a probability distribution function $B(x)$. Let us denote its hazard rate function by $\mu(y) = B^{'}(y)(1 - B(y))^{-1}$ and Laplace -Stieltjes transform by $B^*(y)$, respectively. If the server is busy, an arriving (primary or repeated) customer involves into collision with customer under service and they both moves into the orbit. The retrial time of requests are exponentially distributed with rate σ/N. All random variables involved in the model construction are assumed to be independent of each other.

The system state at time t is denoted by $\{l(t), i(t), y(t)\}$, where $i(t)$ is the number of customers in the system at time t, that is, the total number of customers in orbit and in service, $l(t)$ describes the server state as follows

$$l(t) = \begin{cases} 0, & \text{if the server is free,} \\ 1, & \text{if the server is busy,} \end{cases}$$

$y(t)$ is the supplementary random variable, equal to the elapsed service time of the customer till the moment t.

Thus, we will investigate the Markov process $\{l(t), i(t), y(t)\}$, which has a variable number of components, depending on the server state, since the component $y(t)$ is determined only in those moments when $l(t) = 1$.

Let us define the probabilities as follows:

$$p_0(i,t) = P\{l(t) = 0, i(t) = i\},$$
$$p_1(i,y,t) = \frac{\partial P\{l(t) = 1, i(t) = i, y(t) < y\}}{\partial y}.$$

Assuming that system is operating in steady state, for the stationary probability distribution $p_0(i)$, $p_1(i,y)$ by using standard methods the following system of Kolmogorov equations can be derived

$$-\left[\frac{N-i}{N}\lambda + \frac{i}{N}\sigma\right]p_0(i) + \int_0^\infty p_1(i+1,y)\mu(y)dy$$

$$+\left(1 - \frac{i-1}{N}\right)\lambda p_1(i-1) + \frac{i-1}{N}\sigma p_1(i) = 0, \qquad (1)$$

$$\frac{\partial p_1(i,y)}{\partial y} = -\left[\frac{N-i}{N}\lambda + \frac{i-1}{N}\sigma + \mu(y)\right]p_1(i,y),$$

with boundary conditions

$$p_1(i,0) = \left(1 - \frac{i-1}{N}\right)\lambda p_0(i-1) + \frac{i}{N}\sigma p_0(i). \qquad (2)$$

Denote the partial characteristic functions

$$H_0(u) = \sum_{i=0}^N e^{jui}p_0(i)\,, \qquad H_1(u,y) = \sum_{i=1}^N e^{jui}p_1(i,y)\,, \qquad (3)$$

where $j = \sqrt{-1}$ is imaginary unit, then system (1) and Eq. (2) can be rewritten as

$$-\lambda H_0(u) + \left[\lambda e^{ju} - \frac{\sigma}{N}\right]H_1(u) + e^{-ju}\int_0^\infty H_1(u,y)\mu(y)dy$$

$$+j\frac{(\sigma-\lambda)}{N}\frac{dH_0(u)}{du} + j\frac{(\lambda e^{ju} - \sigma)}{N}\frac{dH_1(u)}{du} = 0\,, \qquad (4)$$

$$\frac{\partial H_1(u,y)}{\partial y} = \left[\frac{\sigma}{N} - \lambda - \mu(y)\right]H_1(u,y) - j\frac{(\lambda-\sigma)}{N}\frac{\partial H_1(u,y)}{\partial u},$$

$$H_1(u,0) = \lambda e^{ju}H_0(u) + j\frac{(\lambda e^{ju} - \sigma)}{N}\frac{dH_0(u)}{du}.$$

3   Asymptotic Analysis

By using asymptotic methods [8] as a consequence of the first order solution to (4) can be obtained as follows

Theorem 1. *Let $i(t)$ be number of customers in a closed retrial queuing system $M/GI/1//N$ with the collisions of customers, then*

$$\lim_{N\to\infty}\mathrm{E}\exp\left\{jw\frac{i(t)}{N}\right\} = \exp\left\{jw\kappa\right\}, \qquad (5)$$

where value of parameter κ is the positive solution of the equation

$$(1-\kappa)\lambda - \delta(\kappa)\frac{B^*(\delta(\kappa))}{2 - B^*(\delta(\kappa))} = 0, \qquad (6)$$

here $\delta(\kappa)$ is

$$\delta(\kappa) = (1 - \kappa)\lambda + \sigma\kappa, \tag{7}$$

and the stationary distributions of probabilities $R_l(\kappa)$ of the service state l are defined as follows

$$R_0(\kappa) = \frac{1}{2 - B^*(\delta(\kappa))}, \quad R_1(\kappa) = \frac{1 - B^*(\delta(\kappa))}{2 - B^*(\delta(\kappa))}. \tag{8}$$

Proof. Denoting $\frac{1}{N} = \varepsilon$, in system (4) let us execute the following substitutions

$$u = \varepsilon w, \quad H_0(u) = F_0(w, \varepsilon), \quad H_1(u, y) = F_1(w, y, \varepsilon), \tag{9}$$

then we will receive system of the equations

$$-\lambda F_0(w, \varepsilon) + \left[\lambda e^{j\varepsilon w} - \varepsilon\sigma\right] F_1(w, \varepsilon) + e^{-j\varepsilon w} \int_0^\infty F_1(w, y, \varepsilon)\mu(y)dy$$

$$+ j(\sigma - \lambda)\frac{\partial F_0(w, \varepsilon)}{\partial w} + j(\lambda e^{j\varepsilon w} - \sigma)\frac{\partial F_1(w, \varepsilon)}{\partial w} = 0, \tag{10}$$

$$\frac{\partial F_1(w, y, \varepsilon)}{\partial y} = [\varepsilon\sigma - \lambda - \mu(y)] F_1(w, y, \varepsilon) - j(\lambda - \sigma)\frac{\partial F_1(w, y, \varepsilon)}{\partial w},$$

$$F_1(w, 0, \varepsilon) = \lambda e^{j\varepsilon w} F_0(w, \varepsilon) + j(\lambda e^{j\varepsilon w} - \sigma)\frac{\partial F_0(w, \varepsilon)}{\partial w}.$$

Taking the limiting transition under conditions $\varepsilon \to 0$ and denoting $\lim_{\varepsilon \to 0} F_0(w, \varepsilon) = F_0(w), \lim_{\varepsilon \to 0} F_1(w, y, \varepsilon) = F_1(w, y)$, then system (10) can be rewritten as

$$\lambda[F_1(w) - F_0(w)] + \int_0^\infty F_1(w, y)\mu(y)dy$$

$$+ j(\lambda - \sigma)\left[\frac{\partial F_1(w)}{\partial w} - \frac{\partial F_0(w)}{\partial w}\right] = 0, \tag{11}$$

$$\frac{\partial F_1(w, y)}{\partial y} = -[\lambda + \mu(y)] F_1(w, y) - j(\lambda - \sigma)\frac{\partial F_1(w, y)}{\partial w},$$

$$F_1(w, 0) = \lambda F_0(w) + j(\lambda - \sigma)\frac{\partial F_0(w)}{\partial w}.$$

The solution of the system (11) can be written in product-form

$$F_0(w) = R_0\Phi(w), \quad F_1(w, y) = R_1(y)\Phi(w), \tag{12}$$

where R_0, $R_1(y)$ are the limiting probability distributions of the server state l under conditions $N \to \infty$ and $\Phi(w)$ is limiting characteristic function of the stationary distribution of random process $\frac{i(t)}{N}$. Substituting this solution into (11) we obtain

$$
\int\limits_0^\infty R_1(y)\mu(y)dy - (R_0 - R_1)\left[\lambda + j(\lambda - \sigma)\frac{\partial\Phi(w)/\partial w}{\Phi(w)}\right] = 0 \ ,
$$

$$
R_1'(y) = -\left[\lambda + \mu(y)\right]R_1(y) - j(\lambda - \sigma)R_1(y)\frac{\partial\Phi(w)/\partial w}{\Phi(w)}, \tag{13}
$$

$$
R_1(0) = \lambda R_0 + j(\lambda - \sigma)R_0\frac{\partial\Phi(w)/\partial w}{\Phi(w)}.
$$

The constant relation of the derivative function to this function allows to write down this function in the following form

$$
\Phi(w) = \exp\left(jw\kappa\right), \tag{14}
$$

coinciding with equality (5). Using the notation (7) the system (13) can be rewritten as

$$
\int\limits_0^\infty R_1(y)\mu(y)dy = \delta(\kappa)\left(R_0 - R_1\right),
$$

$$
R_1'(y) = -\left[\delta(\kappa) + \mu(y)\right]R_1(y), \tag{15}
$$

$$
R_1(0) = \delta(\kappa)R_0.
$$

From the second and third equations, taking into account the normalization condition, it is not difficult to obtain expressions for R_l of the form (8).

Let us return to system (10). Integrating the second equation of the system on y from 0 to ∞, subtracting it from the first equation, substituting the decomposition (12) and taking into account the explicit form (14) of the function $\Phi(w)$, we obtain an equation of the form

$$
\int\limits_0^\infty R_1(y)\mu(y)dy = \lambda(1 - \kappa). \tag{16}
$$

From (16) and the first equations of the system (15) it is obviously follows Eq. (6) for κ.

The theorem is proved.

It is interesting that Eq. (6) can have one, two or three roots $0 < \kappa < 1$. For example, for the gamma distribution function $B(x)$ with a shape parameter α and scale β with parameter values $\alpha = \beta = 2$, $\lambda = 0.29$, $\sigma = 20$, Eq. (6) has three roots, namely $\kappa_1 = 0.031$, $\kappa_2 = 0.188$ and $\kappa_3 = 0.549$.

As a rule the Eq. (6) has three roots in exceptional cases at special values of parameters and such situation arises extremely seldom. Therefore, in the following let us consider some properties of the system when the main Eq. (6) has a single root $0 < \kappa < 1$.

4   Non-ordinary Values of Throughput

First of all, let us define a measure S, called throughput, which is important for any closed retrial queuing systems.

Definition 1. *Let S be defined as the average number of customers, which are successfully served per unit time.*

Since all primary customers sooner or later will be successfully served, the throughput S naturally coincides with the intensity of the incoming (generated by the primary sources) flow. Thus, the throughput S of the considered system in the limiting condition $N \to \infty$ is equal to

$$S = \lambda(1 - \kappa). \tag{17}$$

For retrial queuing systems with collisions of customers value S can take non-ordinary values.

Let us consider the closed retrial queuing system where the service time is gamma distributed with shape parameter α and scale parameter β, with Laplace-Stieltjes transform $B^*(x)$ of the form

$$B^*(x) = \left(1 + \frac{x}{\beta}\right)^{-\alpha}. \tag{18}$$

Notice, that the average service time is α/β and in further examples we will consider a case when $\alpha = \beta$. Therefore, the average service time will be equal to unit and intensity of service as an inverse value to average time, will be also equal to unit.

Let us assume $\sigma = 20$ and the values of parameters λ and $\alpha = \beta$ indicated in the Table 1, in which the found values of throughput S of system with the collision of customers are given.

Table 1. System throughput S at various values of λ and $\alpha = \beta$

		$\alpha = \beta$							
		0.1	0.3	0.5	0.8	1	2	3	5
λ	0.5	0.494	0.486	0.478	0.458	0.432	0.099	0.024	0.0032
	1	0.968	0.909	0.827	0.633	0.478	0.089	0.023	0.0032
	5	3.911	2.441	1.503	0.748	0.487	0.084	0.022	0.0032
	10	6.101	2.961	1.628	0.760	0.488	0.083	0.022	0.0032
	15	7.441	3.173	1.672	0.764	0.488	0.083	0.022	0.0032

Table 1 demonstrates a surprising phenomenon of retrial queues with collision of customers having a throughput, which significantly greater than service intensity, and also at $\alpha > 1$ significantly less than unity.

Let us note, that throughput becomes greater than service intensity only in a case $\alpha < 1$ with increase of intensity $\lambda > 1$. For a case $\alpha > 1$, firstly, values of throughput depends weakly on value of parameter λ and, secondly, with increasing value of α it becomes close to zero.

5   Mean Sojourn Time of the Customer Under Service

For systems with collisions the sojourn time of the customer at the server has a rather complex structure, since it contains terms of zero duration when the customer from the orbit finds the server busy by another customer and a non-zero terms of the services interrupted by collisions and finally a single term of successfully completed service, after which the customer leaves the system. Let us denote by V the total mean sojourn time of the customer in the server.

Also, we will denote by $V(t)$ the residual mean sojourn time of the customer under service. For a further research we will enter the following supplementary random variable $z(t)$, equal to the residual service time, that is time interval from moment t until the end of successful service. The investigation will be carried out under limiting condition of unlimited growing number of sources, i.e. $N \to \infty$. We have the following statement

Theorem 2. *Mean sojourn time of the customer under service in a closed retrial queuing system $M/GI/1//N$ with the collisions of customers can be defined as follows*

$$V = \frac{1 - B^*(\delta)}{\delta B^*(\delta)}, \tag{19}$$

where δ is

$$\delta = (1 - \kappa)\lambda + \sigma\kappa.$$

Proof. Let us introduce the following function of conditional mean residual sojourn time of the customer under service

$$g(z) = E\{V(t)|z(t) = z\}.$$

Using the law of total probability we obtain the following equation

$$g(z) = (1 - \delta\Delta t)(\Delta t + g(z - \Delta t)) + \delta\Delta t V + o(\Delta t). \tag{20}$$

Executing the limiting transition under conditions $\Delta t \to 0$, Eq. (20) can be rewritten as

$$g'(z) = -\delta g(z) + 1 + \delta V.$$

We obtained the Cauchy problem with the initial condition $g(0) = 0$. Let us write the solution in the form

$$g(z) = e^{-\delta z} \int_0^z e^{\delta x}(1 + \delta V)\,dx = (1 + \delta V)\frac{1}{\delta}\left(1 - e^{-\delta z}\right). \tag{21}$$

Using the law of total probability we have

$$V = (1 - R_0) V + R_0 \int_0^\infty g(z) dB(z).$$

Taking into account the explicit form (21) of the function $g(z)$ and performing simple transformations, we obtain

$$V = \frac{1 - B^*(\delta)}{\delta B^*(\delta)},$$

coinciding with (19).

The theorem is proved.   □

We have received a formula (19) for calculating the total mean sojourn time V of the customer under service. Let us consider the influence of the system parameters on the total mean sojourn time V of the customer in the server. We will choose the same parameters of system which have been considered in the previous example when we calculating throughput S of system, namely $\sigma = 20$ and the values of parameters λ and $\alpha = \beta$ are specified in the Table 2

Table 2. Total mean sojourn time V of the customer under service at various values of λ and $\alpha = \beta$

		$\alpha = \beta$							
		0.1	0.3	0.5	0.8	1	2	3	5
λ	0.5	0.324	0.546	0.682	0.857	1	5.038	20.816	153.239
	1	0.204	0.367	0.488	0.727	1	5.576	21.695	154.766
	5	0.067	0.165	0.301	0.640	1	5.937	22.360	155.975
	10	0.046	0.140	0.280	0.632	1	5.979	22.441	156.125
	15	0.039	0.131	0.273	0.629	1	5.993	22.468	156.175

As we can see, at $\alpha < 1$ the values of total mean sojourn time V of the customer under service takes values less than unity. For $\alpha < 1$ there is a high probability of emergence of small values of service time and this fact undoubtedly influences to the total mean sojourn time V of the customer at the server and it takes rather small values. Moreover, let us remark, that with increasing values of parameter λ the values of V decrease.

In the case of $\alpha > 1$ the values of V become greater than unity. Table 2 illustrates that with increase of service parameter α the values of total mean sojourn time V of the customer under service considerably increases and reaches very large values. Let us note that in this situation parameter λ practically doesn't influence on V and with increasing parameter of service α this influence becomes less and less.

Table 2 also shows that in case of exponential service time, i.e. $\alpha = \beta = 1$, the values of total mean sojourn time V of the customer at the server is equal to unit.

Now we will return to Table 1 and we will analyze extraordinary values of throughput S of system with the help of the obtained values of total mean sojourn time V of the customer under service.

First of all, let us note, that in the case of exponential service time at any values of λ, the throughput S takes values nearly 0.5. Further, in Tables 1 and 2 it is demonstrated how much the change of total mean sojourn time V of the customer under service affects the throughput S. We can see, that with increasing the values of V, the throughput S of the system decreases and vice versa.

Previously, throughput S was defined as the average number of customers, which are successfully served per unit time. If the total service time of one customer takes a small value, then in a unit of time the server can serve several customers, hence, the system's throughput will be more unit. For example, in the case $\alpha < 0.5$ and for $\lambda > 5$ the total mean sojourn time V of the customer under service becomes small that involves increase in value of throughput S.

In the case $\alpha > 1$ Tables 1 and 2 also shows how much the big values of V affects on the values of throughput S.

6   Distribution of the Number of Retrials/Transitions of the Tagged Customer into the Orbit

Let us define the server states as follows

$$k(t) = \begin{cases} 0, & \text{server is free,} \\ 1, & \text{server is busy, but not by tagged customer,} \\ 2, & \text{server is busy by tagged customer.} \end{cases}$$

Let us denote by ν the number of transitions of the tagged customer into the orbit in the considered retrial queuing system. We should note, that random variable ν depend on N but for the simplification of notations is not shown explicitly. Applying method of the asymptotic analysis under condition of unlimited growing number of sources, we will find the probability distribution of ν. We have the following statement

Theorem 3. *Let ν be the number of transitions of the tagged customer into the orbit, then*

$$\lim_{N \to \infty} \mathsf{E}\, z^\nu = \frac{1-q}{1-qz}, \tag{22}$$

where value of parameter q has a form

$$q = 1 - R_0 B^*(\delta), \tag{23}$$

here

$$\delta = (1 - \kappa)\lambda + \sigma\kappa.$$

Proof. Denote by $\nu(t)$ the residual number of transitions of the tagged customer into the orbit, that is number of transitions into the orbit of the tagged customer from moment t till the end of its successful service.

Let us introduce the conditional generating functions

$$
\begin{aligned}
G_0(z, i) &= \mathsf{E}\left\{z^{\nu(t)}|k(t) = 0, i(t) = i\right\}, \\
G_k(z, i, y) &= \mathsf{E}\left\{z^{\nu(t)}|k(t) = k, i(t) = i, y(t) = y\right\}, \quad k(t) = 1, 2.
\end{aligned}
\tag{24}
$$

Assuming that system is operating in steady state for conditional generating functions $G_0(z, i), G_k(z, i, y), k = 1, 2$ it is easy to obtain the following system of Kolmogorov equations

$$
\begin{aligned}
-\left[\frac{N-i}{N}\lambda + \frac{i}{N}\sigma\right] &G_0(z, i) + \frac{N-i}{N}\lambda G_1(z, i+1, 0) \\
&+ \frac{i-1}{N}\sigma G_1(z, i, 0) + \frac{\sigma}{N}G_2(z, i, 0) = 0, \\
\frac{\partial G_1(z, i, y)}{\partial y} - &\left[\frac{N-i}{N}\lambda + \frac{i-1}{N}\sigma + \mu(y)\right]G_1(z, i, y) \\
+ \frac{N-i}{N}&\lambda G_0(z, i+1) + \frac{i-2}{N}\sigma G_0(z, i) \\
&+ \frac{\sigma}{N}zG_0(z, i) + \mu(y)G_0(z, i-1) = 0, \\
\frac{\partial G_2(z, i, y)}{\partial y} - &\left[\frac{N-i}{N}\lambda + \frac{i-1}{N}\sigma + \mu(y)\right]G_2(z, i, y) \\
+ \frac{N-i}{N}&\lambda zG_0(z, i+1) + \frac{i-1}{N}\sigma zG_0(z, i) + \mu(y) = 0.
\end{aligned}
\tag{25}
$$

Denoting $\dfrac{1}{N} = \varepsilon$ and executing the following substitutions

$$
\begin{aligned}
i\varepsilon = x, \quad \delta(x) &= (1-x)\lambda + \sigma x, \\
G_0(z, i) = F_0(z, x, \varepsilon), \quad G_k(z, i, y) &= F_k(z, x, y, \varepsilon), k = 1, 2
\end{aligned}
\tag{26}
$$

we obtain the system of equations

$$
\begin{aligned}
-\delta(x)F_0(z, x, \varepsilon) &+ (1-x)\lambda F_1(z, x+\varepsilon, 0, \varepsilon) \\
&+ (x-\varepsilon)\sigma F_1(z, x, 0, \varepsilon) + \varepsilon\sigma F_2(z, x, 0, \varepsilon) = 0, \\
\frac{\partial F_1(z, x, y, \varepsilon)}{\partial y} - &[\delta(x) + \mu(y) - \varepsilon\sigma]F_1(z, x, y, \varepsilon) \\
+ (1-x)\lambda F_0(z, x+\varepsilon, \varepsilon) &+ (x-2\varepsilon)\sigma F_0(z, x, \varepsilon) \\
&+ \varepsilon\sigma zF_0(z, x, \varepsilon) + \mu(y)F_0(z, x-\varepsilon, \varepsilon) = 0, \\
\frac{\partial F_2(z, x, y, \varepsilon)}{\partial y} - &[\delta(x) + \mu(y) - \varepsilon\sigma]F_2(z, x, y, \varepsilon) \\
+ (1-x)\lambda zF_0(z, x+\varepsilon, \varepsilon) &+ (x-\varepsilon)\sigma zF_0(z, x, \varepsilon) + \mu(y) = 0.
\end{aligned}
\tag{27}
$$

Carrying out the limiting transition under condition $\varepsilon \to 0$ in the system (27) and denoting $\lim\limits_{\varepsilon \to 0} F_0(z, x, \varepsilon) = F_0(z, x)$ and $\lim\limits_{\varepsilon \to 0} F_k(z, x, y, \varepsilon) = F_k(z, x, y)$, $k = 1, 2$, we obtain

$$
\begin{aligned}
&-\delta(x)F_0(z, x) + \delta(x)F_1(z, x, 0) = 0, \\
&\frac{\partial F_1(z, x, y)}{\partial y} - [\delta(x) + \mu(y)]\,(F_1(z, x, y) - F_0(z, x)) = 0, \\
&\frac{\partial F_2(z, x, y)}{\partial y} - [\delta(x) + \mu(y)]\,F_2(z, x, y) + \delta(x)zF_0(z, x) \\
&\hspace{6cm} +\mu(y) = 0.
\end{aligned}
\tag{28}
$$

From the first and second equations of system (28) it is easy to show that functions $F_0(z, x)$, $F_1(z, x, 0)$ and $F_1(z, x, y)$ coincide, and designating by $F(z, x)$ their common value, we can write

$$
F_0(z, x) = F_1(z, x, 0) = F_1(z, x, y) = F(z, x).
$$

Let us consider in more detail the third equation of the system (28). The solution of this equation has the form

$$
F_2(z, x, y) = e^{\int_0^y [\delta(x)+\mu(v)]dv} \left\{ F_2(z, x, 0) \right.
$$
$$
\left. - \int_0^y e^{-\int_0^v [\delta(x)+\mu(u)]du}\,[\delta(x)zF(z, x) + \mu(v)]\,dv \right\}.
\tag{29}
$$

Executing the limiting transition at $y \to \infty$ and taking into account that the first factor of the right part of equality (29) in a limiting condition tends to infinity, we can conclude that the expression in curly brackets will be equal to zero, that is

$$
F_2(z, x, 0) = \int_0^\infty e^{-\int_0^v [\delta(x)+\mu(u)]du}\,[\delta(x)zF(z, x) + \mu(v)]\,dv.
\tag{30}
$$

Performing simple transformations, we will receive

$$
F_2(z, x, 0) = [1 - B^*(\delta(x))]\,zF(z, x) + B^*(\delta(x)).
\tag{31}
$$

In order to find the function $F(z, x)$ we introduce the solution $F_0(z, x, \varepsilon)$, $F_k(z, x, y, \varepsilon)$ $k = 1, 2$ of system (27) in the form of the following decomposition

$$
\begin{aligned}
F_0(z, x, \varepsilon) &= F(z, x) + \varepsilon f_0(z, x) + o(\varepsilon^2), \\
F_1(z, x, y, \varepsilon) &= F(z, x) + \varepsilon f_1(z, x) + o(\varepsilon^2), \\
F_2(z, x, y, \varepsilon) &= F_2(z, x, y) + \varepsilon f_2(z, x, y) + o(\varepsilon^2).
\end{aligned}
$$

Substituting these decompositions into the first and second equations of system (27), we obtain equalities

$$-\delta(x)\{F(z,x) + \varepsilon f_0(z,x)\} - \varepsilon\sigma F(z,x)$$
$$+(1-x)\lambda\left\{F(z,x) + \varepsilon\frac{\partial F(z,x)}{\partial x} + \varepsilon f_1(z,x,0)\right\}$$
$$+\varepsilon\sigma F_2(z,x,0) + x\sigma\{F(z,x) + \varepsilon f_1(z,x,0)\} = o(\varepsilon^2),$$
$$\varepsilon\frac{\partial f_1(z,x,y)}{\partial y} - [\delta(x) + \mu(y)]\{F(z,x) + \varepsilon f_1(z,x,y)\} + \varepsilon\sigma F(z,x)$$
$$+(1-x)\lambda\left\{F(z,x) + \varepsilon\frac{\partial F(z,x)}{\partial x} + \varepsilon f_0(z,x)\right\} + \varepsilon\sigma z F(z,x)$$
$$+x\sigma\{F(z,x) + \varepsilon f_0(z,x)\} - 2\varepsilon\sigma F(z,x)$$
$$+\mu(y)\left\{F(z,x) - \varepsilon\frac{\partial F(z,x)}{\partial x} + \varepsilon f_0(z,x)\right\} = o(\varepsilon^2).$$

Equating here coefficients at identical degrees ε, for functions $f_0(z,x)$, $f_1(z,x,y)$ and $f_2(z,x,y)$ we obtain following system

$$-\delta(x)f_0(z,x) + \delta(x)f_1(z,x,0)$$
$$= \sigma\{F(z,x) - F_2(z,x,0)\} - (1-x)\lambda\frac{\partial F(z,x)}{\partial x},$$
$$-[\delta(x) + \mu(y)]f_1(z,x,y) + [\delta(x) + \mu(y)]f_0(z,x) + \frac{\partial f_1(z,x,y)}{\partial y} \qquad (32)$$
$$= (1-z)\sigma F(z,x) + [\mu(y) - (1-x)\lambda]\frac{\partial F(z,x)}{\partial x}.$$

Multiplying the first equality on $R_0(x)$, second on $R_1(x,y)$, adding this products and integrating the received equality on y from 0 to ∞ we obtain

$$\left\{\delta(x)[R_1(x) - R_0(x)] + \int_0^\infty \mu(y)R_1(x,y)dy\right\}f_0(z,x)$$
$$+\delta(x)R_0(x)f_1(z,x,0) - \int_0^\infty [\delta(x) + \mu(y)]f_1(z,x,y)R_1(x,y)dy$$
$$+\int_0^\infty \frac{\partial f_1(z,x,y)}{\partial y}R_1(x,y)dy = (1-z)\sigma F(z,x)R_1(x) \qquad (33)$$
$$+\sigma R_0(x)[F(z,x) - F_2(z,x,0)]$$
$$+\left[\int_0^\infty \mu(y)R_1(x,y)dy - (1-x)\lambda\right]\frac{\partial F(z,x)}{\partial x}.$$

Substituting into (33) $x = \kappa$ let us denote $R_0(x) = R_0$, $R_1(x,y) = R_1(y)$, $F_0(z,x) = F_0(z)$, $F_2(z,x,y) = F_2(z,y)$. Taking into account equalities (15) from Theorem 1, it is not difficult to obtain that the left part of equality (33) is equal to zero. Owing to (16), we obtain

$$(1-z)\sigma F(z) + \sigma R_0[zF(z) - F_2(z,0)] = 0,$$

which, taking into account equality (31), can be rewritten as

$$\left\{1 - z\left[1 - R_0 B^*(\delta)\right]\right\} F(z) = R_0 B^*(\delta). \tag{34}$$

Denote

$$q = 1 - R_0 B^*(\delta),$$

coinciding with (23), we have received that function $F(z)$ has the form

$$F(z) = \frac{1-q}{1-qz},$$

which coincide with (22).

The theorem is proved.

From the proved theorem it is obviously follows that the probability distribution $P\{\nu = n\}$, $n = \overline{0,\infty}$ of the number of transitions of the tagged customer into the orbit is geometric and has the form

$$P\{\nu = n\} = (1-q)q^n, \quad n = \overline{0,\infty}. \tag{35}$$

Let us consider the influence of the system parameters on the values of mean number of retrials ν. We will choose the same parameters of system which have been considered in the previous examples, namely $\sigma = 20$ and the values of parameters λ and $\alpha = \beta$ are specified in the Table 3.

Table 3. Mean number of retrials for various values of λ and $\alpha = \beta$

		$\alpha = \beta$							
		0.1	0.3	0.5	0.8	1	2	3	5
λ	0.5	0.470	1.131	1.86	3.65	6.32	162.7	793	6091
	1	0.656	2.002	4.19	11.57	21.83	204.1	848	6172
	5	1.114	4.192	9.31	22.73	37.07	234.4	891	6236
	10	1.278	4.755	10.28	24.30	39.02	238.1	896	6244
	15	1.355	4.969	10.63	24.83	39.67	239.3	898	6247

From Table 3 it is shown that the mean number of retrials obviously depends on service parameter α that is the expected and logical result. It should be noted that for small values of parameter α influence of parameter λ on mean number of retrials is significant and obvious. But with the increase of parameter α this influence decreases and already for great values of α practically disappears.

The following conclusions can be drawn from the Tables 1, 2 and 3. The presence of extraordinary throughput values S, values of mean number of retrials and mean sojourn time of the customer under service is a consequence of the collision of customers and the admissibility of repeated attempts of service the same customer. Duration of the customer service for repeated attempts has the same probability distribution $B(x)$, but its repeated realization, naturally, accepts various values. If for the distribution $B(x)$ there is a high probability of emergence of small values of service time as in the gamma distribution with the shape parameter $\alpha < 1$, then a small number of retries is sufficient to realize a small value of the service time which will be successful and, as shown in the Table 1, the throughput will be greater than intensity of service.

If the small values of the service time are unlikely for the probability distribution $B(x)$, as in the gamma distribution with the shape parameter $\alpha > 1$, then the number of unsuccessful attempts of service becomes big, as we can see in the Table 3, the server works without results, the mean sojourn time of the customer under service is increase (Table 2) and the throughput S becomes close to zero.

Let us denote by W the sojourn/waiting time of the tagged customer in the orbit. On the basis of the Theorem 3 we can formulate the following statement

Theorem 4. *Characteristic function of the sojourn/waiting time W of the tagged customer in an orbit has the form*

$$\mathsf{E}e^{juW} = (1-q) + q\frac{\sigma(1-q)}{\sigma(1-q) - juN}. \tag{36}$$

Proof. The proof is trivial.

7   Conclusions

In this paper, a finite-source retrial queuing system with collisions of customers was considered. It was shown that at some combinations of system parameter values the throughput takes on exotic values. In addition, in the present paper the sojourn time analysis of the considered system was presented. The research has been conducted by method of asymptotic analysis under condition that number of sources tends to infinity while the primary request generation rate, retrial rate tend to zero. As the result of the investigation it was shown that probability distribution of the number of retrials/transitions of the customer into the orbit is geometric with given parameters, and the normalized sojourn time of the customer in the orbit has Generalized Exponential distribution. The mean sojourn time of the customer under service was obtained. Examples and tables demonstrated the novelty of the investigations.

Acknowledgments. The publication was financially supported by the Ministry of Education and Science of the Russian Federation (Agreement number 02.a03.21.0008) and by Peoples Friendship University of Russia (RUDN University).

References

1. Sztrik, J., Almási, B., Roszik, J.: Heterogeneous finite-source retrial queues with server subject to breakdowns and repairs. J. Math. Sci. **132**, 677–685 (2006)
2. Dragieva, V.I.: System state distributions in one finite source unreliable retrial queue. http://elib.bsu.by/handle/123456789/35903
3. Choi, B.D., Shinand, Y.W., Ahn, W.C.: Retrial queues with collision arising from unslotted CSMA/CD protocol. J. Control Comput. Sci. **11**(4), 335–356 (1992). Queueing Systems 11 Herald of Tomsk State University
4. Nazarov, A., Kvach, A., Yampolsky, V.: Asymptotic analysis of closed Markov retrial queuing system with collision. In: Dudin, A., Nazarov, A., Yakupov, R., Gortsev, A. (eds.) ITMM 2014. CCIS, vol. 487, pp. 334–341. Springer, Cham (2014). doi:10. 1007/978-3-319-13671-4_38
5. Kvach, A.S., Nazarov, A.A.: The research of a closed RQ-system M/GI/1//N with collision of the customers in the condition of an unlimited increasing number of sources. In: Probability Theory, Random Processes, Mathematical Statistics and Applications: Materials of the International Scientific Conference Devoted to the 80th Anniversary of Professor Gennady Medvedev, Doctor of Physical and Mathematical Sciences, Minsk, February 23-26, 2015 Minsk, pp. 65–70 (2015). (In Russian) http://vital.lib.tsu.ru/vital/access/manager/Repository/vtls:000535452
6. Gòmez-Corral, A.: On the applicability of the number of collisions in p-persistent CSMA/CD protocols. Comput. Oper. Res. **37**(7), 1199–1211 (2010)
7. Zhang, F., Wang, J.: Performance analysis of the retrial queues with finite number of sources and service interruptions. J. Korean Stat. Soc. **42**(1), 117–131 (2013). doi:10.1016/j.jkss.2012.06.002
8. Nazarov, A.A., Moiseeva, S.P.: Methods of Asymptotic Analysis in Queueing Theory. NTL Publishing House of Tomsk University, Tomsk (2006). (In Russian)

Infinite–Server Tandem Queue with Renewal Arrivals and Random Capacity of Customers

Ekaterina Lisovskaya[1(✉)], Svetlana Moiseeva[1], and Michele Pagano[2]

[1] Tomsk State University, Tomsk, Russia
ekaterina_lisovs@mail.ru, smoiseeva@mail.ru
[2] University of Pisa, Pisa, Italy
m.pagano@iet.unipi.it

Abstract. A tandem of two queues with infinite number of servers is considered. Customers arrive at the first stage of the tandem according to a renewal process, and, after the completion of their services, go to the second stage. Each customer carries a random quantity of work (capacity of the customer). In this study service time does not depend on the customer capacities; the latter are used just to fix some additional features of the system evolution. It is shown that the two-dimensional probability distribution of the total capacities at the stages of the system is two-dimensional Gaussian under the asymptotic condition of a high arrival rate. Numerical experiments and simulations allow us to determine the applicability area for the asymptotic result.

Keywords: Infinite–server queueing system · Random capacity of customer · Renewal process

1 Introduction

Queueing theory plays a very relevant role in modelling and performance evaluation of information systems. For instance, in computer networks messages exchanged by end-systems have some random size (depending on the protocol stack as well as on the physical medium) and hence require the corresponding amount of memory in all the routers along the path from source to destination [8]. Each router can be modelled as a queueing system and the entire path by a tandem queue, where the number of nodes is equal to the number of hops along the path itself [6].

In more detail, in packet-switching networks routers work according to the *store-and-forward* principle: packets are stored in the buffer, processed (typically some fields in the header are changed – for instance in IP networks the "checksum" is verified and, if it is correct, "time to live" is decremented and the "checksum" is correspondingly updated) and transmitted on the suitable output port depending on the destination address and the forwarding table built by the underlying routing protocol.

Each router can be modelled as a queue with a limited number of servers (depending on the amount of output ports) and a finite amount of memory [9].

© Springer International Publishing AG 2017
V.M. Vishnevskiy et al. (Eds.): DCCN 2017, CCIS 700, pp. 201–216, 2017.
DOI: 10.1007/978-3-319-66836-9_17

However, quite often it is easier to deal with the corresponding infinite–server queueing system [3] and then, solving the underlying optimization problem, find the required number of servers and amount of memory in each node [2].

In this work, we consider a tandem of two queues with infinite number of servers and derive asymptotics for the number of customers (i.e., busy servers) and the buffer occupancy (i.e., the overall volume of the packets in the system) in both phases of the system.

2   Matematical Model

Consider a tandem queue with two stages and an infinite number of servers at each stage. Customers, characterized by some random capacity $\nu > 0$ with distribution function $G(y)$, arrive according to a renewal process. Each arriving customer instantly occupies a server at the first stage, with service time distribution $B_1(x)$. When the service is complete, the customer moves to the second stage for the further service with service time distribution $B_2(x)$. When the service is complete at the second stage, the customer leaves the system. Customer capacities and service times are not dependent on each other and are independent of the epochs of customer arrivals.

Denote the number of customers at the first and at the second stage of the system at moment t by $i_1(t)$ and $i_2(t)$, and denote the total capacities of all customers at the first and at the second stage by $V_1(t)$ and $V_2(t)$, respectively. Our goal is to derive the probabilistic characterization of the multi-dimensional process $\{i_1(t), V_1(t), i_2(t), V_2(t)\}$. This process is not Markovian and, therefore, we use the dynamic screening method for its investigation.

Consider three time axes that are numbered from 0 to 2 (Fig. 1). Let axis 0 show the epochs of customers' arrivals. Axes 1 and 2 correspond to the two stages of the system.

We introduce a set of two functions $S_1(t)$, $S_2(t)$ (dynamic probabilities) that satisfy the conditions

$$0 \leq S_1(t) \leq 1, 0 \leq S_2(t) \leq 1, S_1(t) + S_2(t) \leq 1.$$

and assume that a customer, arriving at the system at time t, may be screened to axis 1 with probability $S_1(t)$, or to axis 2 with probability $S_2(t)$, or may be not screened anywhere with probability $S_0(t) = 1 - S_1(t) - S_2(t)$.

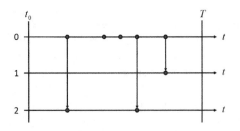

Fig. 1. Screening of the customers arrivals

Let the system be empty at moment t_0, and let us fix some arbitrary moment T in the future. $S_1(t)$ and $S_2(t)$ denote the probability that a customer, arrived at the moment $t > t_0$, will be serviced by time T at the first and second stage, respectively. It is easy to show that

$$S_1(t) = 1 - B_1(T - t), \quad S_2(t) = B_1(T - t) - B_2^*(T - t)$$

for $t_0 \le t \le T$, where

$$B_2^*(\tau) = (B_1 * B_2)(\tau) = \int_0^\tau B_2(\tau - y) dB_1(y)$$

is the convolution of $B_1(x)$ and $B_2(x)$.

Denote the number of arrivals screened before the moment t on axes 1 and 2 by $n_1(t)$ and $n_2(t)$, and denote the total capacities of customers screened on axis 1 and 2 by $W_1(t)$ and $W_2(t)$, respectively. As shown in [5], the multi-dimensional joint probability distribution of the number of customers at the stages of the tandem system at the moment T coincides with the multi-dimensional joint probability distribution of the number of screened arrivals on the respective axes:

$$P\{i_1(T) = m_1, i_2(T) = m_2\} = \{n_1(T) = m_1, n_2(T) = m_2\}$$

for all $m_1, m_2 = 0, 1, 2, \ldots$. It is easy to prove the same property for the extended process $\{i_1(t), V_1(t), i_2(t), V_2(t)\}$:

$$P\{i_1(T) = m_1, V_1(T) < z_1, i_2(T) = m_2, V_2(T) < z_2\}$$

$$= \{n_1(T) = m_1, W_1(T) < z_1, n_2(T) = m_2, W_2(T) < z_2\} \quad (1)$$

for all $m_1, m_2 = 0, 1, 2, \ldots$ and $z_1, z_2 \ge 0$. We use Equalities (1) to investigate the process $\{i_1(t), V_1(t), i_2(t), V_2(t)\}$ via the analysis of the process

$$\{n_1(t), W_1(t), n_2(t), W_2(t)\}.$$

3   Kolmogorov Differential Equations

Let us consider the five-dimensional Markovian process

$$\{z(t), n_1(t), W_1(t), n_2(t), W_2(t)\},$$

where $z(t)$ is the residual time from t to the next arrival (in the renewal input process). Denoting the probability distribution of this process by

$$P(z, n_1, w_1, n_2, w_2, t)$$

$$= P\{z(t) < z, n_1(t) = n_1, W_1(t) < w_1, n_2(t) = n_2, W_2(t) < w_2\}$$

and taking into account the formula of total probability, we can write the following system of Kolmogorov differential equations:

$$\frac{\partial P(z, n_1, w_1, n_2, w_2, t)}{\partial t} = \frac{\partial P(z, n_1, w_1, n_2, w_2, t)}{\partial z}$$

$$+ \frac{\partial P(0, n_1, w_1, n_2, w_2, t)}{\partial z} (A(z) - 1)$$

$$+ S_1(t) A(z) \left[\int_0^{w_1} \frac{\partial P(0, n_1, w_1, n_2, w_2, t)}{\partial z} dG(y) - \frac{\partial P(0, n_1, w_1, n_2, w_2, t)}{\partial z} \right]$$

$$+ S_2(t) A(z) \left[\int_0^{w_2} \frac{\partial P(0, n_1, w_1, n_2 - 1, w_2 - y, t)}{\partial z} dG(y) - \frac{\partial P(0, n_1, w_1, n_2, w_2, t)}{\partial z} \right],$$

$$z > 0; n_1, n_2 = 0, 1, 2 \ldots; \; w_1, w_2 > 0,$$

with the initial condition

$$P(z, n_1, w_1, n_2, w_2, t_0) = \begin{cases} R(z), & n_1 = n_2 = 0 = w_1 = w_2 = 0 \\ 0, & \text{otherwise} \end{cases},$$

where $R(z)$ denotes the stationary probability distribution of the values of the random process $z(t)$.

We introduce the partial characteristic function:

$$h(z, u_1, v_1, u_2, v_2, t)$$

$$= \sum_{n_1=0}^{\infty} e^{ju_1 n_1} \int_0^{\infty} e^{jv_1 w_1} \sum_{n_2=0}^{\infty} e^{ju_2 n_2} \int_0^{\infty} e^{jv_2 w_2} P(z, n_1, dw_1, n_2, dw_2, t),$$

$$z > 0; \; n_1, \, n_2 = 0, 1, 2 \ldots; \; w_1, \; w_2 > 0,$$

where $j = \sqrt{-1}$ is the imaginary unit. Then, we can write the following equations:

$$\frac{\partial h(z, u_1, v_1, u_2, v_2, t)}{\partial t} = \frac{\partial h(z, u_1, v_1, u_2, v_2, t)}{\partial z}$$

$$+ \frac{\partial h(0, u_1, v_1, u_2, v_2, t)}{\partial z} \{A(z) - 1$$

$$+ S_1(t) A(z) \left(e^{ju_1} G^*(v_1) - 1 \right) + S_2(t) A(z) \left(e^{ju_2} G^*(v_2) - 1 \right) \}, z > 0, \quad (2)$$

where

$$G^*(v) = \int_0^{\infty} e^{jvy} dG(y),$$

with the initial condition

$$h(z, u_1, v_1, u_2, v_2, t_0) = R(z). \tag{3}$$

4 Asymptotic Analysis

In general, the exact solution of Eq. (2) is not available, but it may be found under asymptotic conditions. In this paper, we consider the case of infinitely growing arrival rate. Let us write the distribution function of the interarrival times as $A(Nz)$, where N is some parameter used for the asymptotic analysis ($N \to \infty$ in theoretical analysis [4]).

Then, the Eq. (2) takes the form

$$\frac{1}{N} \frac{\partial h(z, u_1, v_1, u_2, v_2, t)}{\partial t} = \frac{\partial h(z, u_1, v_1, u_2, v_2, t)}{\partial z}$$

$$+ \frac{\partial h(0, u_1, v_1, u_2, v_2, t)}{\partial z} [A(z) - 1]$$

$$+ S_1(t)A(z) \left(e^{ju_1} G^*(v_1) - 1 \right) + S_2(t)A(z) \left(e^{ju_2} G^*(v_2) - 1 \right)], \qquad (4)$$

with the initial condition (3).

We solve problem (3) and (4) under such asymptotic condition and we obtain approximate solutions with different order of accuracy, named as "first-order asymptotic" $h(z, u_1, v_1, u_2, v_2, t) \approx h^{(1)}(z, u_1, v_1, u_2, v_2, t)$ and "second-order asymptotic" $h(z, u_1, v_1, u_2, v_2, t) \approx h^{(2)}(z, u_1, v_1, u_2, v_2, t)$.

4.1 The First-Order Asymptotic Analysis

We formulate and prove the following statement.

Lemma. *The first-order asymptotic characteristic function of the probability distribution of the process $\{z(t), n_1(t), W_1(t), n_2(t), W_2(t)\}$ has the form*

$$h^{(1)}(z, u_1, v_1, u_2, v_2, t) = R(z) \exp \left\{ N\lambda \left(ju_1 + jv_1 a_1 \right) \int_{t_0}^{t} S_1(\tau)d\tau \right.$$

$$\left. + N\lambda \left(ju_2 + jv_2 a_1 \right) \int_{t_0}^{t} S_2(\tau)d\tau \right\},$$

where $\lambda = \left(\int_0^{\infty} (1 - A(x)) \, dx \right)^{-1}$ and a_1 is the mean customer capacity.

Proof. By performing the substitutions

$$\varepsilon = \frac{1}{N}, u_1 = \varepsilon x_1, v_1 = \varepsilon y_1, u_2 = \varepsilon x_2, v_2 = \varepsilon y_2,$$

$$h(z, u_1, v_1, u_2, v_2, t) = f_1(z, x_1, y_1, x_2, y_2, t, \varepsilon), \qquad (5)$$

in expressions (3) and (4), we obtain

$$\varepsilon\frac{\partial f_1(z,x_1,y_1,x_2,y_2,t,\varepsilon)}{\partial t} = \frac{\partial f_1(z,x_1,y_1,x_2,y_2,t,\varepsilon)}{\partial z}$$

$$+ \frac{\partial f_1(0,x_1,y_1,x_2,y_2,t,\varepsilon)}{\partial z}\left[A(z) - 1 + S_1(t)A(z)\left(e^{j\varepsilon x_1}G^*(\varepsilon y_1) - 1\right)\right.$$

$$\left. + S_2(t)A(z)\left(e^{j\varepsilon x_2}G^*(\varepsilon y_2) - 1\right)\right], \tag{6}$$

with the initial condition

$$f_1(z,x_1,y_1,x_2,y_2,t_0,\varepsilon) = R(z). \tag{7}$$

Let us find the asymptotic solution of problem (6) and (7), i.e. the function $f_1(z,x_1,y_1,x_2,y_2,t) = \lim\limits_{\varepsilon\to 0} f_1(z,x_1,y_1,x_2,y_2,t,\varepsilon)$ in two steps.

Step 1. Let $\varepsilon \to 0$ in (6) and (7), then we obtain the following equation:

$$\frac{\partial f_1(z,x_1,y_1,x_2,y_2,t)}{\partial z} + \frac{\partial f_1(0,x_1,y_1,x_2,y_2,t)}{\partial z}(A(z) - 1) = 0.$$

We can conclude that $f_1(x_1,y_1,x_2,y_2,t)$ can be expressed as

$$f_1(z,x_1,y_1,x_2,y_2,t) = R(z)\Phi_1(x_1,y_1,x_2,y_2,t), \tag{8}$$

where $\Phi_1(x_1,y_1,x_2,y_2,t)$ is some scalar function that satisfies the condition

$$\Phi_1(x_1,y_1,x_2,y_2,t_0) = 1.$$

Step 2. Let $z \to \infty$ in (6). We obtain

$$\varepsilon\frac{\partial f_1(\infty,x_1,y_1,x_2,y_2,t,\varepsilon)}{\partial t} = \frac{\partial f_1(0,x_1,y_1,x_2,y_2,t,\varepsilon)}{\partial z}.$$

$$\left[S_1(t)\left(e^{j\varepsilon x_1}G^*(\varepsilon y_1) - 1\right) + S_2(t)\left(e^{j\varepsilon x_2}G^*(\varepsilon y_2) - 1\right)\right].$$

We substitute here the expression (8), use the expansion

$$e^{j\varepsilon s} = 1 + j\varepsilon s + O(\varepsilon^2),$$

divide by ε and perform the limit as $\varepsilon \to 0$. Taking into account that $R'(0) = \lambda$, we obtain the following differential equation:

$$\frac{\partial \Phi_1(x_1,y_1,x_2,y_2,t)}{\partial t} = \Phi_1(x_1,y_1,x_2,y_2,t)\cdot$$

$$\lambda\left[S_1(t)(jx_1 + jy_1a_1) + S_2(t)(jx_2 + jy_2a_1)\right], \tag{9}$$

where $a_1 = \int\limits_0^\infty y\,dG(y)$ is the mean customer capacity.

Taking into account the initial condition, the solution of (9) is

$$\Phi_1(x_1, y_1, x_2, y_2, t) = \exp\left\{\lambda\left(jx_1 + jy_1 a_1\right) \int_{t_0}^{t} S_1(\tau)d\tau \right.$$

$$\left. + \lambda\left(jx_2 + jy_2 a_1\right) \int_{t_0}^{t} S_2(\tau)d\tau \right\}.$$

By substituting $\Phi_1(x_1, y_1, x_2, y_2, t)$ from (8), we obtain

$$f_1(x_1, y_1, x_2, y_2, t) = R(z)\exp\left\{\lambda\left(jx_1 + jy_1 a_1\right) \int_{t_0}^{t} S_1(\tau)d\tau \right.$$

$$\left. + \lambda\left(jx_2 + jy_2 a_1\right) \int_{t_0}^{t} S_2(\tau)d\tau \right\}.$$

Therefore, we can write

$$h(z, u_1, v_1, u_2, v_2, t) = f_1(z, x_1, y_1, x_2, y_2, t, \varepsilon)$$

$$\approx f_1(z, x_1, y_1, x_2, y_2, t) = R(z)\Phi_1(x_1, y_1, x_2, y_2, t)$$

$$= R(z)\exp\left\{\lambda\left(jx_1 + jy_1 a_1\right) \int_{t_0}^{t} S_1(\tau)d\tau + \lambda\left(jx_2 + jy_2 a_1\right) \int_{t_0}^{t} S_2(\tau)d\tau \right\}$$

$$= R(z)\exp\left\{N\lambda\left(ju_1 + jv_1 a_1\right) \int_{t_0}^{t} S_1(\tau)d\tau + \lambda\left(ju_2 + jv_2 a_1\right) \int_{t_0}^{t} S_2(\tau)d\tau \right\}.$$

The proof is complete.

4.2   The Second-Order Asymptotic Analysis

The main result is represented by the following theorem.

Theorem. *The second-order asymptotic characteristic function of the probability distribution of the process $\{z(t), n_1(t), W_1(t), n_2(t), W_2(t)\}$ has the form*

$$h^{(2)}(z, u_1, v_1, u_2, v_2, t) = R(z) \exp \left\{ N\lambda \left(ju_1 + jv_1a_1 \right) \int\limits_{t_0}^{t} S_1(\tau)d\tau \right.$$

$$+ N\lambda \left(ju_2 + jv_2a_1 \right) \int\limits_{t_0}^{t} S_2(\tau)d\tau + \frac{(ju_1)^2}{2} \left(N\lambda \int\limits_{t_0}^{t} S_1(\tau)d\tau + N\kappa \int\limits_{t_0}^{t} S_1^2(\tau)d\tau \right)$$

$$+ \frac{(jv_1)^2}{2} \left(N\lambda a_2 \int\limits_{t_0}^{t} S_1(\tau)d\tau + N\kappa a_1^2 \int\limits_{t_0}^{t} S_1^2(\tau)d\tau \right)$$

$$+ (ju_1)(jv_1) \left(N\lambda a_1 \int\limits_{t_0}^{t} S_1(\tau)d\tau + N\kappa a_1 \int\limits_{t_0}^{t} S_1^2(\tau)d\tau \right)$$

$$+ \frac{(ju_2)^2}{2} \left(N\lambda \int\limits_{t_0}^{t} S_2(\tau)d\tau + N\kappa \int\limits_{t_0}^{t} S_2^2(\tau)d\tau \right)$$

$$+ \frac{(jv_2)^2}{2} \left(N\lambda a_2 \int\limits_{t_0}^{t} S_2(\tau)d\tau + N\kappa a_1^2 \int\limits_{t_0}^{t} S_2^2(\tau)d\tau \right)$$

$$+ (ju_2)(jv_2) \left(N\lambda a_1 \int\limits_{t_0}^{t} S_2(\tau)d\tau + N\kappa a_1 \int\limits_{t_0}^{t} S_2^2(\tau)d\tau \right)$$

$$+ (ju_1)(ju_2)N\lambda \int\limits_{t_0}^{t} S_1(\tau)S_2(\tau)d\tau + (jv_1)(jv_2)N\kappa a_1^2 \int\limits_{t_0}^{t} S_1(\tau)S_2(\tau)d\tau$$

$$+ \left. ((ju_1)(jv_2) + (ju_2)(jv_1))N\kappa a_1 \int\limits_{t_0}^{t} S_1(\tau)S_2(\tau)d\tau \right\}, \qquad (10)$$

where $a_2 = \int\limits_{0}^{\infty} y^2 dG(y)$ and $\kappa = \lambda^3 \left(\sigma^2 - a^2 \right)$, a and σ^2 being the mean and the variance of the random variable with distibution function $A(x)$, respectively.

Proof. Let $h_2(u_1, v_{,1}, u_2, v_2, t)$ be a function that satisfies the equation

$$h(z, u_1, v_1, u_2, v_2, t) = h_2(z, u_1, v_1, u_2, v_2, t) \exp \left\{ N\lambda \left(ju_1 + jv_1 a_1 \right) \int_{t_0}^{t} S_1(\tau) d\tau \right.$$

$$\left. + N\lambda \left(ju_2 + jv_2 a_1 \right) \int_{t_0}^{t} S_2(\tau) d\tau \right\}. \tag{11}$$

Substituting these expressions into (3) and (4), we obtain the following problem:

$$\frac{1}{N} \frac{\partial h_2(z, u_1, v_1, u_2, v_2, t)}{\partial t} + \lambda \left[(ju_1 + jv_1 a_1) S_1(t) \right.$$

$$+ (ju_2 + jv_2 a_1) S_2(t) \right] h_2(z, u_1, v_1, u_2, v_2, t)$$

$$= \frac{\partial h_2(z, u_1, v_1, u_2, v_2, t)}{\partial z} + \frac{\partial h_2(0, u_1, v_1, u_2, v_2, t)}{\partial z} \left[A(z) - 1 \right.$$

$$+ S_1(t)A(z) \left(e^{ju_1} G^*(v_1) - 1 \right) + S_2(t)A(z) \left(e^{ju_2} G^*(v_2) - 1 \right) \right], \tag{12}$$

with the initial condition

$$h_2(z, u_1, v_1, u_2, v_2, t_0) = R(z). \tag{13}$$

Let us perform the following substitutions:

$$\varepsilon^2 = \frac{1}{N}, u_1 = \varepsilon x_1, v_1 = \varepsilon y_1, u_2 = \varepsilon x_2, v_2 = \varepsilon y_2,$$

$$h_2(z, u_1, v_1, u_2, v_2, t) = f_2(z, x_1, y_1, x_2, y_2, t, \varepsilon). \tag{14}$$

Substituting this expression into (12) and (13), we obtain the following problem:

$$\varepsilon^2 \frac{\partial f_2(z, x_1, y_1, x_2, y_2, t, \varepsilon)}{\partial t} + f_2(z, x_1, y_1, x_2, y_2, t, \varepsilon) \lambda \left[S_1(t)(j\varepsilon x_1 + j\varepsilon y_1 a_1) \right.$$

$$+ S_2(t)(j\varepsilon x_2 + j\varepsilon y_2 a_1) \right] = \frac{\partial f_2(z, x_1, y_1, x_2, y_2, t, \varepsilon)}{\partial z}$$

$$+ \frac{\partial f_2(0, x_1, y_1, x_2, y_2, t, \varepsilon)}{\partial z} \left[A(z) - 1 + S_1(t)A(z) \left(e^{j\varepsilon x_1} G^*(\varepsilon y_1) - 1 \right) \right.$$

$$+ S_2(t)A(z) \left(e^{j\varepsilon x_2} G^*(\varepsilon y_2) - 1 \right) \right], \tag{15}$$

with the initial condition

$$f_2(z, x_1, y_1, x_2, y_2, t_0, \varepsilon) = R(z). \tag{16}$$

Let us find the asymptotic solution of this problem $f_2(z, x_1, y_1, x_2, y_2, t) = \lim_{\varepsilon \to 0} f_2(z, x_1, y_1, x_2, y_2, t, \varepsilon)$ in three steps.

Step 1. Letting $\varepsilon \to 0$ in (15), we obtain the following equation:

$$\frac{\partial f_2(z, x_1, y_1, x_2, y_2, t)}{\partial z} + \frac{\partial f_2(0, x_1, y_1, x_2, y_2, t)}{\partial z}(A(z) - 1) = 0.$$

Then, we can write

$$f_2(z, x_1, y_1, x_2, y_2, t) = R(z)\Phi_2(x_1, y_1, x_2, y_2, t), \qquad (17)$$

where $\Phi_2(x_1, y_1, x_2, y_2, t)$ is some scalar function, which satisfies the condition

$$\Phi_2(x_1, y_1, x_2, y_2, t) = 1.$$

Step 2. The solution $f_2(z, x_1, y_1, x_2, y_2, t)$ can be represented in the expansion form

$$f_2(z, x_1, y_1, x_2, y_2, t) = \Phi_2(x_1, y_1, x_2, y_2, t)\{R(z)$$

$$+ [(j\varepsilon x_1 + j\varepsilon y_1 a_1)S_1(t) + (j\varepsilon x_2 + j\varepsilon y_2 a_1)S_2(t)] f(z)\} + O(\varepsilon^2), \qquad (18)$$

where $f(z)$ is a suitable function. By substituting the previous expression and the Taylor-Maclaurin expansion

$$e^{j\varepsilon s} = 1 + j\varepsilon s + O(\varepsilon^2)$$

in (15), taking into account that $R'(z) = \lambda(1 - A(z))$, it is easy to show that

$$f'(0) = \lambda f(\infty) + \frac{\kappa}{2}$$

and $\kappa = \lambda^3(\sigma^2 - a^2)$, where a and σ_2 are the mean and the variance of the random variable with distribution function $A(x)$.

Step 3. Letting $z \to \infty$ in (15) and taking advantage of the definition of the function $f_2(z, x_1, y_1, x_2, y_2, t, \varepsilon)$, we can write

$$\lim_{z \to \infty} \frac{\partial f_2(z, x_1, y_1, x_2, y_2, t, \varepsilon)}{\partial z} = 0.$$

Using this and

$$e^{j\varepsilon s} = 1 + j\varepsilon s + \frac{(j\varepsilon s)^2}{2} + O(\varepsilon^3),$$

we obtain

$$\varepsilon^2 \frac{\partial f_2(\infty, x_1, y_1, x_2, y_2, t, \varepsilon)}{\partial t} + f_2(\infty, x_1, y_1, x_2, y_2, t, \varepsilon) \cdot$$

$$\lambda[S_1(t)(j\varepsilon x_1 + j\varepsilon y_1 a_1) + S_2(t)(j\varepsilon x_2 + j\varepsilon y_2 a_1)]$$

$$= \frac{\partial f_2(0, x_1, y_1, x_2, y_2, t, \varepsilon)}{\partial z} \cdot$$

$$\left\{S_1(t)\left[\left(1 + j\varepsilon x_1 + \frac{(j\varepsilon x_1)^2}{2}\right)\left(1 + j\varepsilon y_1 a_1 + \frac{(j\varepsilon y_1)^2}{2}a_2\right) - 1\right]\right.$$

$$\left. + S_2(t)\left[\left(1 + j\varepsilon x_2 + \frac{(j\varepsilon x_2)^2}{2}\right)\left(1 + j\varepsilon y_2 a_1 + \frac{(j\varepsilon y_2)^2}{2}a_2\right) - 1\right]\right\} + O(\varepsilon^3)$$

where $a_2 = \int\limits_0^\infty y^2 dG(y)$.

By substituting here the expansion (18) and considering the limit as $z \to \infty$, we can write

$$\varepsilon^2 \frac{\partial \Phi_2(x_1, y_1, x_2, y_2, t)}{\partial t} + \Phi_2(x_1, y_1, x_2, y_2, t)\lambda \left[(j\varepsilon x_1 + j\varepsilon y_1 a_1)S_1(t)\right.$$

$$+ (j\varepsilon x_2 + j\varepsilon y_2 a_1)S_2(t)] \left[(j\varepsilon x_1 + j\varepsilon y_1 a_1)S_1(t) + (j\varepsilon x_2 + j\varepsilon y_2 a_1)S_2(t)\right] f(\infty)$$

$$= \Phi_2(x_1, y_1, x_2, y_2, t)\lambda \left\{ S_1(t) \left[\left(1 + j\varepsilon x_1 + \frac{(j\varepsilon x_1)^2}{2}\right)\right.\right.$$

$$\left.\cdot \left(1 + j\varepsilon y_1 a_1 + \frac{(j\varepsilon y_1)^2}{2}a_2\right) - 1\right]$$

$$+ S_2(t) \left[\left(1 + j\varepsilon x_2 + \frac{(j\varepsilon x_2)^2}{2}\right)\left(1 + j\varepsilon y_2 a_1 + \frac{(j\varepsilon y_2)^2}{2}a_2\right) - 1\right]\right\}$$

$$+ \Phi_2(x_1, y_1, x_2, y_2, t)f'(0)\left[(j\varepsilon x_1 + j\varepsilon y_1 a_1)S_1(t)\right.$$

$$+ (j\varepsilon x_2 + j\varepsilon y_2 a_1)S_2(t)] \left\{S_1(t)\left[\left(1 + j\varepsilon x_1 + \frac{(j\varepsilon x_1)^2}{2}\right)\cdot\right.\right.$$

$$\left.\left(1 + j\varepsilon y_1 a_1 + \frac{(j\varepsilon y_1)^2}{2}a_2\right) - 1\right] + S_2(t)\left[\left(1 + j\varepsilon x_2 + \frac{(j\varepsilon x_2)^2}{2}\right)\cdot\right.$$

$$\left.\left.\left(1 + j\varepsilon y_2 a_1 + \frac{(j\varepsilon y_2)^2}{2}a_2\right) - 1\right]\right\} + O(\varepsilon^3).$$

As a result of simple transformations and using the notation

$$\kappa = 2\left(f'(0) - f(\infty)\right),$$

we obtain the following differential equation for the function $\Phi_2(w_1, w_2, t)$:

$$\frac{\partial \Phi_2(x_1, y_1, x_2, y_2, t)}{\partial t} = \Phi_2(x_1, y_1, x_2, y_2, t)\left[\frac{(jx_1)^2}{2}\left(\lambda S_1(t) + \kappa S_1^2(t)\right)\right.$$

$$+ \frac{(jy_1)^2}{2}\left(\lambda a_2 S_1(t) + \kappa a_1^2 S_1^2(t)\right) + (jx_1)(jy_1)\left(\lambda a_1 S_1(t) + \kappa a_1 S_1^2(t)\right)$$

$$+ \frac{(jx_2)^2}{2}\left(\lambda S_2(t) + \kappa S_2^2(t)\right) + \frac{(jy_2)^2}{2}\left(\lambda a_2 S_2(t) + \kappa a_1^2 S_2^2(t)\right)$$

$$+ (jx_2)(jy_2)\left(\lambda a_1 S_2(t) + \kappa a_1 S_2^2(t)\right) + (jx_1)(jx_2)S_1(t)S_2(t)\kappa$$

$$+ (jy_1)(jy_2)S_1(t)S_2(t)\kappa a_1^2 + ((jx_1)(jy_2) + (jy_1)(jx_2))\kappa a_1 S_1(t)S_2(t)\right\}.$$

The solution of the latter equation with the available initial condition gives the expression

$$
\begin{aligned}
\Phi_2(x_1, y_1, x_2, y_2, t) = \exp \Bigg\{ & \frac{(jx_1)^2}{2} \left(\lambda \int_{t_0}^{t} S_1(\tau)d(\tau) + \kappa \int_{t_0}^{t} S_1^2(\tau)d(\tau) \right) \\
& + \frac{(jy_1)^2}{2} \left(\lambda a_2 \int_{t_0}^{t} S_1(\tau)d(\tau) + \kappa a_1^2 \int_{t_0}^{t} S_1^2(\tau)d(\tau) \right) \\
& + (jx_1)(jy_1) \left(\lambda a_1 \int_{t_0}^{t} S_1(\tau)d(\tau) + \kappa a_1 \int_{t_0}^{t} S_1^2(\tau)d(\tau) \right) \\
& + \frac{(jx_2)^2}{2} \left(\lambda \int_{t_0}^{t} S_2(\tau)d(\tau) + \kappa \int_{t_0}^{t} S_2^2(\tau)d(\tau) \right) \\
& + \frac{(jy_2)^2}{2} \left(\lambda a_2 \int_{t_0}^{t} S_2(\tau)d(\tau) + \kappa a_1^2 \int_{t_0}^{t} S_2^2(\tau)d(\tau) \right) \\
& + (jx_2)(jy_2) \left(\lambda a_1 \int_{t_0}^{t} S_2(\tau)d(\tau) + \kappa a_1 \int_{t_0}^{t} S_2^2(\tau)d(\tau) \right) \\
& + (jx_1)(jx_2)\lambda \int_{t_0}^{t} S_1(\tau)S_2(\tau)d(\tau) + (jy_1)(jy_2)\kappa a_1^2 \int_{t_0}^{t} S_1(\tau)S_2(\tau)d(\tau) \\
& + ((jx_1)(jy_2) + (jx_2)(jy_1))\kappa a_1 \int_{t_0}^{t} S_1(\tau)S_2(\tau)d(\tau) \Bigg\}.
\end{aligned}
$$

Substituting this expression into (17) and performing the substitutions that are inverse to (14) and (11), we obtain the expression (10) for the asymptotic characteristic function of the process $\{z(t), n_1(t), W_1(t), n_2(t), W_2(t)\}$.

The proof is complete.

Corollary. *When $t = T$ and $t_0 \to -\infty$ we obtain the characteristic function of the process $\{V_1(t), V_2(t)\}$ in the steady state regime*

$$
\begin{aligned}
h(v_1, v_2) = \exp \Bigg\{ & N\lambda j v_1 a_1 b_1 + N\lambda j v_2 a_1 b_2 + \frac{(jv_1)^2}{2}(N\lambda a_2 b_1 + N\kappa a_1^2 \beta_1) \\
& + \frac{(jv_2)^2}{2}(N\lambda a_2 b_2 + N\kappa a_1^2 \beta_2) + (jv_1)(jv_2)N\kappa a_1^2 b \Bigg\},
\end{aligned}
\tag{19}
$$

where

$$b_1 = \int_0^\infty (1 - B_1(\tau)) d\tau, \; \beta_1 = \int_0^\infty (1 - B_1(\tau))^2 d\tau,$$

$$b_2 = \int_0^\infty (B_1(\tau) - B_2^*(\tau)) d\tau, \; \beta_2 = \int_0^\infty (B_1(\tau) - B_2^*(\tau))^2 d\tau,$$

$$b = \int_0^\infty (1 - B_1(\tau))(B_1(\tau) - B_2^*(\tau)) d\tau.$$

From the form of function (19) it is clear that the two-dimensional process $\{V_1(t), V_2(t)\}$ is asymptotically Gaussian with mean

$$\mathbf{a} = [N\lambda a_1 b_1, N\lambda a_1 b_2]$$

and covariance matrix

$$\mathbf{K} = \begin{bmatrix} N\lambda a_2 b_1 + N\kappa a_1^2 \beta_1 & N\kappa a_1^2 b \\ N\kappa a_1^2 b & N\lambda a_2 b_2 + N\kappa a_1^2 \beta_2 \end{bmatrix}.$$

5   Numerical Example

Result (19) is obtained under the asymptotic condition $N \to \infty$. Therefore, it may be used just as an approximation when N is large enough. To test its practical applicability, we considered several numerical examples, varying all the system parameters (including the distributions of the interarrival and service times and of the customer capacity). Since all the different simulation sets led to similar results, for sake of brevity, in the following we discuss in detail just one of them. In particular, we assume that the input renewal process is characterized by the following distribution function:

$$A(x) = \begin{cases} 0, & x < 0.5; \\ x - 0.5, & x \in [0.5, 1.5]; \\ 1, & x > 1.5. \end{cases}$$

Hence, the fundamental rate of arrivals is $\lambda = 1$ customers per time unit. We also assume that customer capacities have uniform distribution in the range $[0; 1]$ and service time at first and second stage have gamma distribution with shape and inverse scale parameters $\alpha_1 = 0.5$, $\beta_1 = 0.5$ and $\alpha_2 = 1.5$, $\beta_2 = 1.5$, respectively.

Our goal is to find a lower bound of parameter N for the applicability of the approximation (19). To this aim, we carried out series of simulation experiments for increasing values of N and compared the asymptotic distributions with the empiric ones by using the Kolmogorov distance

$$\Delta = \sup_x |F(x) - A(x)|$$

as an accuracy measure. Here $F(x)$ is the cumulative distribution function build on the basis of simulation results, and $A(x)$ is the Gaussian approximation based on (19).

In the following we consider the marginal distributions of the total capacity in the two stages of the system.

For the first queue, the asymptotic values of mean and variance are equal to $0.5N$ and $0.25N$ respectively, and the corresponding values of the Kolmogorov distance for increasing values of parameter N are presented in Table 1. Similarly, for the total capacity in the second stage mean and variance are equal to $0.5N$ and $0.272N$ respectively, and Table 2 shows the Kolmogorov distance.

Table 1. Kolmogorov distance for the first queue

N	1	3	5	6	7	10	15	20	50	100
Δ	0.304	0.066	0.037	0.032	**0.028**	**0.022**	**0.017**	**0.015**	**0.009**	**0.007**

Table 2. Kolmogorov distance for the second queue

N	1	3	5	6	7	10	15	20	50	100
Δ	0.316	0.071	0.039	0.034	**0.029**	**0.023**	**0.018**	**0.015**	**0.009**	**0.007**

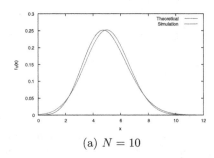
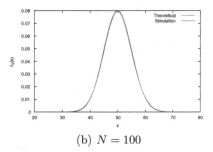

(a) $N = 10$ (b) $N = 100$

Fig. 2. Distributions of the total capacity in the first stage

We can notice that the asymptotic results become more accurate when the parameter N increases. This conclusion is confirmed by Figs. 2 and 3, which compare the asymptotic approximations with the empirical results for the total capacity in each stage of the system for two different values of N.

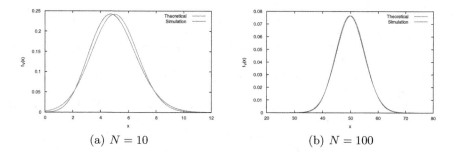

(a) $N = 10$ (b) $N = 100$

Fig. 3. Distributions of the total capacity in the second stage

As typically done in the literature, we suppose that an approximation is applicable if its Kolmogorov distance is less than 0.03 [1,7]. Hence, we can conclude that the asymptotic results are suitable for values of the parameter N equal to 7 or more (marked by boldface in Tables 1 and 2).

6   Conclusions

In this paper a tandem of two infinite–server queues has been analysed under the assumption that customers arrive at the first stage of the tandem according to a renewal process and carry a random quantity of work (capacity of the customer), which is independent of their service time. Since the Kolmogorov differential equations that describe the system cannot be solved analytically, we derived first and second-order asymptotic approximations under the assumption of infinitely growing arrival rate and showed that the two-dimensional probability distribution of the total capacities at the stages of the system is two-dimensional Gaussian under the asymptotic condition of a high arrival rate. Numerical experiments and simulations allowed us to verify the applicability of the Gaussian approximation already for small values of the asymptotic parameter N (in the presented example $N \geq 7$, but similar results have been obtained also for different input parameters).

References

1. Fedorova, E.: The second order asymptotic analysis under heavy load condition for retrial queueing system MMPP/M/1. In: Dudin, A., Nazarov, A., Yakupov, R. (eds.) ITMM 2015. CCIS, vol. 564, pp. 344–357. Springer, Cham (2015). doi:10. 1007/978-3-319-25861-4_29
2. Lisovskaya, E., Moiseeva, S.: Study of the queuing systems $M|GI|N|\infty$. In: Dudin, A., Nazarov, A., Yakupov, R. (eds.) ITMM 2015. CCIS, vol. 564, pp. 175–184. Springer, Cham (2015). doi:10.1007/978-3-319-25861-4_15
3. Moiseev, A., Moiseeva, S., Sinyakova, I.: Modeling of insurance company as infinite-servers queueing system. In: Proceeding International Conference on Application of Information and Communication Technology and Statistics in Economy and Education (ICAICTSEE-2012), pp. 161–163 (2012)

4. Moiseev, A., Nazarov, A.: Asymptotic analysis of the infinite-server queueing system with high-rate semi-Markov arrivals. In: 2014 6th International Congress on Ultra Modern Telecommunications and Control Systems and Workshops (ICUMT), pp. 507–513, October 2014
5. Moiseev, A., Nazarov, A.: Queueing network with high-rate arrivals. Eur. J. Oper. Res. **254**(1), 161–168 (2016)
6. Moiseev, A., Nazarov, A.: Tandem of infinite-server queues with Markovian arrival process. In: Vishnevsky, V., Kozyrev, D. (eds.) DCCN 2015. CCIS, vol. 601, pp. 323–333. Springer, Cham (2016). doi:10.1007/978-3-319-30843-2_34
7. Nazarov, A., Chernikova, Y.: The accuracy of Gaussian approximations of probabilities distribution of states of the retrial queueing system with priority of new customers. In: Dudin, A., Nazarov, A., Yakupov, R., Gortsev, A. (eds.) ITMM 2014. CCIS, vol. 487, pp. 325–333. Springer, Cham (2014). doi:10.1007/978-3-319-13671-4_37
8. Tikhonenko, O.: Queueing systems with common buffer: a theoretical treatment. In: Kwiecień, A., Gaj, P., Stera, P. (eds.) CN 2011. CCIS, vol. 160, pp. 61–69. Springer, Heidelberg (2011). doi:10.1007/978-3-642-21771-5_8
9. Tikhonenko, O., Kempa, W.M.: Performance evaluation of an M/G/N-type queue with bounded capacity and packet dropping. Int. J. Appl. Math. Comput. Sci. **26**(4), 841–854 (2016)

Application of Splitting to Failure Estimation in Controllable Degradation System

Alexandra Borodina[1], Dmitry Efrosinin[2], and Evsey Morozov[1(✉)]

[1] Institute of Applied Mathematical Research, Karelian Research Center RAS
and Petrozavodsk State University, Petrozavodsk, Russia
borodina@krc.karelia.ru, emorozov@karelia.ru
[2] Institute of Control Sciences, RAS and Peoples' Friendship University of Russia,
Moscow, Russia
dmitry.efrosinin@jku.at

Abstract. We consider a regenerative degradation process composed by
a sum of the successive phases, where preventive repair is used to prevent an instantaneous failure. For an optimal control of such a systems,
calculation of the failure probability, the average length of the regeneration cycle with or without failure, etc., is critically important. If the
degradation process is Markovian, then the required steady-state performance measures are analytically available, however in is not the case if
the process is non-Markov, in which case simulation is used to estimate
the unknown parameters of the system. In this work, the regenerative
structure of the degradation process is used to calculate the mentioned
above steady-state parameters. Moreover, provided the failure within a
regeneration cycle is a rare event, we apply a regenerative variant of
the *splitting method* to estimate the failure probability. It is shown that
this approach is much less time-consuming in comparison with crude
Monte Carlo simulation. The efficiency of the approach is demonstrated
by a detailed analysis of the degradation process generated by the i.i.d.
exponential phases. The explicit analytical results for this case are then
compared with the corresponding simulation results obtained by crude
Monte Carlo and splitting method.

Keywords: Rare event simulation · Splitting method · Regenerative
approach · Degradation process

1 Introduction

During last years an intensive attention to the aging and degradation models
for technical and biological objects has been attracted. Some wearing and aging
models were in focus of many investigators in the framework of shock and damage
models. An excellent review and contribution to the earlier papers devoted to
the topic one can be found in [4,16,18]. The aging and degradation models
suppose the study of systems with gradual failures, for which a large number of
multi-state reliability models have been elaborated, see e.g. [17,19].

© Springer International Publishing AG 2017
V.M. Vishnevskiy et al. (Eds.): DCCN 2017, CCIS 700, pp. 217–230, 2017.
DOI: 10.1007/978-3-319-66836-9_18

In this paper, we consider a degradation process in the system with gradual and instantaneous failures proposed in [3]. The main purpose of the work is to develop estimation a small failure probability, in which case crude Monte Carlo simulation turns out to be time-consuming. Following [3], we use the regenerative structure of the basic degradation process, but unlike [3], now we allow the basic process to be non-Markovian, in which case a speed-up simulation, the so-called *splitting method*, is applied to estimate the failure probability with a given accuracy in an acceptable simulation time. The standard splitting algorithm [5, 7, 12] accelerates the rare event (failure) by means of the splitting each trajectory of the process, upon reaching the predefined thresholds, onto a few independent copies.

The main contribution of this work is to apply a regenerative variant of the splitting, developed in [1, 2], to estimate the failure probability in the non-Markov analogue of the system considered in [3]. This approach also allows to avoid inversion of the Laplace transform of a convolution of the (degradation) phases. This inversion becomes highly non-accurate even for a moderate number of the exponential degradation phases when we deals with *asymptotic reliability function*, see Sect. 2.4. The numerical results demonstrate a significant gain in simulation time of the speed-up simulation over the crude (standard) Monte-Carlo simulation. Moreover, the proposed approach also allows to effectively estimate the average length of regeneration cycle with and without failure, the mean (unconditional) cycle length and also to calculate the asymptotic reliability function.

It is worth mentioning that, by the structure of the basic process, the splitting method turns out to be highly adaptive to estimate the required performance measures in the model under consideration. It is in a strong contrast to the application of this method to majority of other models. The main advantage of our model is that each path of degradation process can only *increases*, while in other models any path can, typically, both increases and decreases, significantly increasing simulation time. Finally, regenerative splitting provides confidence estimation of the required (small) failure probability with a given accuracy.

The paper is organized as follows. In Sect. 2.1, the basic degradation process is described in detail, including description its regenerative structure. In Sect. 2.2, the expressions for the mean regeneration cycle, with and without failure, are deduced. In Sect. 2.3, we focus on i.i.d. exponential degradation phases, in which case analytical results for main parameters are obtained. In Sect. 2.4, the asymptotic reliability function is briefly discussed. In Sect. 3, a brief introduction to the splitting method is given. Finally, simulation results and their comparison with the analytical results, where exist, are given in Sect. 4. We conclude our analysis with Sect. 5.

2 Main Results

2.1 Degradation Process

In this section, we describe the degradation of the system by a random process $X := \{X(t)\}_{t \geq 0}$, with a finite state space $E = \{0, 1, \ldots, L, \ldots, M, \ldots, K; F\}$ describing the *degradation stage* of the system. In other words, $X(t) = i$ means that the the system is in the ith stage at instant t. As we show, the states (stages) L, K, M and F play a specific role. We assume that the process starts in state $X(0) = 0$ and crosses (step-by-step) $K - 1$ intermediate *degradation states*. Upon reaching the state M, the following two events are possible. The process X visits the state K, where preventive repair is performed during random time U_K with a given distribution. Afterwards the process returns to stage L. Otherwise, if during a random time V, the process is still at some intermediate stage $j \in [M, K)$, then an instantaneous failure occurs. As a result, the process X jumps to a *complete failure state* F. Then the system is unworkable during a random *repair time* U_F, with a given distribution. After repair, the process returns to initial state 0.

This description is motivated by the application of the controllable model based on recovery policy defined by two parameters (L, K) [3]. In this paper, in particular, the control principles have been illustrated by means of a number of real examples, such as corrosion of the protective covering, damage process due to the fatigue crack growth, wear of a plane bearing, and so on.

For the further analysis, we need the following notations from [3]. Denote by T_i the transition time from state i to $i + 1, i \in E \setminus \{K, F\}$. We assume $\{T_i\}$ to be independent, and call the corresponding degradation process *homogeneous*, if $\{T_i\}$ are independent, identically distributed (i.i.d.). Namely this case is studying in the present paper. Note that

$$S_{ij} := \sum_{k=i}^{j-1} T_k, \quad 0 \le i \le K - 1, j > i,$$

is the transition time from state j to state i. We define the following distribution functions:

$$F_i(t) = \mathbb{P}(T_i \le t); \quad F_V(t) = \mathbb{P}(V \le t); \tag{1}$$

$$F_{ij}(t) = \mathbb{P}(S_{ij} \le t) = F_{i, j-1} * F_j(t) = \int_{0-}^{t} F_{i, j-1}(t - v) dF_j(v); \tag{2}$$

$$F_{i, i+1}(t) = F_i(t), \quad F_{ii}(t) \equiv 0, \tag{3}$$

where $*$ means convolution.

2.2   Basic Relations for General Case

By construction the process X has a regenerative structure. We will consider the regeneration points $\{\tau_n\}$ as the return times to state M. Denote by Z_k, $k \ge 1$, the transition instants of the process X. Then the regeneration points are recursively defined as

$$\tau_{n+1} = \{Z_i > \tau_n : X(Z_i^+) = M\}, \; n \ge 0, \; \tau_0 := 0.$$

and $Y_k = \tau_{k+1} - \tau_k$, $k \ge 1$, are the i.i.d. lengths of the regeneration cycles. As it is seen in Fig. 1, after the process X hits state M, an instantaneous failure

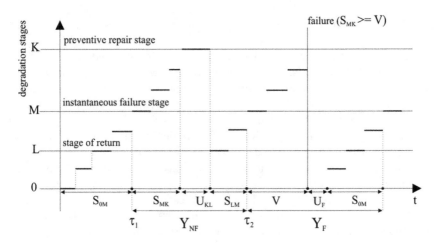

Fig. 1. Realization of the degradation process with two types of cycle

may happen during period V, otherwise, a preventive repair occurs during time S_{MK}. Thus, there are two types of the regeneration cycles of the process X.

Denote by $Y_F = V + U_F + S_{0M}$, the length of the (generic) regeneration cycle with failure, and $Y_{NF} = S_{MK} + U_{KL} + S_{LM}$ the length of cycle ended by the preventive repair. Thus, an unconditional (generic) length of a regenerative cycle Y is

$$Y = Y_F \cdot I_{\{V \leq S_{MK}\}} + Y_{NF} \cdot I_{\{S_{MK} < V\}}, \qquad (4)$$

where I denotes indicator function. Now we calculate the main performance measures of the stationary process X. To find the mean regeneration cycle length $\mathbb{E}[Y]$, consider the probability of a failure during regeneration cycle,

$$p_F = \mathbb{P}(S_{MK} \geq V) = \int_0^\infty (1 - F_{MK}(t)) dF_V(t). \qquad (5)$$

By definition (4), it immediately follows that

$$\begin{aligned} \mathbb{E}[Y] &= \mathbb{E}[V \cdot I_{\{V \leq S_{MK}\}}] + \mathbb{E}[S_{MK} \cdot I_{\{V > S_{MK}\}}] \\ &\quad + \mathbb{E}[(U_F + S_{0M}) \cdot I_{\{V \leq S_{MK}\}}] + \mathbb{E}[(U_{KL} + S_{LM}) \cdot I_{\{V > S_{MK}\}}]. \end{aligned}$$

Further, by the between U_F, S_{0M}, $I_{\{V \leq S_{MK}\}}$, and also by independence between U_{KL}, S_{LM}, $I_{\{V > S_{MK}\}}$, we obtain

$$\begin{aligned} \mathbb{E}[Y] &= \mathbb{E}[\min\{V, S_{MK}\}] + (\mathbb{E}[U_F] + \mathbb{E}[S_{0M}])\mathbb{P}(V \leq S_{MK}) \qquad (6) \\ &\quad + (\mathbb{E}[U_{KL}] + \mathbb{E}[S_{LM}])\mathbb{P}(V > S_{MK}) \\ &= \mathbb{E}[\min\{V, S_{MK}\}] + (\mathbb{E}[U_F] + \mathbb{E}[S_{0M}])p_F \\ &\quad + (\mathbb{E}[U_{KL}] + \mathbb{E}[S_{LM}])(1 - p_F). \end{aligned}$$

Also by (3),

$$\mathbb{E}[\min\{V, S_{MK}\}] = \int_0^\infty (1 - F_V(t))(1 - F_{MK}(t))dt,$$

$$F_{MK}(t) = \int_0^t F_{M,K-1}(t - v)dF_K(v).$$

Denote T_F the length of cycle up to an instantaneous failure, provided it happens, and assume that the density $F_V'(x) = f_V(x)$ exists. The following result is immediate:

$$\mathbb{E}[T_F] = \mathbb{E}[V|V \le S_{MK}] = \int_0^\infty \frac{y\, f_V(y)\, \mathbb{P}(S_{MK} \ge y)}{\mathbb{P}(V \le S_{MK})} \, dy$$

$$= \frac{1}{p_F} \int_0^\infty y f_V(y)(1 - F_{MK}(y))dy. \tag{7}$$

Now we calculate the length of a regeneration cycle with failure. By the independence between U_F, S_{0M}, $I_{\{V \le S_{MK}\}}$, and by (4):

$$\mathbb{E}[Y_F] = \mathbb{E}[Y|V \le S_{MK}] = \mathbb{E}[T_F] + \mathbb{E}[U_F] + \mathbb{E}[S_{0M}]. \tag{8}$$

Assume that the density $F_{S_{MK}}'(x) := f_S(x)$ exists, and find the mean length of the regeneration cycle without failure. Using (3), (4) and the independence between U_{KL}, S_{LM}, $I_{\{V > S_{MK}\}}$, we obtain

$$\mathbb{E}[Y_{NF}] = \mathbb{E}[Y|S_{MK} < V] = \mathbb{E}[(S_{MK} + U_{kl} + S_{LM})|S_{MK} < V]$$

$$= \frac{1}{1 - p_F} \int_0^\infty y f_S(y)\, (1 - F_V(y))\, dy + \mathbb{E}[U_{KL}] + \mathbb{E}[S_{LM}]. \tag{9}$$

These performance measures include convolutions, and can be explicitly calculated in a few special cases only. Moreover, solution normally includes Laplace-Stieltjes transform, and to obtain explicit expression, inversion of the transform is required. However a low accuracy is the weakest point of this numerical approach. Nevertheless the model with i.i.d. exponential degradation stages allows an explicit solution. Below we use it to verify the accuracy and efficiency of the splitting simulation method, which can be applied to more general setting, where analytical solution is unavailable.

2.3   The Exponential Degradation Stages

Recall that we consider the homogeneous case, with i.i.d. exponential $\{T_i\}$. It allows to find $\mathbb{E}[T_F]$, $\mathbb{E}[Y_{NF}]$ analytically, provided V and U_{KL} are exponential as well.

Theorem 1. *For the i.i.d. exponential $\{T_i\}$ with parameter λ and exponential V with parameter ν,*

$$\mathbb{E}[T_F] = \frac{1}{\nu p_F}\left[1 - \frac{\lambda + (K - M + 1)\nu}{\lambda + \nu}\left(\frac{\lambda}{\lambda + \nu}\right)^{K-M}\right], \tag{10}$$

where

$$p_F = 1 - \left(\frac{\lambda}{\lambda + \nu}\right)^{K-M}. \tag{11}$$

Proof. In this case S_{MK} has Erlang distribution (*Erlang* $(\lambda, K - M)$)

$$F_{MK}(x) = 1 - e^{-\lambda y} \sum_{i=0}^{K-M-1} \frac{(\lambda y)^i}{i!}. \tag{12}$$

Then it follows from (7) that

$$\mathbb{E}[T_F] = \frac{\nu}{p_F} \int_0^\infty y e^{-\nu y} e^{-\lambda y} \sum_{i=0}^{K-M-1} \frac{(\lambda y)^i}{i!} dy$$

$$= \frac{\nu}{p_F} \sum_{i=0}^{K-M-1} \frac{\lambda^i}{i!} \int_0^\infty y^{i+1} e^{-(\lambda+\nu)y} dy. \tag{13}$$

Because, see [13],

$$\int_0^\infty y^n e^{-\mu y} dy = n! \mu^{-n-1}, \tag{14}$$

then we obtain from (13)

$$\mathbb{E}[T_F] = \frac{\nu}{p_F(\lambda+\nu)^2} \sum_{i=0}^{K-M-1} \left(\frac{\lambda}{\lambda+\nu}\right)^i (i+1).$$

It is known [13] that,

$$\sum_{i=0}^{n-1} (a + ir)\rho^a = \frac{a - (a + (n-1)r)\rho^n}{1-\rho} + \frac{r\rho(1 - \rho^{n-1})}{(1-\rho)^2}. \tag{15}$$

Taking $n = K - M$, $a = 1$, $r = 1$, we arrive to

$$\mathbb{E}[T_F] = \frac{1}{p_F} \left[\frac{1}{\nu} - \left(\frac{\lambda}{\lambda+\nu}\right)^{K-M} \left[\frac{(K-M)\nu + \lambda + \nu}{\nu(\lambda+\nu)}\right]\right]. \tag{16}$$

Because V is exponential, it follows from (5) and (12) the followinf equality

$$p_F = \nu \int_0^\infty e^{-(\lambda+\nu)y} \sum_{i=0}^{K-M-1} \frac{(\lambda y)^i}{i!} dy = \frac{\lambda}{\lambda+\nu} \sum_{i=0}^{K-M-1} \left(\frac{\lambda}{\lambda+\nu}\right)^i, \tag{17}$$

which is equivalent to (11).

Theorem 2. *For the i.i.d. exponential $\{T_i\}$ with parameter λ and exponential U_{KL} with parameter μ,*

$$\mathbb{E}[Y_{NF}] = \frac{K-M}{(1-p_F)(\lambda+\nu)}\left(\frac{\lambda}{\lambda+\nu}\right)^{K-M} + \frac{1}{\mu} + \frac{M-L}{\lambda}, \qquad (18)$$

where

$$p_F = 1 - \left(\frac{\lambda}{\lambda+\nu}\right)^{K-M}. \qquad (19)$$

Proof. Since S_{LM} is *Erlang* $(\lambda, M-L)$ and U_{KL} is exponential, then

$$\mathbb{E}[U_{KL}] + \mathbb{E}[S_{LM}] = \frac{1}{\mu} + \frac{M-L}{\lambda}. \qquad (20)$$

As above, using (9) and (14), we obtain

$$\int_0^\infty y \frac{f_S(y)(1-F_V(y))}{1-p_F} dy$$
$$= \int_0^\infty y^{K-M} \frac{\lambda^{K-M}}{(K-M-1)!} e^{-(\lambda+\nu)y} dy = \frac{K-M}{\lambda+\nu}\left(\frac{\lambda}{\lambda+\nu}\right)^{K-M}. \qquad (21)$$

By (9), (20) and (21), now (18) follows immediately.

For this setting (and exponential U_F with parameter μ_F), the average unconditional regeneration cycle length is

$$\mathbb{E}[Y] = \frac{1}{\nu}\left(1 - \left(\frac{\lambda}{\lambda+\nu}\right)^{K-M}\right)$$
$$+ \left(\frac{1}{\mu} + \frac{M-L}{\lambda}\right)(1-p_F) + \left(\frac{1}{\mu_F} + \frac{M}{\lambda}\right)p_F, \qquad (22)$$

where p_F satisfies (11). Finally note that, by (10),

$$\mathbb{E}[\min\{V, S_{MK}\}] = \int_0^\infty e^{-(\lambda+\nu)y} \sum_{i=0}^{K-M-1} \frac{(\lambda y)^i}{i!} dy = \frac{p_F}{\nu}. \qquad (23)$$

2.4   Asymptotic Reliability Function

The most important reliability descriptor of operation quality is a reliability function

$$R(t) = \mathbb{P}[T > t | X(0) = 0], \ t \geq 0,$$

where T stands for the life time of the system. In our model, the mean life time can be written as

$$\mathbb{E}[T] = \mathbb{E}[Y_{NF}](\mathbb{E}[N] - 1) + \mathbb{E}[T_F],$$

where $\mathbb{E}[N] := 1/p_F$ is the mean number of cycles until complete failure. This number follows geometric distribution with parameter $1 - p_F$. For exponential (pure Markovian) reliability model, the classical evaluation method of $R(t)$ includes solution of the Kolmogorov differential equations for the state probabilities

$$\pi_i(t) = \mathbb{P}[\hat{X}(t) = i | \hat{X}(0) = 0],$$

of an auxiliary Markov process $\{\hat{X}(t), \, t \geq 0\}$ with an absorption state F. In this case,

$$R(t) = \sum_{x \in E \setminus \{F\}} \pi_i(t).$$

Another way is to evaluate the density function

$$r_x(t) = \frac{1}{dt} \mathbb{P}[T \in [t, t + dt] | \hat{X}(0) = x]$$

of the remaining life time given the initial state is x, see [3], in which case the reliability function becomes,

$$R(t) = 1 - \int_0^t r_0(u) du, \quad t \geq 0.$$

In both methods the results are obtained in terms of Laplace transforms (LT). However, these results in general are not applicable, provided a (natural) assumption that the failure is a rare event. In this case the required inversion of corresponding LT is hardly available, while numerical methods lead to highly incorrect result. Alternatively, it is possible to use the asymptotic of reliability function $R(t)$. This analysis is based on the renewal theory and the limit properties of regenerative processes, see, for instance, [14,15,20,21]. The main result for our model can be formulated as the following limit statement provided failure becomes rarer [21]:

$$R\left(\frac{\mathbb{E}[Y_{NF}]}{p_F} t\right) = \mathbb{P}\left(T \geq \frac{\mathbb{E}[Y_{NF}]}{p_F} t\right) \to e^{-t}, \quad p_F \to 0, \tag{24}$$

uniformly in t. It can be rewritten as

$$R(t) \to e^{-t \frac{p_F}{\mathbb{E}[Y_{NF}]}}, \quad p_F \to 0. \tag{25}$$

In other words, the reliability function is *asymptotically exponential*, and the mean life time becomes (approximately)

$$\mathbb{E}[T] = \int_0^\infty R(t) dt = \frac{\mathbb{E}[Y_{NF}]}{p_F}. \tag{26}$$

This result is intuitive because, as we mention above, the number of cycles up to the (first) failure is geometric with parameter $1 - p_F$.

To determine the reliability function for moderate p_F, neither analytic formulas nor asymptotic relations can be effectively used. To overcome this difficulty, a speed-up simulation (splitting) method is applied below, to effectively estimate the reliability function for arbitrary p_F.

3   Splitting Method for Degradation Process

In this section, we briefly describe the so-called *splitting method* developed for the accelerated estimation of rare events, see [5–7,11,12]. We start the splitting upon the process X reaches stage M, after which an instantaneous failure becomes possible, see Fig. 1.

First, we estimate p_F and cycle parameters $\mathbb{E}[Y]$, $\mathbb{E}[Y_{NF}]$, $\mathbb{E}[Y_F]$ by means of a combination of splitting and regenerative simulation. Then we compare obtained estimates with result of Crude Monte-Carlo simulation, and with analytical results, when available.

Consider a modification of the standard splitting, which starts at the regeneration instants τ_k, when the process reaches stage M, see Fig. 1 The hitting each next degradation stage i corresponds to the ith splitting threshold in the standard splitting, when R_i copies of the random variables (r.v.) T_i is realised, $M \leq i \leq K - 1$. (We take $R_M = 1$). Recall that all new i degradation processes evolve independently. Then each original path generates $D = R_M \cdots R_{K-1}$ (dependent) regeneration cycles called *group of cycles*. The dependence is generated by the same pre-history of realizations of S_{MK} before splitting point at each stage. Thus, we obtain D realizations of S_{MK} for one group of cycles. The cycles belonging to different groups are independent by construction. The total number of groups is denoted by R_{M-1}. The total number of the failures in the ith group is

$$A_i = \sum_{j=(i-1)\cdot D+1}^{i\cdot D} I^{(j)}, i = 1,\ldots,R_{M-1},$$

where indicator $I^{(j)} = 1$ for the cycle with failure ($I^{(j)} = 0$, otherwise), and the groups are i.i.d. The sequence $\{I^{(j)}, j \geq 1\}$ is discrete regenerative with constant cycle length $\beta_i = D$ and regeneration instants $i \cdot D$, $i \in [1, R_{M-1}]$. In particular, the unbiased point estimator $\hat{p}_F$ of p_F is strongly consistent: as $R_{M-1} \to \infty$, w.p.1

$$\hat{p}_F = \frac{\sum_{j=1}^{R_{M-1}} A_j}{R_{M-1} \cdot D} \to \frac{\mathbb{E}\sum_{j=1}^{D} I^{(j)}}{D} = p_F. \tag{27}$$

To construct a confidence interval for the required parameters the regenerative approach is also applicable [8–10]. It follows from [9] that the asymptotic $100(1 - \delta)\%$ confidence interval for p_F is

$$\left[\hat{p}_F \pm \frac{z(\delta)\sqrt{v_n}}{\sqrt{n}}\right], \tag{28}$$

where $z(\delta)$ is defined by $\mathbb{P}[N(0,1) \leq z(\delta)] = 1 - \delta/2$, N is standard normal variable, and

$$v_n = \frac{n^{-1}\sum_{i=1}^{n}[A_i - \hat{p}_F D]^2}{D^2} \tag{29}$$

is a weakly consistent estimator of $\sigma^2 = \mathbb{E}[A_1 - p_F D]^2/D^2$, that is

$$v_n \Rightarrow \sigma^2, \ n \to \infty,$$

in probability.

4 Simulation Results

In general, $\{T_i\}$ not need to be identical. In particular, independent exponential $\{T_i\}$ with different parameters λ_i are considered in [3]. Another distributions are not considered in [3] because in this case analytical results are hardly available, if possible. Conversely, in this paper we obtain numerical results for more general setting based on the slitting simulation method. To illustrate this method, we again consider the simplest exponential setting with $\lambda_i = \lambda$.

Example 1. First we find the main performance measures (cycle characteristics).

(a) Assume the i.i.d. $\{T_i\} \sim exp(\lambda)$, $\nu = 2.5$, $M = 5$, $K = 17$, $L = 1$, $\mu_F = 1.5$, $\mu = 2$, and let n be the number of simulated regeneration cycles. By (11), we have the following exact values of p_F for different values of λ:

$$\lambda = 50 : p_F = 0.443; \quad \lambda = 70 : p_F = 0.343;$$
$$\lambda = 90 : p_F = 0.280; \quad \lambda = 110 : p_F = 0.236. \tag{30}$$

We compare simulation results based on Crude Monte Carlo (MC), and results obtained by regenerative splitting (RS). All tests were executed on a Intel(R) Xeon(R) CPU E5-2630 2.30 GHz processor with 4 GB of RAM, running Linux openSUSE 42.2. We use 50 samples to find (in evident notation) the estimates the variances of $\hat{p}_F$, respectively, Var_{MC}, and Var_{RS}. The results are given in Table 1. (simulation time t_{MC} and t_{RS} is measured in seconds) shows the estimates of p_F.

Note that both methods are well consistent with analytical expression in (11), while for the same number of cycles n, the RS is much less time-consuming than MC.

Note that Var_{RS} is enough large, and $Var_{RS} > Var_{MC}$, perhaps because p_F is not small. To construct $100(1 - \delta)\%$ confidence interval for p_F, we use (28), and consider the case when the instantaneous failure is not very rare. We take $\lambda = 110$, in which case exact value $p_F = 0.236$. Figure 2 represents the 95% confidence intervals and the point RS estimates of p_F.

Table 1. Comparison of MC and RS estimators for exponential T_i

λ	n	$\hat{p}_{F_{MC}}$	$\hat{p}_{F_{RS}}$	Var_{MC}	Var_{RS}	t_{MC}	t_{RS}
50	12597120	0.443	0.445	1.7e-08	6.4e-03	46.7	1.3
70	16796160	0.3436	0.3436	1.2e-08	4.5e-03	62.0	4.1
90	23328000	0.280	0.282	8.5e-08	6.8e-03	86.6	5.5
110	33177600	0.236	0.237	5.5e-08	1.3e-03	123.3	7.8

(b) Table 2 gives the estimates of the mean cycle length $\mathbb{E}[Y]$. For both methods the number of cycles with failure, n_R, is close for given number of cycles n. The number n_R for MC method is given in Table 2. Table 2 shows that both methods provide a good approximation of exact value $\mathbb{E}[Y]$ in (22).

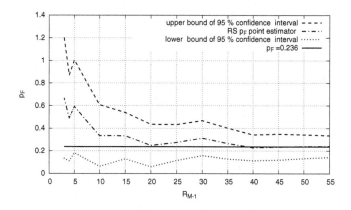

Fig. 2. 95% confidence interval for p_F

Table 2. Comparison of MC and RS estimators for $\mathbb{E}[Y]$

λ	n	$n_R(MC)$	$\hat{\mathbb{E}}[Y]_{MC}$	$\hat{\mathbb{E}}[Y]_{RS}$	$\mathbb{E}[Y]$
50	1259712	557983	0.840	0.847	0.839
70	16796160	5772944	0.756	0.766	0.756
90	23328000	6537331	0.7063	0.7062	0.706
110	33177600	7843054	0.672	0.677	0.672

(c) The estimates of $\mathbb{E}[Y_{NF}]$, the mean length of cycle without failure are given in Table 3. Again, the RS and MC estimates are very close to the exact value given by (18).

Table 3. Comparison of MC and RS estimators for $\mathbb{E}[Y_{NF}]$

λ	n	$\hat{\mathbb{E}}[Y_{NF}]_{MC}$	$\hat{\mathbb{E}}[Y_{NF}]_{RS}$	$\mathbb{E}[Y_{NF}]$
50	1259712	0.808	0.807	0.808
70	16796160	0.722	0.718	0.722
90	23328000	0.674	0.685	0.674
110	33177600	0.643	0.653	0.643

Example 2. To calculate the reliability function $R(t)$ we instead use the asymptotic relation (25)

$$R_a(t) := e^{-D_F\,t},$$

where $D_F = p_F/\mathbb{E}[Y_{NF}]$, assuming that a failure within cycle is to be rare event. We use both methods to obtain the estimate $\hat{D}_F = \hat{p}_F/\hat{\mathbb{E}}[Y_{NF}]$, and use

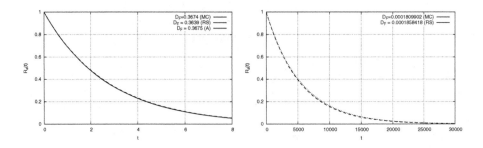

Fig. 3. (a) $R_a(t)$ for $p_F = 0.236$; (b) $R_a(t)$ for $p_F = 0.000118$

approximation (25) in the case of not rare failure, $p_F = 0.236$, $\lambda = 110$. Other parameters are taken as in *Example 1*. In Fig. 3(a) presents plots of function $R_a(t)$ with parameter $\hat{D}_F$ obtained by MC, RS, and D_F with exact values p_F and $\mathbb{E}[Y_{NF}]$. Now we take parameters $\nu = 0.001$, $\lambda = 110$, $\mu = 2$, $\mu_F = 1.5$, $K = 18$, $L = 1$, $M = 5$. Then $p_F = 0.000118$ by (11). In this case we are unable to calculate function $R(t)$ analytically, because of a problem to calculate Laplace transform. However, estimates obtained by MC and RS simulation are available and given by Fig. 3(b). In some cases, the value of function $R(t)$ at a given point t is required. This problem can be solved by means of simulation as well. For instant, it is difficult to calculate the value $R(30000)$ numerically, while both simulation methods gives satisfactory closeness:

$$R_a(30000)[MC] = 0.0043, \quad R_a(30000)[RS] = 0.0037. \tag{31}$$

In both examples, the proximity of estimated obtained by both methods is observed. Moreover, the first example demonstrates a high consistency between the exact value of D_F and the estimates obtained by both simulation methods. It is rather surprising because in this case the probability is not small. It is worth mentioning that, for the i.i.d exponential $\{T_i\}$, the exact value $R(t)$ is available by the numerical inversion of the following Laplace transform expression of $R(t)$:

$$\tilde{R}(s) = \frac{\frac{\nu}{\lambda+\nu+s}\left(\frac{\lambda}{\lambda+s}\right)^M \sum_{i=0}^{K-M-1}\left(\frac{\lambda}{\lambda+\nu+s}\right)^i}{1 - \frac{\mu}{\mu+s}\left(\frac{\lambda}{\lambda+\nu+s}\right)^{K-M}\left(\frac{\lambda}{\lambda+s}\right)^{M-1}}. \tag{32}$$

In particular, it gives $R(30000) = 0.0044425$, which is consistent with (31). For the non-identical $\{T_i\}$ (even exponential) the Laplace transform inversion becomes labour-consuming and highly inaccurate as the number of phases increases.

Example 3. Finally, we consider the i.i.d. $\{T_i\}$ with light-tailed Weibull distribution

$$F_T(x) = 1 - e^{-3x^4}, x \geq 0,$$

implying $\mathbb{E}[T] = 0.6886$. Assume that time $V = 1/\nu$, that is constant. In this case it seems quite hard (if possible) to find the probability p_F analytically, while

Table 4. Simulation results for T_i Weibull

λ	$\hat{p}_{F_{MC}}$	$\hat{p}_{F_{RS}}$	Var_{MC}	Var_{RS}	t_{MC}	t_{RS}
50	0.156	0.153	4.24e-09	8.28e-05	137.7	13.67
70	5.43e-03	5.29e-03	2.38e-10	1.33e-05	69.64	6.53
90	6.43e-05	6.02e-05	1.29e-12	1.58e-10	234.9	17.33
110	—	2.07e-07	—	3.99e-15	>2 h	628.33

MC and RS estimates are available being close. At that, the RS method allows to reduce simulation time significantly. In particular, the estimation of probability of order 10^{-7} by MC simulation is extremely long and often gives no results in an acceptable time (see empty cells in Table 4).

5 Conclusion

In this paper, we apply a speed-up simulation technique, called *multilevel regenerative splitting*, to estimate a small failure probability related to a degradation process. The dynamics of the degradation process is highly adapted to the application of this method. The key idea of the method is to treat the group of the degradation process paths, obtained by the splitting and having the same root, as a regenerative cycle. This approach leads to a modified splitting procedure to preserve properties of the estimates obtained by the standard splitting. This approach allows to compensate a lack or inefficiency of analytical methods discussed in [3]. In particular, simulation can be applied far outside exponential models. As experiments show the RS method gives a significant reduction of simulation time in comparison with the MC simulation. Both methods provide a good approximation of the analytical results, when they are available.

Acknowledgments. The work is supported by the RFBR, projects 15-07-02341, 15-07-02354, 15-07-02360, 16-37-60072, 16-37-60072 mol_a_dk. The work was supported by the Ministry of Education of the Russian Federation (the Agreement number 02.a03.21.0008 of 24 June 2016).

References

1. Borodina, A., Morozov, E.: Accelerated consistent estimation of a high load probability in M/G/1 and GI/G/1 queues. J. Math. Sci. **200**(4), 401–410 (2014)
2. Borodina, A.V.: Ph.D. Thesis. Regenerative modification of the splitting method for estimating the overload probability in queuing systems. Petrozavodsk State University (2008). [in Russian]
3. Efrosinin, D.V., Farhadov, M.P.: Optimal management of the system with the gradual and instantaneous failures. Dependability **28**(1), 27–42 (2009)
4. Esary, J.D., Marshall, A.W., Proshan, F.: Shock models and wear processes. Ann. Probab. **1**(1), 627–649 (1973)

5. Garvels, M.: Ph.D. Thesis. The splitting method in rare event simulation. The University of Twente, The Netherlands, May 2000

6. Glasserman, P., Heidelberger, P., Shahabuddin, P., Zajic, T.: A look at multilevel splitting. In: Niederreiter, H., Hellekalek, P., Larcher, G., Zinterhof, P. (eds.) Monte Carlo and Quasi-Monte Carlo Methods 1996. LNS, vol. 127, pp. 98–108. Springer, New York (1998). doi:10.1007/978-1-4612-1690-2_5

7. Glasserman, P., Heidelberger, P., Shahabuddin, P., Zajic, T.: Splitting for rare event simulation: analysis of simple cases. In: Proceedings of the 1996 Winter Simulation Conference, pp. 302–308 (1996)

8. Glynn, P.W.: Some topics in regenerative steady state simulation. Acta Appl. Math. **34**, 225–236 (1994)

9. Glynn, P.W., Iglehart, D.L.: Conditions for the applicability of the regenerative method. Manage. Sci. **39**, 1108–1111 (1993)

10. Glynn, P.W., Iglehart, D.L.: A joint central limit theorem for the sample mean and regenerative variance estimator. Ann. Oper. Res. **8**, 41–55 (1987)

11. Heegaard, P.E.: A survey of Speedup simulation techniques. In: Workshop tutorial on Rare Event Simulation, Aachen, Germany (1997)

12. Heidelberger, P.: Fast simulation of rare events in queueing and reliability models. In: Donatiello, L., Nelson, R. (eds.) Performance/SIGMETRICS 1993. LNCS, vol. 729, pp. 165–202. Springer, Heidelberg (1993). doi:10.1007/BFb0013853

13. Gradshtein, I., Ryzhik, I.: Tables of Integrals, Sums, Series and Products, 4th edn., 1100 p. Nauka, Moscow (1963). [in Russian]

14. Kalashnikov, V.: Topics on Regenerative Processes. CRC Press, Boca Raton (1994)

15. Keilson, J.: Markov Chain Models - Rarity and Exponentiality. Springer, New York (1979)

16. Kopnov, V.A., Timashev, S.A.: Optimal deatch process control in two-level policies. In: Proceedings of the 4th Vilnius Conference on Probability Theory and Statistics, vol. 4, pp. 308–309 (1985)

17. Lisnuansky, A., Levitin, G.: Multi-state System Reliability. Assessment, Optimization and Application. World Scientific, New Jersey, London, Singapore, Hong-Kong (2003)

18. Murphy, D.N.P., Iskandar, B.P.: A new shock damage model: part II-Optimal maintenance policies. Reliab. Eng. Syst. Saf. **31**, 211–231 (1991)

19. Rykov, V., Dimitrov, B.: On multi-state reliability systems. In: Proceedings of Seminar Applied Stochastic Models and Information Processes, pp. 128–135 (2002)

20. Solovyev, A.: Asymptotic behaviour of the time of the first occurrence of rare event. Eng. Cybern. **9**, 1038–1048 (1971)

21. Kovalenko, I.N.: Analysis of Rare Events to Evaluate Effectiveness and Reliability of Systems Estimation. Soviet Radio, Moscow (1980). [in Russian]

A Token Based Parallel Processing Queueing System with Priority

A. Krishnamoorthy$^{(\boxtimes)}$, V.C. Joshua, and Dhanya Babu

Department of Mathematics, CMS College, Kottayam, India
achyuthacusat@gmail.com, vcjcms@gmail.com, dhanyaabygeorge@gmail.com

Abstract. We consider a single server queueing model with two parallel queues in which one is finite buffer and the other is infinite. Customers arrive according to two independent Poisson processes and service time follows phase type distribution. Customers receive service on the basis of a token system. Customers in the infinite queue are ordinary customers and the customers in the finite queue are priority customers. Customer priority may be either by paying a cost or by any other means. Priority customers have the right to make the strategic decision regarding the queue to which he or she may join. Priority customers are provided service on the basis of token issued to such customers to access the service according to the rule: when $N-1$ customers of lower priority are consecutively served, the next to be served is from the priority line, if there is one waiting, thus ensuring there reduced waiting time. However they can join the lower priority queue in the case they find the waiting time less. This strategy will be discussed in the paper. We perform the steady state analysis and establish the stability condition of the queueing model. Some performance measures are also evaluated. Control problem has been discussed. Some numerical illustrations are provided.

Keywords: Queues · Token · Priority customers · Strategy

1 Introduction

A model is a simplified version of something that is real. If we use mathematical language to make models, they are called mathematical models. Modelling tools play a key role in the study of queueing networks. Mathematical analysis of different variants of queues have been under study by a group of eminent Mathematicians all over the world.

Single server queue with negative arrival is disussed effectively by Artalejo et al. [1]. Chakravarthy et al. analysed two parallel queues with simultaneous services in [3]. But in [12] two M/M/1 queues with customer transfer is studied. Self generation of priorities by waiting customers is a typical situation in clinics/hospitals. Literature of priority in classical queue have been found in [2,15]. In queueing system there are variety of priority schemes mainly preemptive and non-preemptive priority. Differnt priority queues are well ordered in [5]. It was originated from the area of healthcare and is well presented in the paper [4].

© Springer International Publishing AG 2017
V.M. Vishnevskiy et al. (Eds.): DCCN 2017, CCIS 700, pp. 231–239, 2017.
DOI: 10.1007/978-3-319-66836-9_19

We intend to analyze its counterparts in retrial queue set up. They have applications in biology, engineering, telecommunication and several other areas. In this paper we are considered a token based priority i.e. at every fixed number of token a priority customer will be served. Priority is on the basis of paying a large cost or by any other means. Network banking based on token system have been discussed in [9,14]. Token based fair queueing algorithm has been proposed by [13]. A simple queueing model for a distributed protocol with a circulating token is presented in [6]. Neuts in his [11] presented applications of markov chains in queueing theory. Steady state probabilities are computed using Matrix Geometric methods [10]. The rate matrix is computed using Ramaswami's Logarithmic reduction Algorithm by [8]. A control problem due to the finiteness of the buffer is discussed and analysed by Krishnamoorthy et al. in [7].

In this problem there are one infinite buffer and one finite buffer. Motivation behind this model is as follows: a reputed hospital near to our institution approaches us with an offer with an agreement that they offer a limited number of platinum cards so that infront of the long queue in doctors clinic the card holder can enter straightly at every fixed number of turn without waiting a long time. This lead us to model this real life problem.

In a queueing system in which customers are distinguished by class, it is usual to assign priorities according to the perceived importance of the customer. Customers within each class are served in first come, first served order. In this model ordinary customers are served first continuously according to the token number numbered from $1, 2, ..., N-1$ and priority customer at the Nth token if any in the queue. If there is no one in the priority queue at the service of the $(N-1)^{th}$ token, then token returns to 1 and head of the ordinary queue will be served continuously from 1 to $N-1$. Here we consider a $M/PH/1$ queue.

The rest of the article is organized as follows. Section 1.1 provides the mathematical modeling of the problem under study. In Sects. 2 and 3 the steady-state analysis of the model is presented. Some system performance measures are evaluated in Sect. 4.

1.1   Model Description

In this model we consider a single server queueing system with two parallel queues of which one is of infinite buffer and the other is finite of size say 'M'. Customers arrive according to two independent Poisson Processes with arrival rates λ_1 and λ_2 respectively and service time follows phase type distribution with representation (β, S) of order k. The vector S^0 is given by $S^0 = -$ Se.

Customers in the infinite queue are ordinary customers and the customers in the finite queue are priority customers. Customer priority may be either by paying a cost or by any other means. Priority customers also have the right to make the strategic decision regarding the queue to which he or she may join, if it reduces their waiting time. This situation may arise when there are more customers to be served before him/her so he/she will choose the queue which serves him/her faster than the other. Customers are provided service on the basis of token issued. Customers receive service in the following way: A token is circulating from $1, 2, ..., N-1, N$. When N-1 customers of ordinary queue are consecutively served, the next to be served is from the priority queue, if there is one waiting, and thus ensuring there reduced waiting time. However they can join the lower priority queue in the case when they find the waiting time less.

This model is a level independent quasi birth and death process (QBD). Quasi birth death process can be conveniently and efficiently solved by the classical matrix analytic method.

Let $M_1(t)$ be the number of customers at time t in ordinary queue and $M_2(t)$ be the number of customers at time t in priority queue, R(t) be the token numbers from $1, 2, ..., N-1, N$ and S(t) be the service phase of customer in service. The customer in service is counted either in first queue or in second queue depending on the status of the token.

Let $\{M_1(t), M_2(t), R(t), S(t); t \geq 0\}$ be the Markov Process on the state space $\cup l(m_1)$ where $l(m_1) = (0,0) \cup \{(m_1, m_2, r, s), m_1 \geq 1, 0 \leq m_2 \leq M, 1 \leq r \leq N, 1 \leq s \leq k\}$ depending on the strategy.

The strategy be defined as

$$S_1 = \lceil \tfrac{m_1-r}{N} \rceil (N-1) \text{ and}$$

$$S_2 = (m_2 - 1)N + (N - r)$$

If $S_1 \leq S_2$ priority customer decides to join in ordinary queue and if $S_1 > S_2$ priority customer decides to join in priority queue as it offers him/her lesser waiting time.

$$Q = \begin{pmatrix} A_{10} & A_{00} & & & & \\ A_{21} & A_{11} & A_{01} & & & \\ & A_{22} & A_{12} & A_{02} & & \\ & & \ddots & \ddots & \ddots & \\ & & & A_{2\;K} & A_{1\;K} & A_{0\;K} \\ & & & & A_2 & A_1 & A_0 \\ & & & & & \ddots & \ddots & \ddots \end{pmatrix} \tag{1}$$

where $K = M(N-1) - 1$.

The matrices A_{1i}, A_{0i}, A_{2i} are of different orders and A_0, A_1, A_2 square matrices of order $(M+1)N - 1$ are as follows

$$A_0 = \begin{bmatrix} U_0 & & & & \\ & U_1 & & & \\ & & \ddots & & \\ & & & U_1 & \\ & & & & U_M \end{bmatrix}$$

where

$$U_0 = \begin{pmatrix} \lambda_1 I & & \\ & \ddots & \\ & & \lambda_1 I \end{pmatrix}$$

$$U_1 = \begin{pmatrix} \lambda_1 I & & \\ & \ddots & \\ & & \ddots \\ & & & \lambda_1 I \end{pmatrix}$$

$$U_M = \begin{pmatrix} (\lambda_1 + \lambda_2)I & & \\ & \ddots & \\ & & (\lambda_1 + \lambda_2)I \end{pmatrix}$$

$$A_1 = \begin{bmatrix} W_{10} & W_{00} & & & \\ W_{12} & W_1 & W_0 & & \\ & W_2 & W_1 & W_0 & \\ & & \ddots & \ddots & \\ & & & W_2 & W_1 \end{bmatrix}$$

where

$$W_{10} = \begin{pmatrix} S - \lambda I & & \\ & \ddots & S - \lambda I \end{pmatrix}$$

$$W_1 = \begin{pmatrix} S - \lambda I & & \\ & \ddots & S - \lambda I \end{pmatrix}$$

$$W_{12} = \begin{pmatrix} \cdots & \cdots \\ \cdots & \cdots \\ S^0\beta & \end{pmatrix}$$

$$W_2 = \begin{pmatrix} & \\ & \\ S^0\beta & \end{pmatrix}$$

$$W_{00} = \begin{pmatrix} \lambda_2 I & & \\ & \ddots & \\ & & \ddots \\ & & & \lambda_2 I \end{pmatrix}$$

$$W_0 = \begin{pmatrix} \lambda_2 I & & \\ & \ddots & \\ & & \lambda_2 I \end{pmatrix}$$

$$A_2 = \begin{bmatrix} V_0 & & & \\ & V & & \\ & & \ddots & \\ & & & V \end{bmatrix}$$

where

$$V_0 = \begin{pmatrix} & & S^0\beta \\ & \cdot^{\cdot^{\cdot}} & \\ S^0\beta & & \end{pmatrix}$$

$$V = \begin{pmatrix} \cdots & S^0\beta & & \\ & & S^0\beta & \\ & & & \ddots & \\ & & & & S^0\beta \\ \cdots & \cdots & & \cdots & \cdots \end{pmatrix}$$

2 The Steady-State Probability Vector

The matrix $A = A_0 + A_1 + A_2$ can be written as

$$A = \begin{pmatrix} B_1^0 & B_0^0 & O & \cdots & \cdots \\ B_2^1 & B_1 & B_0 & \cdots & \cdots \\ 0 & B_2 & B_1 & B_0 & \cdots \\ 0 & \ddots & \ddots & \ddots & 0 \\ \cdots & \cdots & B_2 & B_1 & B_0 \\ \cdots & \cdots & \cdots & B_2 & B_1^M \end{pmatrix}$$

where the matrices B_1^0, B_0^0, B_2^1 are of different orders and B_0, B_1, B_2, B_1^M are of same orders.

$$B_1^0 = \begin{pmatrix} S - \lambda_2 I & & S^0\beta \\ & \cdot^{\cdot^{\cdot}} & \\ S^0\beta & & S - \lambda_2 I \end{pmatrix}$$

$$B_0^0 = \begin{pmatrix} \lambda_2 I & & & \\ & \lambda_2 I & & \\ & & \ddots & \\ & & & \lambda_2 I \end{pmatrix}$$

$$B_2^1 = \begin{pmatrix} & \\ & \\ & S^0\beta \end{pmatrix}$$

$$B_1 = \begin{pmatrix} S - \lambda_2 I & S^0\beta & \\ & S - \lambda_2 I & S^0\beta \\ & & \ddots & S - \lambda_2 I \end{pmatrix}$$

$$B_0 = \begin{pmatrix} \lambda_2 I & & & \\ & \lambda_2 I & & \\ & & \ddots & \\ & & & \lambda_2 I \end{pmatrix}$$

$$B_2 = \begin{pmatrix} 0 & 0 & \dots & 0 \\ 0 & 0 & \dots & 0 \\ \dots & \dots & \dots & \dots \\ S^0\beta & 0 & 0 & 0 \end{pmatrix}$$

$$B_1^M = \begin{pmatrix} S & S^0\beta & \\ & S & S^0\beta \\ & & \ddots & S \end{pmatrix}$$

We see that A is an irreducible infinitesimal generator matrix and so there exists the stationary vector π of A such that

$$\pi A = 0 \tag{2}$$

and

$$\pi e = 1 \tag{3}$$

where $\pi_0 = (\pi_{00}, \pi_{01}, \dots, \pi_{0N-1})$ and $\pi_i = (\pi_{i1}, \pi_{i2} \dots \dots, \pi_{iM})$ for $1 \le i \le M$.

Solving these equations we get

$$\pi_i = \pi_M \prod_{j=0}^{M-1-i} H_{M-1-j} \tag{4}$$

for $i = 0, 1, \dots \dots K - 1$ where the sequence of matrices H_i are defined as $H_i = -B_2[H_{i-1}B_0 + B_1]^{-1}$ for $i = 1, 2, \dots \dots, M - 1$ and $H_0 = -B_2[B_1^0]^{-1}$.

The vector π_M is obtained from the equation $\pi e = 1$. The stability condition is given by

$$\pi A_0 e < \pi A_2 e$$

$$\pi A_0 e = \pi_0 U_0 e + [\pi_1 + \pi_2 + \dots \pi_{M-1}] U_1 e + \pi_M U_M e$$

$$\pi A_2 e = \pi_0 V_0 e + [\pi_1 + \pi_2 + \dots \pi_M] V_e$$

3   Matrix Analytic Solution

The Quasi-birth-death processes can be conveniently and efficiently solved using the Matrix Analytic Method.

The stationary distribution of the Markov process under consideration is obtained by solving the set of equations $\mathbf{x}Q = 0$, $\mathbf{x}e = 1$.

Let $\mathbf{x}$ be the steady- state probability vector of Q.

Partition this vector as: $\mathbf{x} = (\mathbf{x_0}, \mathbf{x_1}, \mathbf{x_2}, \ldots \ldots)$ where $\mathbf{x}_i = (\mathbf{x}_{i0}, \mathbf{x}_{i1}, \ldots \ldots \mathbf{x}_{iM})$

$$\mathbf{x}_{ij} = (\mathbf{x}_{ij1}, \mathbf{x}_{ij2}, \mathbf{x}_{ij3}, \ldots \mathbf{x}_{ijN})$$

for

$$j = 1, 2, \ldots \ldots, M$$

whereas for

$$r = 1, 2 \ldots N$$

the vectors

$$\mathbf{x}_{ijr} = (\mathbf{x}_{ijr1}, \mathbf{x}_{ijr2} \ldots \ldots \ldots, \mathbf{x}_{ijrk})$$

$\mathbf{x}_{ijrs}$ is the probability of being in state $(ijrs)$ for $r = 1, 2, \ldots N, i \geq 0, j = 1, 2 \ldots \ldots M, s = 1, 2, \ldots \ldots k$ and $\mathbf{x}_{i0}$ is the probability of being in state (i, j, r, s) where

$$\mathbf{x}_i = \mathbf{x}_{i-1} B_i$$

where

$$B_i = -A_{0(i-1)}[A_{1i} + B_{i+1}A_{2(i+1)}]^{-1}, 1 \leq i \leq K$$

are of different dimensions and $\mathbf{x}_{K+1}, \mathbf{x}_{K+2}, \ldots$ are of dimension $k\,[(M+1)N - 1]$
Under the stability condition the steady-state probability vector is obtained as

$$\mathbf{x}_{M(N-1)+i} = \mathbf{x}_{M(N-1)}R^i, i \geq 1$$

where the matrix R is the minimal nonnegative solution to the matrix quadratic equation

$$R^2 A_2 + RA_1 + A_0 = O \tag{5}$$

R can be obtained by successive substitution procedure $R_0 = 0$ and

$$R_{k+1} = -V - R_k^2 W$$

where $V = A_2 A_1^{-1}$, $W = A_0 A_1^{-1}$ by *Logarithmic Reduction Algorithm* developed by Latouche and Ramaswamy. $\mathbf{x}_0$ can be evaluated using $\mathbf{x}e = 1$ and the vectors $\mathbf{x}_0, \ldots \mathbf{x}_{K+1}$ are obtained by solving

$$\mathbf{x}_0 A_{10} + \mathbf{x}_1 A_{21} = 0$$

$$\mathbf{x}_{i-1} A_{0(i-1)} + \mathbf{x}_i A_{1i} + \mathbf{x}(i+1) A_{2(i+1)} = 0,$$
$$1 \leq i \leq K - 1$$

$$\mathbf{x}_{K-1}A_{0K-1} + \mathbf{x}_K A_K + \mathbf{x}_{K+1}A_2 = 0$$
$$\mathbf{x}_K A_{0K} + \mathbf{x}_{K+1}A_1 + \mathbf{x}_{K+2}A_2 = 0$$
$$\mathbf{x}_{i-1}A_0 + \mathbf{x}_i A_1 + \mathbf{x}_{i+1}A_2 = 0, i \geq K + 2$$

subject to the normalising condition

$$\mathbf{x}_0 \left[I + \sum_{j=2}^{M}(N-1) \prod_{i=1}^{j} B_i + \prod_{j=1}^{M(N-1)} B_j(I-R)^{-1} \right] \mathbf{e} = 1.$$

4   Some Performance Measures of the System

Measures of effectiveness are the most purposeful factors of any Queueing system. Some performance measures are evaluated in this paper.

1. Expected Number of customers in the system

$$E[N] = \sum_{i=0}^{\infty} i \sum_{j=0}^{M} j \sum_{r=1}^{N} \sum_{s=1}^{k} \sum_{j=0}^{K} \mathbf{x}_{ijrs}$$

2. Expected Number of ordinary customers in the system

$$E[M_1] = \sum_{i=0}^{\infty} i \mathbf{x}_i \mathbf{e}$$

3. Expected Number of priority customers in the system

$$E[M_2] = \sum_{j=1}^{M} j \sum_{i=0}^{\infty} \sum_{r=1}^{N} \sum_{s=1}^{k} \mathbf{x}_{ijrs}$$

4. Probability that the server is idle

$$t_0 = x_{00}$$

5. Probability that the server is busy with ordinary customer

$$t_1 = \sum_{i=0}^{\infty} \sum_{j=0}^{M} \sum_{r=1}^{N-1} \sum_{s=1}^{k} \mathbf{x}_{ijrs}$$

6. Probability that the server is busy with priority customer

$$t_2 = \sum_{i=0}^{\infty} \sum_{j=0}^{M} \sum_{s=1}^{k} \mathbf{x}_{ijNs}$$

7. The probability that an ordinary/priority customer is blocked from entering the system upon arrival

$$P_b = \sum_{i=0}^{\infty} \sum_{r=1}^{N} \sum_{s=1}^{k} \mathbf{x}_{iMrs}$$

4.1   Conclusion

We are considered a single server parallel queues with priority based on token system and have seen that the this continuous time markov chain have the QBD structure and have found the stationary distribution. The ergodic condition of the Markov chain is derived. The cost analysis of the particular problem also have to study. We can also evaluate the waiting time distribution of the tagged customers in both the queues.

Acknowledgement. The work of the third author is supported by the Maulana Azad National fellowship $[F1 - 17.1/2015 - 16/MANF - 2015 - 17 - KER - 65493]$ of University Grants commission, India.

References

1. Artalejo, J.R., Gomez-Corral, A.: On a single server queue with negative arrivals and request repeated. J. Appl. Probab. **36**, 907–918 (1999)
2. Chakravarthy, S.: A finite capacity dynamic priority queueing model. Comput. Ind. Eng. **22**, 369–385 (1992)
3. Chakravarthy, S., Thiagarajan, S.: Two parallel queues with simultaneous services and Markovian arrivals. J. Appl. Math. Stoch. Anal. **10**, 383–405 (1997)
4. Green, L.: Queueing analysis in healthcare. Patient flow reducing delay in healthcare delivery (2006). Edited by Randolph Hall
5. Jaiswal, N.K.: Priority Queues. Academic press, New York (1968)
6. Keams, P., Peterson, S.: Performance analysis of a token based distributed mutual exclusion protocol. In: Proceedings Southeast Conference (1993)
7. Krishnamoorthy, A., Deepak, T.G., Joshua, V.C.: Queues with postponed work. Top **12**, 375–398 (2004)
8. Latouche, G., Ramaswami, V.: Introduction to Matrix Analytic Methods in Stochastic Modelling. SIAM (1999)
9. Lyan Mark, A., Peha Jon, M.: The priority token bank in a network of queues. In: Proceedings ICC (1997)
10. Neuts, M.F.: Matrix-Geometric Solutions in Stochastic Models - An Algorithmic Approach. The Johns Hopkins University Press, Baltimore (1981)
11. Neuts, M.F.: Markov chains with applications in queueing theory which have a matrix-geometric invariant probability vector. Adv. Appl. Probab. **10**, 185–212 (1978)
12. Neuts, M.F., He, Q.-M.: Two M/M/1 queue with transfer of customers. Queueing Syst. **42**, 377–400 (2002)
13. Patil, R.R.: Token based fair queueing algorithms for wireless networks. IJSEAT **2**, 121–124 (2014)
14. Peha, J.M.: The priority token: integrated scheduling and admission control for an integrated services network. In: ICC(1993)
15. Sharma, V., Virtamo, J.T.: A finite buffer queue with priority. Perform. Eval. **47**, 1–22 (2002)

Optimal Method for Uplink Transfer of Power and the Design of High-Voltage Cable for Tethered High-Altitude Unmanned Telecommunication Platforms

Vladimir Vishnevsky, Boris Tereschenko, Dmitriy Tumchenok, and Artem Shirvanyan$^{(\boxtimes)}$

V. A. Trapeznikov Institute of Control Sciences of Russian Academy of Sciences, 65 Profsoyuznaya Street, 117997, Moscow, Russia
{borist,artshirvanyan}@mail.ru, dtumchenok@gmail.com
http://www.ipu.ru/en

Abstract. The article proposes a method to calculate the design of a high-voltage conducting tether intended for transmission of power from the ground to a tethered high-altitude telecommunication platform. It also describes an approach to the evaluation of wave impedance and selection of an optimal number of lines in the conducting tether to ensure a maximum level of transmitted power. Besides, it provides an example of numerical calculations.

Keywords: Autonomous unmanned aerial vehicles · Tethered high-altitude unmanned telecommunication platforms · Uplink transfer of large volumes of energy · Cable design

1 Introduction

At present, autonomous unmanned aerial vehicles (UAVs) have become widely used. The main drawback of the UAVs is a short time of their operation due to a limited energy resource of batteries or fuel-carrying capacity [1]. Therefore, in recent years we witness persistent efforts aimed at developing tethered UAVs with prolonged operation which are supplied with power from the ground power sources via copper wires.

Uplink transmission of large volumes of energy from industrial sources of 220/380 V 50 Hz while lifting to a height of up to 300 m requires not only copper wires heavy both in terms of weight and gauge but also heavy and bulky on-board voltage converters for powering UAVs electric motors and equipment. That is why the indicated method of transferring energy is inappropriate for tethered high-altitude platforms [2].

V.M. Vishnevskiy et al. (Eds.): DCCN 2017, CCIS 700, pp. 240–247, 2017.
DOI: 10.1007/978-3-319-66836-9_20

2 Calculation of the Optimal Number of Wires in the Conducting Tether, Maximizing the Power Transmitted on Board a High-Altitude Platform

In light of the above, in their papers [3,4] authors described a technology and developed a high-capacity power transfer system via wire lines of a light-gauge conducting tether by means of a high-voltage and high-frequency signal. In this case, the AC transmission system makes it possible to use a multi-phase wire system with a random number of wires and a virtual neutral wire. Systems with 3, 4, 5 and many more wires can be designed.

In a three-phase system an angle between the vectors of voltage and current amounts to 120°, in a 4-phase system - 90°, in a 6-phase system - 60°, etc. With the same load in every phase, the overall current of all phases is equal to zero due to symmetry. It eliminates the need to use a common neutral wire and to transfer the phase power via one wire, accordingly. So, the overall phase power (ground-to-board) is in proportion to the number of phases. On the other hand, the propagation of a signal via a long wire line (conducting tether) is described by the Heaviside differential equation. The solution of this equation shows that the maximum level of the transmitted power is achieved when load impedance R and wave impedance of the long line $\widetilde{R} = \sqrt{\frac{L}{C}}$ are equal, where L and C - line inductance and capacity. Mismatching of these impedances results in the occurrence of current and voltage waves reflected from the end of the line; generation of standing waves with numerous negative consequences (decrease in transmitted power, insulation fault, etc.). Maximum power transmitted via the line is equal to the line wave impedance multiplied by squared maximum permissible current in the line: $W = I^2 \widetilde{R}$.

The increase in the number of phases in the conducting tether results in the growth of wire-to-wire capacity c that leads to the reduction in the line wave impedance. Thereby, while increasing the number of phases (wires) in the conducting tether two opposite trends are observed: increase in the transmitted power with the growth of the number of phases and decrease in the wave impedance and corresponding reduction in power. It means that there can be an optimal number of wires in the conducting tether which will ensure the maximum transfer of power. This article concentrates on the search for an optimal number of wires, maximizing the power transmitted from the ground to the aerial vehicle. Let us consider an n-conductor (n-pole) cable, where the conductors are laid side by side and in a circle (Fig. 1) in such a way that the centers of the conductors form a regular polygon. Let us assume that we know the values of the current-carrying conductor diameter, insulation diameter and inductance L, which is scarcely dependent on the number of conductors in the cable.

Then power transmitted from the ground to the tethered high-altitude platform is defined by the formula

$$W = I_{max}^2 * \widetilde{R} * n, \tag{1}$$

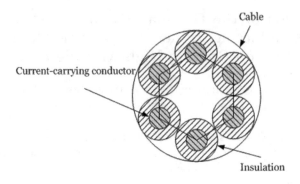

Fig. 1. Section view of 6-conductor cable

where: I_{max} - max. strength of current transmitted via one conductor; $\widetilde{R}$ - conductor wave impedance, being the same for every conductor due to symmetry. Taking into account that $\widetilde{R} = \sqrt{\frac{L}{C}}$ the n-pole cable wave impedance $\widetilde{R}_n$ is written as

$$\widetilde{R}_n = \widetilde{R}_0 \sqrt{\frac{C_0}{C_n}}, \tag{2}$$

where: $\widetilde{R}_0$ and C_0 - two-pole cable wave impedance and capacity, correspondingly. To calculate C_n we should first examine a quadripole (Fig. 2), with capacity formed between every pair of conductors and schematically shown in this figure.

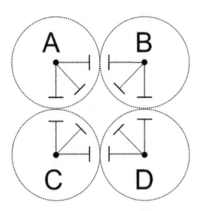

Fig. 2. Quadripole

The total capacity of the conductor (with center A relative to the conductor with center B) can be defined as a sum of capacities of parallel sections, namely, AB, ACB, ADB, ACDB. As a result,

$$C_4 = C_{AB} + C_{ACB} + C_{ADB} + C_{ACDB}.$$

or

$$C_4 = C_{AB} + \cfrac{1}{\cfrac{1}{C_{AC}} + \cfrac{1}{C_{CB}}} + \cfrac{1}{\cfrac{1}{C_{AD}} + \cfrac{1}{C_{DB}}} + \cfrac{1}{\cfrac{1}{C_{AC}} + \cfrac{1}{C_{CD}} + \cfrac{1}{C_{DB}}}. \tag{3}$$

Figure 3 shows a double-wire line the capacity of which is calculated according to the formula:

$$C = \frac{\pi \cdot \varepsilon \cdot \varepsilon_0 \cdot l}{ln\frac{d}{r_{cc}}}, \tag{4}$$

where l - line length, ε and ε_0 - permittivity of external medium and vacuum, d - distance between conductor centers, r_{cc} - current-carrying conductor radius.

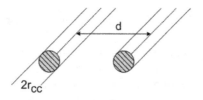

Fig. 3. Double-wire line

By substituting (4) to the formula (3), we will get:

$$C_4 = \pi \cdot \varepsilon \cdot \varepsilon_0 \cdot l \cdot \left(\frac{1}{ln\frac{d_{AB}}{r_{cc}}} + \frac{1}{ln\frac{d_{AC}}{r_{cc}} + ln\frac{d_{CB}}{r_{cc}}} + \frac{1}{ln\frac{d_{AD}}{r_{cc}} + ln\frac{d_{DB}}{r_{cc}}} \right.$$
$$\left. + \frac{1}{ln\frac{d_{AC}}{r_{cc}} + ln\frac{d_{CD}}{r_{cc}} + ln\frac{d_{DB}}{r_{cc}}} \right).$$

For the n-pole cable, correspondingly:

$$C_n = \pi \cdot \varepsilon \cdot \varepsilon_0 \cdot l \cdot \sum_2^n \frac{1}{\sum ln\frac{d_x}{r}}.$$

Taking (2) into consideration, we will get

$$\widetilde{R}_n = \widetilde{R}_0 \cdot \sqrt{\frac{1}{ln(\frac{d_{\text{ins}}}{r_{cc}} \cdot \sum_2^n \frac{1}{\sum ln\frac{d_x}{r_{cc}}})}},$$

where d_{ins} - wire insulation diameter, r_{cc} - current-carrying conductor radius, and d_x - distance between conductor centers. So, let us express the wave impedance of the n-pole cable in terms of the wave impedance of two-pole cable, we get that

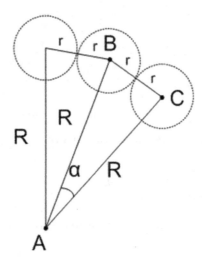

Fig. 4. Three conductors of the n-conductor cable

it will depend only on the distance between the conductor centers, insulation diameter, and current-carrying conductor radius. Therefore, we need to find only the distance between conductor centers.

Let us assume that r is wire insulation radius. Then the distance between the centers of conductors for every pair of adjacent conductors will be equal to $2 \cdot r$. Let us examine Fig. 4, which shows 3 conductors of the n-conductor cable.

Let R be a radius of a circle circumscribed about n-gon, created by the conductor centers, and α - central angle between the centers of adjacent conductors. Then for the ABC triangle, we have:

$$4 \cdot r^2 = 2 \cdot R^2 - 2 \cdot R^2 \cdot cos\alpha \tag{5}$$

Considering that $\alpha = \frac{360°}{n}$ and $sin^2(\frac{\alpha}{2}) = \frac{1-cos(\alpha)}{2}$, then

$$R = \frac{r}{|sin(\frac{180}{n})|}$$

Let us examine Fig. 5. Let x be a minimum number of sides of the polygon created by conductor centers, which we should cross for the two conductor centers under consideration. That is, $x = 1$ for adjacent conductors, $x = 2$ for selected 1 and 3 conductors, etc.

Let us find the distance between such vertices. In the ABC triangle:

$$AB = \sqrt{AC^2 + BC^2 - 2 \cdot AC \cdot BC \cdot cos\alpha}.$$

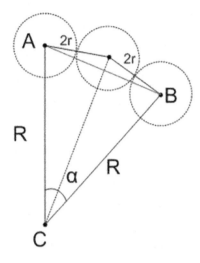

Fig. 5. Distance between the centers of conductors

Considering that $AC = BC = R$ and that $\alpha = \frac{360° \cdot x}{n}$,

$$AB = \sqrt{R^2 + R^2 - 2 \cdot R \cdot R \cdot cos(\frac{360° \cdot x}{n})}.$$

it follows that

$$AB = R \cdot \sqrt{2 - 2 \cdot cos(\frac{360° \cdot x}{n})}.$$

Considering (3) and (4), we obtain

$$AB = \frac{2 \cdot r \cdot |sin(\frac{180 \cdot x}{n})|}{|sin(\frac{180}{n})|}.$$

As $1 \leq x \leq \frac{n}{2}(n, x \in Z)$, then

$$d_x = \frac{2 \cdot r \cdot sin(\frac{180 \cdot x}{n})}{sin(\frac{180}{n})}. \tag{6}$$

Figure 6 shows a schematic representation of the n-pole cable.

To calculate the total capacity of the conductor (with center A relative to the conductor with center B) we should sum up all capacities, generated by 3 conductors (AXB, X - any vertex, except for A and B), 4 conductors (AXYB, XY - any pair of vertices not including A and B) and so on until the n conductors. Thus, a number of various alternatives of k-gons created by the conductor centers

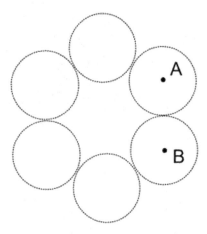

Fig. 6. The n-pole cable

(and accordingly capacities to be taken into account while calculating the total capacity) in the n-pole cable will be defined by the formula:

$$C_{n-2}^{k-2} = \frac{(n-2)!}{(k-2)! \cdot (n-k)!} \qquad (7)$$

The authors of the article wrote a program which can calculate both the conductor wave impedance in the n-conductor cable and overall wave impedance of the cable that characterizes maximum power transmitted with its help. For $R_{\text{imp}\,0}$ we assumed a value of $180\,\Omega$ obtained from experiments. Numerical results are provided in Figs. 7 and 8.

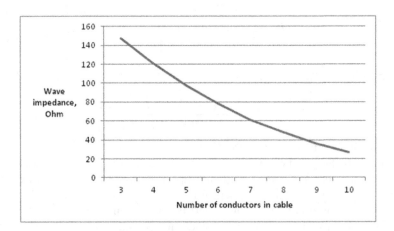

Fig. 7. Wave impedance between a pair of conductors (Ohm) vs. number of conductors

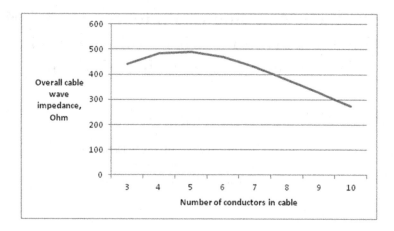

Fig. 8. Overall cable wave impedance (Ohm) vs. number of conductors

3 Conclusions

Therefore, it can be concluded that the maximum power will be achieved with the use of a five-conductor cable.

We would note that the values of the maximum transmitted power for 4-core and 5-core conducting tethers, as seen in Fig. 8, are almost the same (489 and 507, correspondingly). However, bearing in mind that in case of the tether with an even number of conductors, as opposed to the option with an uneven number, we need significantly fewer ground and onboard converters of a lighter weight and smaller dimensions, the optimal solution will be the 4-conductor design.

Acknowledgments. The paper is prepared with financial support from the Russian Science Foundation and DST (India) (grant No. 16-49-02021) for the purposes of joint research and development project of V. A. Trapeznikov Institute of Control Sciences of Russian Academy of Sciences and CMS College Kottayam.

References

1. Foong, K.L.: Autonomous landing of an unmanned aerial vehicle (UAV). Presented at Discover: URECA, NTU poster exhibition and competition, Nanyang Technological University, Singapore, March 2016
2. Widiawan, A.K., Tafazolli, R.: High Altitude Platform Station (HAPS): a review of new infrastructure development for future wireless communications. Wirel. Pers. Commun. **42**, 387–404 (2007). doi:10.1007/s11277-006-9184-9
3. Vishnevsky V.M., Tereschenko B.N.: Research and development of new generation of telecommunication tethered high-altitude platforms. In: T-COMM., vol. 7, pp. 20–24 (2013)
4. Vishnevsky, V.M., Tereschenko, B.N.: Russian Federation Patent - 2572822 "Method of remote power for objects by wire". The patent is registered in the state register of inventions of the Russian Federation, 16 December 2015

Performance Modeling of Finite-Source Retrial Queueing Systems with Collisions and Non-reliable Server Using MOSEL

Tamás Bérczes[1]([⊠]), János Sztrik[1], Ádám Tóth[1], and Anatoly Nazarov[2]

[1] Faculty of Informatics, University of Debrecen, Debrecen, Hungary
berczes.tamas@inf.unideb.hu
[2] Tomsk State University, Tomsk, Russia

Abstract. In this paper we investigate a single-server retrial queueing system with collision of the customer and an unreliable server. If a customer finds the server idle, he enters into service immediately. The service times are independent exponentially distributed random variables. During the service time the source cannot generate a new primary call. Otherwise, if the server is busy, an arriving (primary or repeated) customer involves into collision with customer under service and they both moves into the orbit. The retrial time of requests are exponentially distributed. We assume that the server is unreliable and could be break down. When the server is interrupted, the call being served just before server interruption goes to the orbit. Our interest is to give the main steady-state performance measures of the system computed by the help of the MOSEL tool. Several Figures illustrate the effect of input parameters on the mean response time.

Keywords: Closed queuing system · Finite-source queuing system · Retrial queue · Collision · Unreliable server

1 Introduction

The performance analysis of computing and communicating systems has always been an important subject of computer science. The goal of this analysis is to make predictions about the quantitative behavior of a system under varying conditions, e.g., the expected response time of a server under varying numbers of service requests, the average utilization of a communication channel under varying numbers of communication requests, and so on.

Retrial queueing systems (RQS) are characterized by the feature that arriving customers finding all the servers busy upon arrival are obliged to leave the service area and repeat their requests for service after some random time [1–3]. This feature plays an important role in modeling many problems in telephone switching systems, telecommunication networks, computer networks, call centers, etc. The main difference between retrial queues and classic queues is that

© Springer International Publishing AG 2017
V.M. Vishnevskiy et al. (Eds.): DCCN 2017, CCIS 700, pp. 248–258, 2017.
DOI: 10.1007/978-3-319-66836-9_21

the classic queueing theory does not take the actually existed retrial customers into account. It assumes these retrial customers are either lost due to congestion or delayed in the waiting line (if any).

Since in practice some components of the systems are subject to random breakdowns it is of basic importance to study reliability of retrial queues with server breakdowns and repairs. Finite-source retrial queues with unreliable server have been investigated in several recent papers, for example, [4–10].

Many times when different data is transmitted and there are only a limited number of available free channels may cause a conflict. This may in many cases result in collisions that lead to data loss. Recent results on retrial queues with collisions can be found in, for example [11–15].

The aim of the present paper is to investigate a single-server retrial queueing system with collision of the customer and an unreliable server.

Because of the fact, that the state space of the describing Markov chain is very large, it is rather difficult to calculate the system measures in the traditional way of writing down and solving the underlying steady-state equations. To simplify this procedure we used the software tool MOSEL (Modeling, Specification and Evaluation Language), see [16], to formulate the model and to obtain the performance measures. The organization of the paper is as follows. Section 2 contains the corresponding queueing model with components. In Sect. 3, we present some numerical examples.

2   System Model

Let us consider (Fig. 1) a closed retrial queuing system of type $M/M/1//N$ with collision of the customers and an unreliable server. The number of sources is N and each of them can generate a primary request with rate λ/N. If a customer finds the server idle, he enters into service immediately. The service times are independent exponentially distributed random variables with parameter μ. During the service time the source cannot generate a new primary call. Otherwise,

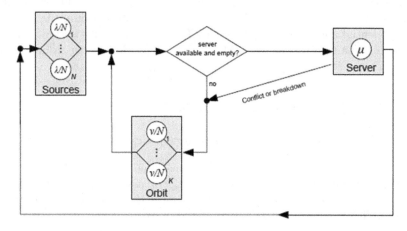

Fig. 1. System

if the server is busy, an arriving (primary or repeated) customer involves into collision with customer under service and they both moves into the orbit. The retrial time of requests are exponentially distribution with rate σ/N. We assume that the server is unreliable and could be break down. The lifetime is supposed to be exponentially distributed with failure rate γ_0 if the server is idle and with rate γ_1 if it is busy. When the server breaks down, it is immediately sent for repair and the recovery time is assumed to be exponentially distributed with rate γ_2. If the server fails in busy state, the interrupted request goes to the orbit.
The server can be in three states:

- *idle state:* If the server is available and it can start serving the arriving requests.
- *busy state:* If the server is available and busy.
- *failed state:* If the server is in failed state, it couldn't start serving any arriving requests until it wouldn't be repaired.

All random variables involved in the model construction are assumed to be independent of each other.

We introduce the following notations (see the summary of the model parameters in Table 1):

- $k(t)$ is the number of active source at time t,
- $o(t)$ is the number of jobs in the orbit at time t.
- $y(t) = 0$ if the server is up and $y(t) = 1$ if the server is failed at time t
- $c(t) = 0$ if the server is idle and $c(t) = 1$ if the server is busy at time t

It is ease to see that:

$$k(t) = \begin{cases} N - o(t), & y(t) = 1 \text{or} c(t) = 0 \\ N - o(t) - 1, & c(t) = 1 \end{cases}.$$

To maintain theoretical manageability, the distributions of inter-event times (i.e., request generation time, service time, retrial time, available state time, sleeping state time, failed state time) presented in the network are by assumption exponential and totally independent. The state of the network at a time t corresponds to a Continuous Time Markov Chain (CTMC) with 3 dimensions:

$$X(t) = (y(t); c(t); o(t)) \tag{1}$$

The steady-state distributions are denoted by

$$P(y, c, o) = \lim_{t \to \infty} P(y(t) = y, c(t) = c, o(t) = o) \tag{2}$$

Note, that the state space of this Continuous Time Markovian Chain is finite, so the steady-state probabilities surely exist. For computing the steady-state probabilities and the system characteristics, we use the MOSEL software tool in this paper. These computations are described in eg. [17, 18].

Table 1. Overview of model parameters

Parameter	Maximum	Value at t
Number of sources	N (population size)	$k(t)$
Generation rate of active sources		λ/N
Total gen. rate	$\lambda_1 N$	$\lambda_1 k(t)$
Service rate		μ
Number of busy servers	1 (number of servers)	$c(t)$
Cust. in service area	N	$c(t) + o(t)$
Requests in Orbit	N (orbit size)	$o(t)$
Retrial rate		σ/N
Server's failure rate (idle case)		γ_0
Server's failure rate (busy case)		γ_1
Server's repair rate		γ_2

When we have calculated the distributions defined above, the most important steady-state system characteristics can be obtained in the following way:

– *Utilization of the server*

$$U_S = \sum_{o=0}^{N} P(0, 1, o) \tag{3}$$

– *Utilization of the repairman*

$$U_r = \sum_{c=0}^{1} \sum_{o=0}^{N} P(1, c, o) \tag{4}$$

– *Availability of the server*

$$A_S = \sum_{c=0}^{1} \sum_{o=0}^{N} P(0, c, o) = 1 - U_r \tag{5}$$

– *Average number of jobs in the orbit*

$$\overline{O} = E(o(t)) = \sum_{y=0}^{1} \sum_{c=0}^{1} \sum_{o=0}^{N} oP(y, c, o) \tag{6}$$

– *Average number of jobs in the server*

$$\overline{C} = E(c(t)) = \sum_{o=0}^{N-1} oP(0, 1, o) \tag{7}$$

– *Average number of jobs in the network*

$$\overline{M} = \overline{O} + \overline{C} = \sum_{y=0}^{1} \sum_{c=0}^{1} \sum_{o=0}^{N} oP(y,c,o)$$
$$+ \sum_{o=0}^{N-1} oP(0,1,o) \tag{8}$$

– *Average number of active sources*

$$\overline{A} = N - M = N - \sum_{y=0}^{1} \sum_{c=0}^{1} \sum_{o=0}^{N} oP(y,c,o)$$
$$- \sum_{o=0}^{N-1} oP(0,1,o) \tag{9}$$

– *Average generation rate of sources:*

$$\overline{\lambda} = \lambda/N\overline{A_1} \tag{10}$$

– *Average waiting time in orbit:*

$$ET_o = \frac{\overline{O}}{\overline{\lambda}} \tag{11}$$

– *Average waiting time in orbit:*

$$ET = \frac{\overline{M}}{\overline{\lambda}} \tag{12}$$

3   Numerical Results

The Table 2 shows the input parameters of the investigated Figures.

Figure 2 shows the steady-state distribution of the three investigated cases. In this figure we can see also the effect of the breakdown of the Server. We can see that the mean number of customers increases as the breakdown intensity are getting larger. From the shape of the curves it is clearly visible that the steady-state distribution of the cases are normally distributed.

Figures 3 and 4 shows the mean response time as a function of the customer generation rate. As we see the mean response time will be greater as we increase the generation rate, but after after λ/N is greater than 1.5 the mean response time starts to decrease.

On Fig. 5 the utilization of the server is displayed as a function of λ. As we see the Utilisation will be greater as we increase the λ/N generation rate. We can understand these property if we take into account that a higher generation rate result more requests in the System.

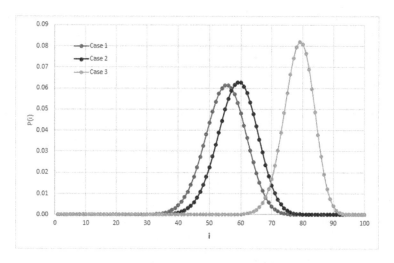

Fig. 2. Steady-state distributions

Table 2. Numerical values of model parameters

Case	N	λ/N	γ_0	γ_1	γ_2	σ/N
Fig. 2 case 1	100	0.01	0.01	0.01	1	0.1
Fig. 2 case 2	100	0.01	0.1	0.1	1	0.1
Fig. 2 case 3	100	0.01	1	1	1	0.1
Fig. 3	100	$0.03 - 8.1$	0.01	0.01	1	0.1
Fig. 4	100	$0.03 - 8.1$	0.1	0.1	1	0.1
Fig. 5	100	$0.03 - 8.1$	0.1	0.1	1	0.1
Fig. 6	100	3	$0.01 - 1.01$	γ_0	1	0.1
Fig. 7	100	3	$0.01 - 1.01$	γ_0	1	0.1
Fig. 8	100	3	0.2	0.2	γ_2	0.1
Fig. 9	100	3	0.2	0.2	γ_0	0.1

On Fig. 6 one can see the effect of the Server's failure rate for the Utilisation. As we see the Utilisation will be smaller as we increase the failure rate.

On Fig. 7 the effect of Server's breakdown with mean response time is displayed. The Mean Response Time is increasing steadily by increasing the failure of the server. On the one hand, this is due to the fact that if the server is breaks down in the busy state, the request is placed in the orbit. On the other hand, the defective server can not accept new requests. Therefore, the increase in the intensity of failure results in an increase in response times.

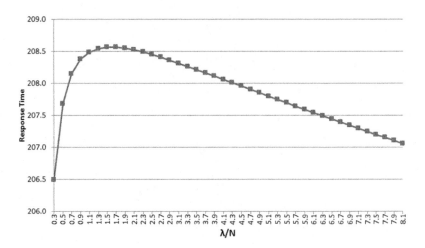

Fig. 3. vs λ/N, $\gamma_0 = \gamma_1 = 0.01$

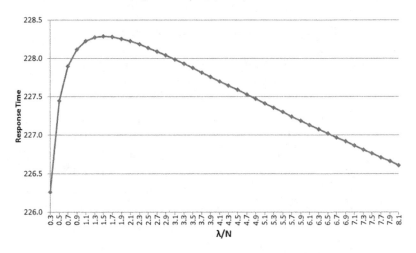

Fig. 4. vs λ/N, $\gamma_0 = \gamma_1 = 0.1$

On Fig. 9 we can see the effect of the Server's repair rate for the Response Time. As we see the Response Time will be smaller as we increase the repair rate. This is because a failed server can not receive or process requests. Thus, the higher repair intensity results in shorter response times.

On Fig. 8 we investigate the effect of the Server's repair rate on Utilisation of the server. As we see the Utilisation will be higher as we increase the repair rate.

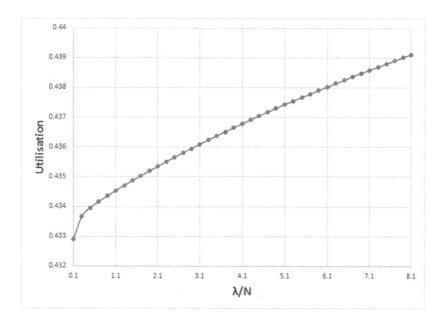

Fig. 5. Utilisation vs λ/N, $\gamma_0 = \gamma_1 = 0.1$, $\gamma_2 == 1$

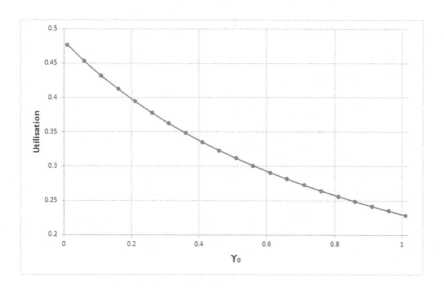

Fig. 6. Utilisation vs γ_0, $\gamma_0 = \gamma_1$, $\lambda/N = 3$

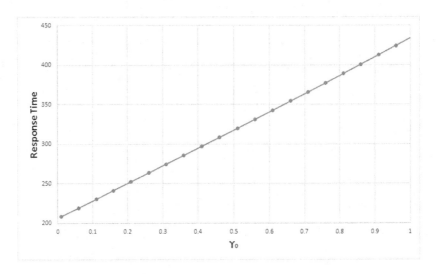

Fig. 7. Mean Response Time vs γ_0, $\gamma_0 = \gamma_1$, $\lambda/N = 3$

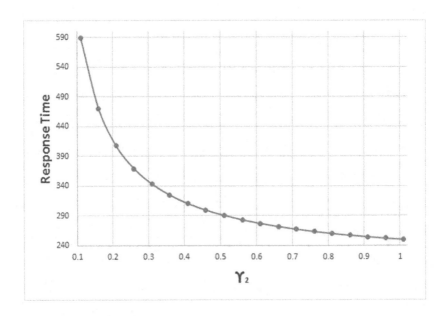

Fig. 8. Response Time vs γ_2, $\gamma_0 = \gamma_1 = 0.2$, $\lambda/N = 3$

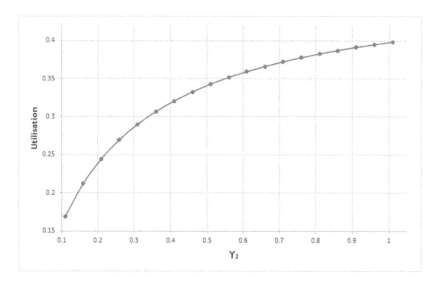

Fig. 9. Utilisation vs γ_2, $\gamma_0 = \gamma_1 = 0.2$, $\lambda/N = 3$

4 Conclusions

In this paper we investigated a single-server retrial queueing system with collision of the customer and an unreliable server. The MOSEL tool was used to formulate and solve the problem, and the main performance and reliability measures were derived and analyzed graphically. We have investigated also the impact of server's failure rate and the server's repair rate on the performance of the system. To our best knowledge, this is the first proposal for the use of single-server retrial queueing system with collision of the customer and an unreliable server.

Acknowledgments. The publication was financially supported by the Ministry of Education and Science of the Russian Federation (the Agreement number 02.a03.21.0008).

The present work of Ádám Tóth and János Sztik was supported by the Austro-Hungarian Cooperation Grant No 96öu8, 2017.

The work of Tamás Bérczes was supported in part by the project EFOP-3.6.2-16-2017-00015 supported by the European Union, co-financed by the European Social Fund.

References

1. Artalejo, J.R., Gómez-Corral, A.: Retrial queueing systems. A computational approach. Springer, Berlin (2008). p. xiii + 318. ISBN 978-3-540-78724-2/hbk;978-3-642-09748-5/pbk; 978-3-540-78725-9/ebook
2. Gómez-Corral, A., Phung-Duc, T.: Retrial queues and related models. Ann. Oper. Res. **247**(1), 1–2 (2016). http://dx.doi.org/10.1007/s10479-016-2305-2

3. Kim, J., Kim, B.: A survey of retrial queueing systems. Ann. Oper. Res. **247**(1), 3–36 (2016). http://dx.doi.org/10.1007/s10479-015-2038-7

4. Almási, B., Roszik, J., Sztrik, J.: Homogeneous finite-source retrial queues with server subject to breakdowns and repairs. Math. Comput. Model. **42**(5–6), 673–682 (2005)

5. Dragieva, V.I.: Number of retrials in a finite source retrial queue with unreliable server. Asia-Pac. J. Oper. Res. **31**(2), 23 (2014)

6. Gharbi, N., Dutheillet, C.: An algorithmic approach for analysis of finite-source retrial systems with unreliable servers. Comput. Math. Appl. **62**(6), 2535–2546 (2011)

7. Roszik, J.: Homogeneous finite-source retrial queues with server and sources subject to breakdowns and repairs. Ann. Univ. Sci. Bp. Sect. Comput. **23**, 213–227 (2004). Rolando Eötvös

8. Wang, J., Zhao, L., Zhang, F.: Performance analysis of the finite source retrial queue with server breakdowns and repairs. In: Proceedings of the 5th International Conference on Queueing Theory and Network Applications, pp. 169–176. ACM (2010)

9. Wang, J., Zhao, L., Zhang, F.: Analysis of the finite source retrial queues with server breakdowns and repairs. J. Ind. Manag. Optim. **7**(3), 655–676 (2011)

10. Zhang, F., Wang, J.: Performance analysis of the retrial queues with finite number of sources and service interruptions. J. Korean Stat. Soc. **42**(1), 117–131 (2013)

11. Ali, A.-A., Wei, S.: Modeling of coupled collision and congestion in finite source wireless access systems. In: Wireless Communications and Networking Conference (WCNC), pp. 1113–1118. IEEE (2015)

12. Balsamo, S., Rossi, G.-L.D., Marin, A.: Modelling retrial-upon-conflict systems with product-form stochastic petri nets. In: Dudin, A., De Turck, K. (eds.) ASMTA 2013. LNCS, vol. 7984, pp. 52–66. Springer, Heidelberg (2013). doi:10.1007/978-3-642-39408-9_5

13. Choi, B.D., Shin, Y.W., Ahn, W.C.: Retrial queues with collision arising from unslotted CSMA/CD protocol. Queueing Syst. **11**(4), 335–356 (1992)

14. Gómez-Corral, A.: On the applicability of the number of collisions in p-persistent CSMA/CD protocols. Comput. Oper. Res. **37**(7), 1199–1211 (2010)

15. Kim, J.-S.: Retrial queueing system with collision and impatience. Commun. Korean Math. Soc. **25**(4), 647–653 (2010)

16. Begain, K., Bolch, G., Herold, H.: Practical Performance Modeling, Application of the MOSEL Language. Kluwer Academic Publisher, Boston (2001)

17. Bolch, G., Greiner, S., de Meer, H., Trivedi, K.: Queueing Networks and Markov Chains, 2nd edn. Wiley, Amsterdam (2006). ISBN 0-471-56525-3

18. Wüchner, P., Sztrik, J., de Meer, H.: Modeling wireless sensor networks using finite-source retrial queues with unreliable orbit. In: Hummel, K.A., Hlavacs, H., Gansterer, W. (eds.) PERFORM 2010. LNCS, vol. 6821, pp. 73–86. Springer, Heidelberg (2011). doi:10.1007/978-3-642-25575-5_7

Estimation of High-Speed Performance of the Transport Protocol with the Mechanism of Forward Error Correction

Pavel Mikheev[✉], Sergey Suschenko, and Roman Tkachyov

National Research Tomsk State University, Lenina str., 36, 634050 Tomsk, Russia
doka.patrick@gmail.com

Abstract. We propose a model of a virtual connection controlled by the transport protocol with a forward error correction mechanism in the group failure mode with restrictions on the values of protocol parameters in the form of a Markov chain with discrete time. The analysis of the impact of protocol parameters window size and the duration of the timeout of waiting confirmation, the probability of distortion of the segments in the individual links of the transmission path data, the duration of the round trip delay, the parameters of the mechanism to restore the distorted segments on the throughput of the transport connection. In the area of protocol parameters, the characteristics of the transmission channel and parameters of the forward error correction mechanism found in the area of superiority of the control procedures of the transport protocol with forward error correction over the classic procedure with decision feedback on the criterion of the throughput of a transport connection. The expediency of applying of the method of forward error correction for transport links with large round-trip delay.

1 Introduction

One of the most important indicators of the quality of subscribers interaction in computer networks is the capacity of transport connections. This operating characteristic is largely determined by the logic of the transport protocol, its parameters—the width of the window and of timeout duration [1], as well as additional mechanisms for increasing the speed by reducing the number of retransmissions of the distorted data [2–4] in channels with a high error rate. Currently, the direct error correction technology [2–4] is widely used as an additional service in transport protocols, along with the method of forward error correction decisive feedback to reduce the amount of retransmitted traffic. The study of the transport connection and analysis of its potential capabilities was performed in [2–10]. But the analytical results were obtained only for a single-link data transmission path [5–7], or with restrictions on protocol parameters [8–10]. Known forward error correction technologies are used by the transport protocol, usually as a service of lower-level network architecture [2–4]. A comprehensive analysis of the advantages and the effectiveness of the methods of forward error

© Springer International Publishing AG 2017
V.M. Vishnevskiy et al. (Eds.): DCCN 2017, CCIS 700, pp. 259–268, 2017.
DOI: 10.1007/978-3-319-66836-9_22

correction were performed only at qualitative level, as well as numerically for some individual cases and did not allow us to highlight areas of possible application of methods in the area of protocol parameters and transport connection parameters. However, the potential of the transport protocol using the methods of forward error correction has not been studied yet. There are no analytical dependencies of overall effect of protocol parameters, characteristics of the data transmission path and parameters of the correction method on the speed of the transport connection. The influence of the relations between the duration of the round-trip delay and the protocol parameters on the throughput of the data transmission path of data controlled by transport protocol was not studied. The paper proposes a mathematical model of the process of data transfer with forward error correction in the phase of information transport in the form of a Markov chain with discrete time, in analytical form we found the stationary distributions of probabilities for different repeat modes [1], with a restriction from below to the protocol parameter of the time-out waiting for end-to-end acknowledgment, analytical relation are obtained for the throughput, on the basis of which the analysis of the of potential possibilities of a transport connection and in the characteristic area of protocol parameters, the data transmission path and the forward error correction mechanism, an area is found in which the application of the methods of forward error correction gives the increase in the performance of the transport connection.

2   Transport Connection Model

Let's consider the process of transferring data between subscribers of transport protocol based on the algorithm with forward error correction decisive feedback. An example of such reliable family protocols is the TCP protocol that dominates modern computer networks [1]. We assume that the interacting subscribers have unlimited data stream for transmission, and the exchange is performed by segments of the transport protocol data of the same length. We believe that the section of hops along the transmission path data have the same speed in both directions, and the length of loop segment in a separate link is t. In general case, the path length from the source to the destination carrying the information flow and the length of the return path through which the confirmation of the received segments are transmitted can be different. We believe that the length of the data transmission path, expressed in number of hops areas in the forward direction is equal to $D_n \geq 1$. The return path, by which the confirmation is sent to the sender about the correct processing of the sequence of blocks of segments, has a length of $D_o \geq 1$. The probabilities of distortion of the segment in the communication channels are given for forward $R_n(d)$, $d = \overline{1, D_n}$ and reverse $R_o(d)$, $d = \overline{1, D_o}$ directions of transmission of each area of hops. Then the reliability of the transmission of segments along the path from the source to the destination and back again will be $F_n = \prod_{d=1}^{D_n}(1 - R_n(d))$ and $F_o = \prod_{d=1}^{D_o}(1 - R_o(d))$ respectively. We believe that the loss of segments due to the absence of buffer memory in the nodes of the path does not occur. We believe that the transmission of data by the sender is realized by blocks containing B segments, of which $1 \leq A \leq B$ —are informative, and $B - A$— additional redundant of the same length. We assume that all segments have checksums to detect errors in each of them. The loss (distortion) to $B - A$ arbitrary segments in the block allows to restore all segments of the block (for example, transfer of duplicates when $A = 1$, $B \geq 2$ the excess segment with the bitwise parity of all information segments using the technology of RAID arrays of the fourth level [11] when $A > 1$, $B = A + 1$ etc.). Flow control is implemented by a sliding window mechanism [1] with a formal parameter of the width of the window $w \geq 1$, expressed in number of blocks. We believe that confirmations about the correctness of the blocks of segments received by the addressee are transmitted in each segment of the counter flow. If it is impossible to restore directly the transmitted block segments (distortion of more $B - A$ segments in the block), the entire block is retransmitted. Then, the process of information transfer in a virtual connection controlled by a transport protocol can be described by a Markov process with discrete time (with the cycle duration t) because the time between receiving acknowledgments has the geometric distribution with parameter F_o. This model is a generalization of the formalizations of the data transfer process proposed in [3–6] in case of a transport connection of random length and the mechanism of direct error correction. The range of possible states of the Markov chain is determined by the duration of the waiting timeout for the confirmation of S, expressed in the number of cycles of duration t. The size of the timeout is related to the length of the path, the width of the window and the size of the block by the inequalities $S \geq wB + 1$, $S \geq D_n + D_o + B - 1$. It is obvious that the sum of the lengths of forward and reverse paths can be interpreted as a circular delay of a single segment $D = D_n + D_o$ in the deterministic transport connection is expressed in the duration t. The circular delay for the block of segments will be $D + B - 1$. The States of Markov chain $i = \overline{0, wB}$ corresponds to a queue size of transmitted but not confirmed segments in the source stream, and states $i = \overline{wB + 1, S - 1}$—the time during which the sender is not active and waits to receive confirmation about the correctness of receiving the transmitted sequence of w blocks of segments. From zero state into $(D + B - 2)$-th sender is moving with every cycle t with a probability of determine event. In states $i \geq D + B - 2$ after the expiration of next discrete cycle t, the sender starts receiving confirmation and, depending on the results of the pre-rate segment blocks, taking into account the technology of forward error correction, the sender transmits new segment blocks (with positive confirmation) or repeatedly—distorted (not responding to direct recovery). Completion of cycle of being in the state $D + B - 2$ corresponds to the time of bringing the first block of segments to the destination and receiving confirmation on it. Further growth of the state number occurs with the probability of distortion of confirmation $1 - F_o$ in the reverse path. In states $i \geq D + B - 2$ in the selective failure mode of confirmation gives rise to $(D - 1)$-th state when $w \geq K + 2$, $K = \lfloor \frac{D-2}{B} \rfloor$, where $\lfloor \cdots \rfloor$—means "integer part" of the fraction, or to the state kB, $k = \overline{1, w-1}$ where $w \leq K+1$, only if the blocks delivered to the destination are successfully transmitted, otherwise the transition to the 0-th state follows. Due to the fact that in states $i \geq wB$ the source suspends the sending of segment blocks, obtaining confirmation when $w \geq K+2$ in the states $i = \overline{(w+k)B-1, (w+k+1)B-2}$, $k = \overline{1, K}$ leads to transition into state $D-kB-1$, $k = \overline{1, K}$ in the selective repeat mode —always, and in the group failure mode— only with successful delivery of data (otherwise—in the 0-th state). When $w \leq K+1$ from states $i = \overline{D+(w-k)B-2, D+(w-k+1)B-3}$, $k = \overline{1, w-1}$ transitions to the states occurs kB, $k = \overline{1, w-1}$ in the selective repeat mode —without conditions, and in the group repeat mode— under the condition of successful delivery of the data blocks (otherwise—to the 0-th state). In states $i = \overline{(w+K+1)B-1, S-2}$ for arbitrary width of the window and any repeat mode, the transition to the zero state is performed, since the size of the queue of transmitted, but not confirmed information segments is zeroed. In the state $S-1$ the waiting timeout of confirmation from the receiver about the correctness of the received blocks of segments expires and the unconditional transition to the zero state occurs.

3   State Probabilities for a Transport Protocol with a Forward Error Correction Mechanism

The transition probabilities π_{ij} from the initial state i to the resultant j Markov chain describing the process of transmission of information flow with forward error correction technology in the group failure mode presented in Table 1.

Table 1. The transition probabilities

π_{ij}	i	j	k	Limits
1	$\overline{0, D+B-3}$	$i+1$		
$1-F_o$	$\overline{D+B-2, S-2}$	$i+1$		
$F_o\Phi^k$	$\overline{D+Bk-2, D+B(k+1)-3}$	$D-1$	$\overline{1, G}$	$w \geq K+2$
$F_o(1-\Phi^k)$	$\overline{D+Bk-2, D+B(k+1)-3}$	0	$\overline{1, G}$	$w \geq K+2$
$F_o\Phi^G$	$\overline{D+B(G+1)-2, B(w+1)-2}$	$D-1$		$w \geq K+2$
$F_o(1-\Phi^G)$	$\overline{D+B(G+1)-2, B(w+1)-2}$	0		$w \geq K+2$
$F_o\Phi^{G+k}$	$\overline{B(w+k)-1, B(w+k+1)-2}$	$D-Bk-1$	$\overline{1, K}$	$w \geq K+2$
$F_o(1-\Phi^{G+k})$	$\overline{B(w+k)-1, B(w+k+1)-2}$	0	$\overline{1, K}$	$w \geq K+2$
F_o	$\overline{B(w+K+1)-1, S-2}$	0		$w \geq K+2$
$F_o\Phi^k$	$\overline{D+Bk-2, D+B(k+1)-3}$	$B(w-k)$	$\overline{1, w-1}$	$w \leq K+1$
$F_o(1-\Phi^k)$	$\overline{D+Bk-2, D+B(k+1)-3}$	0	$\overline{1, w-1}$	$w \leq K+1$
F_o	$\overline{D+Bw-2, S-2}$	0		$w \leq K+1$
1	$S-1$	0		

Here $G = \left\lfloor \frac{B(w+1)-2-(D+B-2)+1}{B} \right\rfloor = w - \left\lfloor \frac{D-1}{B} \right\rfloor$ the distance between the beginning of the termination of sender's activity $B(w+1)-2$ (complete transmission of w blocks) and the beginning of the receipt's acknowledgement $D+B-2$, expressed in the dimensions B when $w \geq K + 2$, $K = \left\lfloor \frac{D-2}{B} \right\rfloor$. Out of the form of the transition probabilities, it is not difficult to see that when $F_n = 1$ we get the transition probabilities for the selective repeat mode.

The variety of the form of the solution of the system of equilibrium equations for the probabilities of states of Markov chain is determined by the relations between the protocol parameters w, S, the block size of the segments B and the total length of the tract D. Since the duration of the timeout must exceed the value of wB and be no shorter than the round trip delay of block segments $(S \geq D + B - 1)$, we can identify different solutions for various areas of changes of protocol parameters. For protocol parameters associated with the total path length by inequalities in the form of $w \geq K + 2$, $S \geq D + (w + 1)B - 2$, the system of local equilibrium equations is written as follows:

$$P_0 = P_{S-1} + F_o \left[\sum_{k=1}^{w-1}(1-\Phi^k) \sum_{i=D+Bk-2}^{D+B(k+1)-3} P_i + (1-\Phi^w) \sum_{i=D+Bw-2}^{B(w+K+1)-2} P_i + \sum_{i=B(w+K+1)-1}^{S-2} P_i \right];$$

$$P_i = P_{i-1}, \; i = \overline{1, D-BK-2}, \overline{D-B(K-k), D-B(K-k-1)-2}, \; k=\overline{0,K};$$

$$P_{D-Bk-1} = P_{D-Bk-2} + F_o \left[\sum_{i=B(w+k)-1}^{D+B(G+k+1)-3} P_i \Phi^{G+k} + \sum_{i=D+B(G+k+1)-2}^{B(w+k+1)-2} P_i \Phi^{G+k+1} \right], \; k=\overline{1,K};$$

$$P_{D-1} = P_{D-2} + F_o \left[\sum_{k=1}^{G} \sum_{i=D+Bk-2}^{D+B(k+1)-3} P_i \Phi^k + \sum_{i=D+B(G+1)-2}^{B(w+1)-2} P_i \Phi^{G+1} \right];$$

$$G = \left| \frac{B(w+1) - 2 - (D + B - 2) + 1}{B} \right| = w - \left\lfloor \frac{D-1}{B} \right\rfloor;$$

$$P_i = P_{i-1}(1 - F_o), \quad i = \overline{D + B - 1, S - 1}.$$

Taking into account the condition of normalization, the solution of this system is determined by the relations:

$$P_i = P_0, \; i = \overline{0, D - BK - 2};$$

$$P_i = \frac{P_0}{Z} \left\{ (1-\Phi)\left[1 - \left(\Phi(1-F_o)^B\right)^{G+k}\right] + \left(1 - \Phi(1-F_o)^B\right)\Phi^{G+k} \right.$$
$$\left. \times (1-F_o)^{B(w+k-1)-D-1} \right\}, \; i = \overline{D - Bk - 1, D - B(K-1) - 2}, \; k=\overline{1,K};$$

$$P_i = P_0\frac{1 - \Phi(1 - F_o)^B}{Z}, \; i = \overline{D - 1, D + B - 2};$$

$$P_i = P_0\frac{1 - \Phi(1 - F_o)^B}{Z}(1 - F_o)^{i-D-B+2}, \quad i = \overline{D + B - 1, S - 1};$$

$$P_0 = ZF_o\left(1 - \Phi(1 - F_o)^B\right) \Bigg/ \Bigg\{ ZF_o(D - BK - 1)\left(1 - \Phi(1 - F_o)^B\right)$$

$$+ \left(1 - \Phi(1 - F_o)^B\right)^2 \left[1 + F_o(B - 1) - (1 - F_o)^{S-D-B+2}\right]$$

$$+ BF_o\left[K(1 - \Phi)\left(1 - \Phi(1 - F_o)^B\right) + \Phi^{G+1}\left(1 - \Phi^K(1 - F_o)^{BK}\right)\right.$$

$$\times \left.\left[(1 - F_o)^{Bw-D+1}\left(1 - \Phi(1 - F_o)^B\right)\right] - (1 - \Phi)(1 - F_o)^{G+1}\right]\Bigg\};$$

$$Z = \left(1 - \Phi(1 - F_o)^B\right)\Phi^{G+K+1}(1 - F_o)^{B(w+K)-D+1}$$

$$+ (1 - \Phi)\left(1 - (\Phi(1 - F_o)^B)^{K+G+1}\right).$$

Hence, for the parameter values of the mechanism forward error correction $A = B = 1$ we obtain known result [8], and for absolutely reliable reverse data transmission path ($F_o = 1$) the probability distribution of states of the Markov chain is determined by two evenly distributed areas of values:

$$P_i = \frac{1 - \Phi}{B + (D - 1)(1 - \Phi)}, \quad i = \overline{0, D - 2};$$

$$P_i = \frac{P_0}{1 - \Phi}, \quad i = \overline{D - 1, D + B - 2};$$

$$P_i = 0, \quad i = \overline{D + B - 1, S - 1}.$$

With restrictions on the protocol parameters of $1 \le w \le K + 1$, $S \ge D + (w + 1)B - 2$ the equilibrium equations are rewritten as:

$$P_0 = P_{S-1} + F_o\left[\sum_{k=1}^{w-1}\sum_{i=D+Bk-2}^{D+B(k+1)-3} P_i(1 - \Phi^k) + \sum_{i=D+Bw-2}^{S-2} P_i\right];$$

$$P_i = P_{i-1}, \quad i = \overline{Bk + 1, B(k + 1) - 1}, \overline{B(w - 1) + 1, D + B - 2}, \quad k = \overline{0, w - 2};$$

$$P_{Bk} = P_{Bk-1} + F_o \sum_{i=D+B(w-k)-2}^{D+B(w-k+1)-3} P_i\Phi^{w-k}, \quad k = \overline{1, w - 1};$$

$$P_i = P_{i-1}(1 - F_o), \quad i = \overline{D + B - 1, S - 1}.$$

The state probabilities in this case have a subset of ($i = \overline{(w - 1)B, D + B - 2}$) values, and invariant to the number of state:

$$P_i = P_0, \quad i = \overline{1, B - 1};$$

$$P_i = P_0 \frac{1 - \Phi + \Phi\left(1 - (1 - F_o)^B\right)\left[\Phi(1 - F_o)^B\right]^{w-1-k}}{1 - \Phi + \Phi\left(1 - (1 - F_o)^B\right)\left[\Phi(1 - F_o)^B\right]^{w-1}}, \quad i = \overline{Bk, B(k + 1) - 1},$$

$$k = \overline{1, w - 1};$$

$$P_i = P_0 \frac{1 - \Phi(1 - F_o)^B}{1 - \Phi + \Phi(1 - (1 - F_o)^B)\left[\Phi(1 - F_o)^B\right]^{w-1}}, \; i = \overline{Bw, D + B - 2};$$

$$P_i = P_0 \frac{1 - \Phi + \Phi(1 - \Phi(1 - F_o)^B)(1 - F_o)^{i-D-B+2}}{1 - \Phi + \Phi(1 - (1 - F_o)^B)\left[\Phi(1 - F_o)^B\right]^{w-1}}, \; i = \overline{D + B - 1, S - 1};$$

$$P_0 = F_o\left(1 - \Phi(1 - F_o)^B\right)\left(1 - \Phi + \Phi(1 - (1 - F_o)^B)\left[\Phi(1 - F_o)^B\right]^{w-1}\right) \Big/ \Big\{ BF_o(w-1)$$

$$\times (1 - \Phi)\left(1 - \Phi(1 - F_o)^B\right) + BF_o\Phi^2(1 - F_o)^B\left(1 - (1 - F_o)^B\right)$$

$$\times \left[1 - (\Phi(1 - F_o)^B)^{w-1}\right] + \left(1 - \Phi(1 - F_o)^B\right)^2\left[1 + F_o(D - B(w - 2) - 2)\right.$$

$$\left. - (1 - F_o)^{S-D-B+2}\right]\Big\}.$$

With an absolutely reliable reverse data path transmission ($F_o = 1$), the distribution of probabilities of states of the Markov chain has two equally probable intervals:

$$P_i = \frac{1 - \Phi}{D - B(w - 2) - 1 + B(w - 1)(1 - \Phi)}, \quad i = \overline{0, B(w - 1) - 1};$$

$$P_i = \frac{P_0}{1 - \Phi}, \quad i = \overline{B(w - 1), D + B - 2};$$

$$P_i = 0, \quad i = \overline{D + B - 1, S - 1}.$$

For $w = 1$ we obtain a distribution that is invariant to the level of errors in the forward path

$$P_i = \frac{F_o}{1 - (1 - F_o)^{S-D-B+2} + F_o(D + B - 2)}, \quad i = \overline{0, D + B - 2};$$

$$P_i = P_0(1 - F_o)^{S-D-B+2}, \quad i = \overline{D + B - 1, S - 1}.$$

4   Analysis of the Speed Performance of a Transport Connection

The most important operating characteristics of a transport connection is its throughput determined by parameters of the transport protocol, the data transmission path and the forward error correction mechanism, and overhead as well as the peculiarities of procedure of transmission control [1]. The normalized speed of the transport connection is determined by the average number of undistorted information segments delivered to the recipient (taking into account the repeat mode and the mechanism of forward recovery of distorted segments) for the average time between two consecutive receipts of confirmation onto the transmitted segment blocks [8]. Since the sender receives confirmation about successfully delivered to the recipient block of segments and confirmation to the sent blocks are transferred to each segment of the counter flow, then to the next block the sender receives B acknowledgement during the transmission of the next block Bt, the probability of error-free delivery of which is $1 - (1 - F_o)^B$. Therefore, the time between receiving acknowledgements to the sent blocks from the information and redundant segments is distributed geometrically with the duration of the period B of clock cycles of duration t, and the distribution parameter equal to the reliability of receiving acknowledgements for the B attempts: $1 - (1 - F_o)^y B$.

Then the average time between the arrival of consecutive acknowledgements per block in the cycles of t will be $\bar{T} = B/(1 - (1 - F_o)^B)$. The capacity of transport connections for the group failure mode taking into account the retransmission of all blocks of segments, starting with the first not received [8], is given by the following dependence of the:

$$Z(D, w, S, A, B) = A\Phi\frac{1 - (1 - F_o)^B}{B}\left\{\sum_{k=1}^{w}\frac{1 - \Phi^k}{1 - \Phi}\sum_{i=D+kB-2}^{D+(k+1)B-3}P_i\right.$$

$$\left. + \frac{1 - \Phi^w}{1 - \Phi}\sum_{i=D+(w+1)B-2}^{S-1}P_i\right\}.$$

In the case of absolutely reliable reverse transmission data path $(F_o = 1)$ throughput when $w \geq K + 2$ is invariant to the window size

$$Z(D, w, S, A, B) = \frac{A\Phi}{B[(D-1)(1-\Phi) + B]},$$

and when $w \leq K + 1$ is variable from

$$Z(D, w, S, A, B) = \frac{A\Phi}{B[D + B - 1 - B\Phi(w - 1)]}.$$

Unlimited window size growth $(w \to \infty)$ leads to the relation

$$Z(D, \infty, \infty, A, B) = \frac{A\Phi(1 - (1 - F_o)^B)}{B[F_o(D-1)(1-\Phi) + (1 - \Phi(1 - F_o)^B)(1 + F_o(B-1))]}.$$

5   The Preferred Area of Application of Forward Error Correction

The effective application of the technology of forward error correction involves searching for the length parameters of the block sequence of the segments B and the number of redundant segments $B - A$ in the error correction block that ensure the maximum speed of the transport connection with the specified characteristics and protocol parameters. The presence of redundant segments in the transmitted sequence increases the probability of delivery of the information segments to the recipient in the group; however, this is achieved by increasing the overhead in the form of time for transferring additional data. In this connection, the problem arises of searching in the multidimensional feature space ranges of values of characteristics of transport connection (D, F_o, F_n), parameters of transport protocol (w, S) and the forward error correction mechanism (A, B), which ensures the superiority of control procedure with forward error correction to the classical protocol procedure with decisive feedback without using error correction. Comparison of control procedures is fulfilled in conditions of equal intensities of subscriber information flows offered for transmission $\lambda = Aw$. We will determine the normalized increase in performance speed from the use of forward error correction mechanism in comparison with the classical protocol procedure with decisive feedback in the form of: $\Delta(D, w, S, A, B) = Z(D, w, S, A, B) - Z(D, Aw, S, 1, 1)$. In general case, the comparative analysis can be carried out only numerically. The most significant benefit is determined by the relation between the width of the window, block size B and duration of round-trip transport connection D. For each set of values parameters of forward error correction mechanisms A and B there is a lower bound of the duration of the round trip delay of a single segment D, beyond which there are ranges of window size values and the reliability of segment transmission in the forward and reverse paths of the transport connection, providing a positive benefit. In a number of cases, by reducing the dimension of the feature space, the area of positive values of benefit can be found in a simple analytical form. Assuming that absolutely reliable reverse data path transfer $F_o = 1$ when $w \geq K + 2$ the benefit takes the simple analytical form, which is invariant to the window size $\Delta(D, w, S, A, B) = \frac{A\Phi}{B[B + (D-1)(1-\Phi)]} - \frac{F_n}{1 + (D-1)(1-F_n)}$. Hence, when $A = 1$, $B = 2$ it is easy to see that the area of positive values of benefit for parameter $F_n \in [0, 1]$ exists on $D \geq 11$ the interval

$$F_n \in \left(\frac{D - 2 - \sqrt{D^2 - 12(D-1)}}{2(D-1)}, \frac{D - 2 + \sqrt{D^2 - 12(D-1)}}{2(D-1)} \right).$$

The area of positive benefit values with unlimited growth of round trip delay of a single segment $(D \to \infty)$ is expanded to $F_n \in (0, 1)$. Numerical studies show that the value of benefit is growing.

6   Conclusions

The paper proposes a model for the process of transferring data segments in a transport connection controlled by a reliable transport protocol with a forward error correction mechanism and confirmation of data received by the recipient in the group repeat mode. The mathematical model is based on the description of the queue of transmitted, but not confirmed, data segments by Markov chain with a finite number of states and discrete time. Stationary distributions of Markov chain states are obtained for different areas of window size change and duration of time-out. Analytical expressions for the transport connection capacity are found. In general, the throughput is largely determined by the ratio between the window width and the duration of the round-trip delay. It is shown that for group repeat mode the implementation of a forward error correction mechanism is suitable for transport connections with large round-trip delay.

References

1. Fall, K., Stevens, R.: TCP/IP Illustrated: The Protocols, vol. 1, 2nd edn., 1017 p. Addison-Wesley Professional Computing Series, Redwood City (2012)
2. Lundqvist, H., Karlsson, G.: TCP with end-to-end FEC. In: 2004 International Zurich Seminar on Communications, pp. 152–156 (2004)
3. Barakat, C., Altman, E.: Bandwidth tradeoff between TCP and link-level FEC. Comput. Netw. **39**, 133–150 (2002)
4. Shalin, R., Kesavaraja, D.: Multimedia data transmission through TCP/IP using Hash Based FEC with AUTO-XOR scheme. ICTACT J. Commun. Technol. **03**(03), 604–609 (2012)
5. Boguslavskij, L.B., Gelenbe, E.: Analytical models of data link control procedures in packet-switching computer networks. Autom. Remote Control **41**(7), 1033–1042 (1980)
6. Gelenbe, E., Labetoulle, J., Pujolle, G.: Performance evaluation of the HDLC protocol. Comput. Netw. **2**(4/5), 409–415 (1978)
7. Kokshenev, V.V., Suschenko, S.P.: Analysis of the asynchronous performance management procedures link transmission data. Comput. Technol. **15**(Special issue 5), 61–65 (2008)
8. Kokshenev, V.V., Mikheev, P.A., Suschenko, S.P.: Comparative analysis of the performance of selective and group repeat transmission models in a transport protocol. Autom. Remote Control **78**(2), 247–261 (2017)
9. Kokshenev, V.V., Mikheev, P.A., Suschenko, S.P.: Transport protocol selective acknowledgements analysis in loaded transmission data path. Vestnik TSU. Series Control Comput. Facil. Comput. Sci. **3**(24), 78–94 (2013)
10. Kokshenev, V.V., Suschenko, S.P.: Modeling sessions with Markov's chains. In: Theory of Probability, Random Processes, Mathematical Statistics and Applications: Proceedings of the International Scientific Conference Dedicated to the 80th Anniversary of Professor G.A. Medvedev. Minsk on 23–26 February 2015, Minsk, RIVS, pp. 311–316 (2015)
11. Tannenbaum, A.: Modern Operating Systems. SPB, Piter (2002)

Optimal Methods of Storage, Transfer and Processing of DICOM Data in Medical Information Systems

Alexandr Golubev$^{(\boxtimes)}$, Peter Bogatencov, and Grigore Secrieru

Research and Educational Networking Association of Moldova,
Str. Academiei 5, office 324, 2028 Chisinau, Republic of Moldova
{galex,bogatencov,secrieru}@renam.md
http://renam.md/

Abstract. This article describes the problems and possible solutions for optimization of medical images storing, providing stable and secure access, based on the distributed warehouse for huge volumes of data with different levels of access providing. The standard for working with medical images is the DICOM format, which allows storing images in good quality. The main problem of data storing in DICOM format is caused by the fact that one investigation can has more than one-gigabyte volume of data and consist of thousands of images. In Moldova in 2015 was launched "DICOM Network" project, whose goal is to provide access to investigations for medical staff with the appropriate access rights, as well as to patients to their personal radiography investigations. Now the system collects and processing more than 500 gigabytes of data every month. In the article, we describe and analyze possible solutions for optimizing data storing and processing workflows.

Keywords: DICOM · Distributed system · HPC · Processing algorithms · Radiology · e-Health · Medical emergency · VI-SEEM · RENAM services

1 Introduction

Modern medical information systems integrate various types of medical equipment. This article describes the actual problems and solutions for optimizing the processing and storage of medical radiography investigations. The standard for working with medical images is the DICOM format, which allows storing studies in good quality with the patient's personal data included. The main problem in storing data in DICOM format [1] are caused due to the fact that one study can take more than 1 gigabyte and consist of thousands of images.

The problem of storing medical investigations archive on national level can be considered as Big Data issue. Solution for this issue should take into account the different data access levels. On the one hand, a medical investigation contains personal patient data, which means that data access should be restricted

© Springer International Publishing AG 2017
V.M. Vishnevskiy et al. (Eds.): DCCN 2017, CCIS 700, pp. 269–280, 2017.
DOI: 10.1007/978-3-319-66836-9_23

and secured. This could be reached by permission-based categories of users and individual investigations access on supervised approval. On the other hand, data should be accessible by any authorized user, like patient or doctor, from any location. One of the main priority is system performance that should allow high speed access of the huge amount of data.

Thus, the process of storing and transferring large volumes of medical data can be divided into several components: archiving and storage, retrieval and accessibility of data, data transfer to the end user's computer and processing the data on the client application. When medical investigation is completed and DICOM image set imported from the equipment raises the problem of data archiving and data storing. If an X-ray photograph usually does not contain more than 2–3 images, a tomographic survey can contain up to 1000 slices and take up to a gigabyte of memory on the physical storage. It is easy to calculate that for a large hospital or a large diagnostic center with a daily flow of 500 or more patients, the data volumes will be terabytes per month, but the archive of investigations should be saved for a minimum of 3 years by law (and even longer period for practical use). As a result, medical institutions could not archive such a volume of data, because in many cases there are no available sufficient capacity for their storage at the level of one institution.

It should be noted that when solving the practical task of importing medical images in DICOM format directly from the device, it is necessary to create local networks between the device and data storage, isolated from global networks, since connecting medical equipment directly to the Internet is too dangerous and not recommended by the existing policies. Thus, for each floor or even a cabinet with equipment, it is assumed that there is a certain processor for importing data and recording them into a global database. Taking into consideration of the above factors, storing data on single storage element is not possible or in any cases will be too costly.

The next problem is the transfer of data via the internet channels directly to the end user. Of course, when image set is 2–3 slices, the investigation data packet does not exceed 1–2 megabytes, but when transferring a full tomographic investigation that can have gigabytes of data and the download process can take tens of minutes, depending on the channel capacity, which is unacceptable for many end users. Even more critical is the use of this data directly in the surgery room, when speed is an important component. In addition to the problem of transferring large amounts of data, should be taken into account the problem of local data processing during visualization, especially if the visualization takes place on mobile devices with limited processing capacity.

This article describes the algorithms and methods for solving the problem of working with large volumes of data. These algorithms and practical solutions based on the "DICOM Network"[1] application are considered. In addition, the

---

[1] This project is supported by the Agency for Innovation and Technology Transfer of Moldova, Grant Contract No186T dated 27 February 2015 and by European Commission H2020 project RE for regional Interdisciplinary communities in Southeast Europe and the Eastern Mediterranean (VI-SEEM) Grant Agreement 675121.

article examines the possibilities of international cooperation to solve the problem of storage and access to data based on using of the VI-SEEM platform.

2   DICOM Network

The "DICOM Network" [2] project was launched in Moldova. In 2015, whose goal is to provide access to investigations for medical staff with the appropriate access rights and as well as patients to the personal radiography investigations. The current realization of the project allowed connecting eleven types of medical equipment to the system. Today the system collects and processes more than 500 gigabytes of data per month.

The system Data Storage and Data Processing components distributed between different processing units and storages, which could be customized using specific interfaces. The general architecture of the DICOM Network system presented in the Fig. 1.

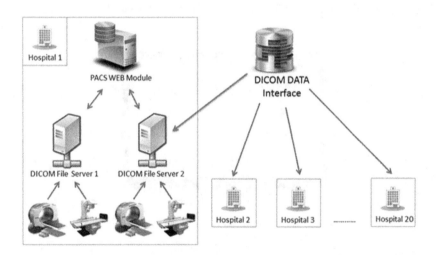

Fig. 1. General scheme of DICOM Network architecture

The system structure comprises the following data servers and modules:

- Data from equipment are collecting by the DICOM Server modules that can be installed in the same location with the used medical equipment or can be distributed through different institutions and even cities or countries.
- All the investigations (DICOM Images) are archiving on DICOM Servers, but the information about investigation is stored in DICOM Portal (like www.dicom.md) database. Usually various DICOM Servers connected to one DICOM Portal.

– DICOM Portal stores all data like users info, access info, system settings, DICOM Server settings and some others, but not DICOM images it salves. Each institution can deploy DICOM Portal internally using one or several own DICOM Servers.

– DICOM DATA Interface collects information about users and investigations from all DICOM Portals and provides functionality for data exchange and unification.

This system covers all necessary workflows for processing and documentation of medical investigations - from collecting images directly from equipment to archiving investigation in the patient medical record [4]. DICOM Network offers extended functionality for enhancing quality of medical management and secured access to investigations. This helps doctors, specialists and penitents to gain access to structured database of medical images, allows documenting images that are collecting from various medical apparatus. At institutional level, the system helps to reduce costs of investigation, raise the quality of provided services.

DICOM Network is already in production operation and can be accessed by link http://www.dicom.md/. The GUI interface is presented in the Fig. 2.

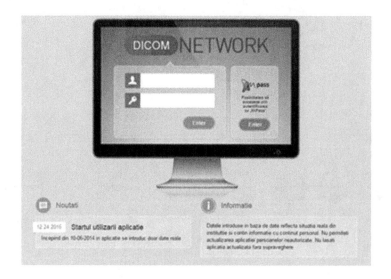

Fig. 2. DICOM Network GUI

The system initially was deployed at the National Centre of Ambulance of Moldova and during the first year of functioning has shown it effectiveness and attractiveness for personnel that is resulting in:

– Two DICOM Portals were set up: http://dicom.md/ national portal for Moldavian medical community and http://viseem.dicom.md/ for research community.

- Three DICOM Servers installed.
- Eleven types of medical equipment were connected to the DICOM Network.
- Sixteen medical doctors from radiology section are using DICOM Network system in 24/7 regime.
- About 250 investigations per day are collecting by the system.
- Over 500 doctors have access to their patients investigations from their working place.
- Over 12 776 investigations were searched and downloaded by doctors and radiology specialists during 12 months period.
- More than 133 950 investigations stored by the system.
- Additional budget savings ensured for hospitals due to refusing of printing investigations using expensive consumables.

"DICOM Network" system is actively developing, as far as doctors are interested in operative accessing radiographic image sets directly from their workplace and institutions wants to save money for consumables. A new equipment connected to the system that increases the number of imported investigations. The main problem faced by the system is the extremely increasing amounts of data that must be preserved.

Today the data is stored on two DICOM servers and one Storage element. Moreover, DICOM servers store an archive for the last month and on a common storage is storing a common archive for the last 2 years. This architecture allows providing the fastest access to more recent data that is used more often, while downloading from Storage archive takes much longer time.

3   DICOM Format Optimisation

First, to describe the solution for DICOM data optimization, here we overview of the algorithms and processes that are necessary for any information system working with radiography. To obtain the source images, DICOM provides functionality for the import of data directly from the equipment. After importing the image, you must process and save it on commonly accessible a storage. As a result, a medical research database available to the user is created. The user accessing the database should receive data through the local or wide area network, and then using the visualization application to work with DICOM images on his workplace.

Thus, the problem of data handle optimization for information systems can be divided into three stages. First, when you import and write the source files on storage, you need to archive the data to minimize it volume saving the quality of the images. Secondly, when accessing data, it is necessary to transfer the data to user as quickly as possible while optimizing the data format to reduce the amount of transmitted images. Thirdly, the data should be optimized to speed up its loading and processing on the local processing unit.

3.1 Data Archiving

As it was mentioned above, one radiographic examination can vary from 1 to 1000 images. Thus, the investigation can consist (has a size) more than of one image. Of course, the most interesting and useful is archiving of large tomographic surveys with a large number of slices. First, let us look at the structure of the DICOM file. In general, each file consists lines of the patient's metadata and the image itself in the raw format. Since during processing all meta-information that is written to the database, when archiving we can discard metadata and save only the image. If necessary, it will be possible to restore this data and generate a DICOM file on client request, or use the new file format when transferring to a specialized application DICOM Viewer. Of course metadata is only a small part of the file and only takes a couple of kilobytes, but it should be taken into consideration, that metadata for one investigation is almost identical, so for a set of data in 1000 slices it will be possible to save significant amounts of physical storage space. On the other hand, excluding the patient's personal data from the original files significantly reduces the possibility of data leakage. This approach also allows the transfer of the impersonated data.

Archiving an image is already a much more complicated task that requires more complex data compression algorithms. It should be taken into account that it is necessary to exclude loss of image quality since each pixel is important for data processing and visualization. Based on analysis were found that archiving an individual slice without loss of image quality does not give a tangible result, but it is worth noting that the proportion of individual slices is small enough and refers mainly to x-rays, when one investigation consists of 1–2 images. The quantity of this data type that is stored does not exceed 1.

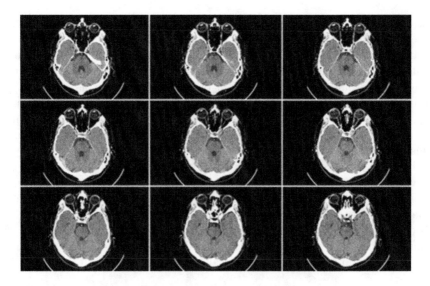

Fig. 3. Similar radiology investigations slices.

As you can see from Fig. 3, the differences in adjacent slices are so small that you can confidently convert these slices into video stream where only the differences for pictures are retained and not the raw format for each image. In this way, you can reduce the amount of images with multiple slices by 10 or more times. Of course, you need to take into account that for archiving you will need considerable computing capacity, which involves the use of high-performance computing systems such as HPC or GRID.

3.2   Data Access Optimization

Another problematic aspect of data storage is the distribution of data among various data storage systems with different levels of security and access speed. As already mentioned above, storing data in one centralized system is not possible for a number of reasons. To solve this problem, we proposed a distributed data storage system showed in Fig. 4.

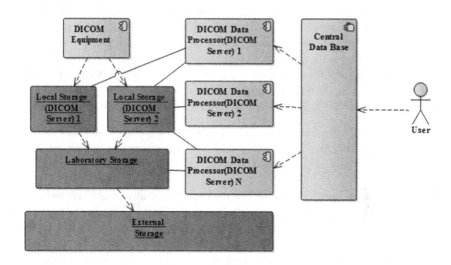

Fig. 4. Distributed big data storage.

The main components of the proposed distributed storage system architecture are:

– Local Storage. Since the equipment connected to front-end DICOM server directly, the server must be located both in the local network and in the global network. The medical equipment should be located only in the local network. For solution of this problem is using a local server, physically located near to the equipment (one per cabinet or laboratory). This server receives the DICOM data and performs the primary data processing. On the one hand, this server provides the fastest access to data; however, a real storage capacity of this server not exceeds 1–2 terabytes, which implies the stored data availability within 1–2 months. This allows high-speed access to the latest data. After a defined period, the data archived and transferred to next instructional storage level ("Laboratory Storage") or to "External Storage" level.

- Laboratory Storage. Storage element for one institution, available in the local network of the organization. This storage has extended parameters, but usually they are limited too and can store data for 1–2 years.
- External Storage. Distributed storage, located partly inside and mainly outside of the organization where the investigations were created and interoperable through the global network infrastructure. This type of storage has the lowest access speed, but it can store a huge distributed archive of data at the national or international level.

3.3   Data Transfer Optimization

Regardless of the storage location for their visualization, images must be uploaded to the end user's computer. At the same time, regardless of the transmission channel, which can be a high-bandwidth local network or a global low-speed Internet network, data packets can be very large, and the speed of downloading may depend on the quality of medical services and the interest of the doctor in using the information System. The main solution is certainly the system of archiving and decompressing the data described above, but it is worth taking into account the following opportunities for transmission optimization:

- Preparing a data packet taking into account the permission of the receiving client application. That is, if a user views a survey on a low-resolution mobile device, there is no point in sending full-screen images at the maximum resolution.
- Ability to load a specific slice in the maximum resolution. At the request of the client, it is necessary to provide the possibility of sending a particular image at the maximum resolution, for example, if necessary, zoom in/zoom out.
- Loading data in background with a separate thread. In this case, the end user can begin to visualize not a complete set of data, while the full set will be loaded asynchronously and displayed as the load is loaded.
- Caching data at the client and server level. All of the above can significantly speed up the transfer of data and make the application for visualization more user friendly.

3.4   Preprocess of Data for Visualization

After uploading data to the end user's computer, the application that implements the DICOM image rendering must process the data for display. In this case, all data must be loaded into the RAM for fast processing. It should be noted that the DICOM Viewer (the application that displays the DICOM images) should not only display the image and change the slices, but also perform more complex operations from drawing to building a 3D model and modeling the tissues. Not every personal computer, much less a mobile device, has satisfactory computing capabilities, so the above operations can take a long time.

The solution for this problem should be to prepare data for visualization on the server. Given the use of high-performance systems, the server can build and transfer ready-made models that will not require complex conversions on the client application.

4   Integration into VI-SEEM Platform

As was shown above, the solution for storing such large volumes of data is the distributed storage of the image archive. One solution is to participate in international projects and use storage elements from the resources of the project. This cooperation is mutually beneficial for both "DICOM Network" and the VI-SEEM platform. VI-SEEM provides project resources while "DICOM Network" datasets for research community. Of course, all the data arriving at the VI-SEEM server is impersonated.

VI-SEEM project deploys and offering user-friendly integrated e-Infrastructure platform for Scientific Communities in Climatology, Life Sciences and Cultural Heritage for the South-Europe and Mediterranean regions by linking compute, data, and visualization resources, as well as generalized services, software and tools. The regional infrastructure deployment concept is presented in Fig. 5.

As far as DICOM Network was selected as pilot application for integration into distributed regional VI-SEEM platform [5], the DICOM Network system architecture was adjusted to the project needs. The updated system architecture presented in the Fig. 6.

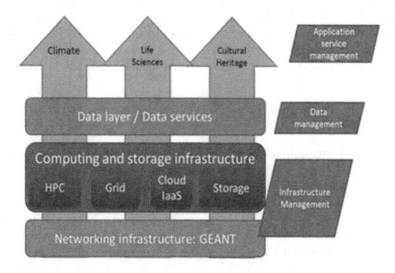

Fig. 5. VI-SEEM technology context

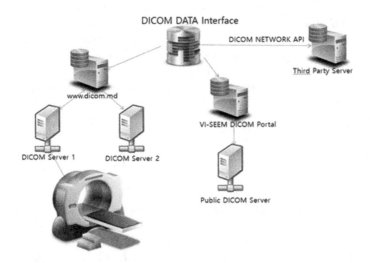

Fig. 6. DICOM Network architecture integration with VI-SEEM platform.

In the Fig. 6 presented the concept of connecting national DICOM Network application, that it is containing existing DICOM Portal http://dicom.md/, with the DICOM Portal installed in VI-SEEM platform resources. DICOM DATA interface grants the interconnectivity for different users of the both portals and allows displaying DICOM investigations using the both portals interfaces. Public DICOM Server grants possibility for any VI-SEEM platform member to pull and retrieve the investigations from DICOM Network application and use the developed facilities based on configured and granted access rules.

VI-SEEM platform will offer possibility to install and configure publically available DOCOM Portal that can be used by any interested institutions to store, access and share medical images. Setting up public DICOM Portal instance will increase the level of access to DICOM investigations and will help to make available DICOM Network services to regional medical research and practicing community.

5   Conclusions

At the moment, "Dicom Network", although actively developing, but still far from realizing the potential built into the system. Taking into account the growing number of medical equipment and the trend towards modernization and computerization of health facilities, the system will be able to receive and will have to process dozens of terabytes of source information. The storage and subsequent transfer of such large amounts of data is an expensive process, impossible without optimization. On the other hand, for the successful development of the system it is necessary to provide the archive not for 3 years, as provided by law, but for tens of years to monitor the patient's condition and maintain a full medical record. It is also necessary to take into account the need for backup copies of such important information. It is easy to calculate that even for such a small country as Moldova, the data volumes are too large to store them in an unprocessed form. Thus, the issue of data optimization and archiving is a key factor for the development of such systems.

The benefits of using archiving algorithms are opening four perspective directions:

- Reduce the costs of medical data storage and maintenance;
- Reduce internet traffic for data access and in such way reduce the costs for data transfer;
- Increase the quality of radiology services for patients.
- Solve the main problem for DICOM images database insufficient space for quickly increased amount of data.

As far as radiology medical investigations services are offered by majority of medical institutions starting from small villages to the huge laboratories centers there algorithms should be in a great demand, because on the one hand it reduce the costs [3] and save organization budget and on other hand increase quality offered services and opens huge opportunities for research and collaboration with other institutes. Of course the most perspective market are huge diagnostically centers in governmental and private sector that have modern equipment and huge number of investigations that should be archived and transferred to other medical institutions. But these solutions will be also interesting for small hospitals that do not have their own equipment but anyway need to have access to the radiology investigations for their patients. Using the proposed algorithms will have possibility to have access to the investigations that were done in the external institution, like private or governmental diagnostics centers.

Datasets collecting in the DICOM Network system will provide new opportunities for researchers. Although the system is now in production stage, functionality of the DICOM Network is permanently enhancing. During the process of the system implementation beneficiaries specified their necessities for providing additional features and services, such as:

- Studying and realization of new methods for optimization of data transfer and archiving.
- Image preprocessing and detection of anomalies.
- Incorporation of expert systems to help making diagnoses for doctors.
- Development of open APIs for "Dicom Network" to collect, archive, access and jointly process medical images at international level.

Algorithms and solutions discussed in this article can be applied in any medical information system and are not tied to a specific project, since the developed approach involves tight integration with the DICOM standard and rather complements it rather than modifies it.

References

1. DICOM format description. http://dicom.nema.org/standard.html
2. Golubev, A., Bogatencov, P., Secrieru, G., Iliuha, N.: DICOM network - solution for medical imagistic investigations exchange. In: International Workshop on Intelligent Information Systems. Proceedings IIS, 13–14 September, IMI ASM, Chisinau, pp. 179–182 (2011). ISBN 978-9975-4237-0-0
3. Anagnostaki, A., Pavlopoulos, S., Kyriakou, E., Koutsouris, D.: Cost and benefits of picture achieving and communication system. J. Am. Med. Inform. Assoc. 1(5), 361–371 (1994)
4. Bogatencov, P., Iliuha, N., Secrieru, G., Golubev, A.: DICOM network for medical imagistic investigations storage, access and processing. In: Networking in Education and Research Proceedings of the 11th RoEduNet IEEE International Conference, Sinaia, Romania, 17–19 January 2013, pp. 38–42. ISSN-L 2068–1038
5. VI-SEEM - Virtual Research Environment (VRE) in Southeast Europe and the Eastern Mediterranean. https://vi-seem.eu

The Augmented Reality Service Provision in D2D Network

M. Makolkina[1,2(✉)], A. Vikulov[1], and A. Paramonov[1,2]

[1] The Bonch-Bruevich Saint-Petersburg State University of Telecommunications,
Saint-Petersburg, Russia
makolkina@list.ru, alex-in-spb@yandex.ru
[2] Peoples Friendship University of Russia (RUDN University),
6 Miklukho-Maklaya St, Moscow 117198, Russian Federation
https://www.sut.ru/

Abstract. This paper is dedicated to the augmented reality service provision by using of D2D communications. We consider the service which may be provided by using of local data source such as panoramic view video camera or any other. Local source provide information useful for nearest users. We propose using of multicast service. Our approach allows reach high users density users near the source. The aim of the paper is to determine an appropriate bitrate for the multicast stream on the base on the distribution of users in the service area.

Keywords: Augmented reality · Multicasting · D2D technology · Users distribution · Data bitrate

1 Introduction

The evolution of mobile networks, including the concept of 5G networks architecture, implies a heterogeneous structure building using various wireless technologies. In particular, this concept assumes the wide use of in-band and out-of-band clustering using device-to-device (D2D) communication technologies, which allows to significantly improve the efficiency of the radio-frequency spectrum use, as well as the quality of service provisioning [1]. The augmented reality service (AR) [1] is based on providing the user with "additional" information in the form of text messages, characters, sounds (sonification) and streaming video. Depending on the purpose of the service, various options for its implementation can be presented. In particular, it is possible to conditionally allocate a group of services, the provision of which has territorial affiliation to a certain geographical position or region. In particular, these are services that provide additional information about objects in the close proximity of the user (tens - hundreds of meters). In this case, the sources of the augmented reality data can be the devices that are located in the immediate proximity to the user. In such cases, the use of D2D communication allows to localize the traffic and offload the mobile network.

V.M. Vishnevskiy et al. (Eds.): DCCN 2017, CCIS 700, pp. 281–290, 2017.
DOI: 10.1007/978-3-319-66836-9_24

2   Related Works

A comprehensive analysis of D2D communications is given in [3,4]. The paper [3] deals with the use of out-of-band D2D communications and the D2D devices clustering when the criterion for selecting the head node of the cluster is the quality of the communication channel with the base station (BS). In [14] the authors showed that clusters built on the principle of in-band D2D increase the spectral efficiency of the system for broadcast messages. Numerical analysis and simulation have shown that this proposal gives 66% gain in bandwidth compared to traditional solutions, in the case when 20% of users use D2D communication. The authors derived an expression for the probability density function (pdf) for the optimal number of repeaters in the cluster and the interlayer interaction scheme. The authors also showed through simulation that their proposed scheme provides an increase of up to 40% in terms of resource efficiency use. Various aspects of out-of-band D2D communication were considered in [6,15,16]. In [6,15], the authors developed an analytical model of network offloading for various D2D scenarios, using stochastic geometry. The authors evaluate the potential of out-of-band D2D communication using both the system level and mathematical analysis. They show that at 30% clustering, the performance and energy efficiency of the network increase four and two times, respectively. The authors study the problems of implementing D2D with network support and social networks focus in [16]. In addition, they use the existing experimental LTE Testbed [17] to implement their proposed D2D system and demonstrate its usability in terms of latency and user satisfaction.

3   The Statement of the Problem

We will consider the use of D2D communications when providing augmented reality services. As the application for such services, for example, it is possible to consider local services, i.e. services for the subscribers that are located in some local area. For example, access to a panoramic view camera, located on the roof of the building, a mast or a UAV in the close proximity of the subscriber. The purpose of such service can be to provide a general view of the terrain in real time, as well as providing additional information about the objects in the panoramic view zone, if the appropriate equipment and software is available. When implementing this service using D2D communications, its delivery scheme may look like it is shown in Fig. 1

The Panoramic View Camera (V) transmits the video stream through the Wi-Fi access point (AP) located directly next to the video camera. This AP can be combined with a video camera within a single device (UAV, external video camera, etc.). In this case, the customer of the service can be located within the AP reach zone. Thus, users of the service can be considered as a cluster in a mobile network when using out-of-band clustering. Since all mobile stations (MS) in the service zone receive the same video stream, it is possible to increase the efficiency of the channel resource utilization with the use of point-to-multipoint

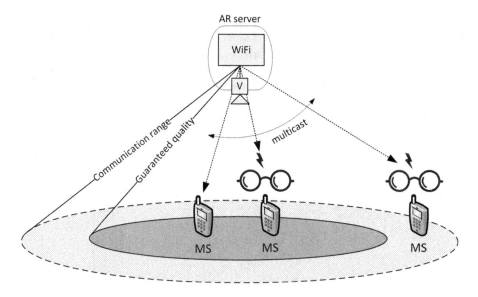

Fig. 1. Panoramic view service scheme.

packet transfer type (multicast). Users are randomly distributed within the AP communication zone, different users are in different signal reception conditions. When using point-to-point (unicast) connections, the AP-MS pair selects the most suitable modulation and coding scheme for packet transmission, depending on the specific signal propagation conditions. However, when using multicast, AP selects a single encoding scheme for transmitting packets to all users. In fact, the encoding scheme determines the bit rate of the video stream. The high speed of the stream allows broadcasting with high quality of the image, but it does not provide a sufficiently low packet loss for users with relatively low RF reception conditions. I.e. when choosing a transmission rate (modulation and coding scheme - MCS), it is necessary to follow a compromise between the quality of the image and the number of users served. For example, choosing a too low MCS based on the requirements for the most remote user will result in a decrease of the quality of service for all users in the receiving area. One possible solution is to fix the transmission rate at the minimum sufficient level. However, in this case it is impossible to improve the quality of the service when the reception conditions of most (or all) users permit the higher MCS usage. In this paper, we propose a method for selecting a coding scheme based on an analysis of the distribution of users in the service area.

4   Throughput Distribution Analysis in the Service Zone

The model. Quality of service provisioning is characterized by such probabilistic and temporal indicators as probability of availability, probability of losses and data delivery time. These indicators depend on throughput and traffic parameters of the communication network. Thus, for given traffic characteristics, the throughput value most fully characterizes the decision taken from the point of view of the quality of the communication services provided. Under the throughput, we will understand the achievable data transfer rate. We will consider the channel throughput between the elements of the network bij as a metric. In the analysis problem, we will assume that the AP service zone is a circle with radius R, centered at the AP location, Fig. 2

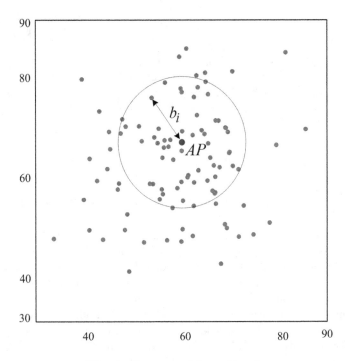

Fig. 2. Cluster model considered

While considering the IEEE 802.11 family of standards as a technology for D2D communication, it is necessary to determine the nature of the bij dependence on users placement in the service area. According to [18], the data transfer rate between the two devices is determined by the selection of one of the implemented MCS according to reception conditions (radio channel quality). These conditions are estimated by the signal strength at the receiver's input s = prx or by signal to noise ratio (SNR) or signal-to-interference-plus-noise ratio (SINR). In this case, the noise power is considered as the total power of all received signals and their interference, except the useful signal. The interference power is understood as the power generated by certain signal sources, in our case by neighboring clusters. According to [18], the typical dependence form of the transmission rate on the signal strength at the receiver input s is a step function that increases with increasing power. Figure 3 shows the example of the transmission rate dependence for the IEEE 802.11n standard, with the 20 MHz channel width (HT20) on the s value.

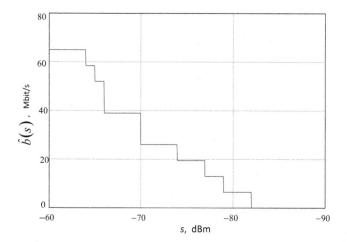

Fig. 3. Transmission rate dependence on the s value

A similar dependence can be constructed for the SNR and SINR values. The signal strength at the receiver input can be described by received channel power indicator (RCPI). According to [18], the value of this indicator is measured with an accuracy of ±5 dB (95 In addition to these metrics, the signal power at the receiver input can be described by the received signal strength indicator (RSSI). For this value, the exact correspondence with the power of the received signal is not determined. According to the IEEE 802.11 standard, it can vary from the minimum to the maximum value, with the minimum value corresponding to the minimum, and the maximum value of the maximum power of the received signal. Each of the parameters above affects the throughput of the channel and can be selected as a metric in the selection of the transmission rate task. The choice of a particular parameter depends on the possibilities of its evaluation and the problem conditions. The choice of SINR in solving the clustering problem allows to take into consideration the interference produced by the devices of neighboring clusters. In this analysis, as the metric, we chose the signal power at the receiver input s, which can be estimated by the RCPI parameter. The throughput value, determined by this model (Fig. 3), depends on the reception conditions and in practice it has a wide spread. Accordingly, it can be assumed that the approximation of the step-like model by a continuous function, for the purpose of bandwidth analysis, will not introduce a significant error in the results, but will greatly simplify the task. Considering the panoramic view service with the use of UAV survey, we believe that the service users are concentrated on the outdoor open terrain, we will describe the attenuation of the signal by the model recommended [19] for outdoor use (1)

$$L(d) = 20lg(4\pi d/\lambda) \qquad (1)$$

where d is the distance (m); λ - wavelength (m).

Taking into account the attenuation model, we will describe the throughput dependence on the distance by a step function (Fig. 4)

From the given drawing, the area of the shaded area can be determined by a formula (2) the Number of new objects in the area can be defined as

$$\hat{b}(d) = \hat{b}(L(d)) \qquad (2)$$

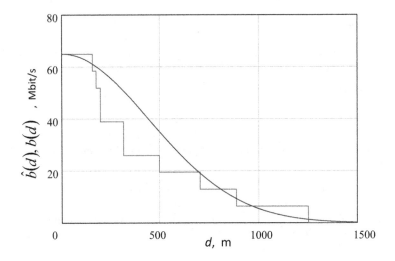

Fig. 4. The throughput dependence on the distance

The step-like function is approximated by a gaussoid [20], which shows the change in the throughput with the distance

The number of the new objects in the field can be defined as (3)

$$b(d) = \begin{cases} 0 & d < 0 \\ b_{max} e^{-\frac{d^2}{2C^2}} & 0 \le d \le R \\ 0 & d > R \end{cases} \qquad (3)$$

where d is the distance (m); D constant; b_{max} maximum possible bandwidth (Mbit/s); c half-width of the curve (m). $R = arg\{\hat{b}(d) = 0\}$ (m)

Generally, mobile stations are randomly distributed over the served territory, hence the distance d and the attenuation of the signal L(d) between them are also a random variables. In this case, we do not take into account the fading of the signal, which also affects the nature of the random attenuation value, and consequently, of the throughput. In this analysis, we will only consider the factor of the relative location of users b(d). Since the bandwidth is a function of the random variable (of coordinates), the distribution function b can be determined according to [21] as

$$F(b) = \iint_{D_b} f(x, y) dx dy \qquad (4)$$

where D_b - is the range of function b. $f(x, y)$ - distribution function of users on disk R. Probability density of b is determined as

$$f(b) = \frac{dF(b)}{db} \qquad (5)$$

The expectation of b

$$M(b) = \int_0^{b_0} b f(b) db \qquad (6)$$

We will consider one type of user distribution across the territory - a uniform distribution.

5   User Distribution in Service Zone

We assume that the uniform distribution is given on a disk whose area is equal to S, radius R ($S = \pi R^2$), on $0 \leq r \leq R$ range. Disk radius R is determined as $R = arg\{\hat{b}(d) = 0\}$ (m). Probability density function $f(r) = const$ for disk

$$f(r) = \frac{1}{S} = \frac{1}{\pi R^2} \qquad (7)$$

If the function expressing the dependence of the bandwidth b on the distance to the base station has the form (3), i.e. $b(d) = b_{max} e^{-\frac{d^2}{2c^2}}$. Where R is the service zone radius, determined from attenuation model can be evaluated from (3) $d = c\sqrt{-2ln(\frac{b}{b_{max}})}$ (m). Throughput distribution function b on the disk R according to the (8)

$$F(b) = \int_0^{2\pi} \int_{c\sqrt{-2ln\left(\frac{b}{b_{max}}\right)}}^{R} \frac{1}{S} r\, dr\, d\theta = \frac{1}{2\pi R^2} r^2 \Big|_{c\sqrt{-2ln\left(\frac{b}{b_{max}}\right)}}^{R} 2\pi =$$

$$\frac{1}{R^2}\left(R^2 + 2c^2 ln\left(\frac{b}{b_{max}}\right)\right) =$$

$$1 + \frac{2c^2}{R^2} ln\left(\frac{b}{b_{max}}\right) \qquad (8)$$

Probability density function, according to (9)

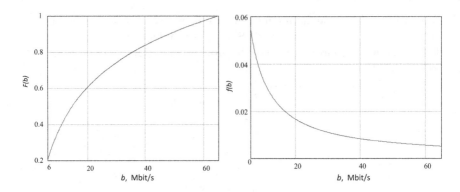

Fig. 5. Throughput probability distribution and probability density function

$$f(b) = \frac{dF_b(r)}{dr} = \frac{d}{dr}\left(1 + \frac{2c^2}{R^2}ln\left(\frac{b}{b_{max}}\right)\right) = \frac{2c^2}{R^2b} \tag{9}$$

They are presented on Fig. 5

For example, for the IEEE 802.11g standard, the expectation M(b) of the throughput in the communication area (in the disk R), according to the (10), will be defined as

$$M(b) = \int_{b_{min}}^{b_{max}} b\frac{2c^2}{R^2b}db = 2\frac{c^2}{R^2}(b_{max} - b_{min}) \tag{10}$$

For example, with the selected approximation of the bandwidth function from the distance and the uniform distribution of users in the communication area, the mathematical expectation of the bandwidth for the IEEE 802.11n standard is 19.51 Mbps. Thus, the choice of the data rate can be realized as a solution of an Eq. (11) with respect to b0.

$$F(b_0) = p(b < b_0) \tag{11}$$

According to (3) the chosen data rate is

$$b_0 = b_{max}\, e^{-\frac{R^2(1-p)}{2c^2}} \tag{12}$$

where R, c and b_{max} are determined as it has been shown above, p probability that the data rate does not exceed the quality reception threshold. Probability p can be interpreted as a part of clients with lower quality of service.

6   Conclusions

1. The provision of augmented reality services can be realized using D2D communication technologies, i.e. Out-of-band clustering of mobile devices. 2. To provide a number of services, for example, the discussed panoramic view service,

a point-to-multipoint connection (multicast) can be used. When using multicast, different users are in different conditions and receive a service with different quality. 3. When multicast is provided, the choice of the AP data rate is required, which will fulfill the compromise condition between the quality of the service, which is determined by the data transfer rate, and the quality of the data delivery to users. 4. With the uniform distribution of users in the service area, the data rate can be calculated from the distribution function (8), according to (12) when specifying the user share for which it will be too high due to the reception conditions. 5. In the case considered, the augmented reality service represented the video stream transmitted directly from the panoramic video camera. However, the service can be integrated with some object data source (symbols and texts).

Acknowledgments. The publication was financially supported by the Ministry of Education and Science of the Russian Federation (the Agreement number 02.a03.21.0008), RFBR according to the research project No. 16-37-00209 mol_a Development of the principles of integration the Real Sense technology and Internet of Things.

References

1. Muthanna, A., Masek, P., Hosek, J., Fujdiak, R., Hussein, O., Paramonov, A., Koucheryavy, A.: Analytical evaluation of D2D connectivity potential in 5G wireless systems. In: Galinina, O., Balandin, S., Koucheryavy, Y. (eds.) NEW2AN/ruSMART -2016. LNCS, vol. 9870, pp. 395–403. Springer, Cham (2016). doi:10.1007/978-3-319-46301-8_33
2. Asadi, A., Mancuso, V.: Network-assisted outband D2D-clustering in 5G cellular networks: theory and practice. IEEE Trans. Mobile Comput. (2016)
3. Asadi, A., Wang, Q., Mancuso, V.: A survey on device-to-device communication in cellular networks. IEEE Commun. Surv. Tutorials **99**, 11 (2014)
4. 3GPP, 3rd generation partnership project;technical specification group services and system aspects; policy and charging control architecture (release 13), // TR 23.203 V13.4.0 (2015)
5. Andreev, S., Pyattaev, A., Johnsson, K., Galinina, O., Koucheryavy, Y.: Cellular traffic offloading onto networka-ssisted device-to-device connections. IEEE Commun. Mag. **52**(4), 20–31 (2014)
6. Sim, G.H., Loch, A., Asadi, A., Mancuso, V., Widmer, J.: 5G millimeter-wave and D2D symbiosis: 60 GHz for proximity based services. IEEE Wireless Commun. Mag. (2016)
7. Zhou, B., Hu, H., Huang, S.-Q., Chen, H.-H.: Intracluster device-to-device relay algorithm with optimal resource utilization. IEEE Trans. Veh. Technol. **62**(5), 2315–2326 (2013)
8. Asadi, A., Mancuso, V.: On the compound impact of opportunistic scheduling and D2D communications in cellular networks. In: Proceeding of ACM MSWIM (2013)
9. Asadi, A., Mancuso, V.: Wifi direct. In: Proceeding of IFIP Wireless Days
10. Wi-Fi Alliance, Wi-Fi peer-to-peer (P2P) technical specification v1.1. www.wi-fi.org/wi-fi-peer-peer-p2p-specification-v11

11. Sim, G.H., Nitsche, T., Widmer, J.: Addressing MAC layer inefficiency and deafness of IEEE802.11ad millimeter wave networks using a multi-band approach. In: Proceeding of IEEE PIMRC (2016)
12. Sim, G.H., Li, R., Cano, C., Malone, D., Patras, P., Widmer, J.: Learning from experience: efficient decentralized scheduling for 60 GHz mesh networks. In: IEEE WoWMoM (2016)
13. Seppala, J., Koskela, T., Chen, T., Hakola, S.: Network controlled device-to-device (D2D) and cluster multicast concept for LTE and LTE-A networks. In: Proceeding of IEEE WCNC (2011)
14. Andreev, S., Galinina, O., Pyattaev, A., Johnsson, K., Koucheryavy, Y.: Analyzing assisted offloading of cellular user sessions onto D2D links in unlicensed bands. IEEE JSAC **33**(1), 67–80 (2015)
15. Andreev, S., Hosek, J., Olsson, T., Johnsson, K., Pyattaev, A., Ometov, A., Olshannikova, E., Gerasimenko, M., Masek, P., Koucheryavy, Y., Mikkonen, T.: A unifying perspective on proximity-based cellular-assisted mobile social networking. IEEE Commun. Mag. **54**(4), 108–116 (2016)
16. Pyattaev, A., Hosek, J., Johnsson, K., Krkos, R., Gerasimenko, M., Masek, P., Ometov, A., Andreev, S., Sedy, J., Novotny, V., et al.: 3GPP LTE-assisted Wi-Fi-direct: trial implementation of live D2D technology. ETRI J. **37**(5), 877–887 (2015)
17. IEEE Std 802.11-2012. Wireless LAN Medium Access Control (MAC) and Physical Layer (PHY) Specifications
18. ITU-R pp. 1238-8, Propagation data and prediction methods for the planning of indoor radiocommunication systems and radio local area networks in the frequency range 300 MHz to 100 GHz
19. Wolfram MathWorld. Gaussian Function. http://mathworld.wolfram.com/GaussianFunction.html

Measurement System Architecture for Measuring Network Parameters of e2e Services

Vyacheslav Kulik[1], Ruslan Kirichek[1,2(✉)], Alexey Borodin[3], and Andrey Koucheryavy[1]

[1] Saint Petersburg State University of Telecommunication,
Saint Petersburg, Russia
ruslan.stk@gmail.com, akouch@mail.ru
[2] RUDN University, Moscow, Russian Federation
[3] PJSC "Rostelecom", Moscow, Russian Federation

Abstract. In this article, we examine possible methods for measuring uplink, downlink throughput, and network latency RTT value between two devices operating at the level of communications providers' "last mile" and propose various methods for measure uplink, downlink throughput, and network delay (RTD) value level based on various network protocols (HTTP / HTTPS, MQTT, XMPP, AMQP, WebSocket, CoAP, QUIC, etc.). Based on ITU-T Rec. Q.3960, the following was developed: the architecture of the hardware and software complex to be used for testing the quality of the provided communication services; a method for measuring network parameters between a last mile device and remote cloud service based on the developed software and hardware complex; a method for measuring network performance parameters between end nodes that operate behind NAT on the basis of the developed software and hardware complex. A protocol of data transmission for the developed software and hardware complex is described. The article also presents a concise study of the reasearch of the measurement system developed.

Keywords: e2e · Measurement · Uplink · Downlink · Latency · QoS · HTTP · STUN · NAT

1 Introduction

At present, information technologies are one of the most actively developing branches of science. The IT sector includes many promising areas, including such areas as M2M-communications [1], D2D-communications, SDN [2], Internet Things [1–6], cloud technologies [7], Tactile Internet [8], Augmented Reality [9], Internet Nanothings [10,11], etc. The use of these technologies allows the introduction of new types of communication services, the support of which is expected to be implemented in future communication networks and which must meet certain requirements for indicators of network parameters in the transmission of data. At the moment, there are many different solutions for measurement

© Springer International Publishing AG 2017
V.M. Vishnevskiy et al. (Eds.): DCCN 2017, CCIS 700, pp. 291–306, 2017.
DOI: 10.1007/978-3-319-66836-9_25

objective indicators of communication quality. These solutions record the main indicators of network parameters for end-users of the "last mile" of communication operators. One of the main problems of these systems is the heterogeneity of measurement methods and the absence of a single software and network interface for measuring network parameters. This problem was identified in ITU-T Recommendation Q.3960 "Framework of Internet related performance measurements" [12]. This standard describes a model for measurement the quality of the provision of telecommunications services by the "last mile" provider and a technique for measurement the network parameters between the program probe (the end node of measurement) installed on the user's computing device and the remote service on which the program probe also functions. Measurement of network parameters for the quality of communication services between these software probes is performed using a remote measurement server that monitors the measurement procedure.

According to ITU-T Q.3960, measurement of the quality of communication services of the telecommunications operator takes place in two scenarios: measurement the quality of communication from the subscriber's probe to the provider edge router; measurement of the quality of communication from the subscriber's probe to the probe operating on the Internet service. This distinction allows us to assess the quality of the provision of telecommunications services within the operator's network. These scenarios assume the presence of probes - software or hardware terminals between which testing is carried out and a measuring server that monitors the measurement process.

In this article, the architecture of the hardware and software complex was developed to measure the quality of the communication services provided. Based on this architecture, a method for measurement network parameters between the last mile device and a remote cloud service was developed and a method for measuring network parameters between terminal nodes that operate behind NAT [13]. The protocol of data transmission was also described and testing was conducted for the developed measurement system (MS).

2   Architecture of the Measurement System

Based on the recommendation of ITU-T Q.3960 "Framework of Internet related performance measurements", a MS architecture was developed to measure the quality of services provided by the operator. This architecture is depicted in Fig. 1.

This measurement system includes the following elements:

1. The probe is a customer of the quality monitoring system of communication services. Includes the following subsystems:
   - Software measurement peer (further SMP);
   - Software measurement peer communication interface (further SMP CI).

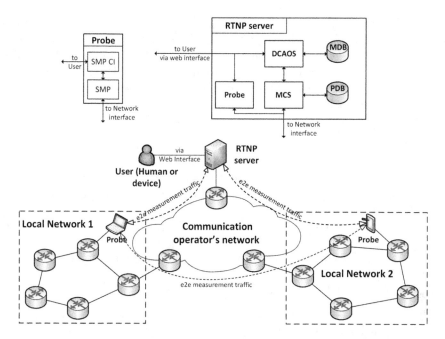

Fig. 1. The architecture of the MS for measuring the quality of the provided communication services.

2. Remote testing network parameters server for network parameters and communication services (further RTNP Server). Includes the following subsystems:
   - Measurement control system (MCS);
   - Data collection, analysis and output system (DCAOS);
   - Measurement database (MDB);
   - Peer database (PDB).

Depending on the network construction features and the tasks assigned to RTNP, all the above-mentioned elements and subsystems can be implemented jointly or separately in various combinations, including on the same hardware.

The probe is a structurally isolated component of the RTNP system, which is a hardware-software device with SMP and SMP CI functions.

The software measurement peer (SMP) is software running in the background and performing measurement scenarios by means of measurement configurations received from the MCS that collects, accumulates, transmits measurements and generates traffic flows on the basis of the received measurement configurations using the selected 3–7 layer protocol, according to the OSI model. SMP implements the following functionality:

- Receiving and transmitting information from / to SMP CI;
- Performs measurement scenarios, according to the measurement configuration received from SMP CI;
- Generates traffic flows based on measurement configuration received from SMP CI.

Software measurement peer communication interface (SMP CI) - an application that performs the functions of the interface with the software measurement peer. This interface can be: graphical user interface - GUI, command line interface - CLI, network interface - NI, application interface - API. This application can perform functions as one of the above interfaces, and several simultaneously, depending on the task.

SMP CI implements the following functionality:

- Receiving and transmitting information from / to SMP;
- Input and output of the preliminary SMP configuration (RTNP address and port, hardware platform type, access levels to SMP, PUID, etc.);
- Selection, pre-configuration and launch of the measurement scenario (including from MCS);
- Input and output configuration of the current measurement (measurement scenario, PUID, IP address, destination SMP port, MUID, etc.);
- Accepts and sends intermediate and final measurement results (including from / to MCS);
- Updating the SMP software and the SMP CI itself.

The server of remote testing network parameters (RTNP) – is a server software, which includes subsystems: MCS, DCAOS, MDB, PDB.

The measurement control subsystem (MCS) is a server software that is a subsystem that provides control over the execution of measurement scenarios.

MCS implements the following functionality:

- Registration of SMP in the system, with the assignment of a unique node identification number to it - PUID;
- Generating measurement configurations for SMP, including a unique identification number for measuring - MUID;
- Receiving measurement configurations from SMP CI and assigning them a unique identification number for measuring - MUID;
- Receiving and transmitting measurement configurations from / to SMP CI;
- Reception and transmission of intermediate and final information about measuring from / to SMP CI;
- Interaction with DCAOS;
- Write and read data about the SMP node from / to the PDB, by the PUID key;
- Continuous monitoring of network parameters and communication services for SMP registered in the PDB;
- If you need to run a script based on peer to peer (p2p), it performs the functions of the STUN server and the TURN server.

Subsystem for data collection, analysis and output (DCAOS) - server software, which is a subsystem that collects, analyzes and outputs data obtained during the execution of a measurement scenario.

DCAOS implements the following functionality:

- Interaction with MCS;
- Input and output of information about measuring;
- Analysis of information received about the measuring;
- The conclusion of useful information about the conducted measuring to the user with the help of the Web-interface;
- Record and read measurement information to / from MDB, by MUID key.

The measuremet database (MDB) is a server software that is a subsystem that records, reads and stores the data on measurement conducted on this RTNP, using the MUID - Measurement Unique IDentificator key.

A node database (PDB) is a server software that is a subsystem that records, reads and stores data about registered SMP hosts on this RTNP Server, by the PUID-Peer Unique IDentificator key.

3   Measurement of Network Parameters Between the End Node and the Tested Remote Internet Service

Based on this MS, a scenario was developed for measuring network parameters between a client node and a remote Internet service with an external IP address (e2s). This scenario involves measuring based on the TCP transport layer protocol. Thus, the process of measuring network parameters for this scenario is carried out using protocols that operate over the TCP protocol.

In this scenario, the following parameters are measured:

- Network latency (RTD);
- Maximum uplink throughput value;
- Maximum downlink throughput value.

The scenario for measuring the network parameters for e2s is shown in Fig. 2. Measure the network the following scenario the following scenario:

1. SMP nodes (SMA - source measurement agent and DMP - destination measurement peer) are first registered with the MCS server;
2. Then, the SMA node sends a request to initialize the measurement on the parameters passed in the body of the request;
3. The MCS server analyzes the measurement parameters received from the SMA request, generates a measuring configuration, and sends it to the SMP hosts;
4. Next, SMP hosts perform one of types of measurement, periodically sending information about the MCS measurement process:
    - RTT latency network latency;
    - Uplink bandwidth measurement;
    - Downlink bandwidth measurement.
5. After measuring, the SMP hosts send information about the MCS measurement results.
6. MCS sends the processed measurement result to the SMA node.

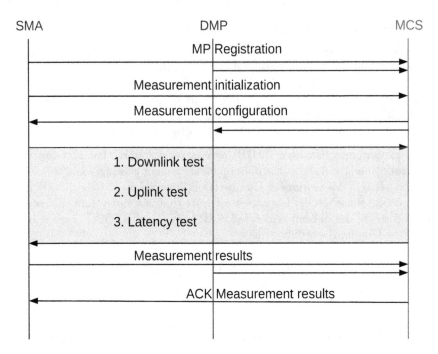

Fig. 2. Scenario for measuring network parameters for e2s.

4   Measurement of Network Parameters Between the End Node and the Tested Remote Internet Service

On the basis of this MS, a scenario was developed for measuring network parameters between the terminal nodes that operate behind NAT (e2e). This scenario involves measuring based on transport layer protocols UDP, TCP. To measure network parameters using the application protocols operating on top of the UDP protocol, the procedure STUN [14] is used. To measure network parameters based on application protocols that operate on top of the TCP protocol, the TURN procedure is used [15].

In this scenario, the following parameters are measured:

- The RTD (round trip delay);
- Maximum uplink throughput value;
- Maximum downlink throughput value.

The general view of the complex for measuring the bandwidth of access to the Internet based on the STUN procedure is shown in Fig. 3.

The STUN procedure allows measuring based on application layer protocols that use the UDP transport protocol to bypass NAT using a logical point-to-point connection (SMA-DMP).

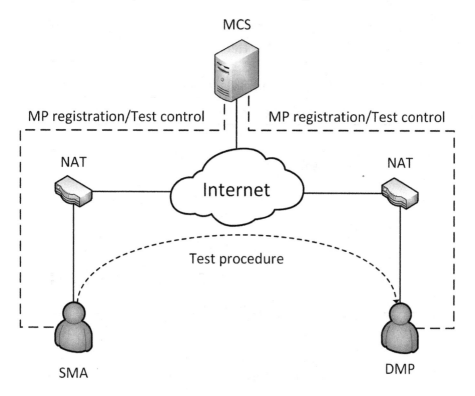

Fig. 3. A general view of the network measurement suite for e2e, based on the STUN procedure.

The scenario for measuring the network parameters for e2e is shown in Fig. 4. The measuring of the network the following scenario the following scenario:

1. First, SMPs (SMA - source measurement agent and DMP - destination measurement peer) register on the MCS server and send unique SMP identifiers (UIDs);
2. Then, the SMA node sends a request to initialize the measurement on the parameters transmitted in the body of the request, including the UID of the DMP;
3. The MCS server analyzes the measurement parameters received from the SMA request, generates a measuring configuration, including the addresses and NAT ports to which the SMP hosts are assigned and sends it to the SMP hosts;
4. Next, SMP hosts perform one of types of measurement, periodically sending information about the MCS measurement process:
   - Measure the RTD (round trip delay);
   - Uplink bandwidth measurement;
   - Downlink bandwidth measurement.

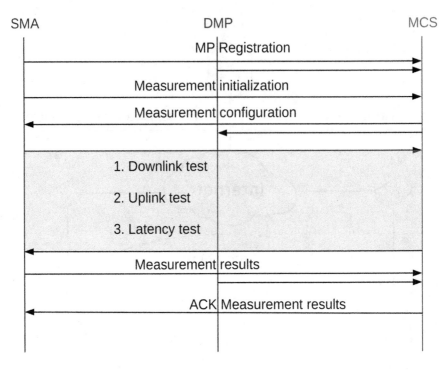

Fig. 4. Scenario for measuring network parameters for e2e, based on the STUN procedure.

5. After measuring, the SMP hosts send information about the MCS measurement results. item MCS sends the processed measurement result to the SMA node.

A general view of the complex for measuring the bandwidth of access to the Internet based on the TURN procedure is shown in Fig. 5.

The TURN procedure allows measurement based on application layer protocols that use the TCP and UDP transport protocols to bypass NAT, using the MCS TURN server as an intermediate node (SMA → MCS → DMP). This procedure is not recommended to measure based on the UDP application layer protocols.

The scenario to measure the network parameters for e2e is shown in Fig. 6. The measuring of the network the following scenario the following scenario:

1. First, SMPs (SMA - source measurement agent and DMP - destination measurement peer) register on the MCS server and send unique SMP identifiers (UIDs);
2. Then, the SMA node sends a request to initialize the measurement on the parameters transmitted in the body of the request, including the UID of the DMP;

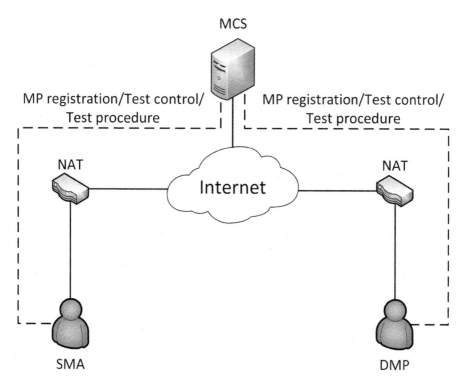

Fig. 5. General view of the results complex for e2e, based on the TURN procedure.

3. The MCS server analyzes the measurement parameters received from the SMA request, generates a measuring configuration, including the addresses and NAT ports to which the SMP hosts are assigned and sends it to the SMP hosts;
4. Next, SMP hosts perform one of types of measurement, periodically sending information about the MCS measurement process:
   - RTD (round trip delay);
   - Uplink bandwidth measurement;
   - Downlink bandwidth measurement.
5. After measuring, the SMP hosts send information about the MCS measurement results;
6. MCS sends the processed measurement result to the SMA node.

5   Data Transfer Protocol for Measurement System

A management protocol based on the HTTP protocol [18,19] and the JSON data representation format is defined for this system. This protocol uses HTTP POST requests to send data in the JSON format [20] between the MS peers. Data conversion functions are performed by a library for the programming language C++

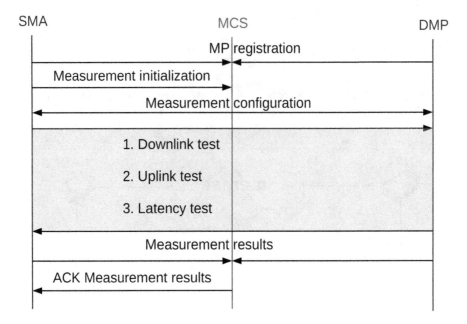

Fig. 6. Scenario of measuring network parameters for e2e, based on the TURN procedure.

- RapidJSON, built into the subsystems SMP CI and MCS MS. The measuring process itself is based on the punctures given during initialization measurement. When registering a node on the measure device, each peer sends the following information:

- Message type (setup);
- Write window throughput for measurement peer;
- Read window throughput for measurement peer;
- Measurement peer address;
- Measurement peer port.

After registering the peer, the measurement management engine returns the following information to the peer:

- Message type (setupACK);
- Measurement peer address;
- Measurement peer port;
- Measurement peer UID.
- When the measure is initialized, the following information is passed to the RTNP server from source SMP:
- Measurement type;
- Message type;
- Timeout;
- Measurement duration;

- Write window throughput for source peer;
- Read window throughput for source peer;
- Threads quantity;
- Package quantity - for measurements with a delay or as an alternative to the parameter measurement duration (optional);
- Size of measurement message body - for uplink and downlink throughput measurements (optional);
- Measurement protocol: HTTP, HTTPS, CoAP, MQTT, AMQP, XMPP, FTP, RTP, etc.;
- Used encryption: SSL/TLS, DTLS, NONE;
- Encryption/decryption method: RSA. NTRUEncrypt, etc.;
- Destination peer address;
- Destination peer port;
- Destination UID.

During the SMP measuring process, the MCS measurement progress data is sent:

- Measurement UID;
- Destination peer UID;
- Source peer UID;
- Measurement type;
- Message type;
- Thread number;
- Latency;
- Uplink data transmission throughput;
- Downlink data transmission throughput;
- Package loose indicator;
- Measurement protocol: HTTP, HTTPS, CoAP, MQTT, AMQP, XMPP, FTP, RTP, etc.;
- Used encryption: SSL/TLS, DTLS, NONE;
- Encryption/decryption method: RSA. NTRUEncrypt, etc.

The measurement results are stored in the MDB database, including:

- Measurement UID;
- Destination peer UID;
- Source peer UID;
- Measurement type;
- Message type;
- Threads quantity;
- Date and time;
- Latency;
- Uplink data transmission throughput;
- Downlink data transmission throughput;
- Package loose indicator;
- Measurement protocol: HTTP, HTTPS, CoAP, MQTT, AMQP, XMPP, FTP, RTP, etc.;

- Used encryption: SSL/TLS, DTLS, NONE;
- Encryption/decryption method: RSA. NTRUEncrypt, etc.;
- Destination peer address;
- Destination peer port;
- Source peer address;
- Source peer port;
- Number of concurrent connections Measurement Agent public IP (Optional);
- Measurement Agent phone status and radio technology (e.g. UMTS, LTE, etc.) and quality (eg.RSSI). (Optional for radio-frequency networks);
- Geo-location and tracking accuracy as well as movement during the measurement (latitude, longitude, altitude, timestamp, speed, bearing, location provider, information to assess GPS accuracy, etc.). (Optional).

6   Measurement System Testing

The developed MS was tested for all the described methods of measuring network parameters (e2s, e2e STUN, e2e TURN) between SMA, DMP and MCS devices. The model on the basis of which the testing was conducted is shown in Fig. 7.

Table 1. The results of the preliminary testing of the MS

Network parameters	SMA-MCS MS	DMP-MCS MS	SMA-MCS IPerf3+Ping	DMP-MCS IPerf3+Ping
Latency RTT (ms)	14,63	12,88	14,98	12,36
Downlink Throughtput (*Mbps*)	47,97	45,93	48,44	46,74
Uplink Throughtput (*Mbps*)	47,30	46,32	48,64	46,81

For each of the methods, the following network indicators were obtained:

- RTD (round trip delay);
- Uplink bandwidth measurement;
- Downlink bandwidth measurement.

In this model there are the following elements:

- MCS is a remote monitoring server for measuring network parameters. This server is a computing device (VPS) with an installed Debian Linux 8 OS, which has an external IPv4 address;
- SMA is a computing device located behind NAT and initiating a procedure for measuring network parameters between it and the DMP node;
- DMP is a computing device, depending on the scenario, located behind NAT or having an external IPv4 address and acting as the destination node for testing.

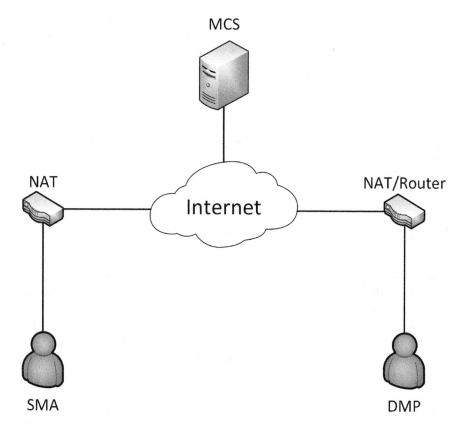

Fig. 7. Model network for testing the developed MS.

Table 1 shows the averaged results of preliminary measuring of network parameters. Also in Table 2, network parameter measurements are performed using public network measuring services.

The data shown in Table 1 indicates that the network parameters for the MSC server are determined by the VPS service provider and are limited in terms of downlink and uplink throughput in the 46–49 Mbps range and are limited in the RTD (round trip delay) within 10–12 ms. Network parameters when using the MS and network testing utilities IPerf3 and Ping show similar parameters.

According to the data displayed in Table 2 it can be seen that the popular network parameter measurement services do not show real indicators of communication services provided by the operator to the user.

Table 3 shows the average results of measuring network parameters for each of the methods described in this article. According to Table 3, the e2s method, which describes measuring network parameters directly to the web service on an external IPv4 address, shows the best network latency, uplink and downlink throughput. This method allows you to measure network parameters using many different application protocols that function both on top of the TCP transport

Table 2. Results of SMP testing using public network measuring services

Network parameters	Speedtest.net SMA	2ip.ru SMA	Speedtest.net DMP	2ip.ru DMP
Latency RTT (ms)	1,98	7,57	9,00	59,17
Downlink Throughtput (*Mbps*)	93,49	67,11	40,80	28,81
Uplink Throughtput (*Mbps*)	91.36	85,30	35,17	50,84

Table 3. Results of testing MS

Network parameters	e2s	e2e STUN	e2e TURN
Latency RTT (ms)	6,31	8,17	33,17
Downlink Throughtput (*Mbps*)	64,19	64,46	45,74
Uplink Throughtput (*Mbps*)	71,69	69,11	46,23

protocol and over UDP. The e2e STUN method also shows a satisfactory performance for the measured network parameters, but nevertheless this method allows measuring only with application protocols running on top of UDP. The e2e TURN method shows a lower throughput and a higher network delay rate than the e2s and e2e STUN methods, since the way the traffic tunneling through the MCS server limits the maximum network parameters by the bandwidth and delay of the MCS server.

7   Conclusions

In this article the hardware and software complex architecture was developed for measurements of the quality of the communication services provided. Based on this architecture, a method for measuring network parameters between a last mile device and remote cloud service was developed and a method for measuring network parameters between the end nodes functioning behind the NAT. The protocol of data transmission was also described and measurements were made for the developed measurement system (MS). Based on the results of the study, it can be argued that the existing methods of measuring network parameters for accessing the Internet by the user are biased and measure the indicators of access to specific measuring services. This software and hardware was developed based on ITU-T Rec. Q.3960, describing the scenarios for measuring network parameters for the end user (e2e). Application of the methods described in ITU-T Rec. Q.3960 and inclusion of the last node of the responsibility zone of the software provider Probe will allow an objective assessment of the quality of the provided communication services and the quality of access to remote devices on the Internet.

Acknowledgments. The publication was financially supported by the Ministry of Education and Science of the Russian Federation (the Agreement number 02.a03.21.0008).

References

1. Paramonov, A., Koucheryavy, A.: M2M traffic models and flow types in case of mass event detection. In: Balandin, S., Andreev, S., Koucheryavy, Y. (eds.) NEW2AN 2014. LNCS, vol. 8638, pp. 294–300. Springer, Cham (2014). doi:10. 1007/978-3-319-10353-2_25
2. Kirichek, R., Vladyko, A., Zakharov, M., Koucheryavy, A.: Model networks for Internet of Things and SDN. In: 18th International Conference on Advanced Communication Technology (ICACT), pp. 76–79 (2016). doi:10.1109/ICACT.2016. 7423280
3. Kirichek, R., Koucheryavy, A.: Internet of Things laboratory test bed. In: Zeng, Q.A. (ed.) Wireless Communications, Networking and Applications. Lecture Notes in Electrical Engineering, vol. 348. Springer, New Delhi (2016). doi:10.1007/ 978-81-322-2580-5_44
4. Kirichek, R., Golubeva, M., Kulik, V., Koucheryavy, A.: The home network traffic models investigation. In: 18th International Conference on Advanced Communication Technology (ICACT), pp. 97–100 (2016). doi:10.1109/ICACT.2016.7423288
5. Iera, A., Floerkemeier, C., Mitsugi, J., Morabito, G.: The Internet of Things. IEEE Wirel. Commun. 17(6), 8–9 (2010). doi:10.1109/MWC.2010.5675772
6. Gubbi, J., Buyya, R., Marusic, S., Palaniswami, M.: Internet of Things (IoT): a vision, architectural elements, and future directions. Fut. Gener. Comput. Syst. 29(7), 1645–1660 (2013). doi:10.1016/j.future.2013.01.010
7. Kirichek, R., Kulik, V.V., Koucheryavy, A.: False clouds for internet of things and methods of protection. In: 18th International Conference on Advanced Communication Technology (ICACT), pp. 201–205 (2016). doi:10.1109/ICACT.2016. 7423328
8. Ateya, A., Vybornova, A., Kirichek, R., Koucheryavy, A.: Multilevel cloud based Tactile Internet system. In: 19th International Conference on Advanced Communication Technology (ICACT), pp. 105–110 (2017). doi:10.23919/ICACT.2017. 7890067
9. Koucheryavy, A., Makolkina, M., Paramonov, A.: Applications of augmented reality traffic and quality requirements study and modeling. In: Distributed Computer and Communication Networks: Control, Computation, Communications (DCCN-2016), vol. 3, pp. 289-300. RUDN University, November 2016
10. Kirichek, R., Pirmagomedov, R., Glushakov, R., Koucheryavy, A.: Live substance in cyberspace -biodriver system. In: Proceedings of the 18th International Conference on Advanced Communication Technology, ICACT, pp. 274–278 (2016). doi:10. 1109/ICACT.2016.7423358
11. Pirmagomedov, R., Hudoev, I., Kirichek, R., Koucheryavy, A., Glushakov, R.: Analysis of delays in medical applications of nanonetworks. In: 8th International Congress on Ultra Modern Telecommunications and Control Systems and Workshops (ICUMT), pp. 80–86 (2016). doi:10.1109/ICUMT.2016.7765231
12. ITU-T Q.3960 Framework of Internet related performance measurements: ITU-T SG11. Series Q: Switching and Signalling, Jule 2016. http://handle.itu.int/11. 1002/1000/12747
13. Srisuresh, P., Holdrege, M.: RFC 2663. IP Network Address Translator (NAT) Terminology and Considerations. IETF Network Working Group, August 1999. https://tools.ietf.org/html/rfc2663
14. Rosenberg, J., Mahy, R., Matthews, P., Wing, D.: RFC 5389 Session Traversal Utilities for NAT (STUN). IETF Network Working Group, October 2008. https:// tools.ietf.org/html/rfc5389

15. Mahy, R., Matthews, P., Rosenberg, J.: RFC 5766 traversal using relays around NAT (TURN): relay extensions to session traversal utilities for NAT (STUN). IETF Network Working Group, April 2010. https://tools.ietf.org/html/rfc5766
16. Fielding, R., Gettys, J., Mogul, J., Frystyk, H., Masinter, L., Leach, P., Berners-Lee, T.: RFC 2616 Hypertext Transfer Protocol – HTTP/1.1. IETF Network Working Group, June 1999. https://tools.ietf.org/html/rfc2616
17. Fielding, R., Reschke, J.: RFC 7230. Hypertext Transfer Protocol (HTTP/1.1): Message Syntax and Routing. IETF Network Working Group, June 2014. https://tools.ietf.org/html/rfc7230
18. Rescorla, E.: RFC 2818 HTTP over TLS. IETF Network Working Group, May 2000. https://tools.ietf.org/html/rfc2818
19. Belshe, M., Peon, R., Thomson, M.: RFC 7540 Hypertext Transfer Protocol Version 2 (HTTP/2). IETF, May 2015. https://tools.ietf.org/html/rfc7540
20. Bray, T.: RFC 7159 The JavaScript Object Notation (JSON) Data Interchange Format. IETF, March 2014. https://tools.ietf.org/html/rfc7159

The Application of Classification Schemes While Describing Metadata of the Multidimensional Information System Based on the Cluster Method

Maxim Fomin[(✉)]

Department of Information Technologies, RUDN University,
Miklukho-Maklaya street 6, Moscow 117198, Russia
fomin_mb@rudn.university

Abstract. Metadata of the multidimensional information system can be described through setting the options for the cells of the multidimensional cube. The cluster method can be used for the description of the sparse data cube structure. The core of this method is the formation of groups of members which are semantically connected with groups of members of other dimensions. Connected groups related to different dimensions describe the cluster of cells. Clusters can be merged into sets of cells. The term where such sets are combined by operations of set theory describes the structure of the multidimensional data cube. Classification schemes can be used while forming a cluster. Every classification scheme is a graph describing the hierarchy of members which are connected with a separate structural component of the observed phenomenon. The coupling between several classification schemes related to different structural components helps to describe the metadata of the multidimensional information system.

Keywords: Multidimensional data model · Sparse cube · Classification scheme · Set of possible member combinations · Cluster of member combinations

1 Introduction

The data cube is characterized by dimensions in the information system where measures related to the observed phenomenon exist in the multidimensional form. Every dimension complies with some certain analysis aspect of this observed phenomenon. The multidimensional cube is significantly sparse and unevenly filled in case the system contains a lot of semantically diverse data [1–8]. As a result, there is a problem of developing an adequate way to describe the structure of an analytical space which use would make it possible to effectively organize the data analysis process [9–18].

Every possible cube cell complies with some fact. The cluster method can be used for the effective description of the multidimensional cube structure. This

© Springer International Publishing AG 2017
V.M. Vishnevskiy et al. (Eds.): DCCN 2017, CCIS 700, pp. 307–318, 2017.
DOI: 10.1007/978-3-319-66836-9_26

method is based on the semantic analysis of different dimensions' members' compatibility in possible cube cells. It allows describing the metadata of the information system as a set of possible member combinations. Possible member combinations comply with possible cells of the multidimensional cube. It can be difficult to analyze member combinations of the whole cube if there are a lot of dimensions while describing the set of possible member combinations. In order to simplify it the problem can be divided into several stages:

- splitting of the observed phenomenon which is described by the information system into structural constituents;
- compatibility analysis of the members characterizing these structural constituents;
- formation of classification schemes which contain the description of possible member combination for every structural constituent;
- coupling of combinations brought from different classification schemes into sets of possible member combinations of all the multidimensional cube dimensions.

In the process of carrying out the steps described above, the characteristics of the observed phenomenon and the relationships among them should be considered from the standpoint of classification, which would reflect the semantics of the observed phenomenon. Classification can be performed using a hierarchical principle. In this case, the detected properties can be represented in the form of a connected acyclic graph. The characteristics of the observed phenomenon are divided into classes, which can belong to different hierarchy levels of the graph. Advantages of the hierarchical method of classification are the possibility of using a large number of characteristics and informational value.

2   The Description of the Sparse Data Cube with the Use of Member Combinations

Each dimension of the multidimensional data cube H complies with an aspect of analysis of the observed phenomenon for which the multidimensional information system is designed. All the dimensions form the set $D(H) = \{D^1, D^2, ..., D^n\}$, where D^i is i-dimension, $n = \dim(H)$ – multidimensional cube dimensionality. The dimension is defined by the set of members: $D^i = \{d_1^i, d_2^i, ..., d_{k_i}^i\}$, where k_i – the quantity of members of i-dimension. The members of D^i are chosen from a set of positions of the basic classifier which complies with the aspect of the observed phenomenon associated with D^i.

The multidimensional data cube is a structured set of cells. Every cell c of the multidimensional cube can be compared with the member combination $c = (d_{i_1}^1, d_{i_2}^2, ..., d_{i_n}^n)$; one member is for one dimension [19]. In the case of the sparse cube not all possible member combinations comply with possible cube cells, i.e. cells describing facts.

If the multidimensional cube contains semantically diverse data it is possible that some dimensions cannot be defined in compliance with the existing set of other dimension members. In this situation the members of some dimensions cannot be defined while describing a possible cell of the multidimensional cube. A special member "Not in use" can be applied for the definition of these semantically undefined dimensions [19]. We will use such an extended method of member definition in cells. In this case the structure of the multidimensional data cube in the information system can be described as the set of possible member combinations. Different values from the classifiers, which complies with the dimensions, and the special value "Not in use" can be applied in the combinations of this set. To refer to the set of possible member combinations we will use the abbreviation "SPMC".

The observed phenomenon is characterized by the measure values defined in possible cells of the multidimensional cube. The full set of measures composes the set $V(H) = \{v_1, v_2, ..., v_p\}$, where v_j is j-measure, p – the quantity of measures in the hypercube. Not all the measures from the $V(H)$ can be defined in the possible cell. This situation can appear in case of semantic inconsistency between the members defining the cell and some measures. While describing SPMC for every possible sell it is necessary to define its own set $V(c) = \{v_1, v_2, ..., v_{p_c}\}$, which consists of certain measures for this cell, $1 \leq p_c \leq p$. We can use the special value "Not in use" for the description of c measures, which are not included in the set $V(c)$. The rule must be hold: a set of measures $V(c)$ defined in a possible cell c can not be empty. Description of measures in cells of multidimensional cube matching the combinations of members not included into the SPMC does not make sense.

3 The Appliance of the Cluster Method for the Description of the Multidimensional Data Cube Structure

The SPMC structure describes the semantic of the observed phenomenon. The information about it is contained in the multidimensional data cube. Taking into account the semantic the succinct description of the SPMC can be obtained with the help of the cluster method based on the analysis of links between members [20]. The cluster method allows identifying the groups of members. The group $G_j^i = \left\{d_1^i, d_2^i, ..., d_{m_j}^i\right\}$ of members in i-dimension includes m_j members $(1 \leq m_j \leq k_i)$, where j is a group number and contains members, which equally coincide in the SPMC with the members from some groups of members of other dimensions.

It is possible to define connected groups of member in different dimensions with the help of the semantic analysis. The cluster of member combinations K is the set of member combinations, which can be obtained with the help of Cartesian product where operands are groups of members or special value "Not in use"; one operand stands for every dimension used in the cluster $\mathrm{SPMC}(K) = G_1 \times G_2 \times ... \times G_n$. Clusters of member combinations can be used for the description of the SPMC.

In the observed phenomenon, it is possible to distinguish different semantic components. In this case, it is possible to form subsets of member combinations, each of which corresponds to its semantic component. A subset of member combinations is the union of clusters of combinations. It can be constructed as a result of the analysis of the compatibility of the values of the characteristics of the observed phenomenon, corresponding to some of its semantic component.

While performing following steps the cluster method allows getting SPMC description for the multidimensional cube [20].

1. N semantic components should be defined in the structure of the observed phenomenon. These components should be contrasted to the subset of combinations Q_k, $k = 1, ..., N$. The expression for the SPMC(H) should be defined in a form, where the subsets Q_k are connected with the help of operations of set theory.

2. The layers of the dimensions $L^i = \{D^{j_1}, D^{j_2}, ..., D^{j_i}\}$ should be defined in every subset Q_k where $i = 1, ..., m_k$ is the layer number in the subset, m_k is the quantity of layers, j_i is the dimension number in a layer, l is the quantity of dimensions in i-layer. The layer of dimensions is the set of dimensions where member combinations in the set do not depend on member combinations of dimensions outside this very particular layer. When all the subsets of member combinations are known for every layer of dimensions the subset of combinations Q_k can be obtained with the help of Cartesian product layer by layer: $Q = \mathrm{SPMC}(L^1) \times \mathrm{SPMC}(L^2) \times ... \times \mathrm{SPMC}(L^m)$. In this expression $\mathrm{SPMC}(L^i)$ is the set of member combinations of the i-th layer.

3. For every layer L^i in the subset Q_k the subset of possible member combinations $\mathrm{SPMC}(L^i)$ should be defined as the set of clusters of member combinations in the layer. Every cluster in the layer is defined by the collection of groups of members G_j^k, there k is the number of dimension in the layer and $j = 1, ..., l$ is the group number: $K = \left\{ G_1^{j_1}, G_2^{j_2}, ..., G_l^{j_l} \right\}$. The clusters of member combinations can be obtained with the help of Cartesian product of the groups of members (or special member "Not in use" instead of a group), one group for every dimension in the layer: $\mathrm{SPMC}(K) = G_1^{j_1} \times G_2^{j_2} \times ... \times G_l^{j_l}$.

In the picture below (Fig. 1) you can see the diagram which describes the relationship among structural elements of the SPMC.

We can distinguish two typical cases of split of observed phenomenon into semantic components and representations of SPMC(H) using several subsets. The first one takes place when different subdivisions of the dimensions onto the layers occur during the analysis of different semantic components, and the second one – when there is a simple way of building a subset describing the SPMC redundantly, and the efficient way to describe the combinations which are to be excluded from this subset to reduce it to the SPMC. Let us consider these cases in more detail.

In the first case, the decomposition of the observed phenomenon on semantic components corresponds to the union of member combinations subsets:

$$\mathrm{SPMC}(H) = Q_1 \cup Q_2 \cup ... \cup Q_l.$$

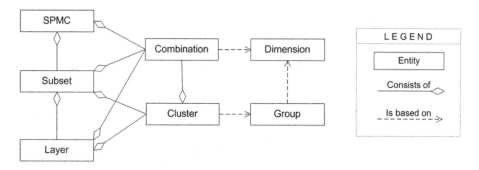

Fig. 1. The diagram of the SPMC structure

Set of analytical space dimensions can be divided into layers in different ways due to differences in semantics of the observed phenomenon components:

$$D(H) = L_i^1 \cup L_i^2 \cup ... \cup L_i^{m_i},$$

there $i = 1...l$ – number of component, and m_i – the quantity of layers in i-component. Each subset Q_i is formed according to its split of set of the dimensions can into layers.

In the second case, set of possible member combinations is represented as the difference of two subsets:

$$\text{SPMC}(H) = R \backslash Q.$$

there R – set of member combinations, described with an excess (set to reduce), and Q – set of combinations to be excluded. Set to reduce may be formed using the following rules. It should include member combinations obtained by the Cartesian product of all members of all dimensions. It must be supplemented with a set of combinations that contain the special value "Not in use" for some dimensions, for which this value is acceptable. From this set it should be excluded those combinations which can be obtained by replacing the special value "Not in use" by the member. This approach can be used in case the set $\text{SPMC}(H)$ has a complex structure and it may be offered a simple algorithm of forming a subset Q.

4   The Application of the Classification Schemes for the Description of the Observed Phenomenon

From the position of semantic the description of the observed phenomenon within the multidimensional data model consists of the classification characteristics (i.e. dimensions of data cube) detecting and establishing the links between them. Meanwhile the observed phenomenon is not considered to be a multi-component object and classification characteristics are not ranked: there are neither major nor minor ones among them. The search for links among the members is the key to establishing the links among dimensions. It can be rather difficult if there are a great number of dimensions. The drawbacks of the approach can be diminished with the introduction of additional objects into the data model of the information system – classification schemes of characteristics for the observed phenomenon (thus the abbreviation CSC stands for a classification scheme). There are certain requirements for CSCs:

1. It is necessary to take into account the component structure of the observed phenomenon while defining CSC. If the observed phenomenon can be semantically divided into separate structural components for which is possible to choose their own sets of aspects for analysis every component should be compared with CSC. The procedure of CSC formation is based on defining and analysis of the characteristics relevant to the chosen aspects of the analysis. The dimensions of the multidimensional cube should be compared with the characteristics.

2. CSC should be formed on the hierarchical principle. Ranking should be established among the characteristics related to CSC. This ranking allocates the dimensions which to some extent convey the essence of the structural component for the observed phenomenon. This component is compared with CSC. It is necessary to define the major dimension which is more likely to reflect the semantic of the structural component relevant to CSC. The hierarchy of characteristics should be formed from other dimensions included into CSC, which are semantically subordinate to the major dimension and express some particular properties of the structural component for the observed phenomenon. The following principle should be observed: the members of the major dimension signify the most important properties of the observed phenomenon, the members of other dimensions which come hierarchically below the major one signify some subordinate properties specifying the essence of the major dimension.

3. While forming the hierarchy of the characteristics for the observed phenomenon in CSC it should be possible to describe the members of the major dimension separately or in groups of members as different members can be connected with different semantic aspects of the structural component for the observed phenomenon. Different hierarchies of characteristics should be formed for the members of the major dimension, which are semantically different.

4. In the hierarchy of the characteristics in CSC there must be the information about the set of measures describing the observed phenomenon in case of choosing some particular members from the hierarchy.

Let us consider "Lending" as the process, which can be divided into separate structural components. There are four components which can be compared with CSCs: "Participants of lending", "Instruments for lending", "Conditions of lending" and "Risk factors for lending".

During the development of the information system classification schemes can be the source of the classification information about the observed phenomenon. At the same time CSC is semantically connected with the structural component of the observed phenomenon and can be the source of information about the characteristics of the structural component presented in the hierarchical form. CSC is technologically connected with the dimensions of data cube and can be the sample during the forming of metadata for the multidimensional information system.

5   Construction of the Classification Scheme as a Tree of Member Combinations

The classification scheme (CSC) is an object of the multidimensional information system which describes the structural component of the observed phenomenon and contains the following information:

- The set of dimensions included into the CSC;
- The set of members of these dimensions included into the CSC;
- The major dimension chosen in the set of dimensions included into the CSC;
- The set of measures included into the CSC;
- The tree of member combinations of CSC which form the hierarchy of the classification characteristics.

The hierarchical principal of CSC forming is realized in the structure of a tree which presents the member combinations in CSC. The tree of combinations can be formed as a result of the semantic analysis of the structural component for the observed phenomenon. The tree can be defined while describing the process of its formation. One should start the formation of the tree from its roots where the groups of members of the major dimension are placed. Then it is necessary to go down passing the hierarchical levels and adding a group of members to each of them. Thus every group reveals the essence of every previous member on the previous level. What is more it is necessary to add the group related to the dimension which is mostly connected with the members of the previous level. As a result different sequences of dimensions can appear in different tree brunches on the way from the roots to leaves.

Moving down through the hierarchical levels of the tree we have the members placed on these levels that express the less and less significant characteristics of the observed phenomenon. Thus, the ranking of the characteristics for the observed phenomenon is set up.

As a result of performing all the steps described above it is possible to form the tree of member combinations in CSC. Following rules are implemented for the tree structure:

1. The root of the tree is the node "Major dimension".
2. The tree itself is a hierarchical structure where the levels are set through alternating such nodes as "Group of members" and "Dimension". At the same time groups of members should be formed in the dimensions relevant to the nodes hierarchically placed one level higher.
3. Leaves of the tree are nodes "Group of members".

4. The node "Group of members" (except the node which is a tree leaf) should
be relevant to the node "Dimension" hierarchically placed on a lower level.
Only one node or several nodes "Group of members" placed on a lower level
can be relevant to the node "Dimension".
5. Moving from the root to a leaf you can see every dimension only once.

There is a tree of member combinations in the picture below (Fig. 2).

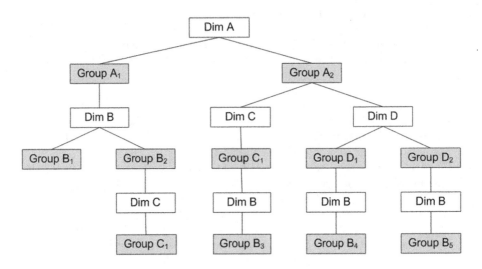

Fig. 2. The example of a tree of member combinations of SPMC

Every way from the root of a tree of member combinations to a leaf contains
a certain collection of groups of member from different dimensions. It means
that the way specifies the cluster of member combinations in CSC. To form the
whole set of clusters it is necessary to traverse the whole tree. It the process of
breadth first search on the trees the number of forming clusters increases every
time when you meet several groups of members on some hierarchical level related
to one unit "Dimension". If there is some dimension in the structure of CSC but
it is absent on the way from the root to a leaf the value of this dimension in
cluster which is relevant to the way considered is "Not in use".

There (Table 1) are clusters of member combinations for CSC that corresponds to a tree you can see in Fig. 2.

From the position of semantic every cluster in a tree of member combinations
in CSC is relevant to some set of properties for the observed phenomenon. In the
information system these properties are described as collection of some measures
included into the classification scheme. There can be different sets of measures in
different clusters. The information about the measures in the cluster of member
combinations for CSC should be described in a tree of member combinations for
CSC as leave attribute.

Table 1. Clusters of member combinations for a classification scheme

N	Dimension A	Dimension B	Dimension C	Dimension D
1	A1	B1	Not in use	Not in use
2	A1	B2	C1	Not in use
3	A2	B3	C1	Not in use
4	A2	B4	Not in use	D1
5	A2	B5	Not in use	D1

6 The Application of Classification Schemes While Forming the Structure of the Multidimensional Data Cube

The important property of CSC is the possibility to use member combinations which CSC describes while forming the metadata of the information system. The set of dimensions for the multidimensional cube is formed in accordance with the following principle: all the characteristics that can influence measures used for the analysis of the observed phenomenon should present here. It is a rather difficult task to form SPMC on such a wide range of dimensions for the multidimensional cube. CSCs which fulfill the classifying function in relation to SPMC help to resolve this problem.

Every CSC related to the observed phenomenon contains the information about the member compatibility in that part of dimensions which is connected with some properties of this observed phenomenon. The task is to join these combinations from different CSC together in SPMC. Joining member combinations from two CSCs you can observe one of the following situations:

– dimensions included into the first CSC are absent from the other and vice versa;
– sets of dimensions in CSCs partially coincide.

In the first situation the compatibility of members from one CSC does not depend on the members from the other CSC. This situation complies with the case when SPMC describing the structure of the multidimensional data cube is divided into layers. In this case for every CSC the set SPMC(CSC) is the description of some layer of SPMC(H).

In case of junction between sets of dimensions related to two meeting CSCs there is a necessity to prolong the combinations from one CSC with the combinations from the other CSC with a partial junction in dimensions while forming SPMC(H). In this case the only possible solution can be absent. In the described situation a business analyst should choose the right variant for SPMC(H) forming based on the semantic analysis. Moreover following issues should be resolved:

– if members in combinations from different CSC in area of intersection of dimensions coincide whether it is right to consider such combinations to be

the continuation of each other or to consider that each of them creates its own combination in SPMC(H);
- if for any member combination from one CSC there are several continuations from the other CSC what variant is to choose while forming the combination in SPMC(H).

7 Conclusion

Metadata of the multidimensional information system created with the help of the cluster method have the structure of sparse and unevenly filled multidimensional cube. The generation of such metadata is a difficult task which can be resolved while considering the observed phenomenon described in the information system to be a set of structural components. Every structural component complies with the classification scheme where the data can be presented as a set of possible member combinations connected with the characteristics of this structural component. In comparison with the metadata of the information system in general classification schemes describe the narrow set of properties for the observed phenomenon and present characteristics of these properties in the hierarchical form. In order to form a classification scheme it is necessary to analyze semantically the characteristics of the structural components for the observed phenomenon, every single structural component individually. A limited quantity of dimensions simplifies the process of forming a hierarchy of values of characteristics.

As a result there is a possibility to identify the inner structure of the multidimensional data cube. Collections of dimensions included in different classification schemes partially intersect. The junction of member combinations from different CSs allows restoring the structure of the multidimensional data cube. This procedure should be performed with the participation of the analyst who decides on the choice of the way to continue one combination to another in case of multivaluedness.

Acknowledgments. The work is partially supported by the Ministry of Education and Science of the Russian Federation (the Agreement number 02.a03.21.0008).

References

1. Thomsen, E.: OLAP Solution: Building Multidimensional Information System. Willey Computer Publishing, New York (2002). ISBN: 0-471-40030-0
2. Hirata, C.M., Lima, J.C.: Multidimensional cyclic graph approach: Representing a data cube without common sub-graphs. Inf. Sci. **181**, 2626–2655 (2011)
3. Karayannidis, N., Sellis, T., Kouvaras, Y.: CUBE file: A file structure for hierarchically clustered OLAP cubes. In: Bertino, E., Christodoulakis, S., Plexousakis, D., Christophides, V., Koubarakis, M., Böhm, K., Ferrari, E. (eds.) EDBT 2004. LNCS, vol. 2992, pp. 621–638. Springer, Heidelberg (2004). doi:10. 1007/978-3-540-24741-8_36

4. Chun, S.-J.: Partial prefix sum method for large data warehouses. In: Zhong, N., Raś, Z.W., Tsumoto, S., Suzuki, E. (eds.) ISMIS 2003. LNCS (LNAI), vol. 2871, pp. 473–477. Springer, Heidelberg (2003). doi:10.1007/978-3-540-39592-8_67

5. Messaoud, R.B., Boussaid, O., Rabaseda, S.L.: A Multiple Correspondence Analysis to Organize Data Cube. In: Databases and Information Systems IV DB&IS 2006, pp. 133–146. IOS Press, Vilnius (2007). ISBN:978-1-58603-715-4

6. Jin, R., Vaidyanathan, J.K., Yang, G., Agrawal, G.: Communication and memory optimal parallel data cube construction. IEEE Trans. Parallel Distrib. Syst. **16**, 1105–1119 (2005)

7. Luo, Z.W., Ling, T.W., Ang, C.H., Lee, S.Y., Cui, B.: Range top/bottom k queries in OLAP sparse data cubes. In: Mayr, H.C., Lazansky, J., Quirchmayr, G., Vogel, P. (eds.) DEXA 2001. LNCS, vol. 2113, pp. 678–687. Springer, Heidelberg (2001). doi:10.1007/3-540-44759-8_66

8. Fu, L.: Efficient evaluation of sparse data cubes. In: Li, Q., Wang, G., Feng, L. (eds.) WAIM 2004. LNCS, vol. 3129, pp. 336–345. Springer, Heidelberg (2004). doi:10.1007/978-3-540-27772-9_34

9. Chen, C., Feng, J., Xiang, L.: Computation of sparse data cubes with constraints. In: Kambayashi, Y., Mohania, M., Wöß, W. (eds.) DaWaK 2003. LNCS, vol. 2737, pp. 14–23. Springer, Heidelberg (2003). doi:10.1007/978-3-540-45228-7_3

10. Salmam, F.Z., Fakir, M., Errattahi, R.: Prediction in OLAP data cubes. J. Inf. Knowl. Manag. **15**, 449–458 (2016)

11. Romero, O., Pedersen, T.B., Berlanga, R., Nebot, V., Aramburu, M.J., Simitsis, A.: Using Semantic web technologies for exploratory OLAP: A survey. IEEE Trans. Knowl. Data Eng. **27**, 571–588 (2015)

12. Gomez, L.I., Gomez, S.A., Vaisman, A.: A generic data model and query language for spatiotemporal OLAP cube analysis. In: Proceedings of the 15th International Conference on Extending Database Technology – EDBT 2012, pp. 300–311, Berlin (2012). ISBN:978-1-4503-0790-1

13. Tsai, M.-F., Chu, W.: A Multidimensional Aggregation Object (MAO) framework for computing distributive aggregations. In: Kambayashi, Y., Mohania, M., Wöß, W. (eds.) DaWaK 2003. LNCS, vol. 2737, pp. 45–54. Springer, Heidelberg (2003). doi:10.1007/978-3-540-45228-7_6

14. Vitter, J.S., Wang, M.: Approximate computation of multidimensional aggregates of sparse data using wavelets,. In: Proceedings of the 1999 International Conference on Management of Data - SIGMOD99, pp. 193–204. ACM, New York (1999). ISBN:1-58113-084-8

15. Leonhardi, B., Mitschang, B., Pulido, R., Sieb, C., Wurst, M.: Augmenting OLAP exploration with dynamic advanced analytics. In: Proceedings of the 13th International Conference on Extending Database Technology - EDBT 2010, pp. 687–692. ACM, New York (2010). ISBN:978-1-60558-945-9

16. Wang, W., Lu, H., Feng, J., Yu, J.X.: Condensed Cube: An Effective approach to reducing data cube size. In: Proceedings of the 18th International Conference on Data Engineering - ICDE02, pp. 155–165. IEEE Computer Society, Washington (2002). ISBN:0-7695-1531-2

17. Goil, S., Choudhary, A.: Design and implementation of a scalable parallel system for multidimensional analysis and OLAP. In: Parallel and Distributed Processing - 11th IPPS/SPDP 1999, pp. 576–581. Springer, Heidelberg (1999) ISBN:978-3-540-65831-3

18. Cuzzocrea, A.: OLAP data cube compression techniques: a ten-year-long history. In: Kim, T., Lee, Y., Kang, B.-H., Ślęzak, D. (eds.) FGIT 2010. LNCS, vol. 6485, pp. 751–754. Springer, Heidelberg (2010). doi:10.1007/978-3-642-17569-5_74

19. Viskov, A.V., Fomin, M.B.: Methods of description of possible combinations of signs and details while using the multidimensional models in infocomm systems. T-Comm. - Telecommun. Transp. **7**, 45–47 (2012)
20. Fomin, M.: Cluster method of description of information system data model based on multidimensional approach. In: Vishnevskiy, V.M., Samouylov, K.E., Kozyrev, D.V. (eds.) DCCN 2016. CCIS, vol. 678, pp. 657–668. Springer, Cham (2016). doi:10.1007/978-3-319-51917-3_56

Yet Another Method for Heterogeneous Data Fusion and Preprocessing in Proactive Decision Support Systems: Distributed Architecture Approach

Van Phu Tran, Maxim Shcherbakov$^{(\boxtimes)}$, and Tuan Anh Nguyen

Department of Computer Aided Design, Volgograd State Technical University,
Lenin Avenue, 28, 400005 Volgograd, Russia
vanphu.vstu.russia@gmail.com, maxim.shcherbakov@vstu.ru,
anhtuank37@gmail.com

Abstract. In the multi-sensors environment, the crucial issue is collecting data from different sources. The paper considers the problem of data gathering from different data sources in the framework of proactive intelligent decision support systems design. The aim is to provide the invariate access to heterogeneous data stored in a data warehouse for further processing. There are two types of data sources are considered in the paper: machine or sensor data and video streams. We propose an 'on-fly' method of heterogeneous data fusion and preprocessing toward to minimisation of execution time of queries. A proposed method is implemented in the five-layer distributed architecture of the system based on Apache Kafka and Spark Streaming technology. The main conclusion is that in case of heterogeneous data (like video and loged data) and functional requirements for query execution over these data, the distributed data preprocessing might be efficient in comparison with batch processing.

Keywords: Distributed computer networks · Heterogeneous data · Data fusion

1 Introduction

Nowadays, development of personal intelligent assistance technologies is a hot topic for academic and business society. Artificial intelligence and machine learning tools help to upgrade existing decision support systems to a new class of decision making systems which generate decisions automatically. Advanced techniques for decision making systems design are based on proactive computing which allows to shift human intervention on the upper level of management [12]. Based on this idea, a new generation of systems called *proactive decision support systems* is the object of exploration in this study. The concept of proactive systems described by the following scheme *detect - forecast - decide - act*. Principles of decision support system design based on data-driven approach and

© Springer International Publishing AG 2017
V.M. Vishnevskiy et al. (Eds.): DCCN 2017, CCIS 700, pp. 319–330, 2017.
DOI: 10.1007/978-3-319-66836-9_27

event processing [3]. A particular problem in proactive computing is a problem of management (or control) using predictive analytics. In control theory, this approach is well-known as model predictive control or MPC [4,15]. In management, the approached is called predictive analytics or proactive maintenance. Generalized, a control scheme is expanded by components implementing predictive analytic technology or by using predicted features. As an example, the domain of urban development is considered in the research. In details, we explore the possibilities of implementation data-driven solutions for improving of urban processes management, e.g. public transport network analysis [6]. Implementation of data-driven solutions using predictive analytics can lead to sufficient cost reduction in urban management due to avoiding negative outcomes. In this case, two problems are considered: heterogeneous data collecting for further analysis and forecasting.

Basically, data collecting and data fusion are core processes in data-driven solutions [11]. The crucial factor here is access data time in decision making procedures. If data kept in distributed data storage and data has different format, then it is necessary to process the data during user's query execution. For instance, if a user requests a details about certain traffic situation on the street described by video streams and logged data, the video data and logged data should be processed and yielded as an output.

To solve the forecasting problem, forecasting methods based on expert knowledge and forecasting methods based on data processing [7] are used. In the latter case, data samples including features describing the object is required. If the search of proper imput feature combinations and model hyperparameters can be reduced to a combinatorial search problem on a grid or described as optimization problem [2], the collection and preliminary processing of diverse data from different sources is an unstructured task and requires significant time-consuming [5]. Variaty of collected data is crucial, as data can be different types: sensors data, social data, video streams, images, etc., so data collecters need to be adopted to this kind of data types.

The study is devoted to solving the problem of efficient access to data obtained from various sources by means of their preliminary processing and transforming to a given structure. The authors propose a new method for collecting and merging different types of data (sensory/log-data and video) for a predetermined real-time storage scheme. This method can be used for solving the task of monitoring the transport situation in the city.

2   Background

Recents studies devoted to design of proactive decision support systems for large-scale heterogeneous data processing [1,6]. We focus on two problems of efficient data processing that need to be solved: (i) design of efficient architecture of data processing systems and (ii) design of data scheme for fast data access.

Nowadays, the development of new video equipments, satellite television, and drones with video cameras increases of video data obtaining in real-time. Large amount of video data is difficult to process efficiently with a single resource, such as CPU and GPU. The graphics processor unit was designed for high-speed image processing, but this is not fast enough for large-scale image processing due to a memory limit.

Distributed File Systems, DFS (particular Hadoop Distributed File System, HDFS) was designed for keep heterogeneous data distributed on many computers in a cluster. However, DFS concept intends that data processing is performed in batch mode using Map-Reduce technology. In case of video data, it is stored in HDFS, and then video is processed after a certain time t. This is called *near real time* image/video processing.

In paper [13] we presented an approach for data processing in distributed architecture based on Kafka cluster. All video streams are sent to Kafka *brokers* through *producers*. Note, presented previously architecture is efficient for a 10 KB data packages. In case, of video streams processing, we observe decreasing of Kafka cluster performance. The following alternative options can be chosen. The first option is to send data including information about storage locations instead video files. The original video files are stores in DFS like NAS, HDFS, S3 etc. The second option is splitting messages into smaller bunches. Firstly, the initial data package is divided into 10 KB data bunches, then data sent to brokers with the similar broker's key to be sure that *consumers* restore initial data back properly. The third option is data compression. For instance, if the original message has XML format, the compression rate is very high. Compression can be performed by using supported. GZIP by managing *compression.codec* and *compressed.topics* settings in *producers*. However, the main idea is not about just sending video files via Kafka Zookeeper, but is about pre-processing of video data, as we intent reduce query run-time. Properties of video stream also need to be extracted. The image processing library HIPI or OpenCV used with the Apache Hadoop MapReduce [8].

Apache Spark is a low latency technology can be used for real time transfer data with. All data including video and images is proceed in memory on the cluster [10]. In [10] the architecture of transport data analytics system is presented. The system uses Flume Agent for data collecting and aggregation, Hadoop HDFS for raw data storage and analysis of archival queries based on Cloudera Impala, and Apache Spark for batch processing. For some cases, Apache Kafka is better than the Flume Agent for data collecting and transfering [9]. Also results in [9] show that performance of Apache Kafka with configuration 1 *producer*, 1 *broker* and 1 *consumer* is approximately 800,000 messages per second (the size of one message is 10 KB). Flume in this case shows 46,000 messages per second only.

3   Yet Another Method

In previously published paper [13], we proposed a method for data fusion in distributed architecture based on Kafka cluster. The framework of EVGEN, was used to generate machine data of vehicles in real time mode [14]. All events are created according the following template

```
{
data  :  [
          "uid"  :  {Integer},
          "eventStart"  :  {Long int},
          "eventEnd"  :  {Long int},
          "long"  :  {Double},
          "lat":  {Double},
          "velocity":  {Double},
          "status":  {String}

]
}
```

The notation is following *uid* – object ID, *eventStart* – time when event was started, *eventEnd* – time when event was terminated, *long* – longitude of the object's location, *lat* – latitude of the object's location, *velocity* the current velocity of the object, *status* – a status of the object. However, the template is a subject of modification and changes.

We measured and evaluated performance in comparison with a method of data access without preliminary pre-processing. The performance indicator is query run-time over data containing video and logged geospatial data [6]. The example of logged file is represented at schemes. The first scheme represents log file obtained from vehicle with id "f49a09f4-5e79-42cb-8f89-f0df33a08a02" and type is "taxi".

```
{
         "VehicleID":" f49a09f4 −5e79 −42cb−8f89 −f0df33a08a02",
         "VehicleType":" taxi",
         "Position"  :
         {
                 "latitude":" ,46.583435",
                 "longitude":"44.892395"
         },
         "timestamp":1465471124373,
         "speed":65.0
}
```

The second scheme includes data about vehicle with type "bus" and id is "42760523-7259-4b6d-b4ea-640a2c3e789c".

```
{
         "VehicleID":"42760523 −7259 −4b6d−b4ea −640a2c3e789c",
         "VehicleType":" bus",
         "Position"  :
         {
                 "latitude":" ,46.94521",
                 "longitude":"44.93591"
         },
         "timestamp":"2017 −06 −10  07:13:55",
         "speed":88.0

}
```

Different configurations of distributed architecture were used for run-time evaluation. The basic method was implemented for street traffic data processing where data contains information about vehicles movement (as a textual logged files) and video frames. Figure 1 represents a frame from video stream containing traffic situation.

Fig. 1. A frame obtained from video stream: a traffic condition on a road

We assumed, that a user needs to obtain information about vehicles movements in defined time period $[t_0, t_k]$ for a certain location. This information should be extracted from logged files and video streams. Data fusion method was implemented based on Apache Kafka distributed architecture. A proposed method (PM) was compared with a method without additional data preprocessing (BM). Experiments were made for the different intensity of incoming data stream. For intensity of 3000 events per second, run-time of PM in the average 14.153 ± 1.04 s, it is 3.24 times less in comparison with BM (45.899 ± 2.56 s). Note, the statistical significance of difference was estimated by non-parametric Siegel test with significant level $\alpha = 0.01$.

In this study, we propose an advanced method based on distributed architecture towards to improvement of the method declared in [8]. We suggest the real-time data merging in the memory according to possibilities of Spark Streaming technology. The advanced method contains following steps.

1. **Determination of the required data scheme.** Observed objects are described by a set of heterogeneous data. The data scheme for storing this kind of data is represented according to the format

$$sD = \langle gId, timestamp, (lat, lon), attrD \rangle,$$

where gId – is the object's global identifier which is unique for every observed object; $timestamp$ - timestamp defining when observation was made; (lat, lon) - coordinates of object location at time $timestamp$ as a pair of latitude and longitude; $AttrD$ – a list (dictionary) of key-value pairs describing object's features and its values, for instance, "speed: 68".

2. **Description of data sources and settings of data collectors.** We use term *data collectors* for software having access to data sources and collecting data about observed objects. High-level description of an arbitrary data sources is specified by the format

$$sC = \langle sId, acs, (lat, lon), attrS \rangle ,$$

where sId – is the unique data source id (for example, the link to web service where data comes from), acs – is a key-value list of access settings to the data source marked as sId, $attrS$ – the internal schema of data received from the data source. We consider that $attrS$ contains the mandatory parameter *type*, which takes the value from the list $<type:$ JSON, $type:$ VIDEO$>$.

3. **Binding data schemas.** In this step, the link between the initial scheme of the data source and required scheme is created. This link is represented as a set R containing pairs of attributes from a set $attrD$ of sD scheme and attributes from a set $attrS$ in a scheme sC.

$$R = \{r_{i,j}\} ; r_{i,j} = \langle attrD_i, attrS_j \rangle ,$$

Note, that $\exists r_{lon} : \langle lon^{(sC)}, lon^{(sD)} \rangle$, and $\exists r_{lat} : \langle lat^{(sC)}, lat^{(sD)} \rangle$.

4. **Implementation of data transformation algorithms.** In accordance with the binding settings R, algorithms $\{\alpha_k\}_{k=1}^{\|R\|}$ for transforming data from initial scheme to desired are defined.

$$\forall r_{i,j} \in R; \exists \alpha_{r_{i,j}} : v(attrD_i) \rightarrow v^*(attrS_j),$$

where v indicates a value of an attribute. It should be noted, that the transformations can be simple and complex. In a simple case, for example, for *type*:JSON univocal correspondence of two data schemes could be established. If *type*:VIDEO the video stream is transformed using frame-by-frame conversion with binding time attribute *timestamp* and geographic coordinates (lat, lon) for each frame. In addition, $attrD$ specifies parameters with video ID, frame ID, the name of a frame and a path to the stored video file or to the stored image (frame). If computer vision tools are used for pattern recognition, $attrD$ contains attributes for its usage. Also, it is necessary to include in $attrD$ attributes declaring recognition performance evaluation.

5. **Splitting data.** The scheme DS for splitting data flows into micro-flows is defined.

$$DS_{\alpha_k} = \left\langle df, \alpha_k, \{mdf_l\}_{l=1}^{L_{\alpha_k}} \right\rangle ,$$

where df is an initial data flow, and mdf_l is l in-memory data flows for the certain algorithm α_k, L_{α_k} – is a number of flows. In this step, data are split into data flows to be processed in distributed architecture according to predefined tasks, e.g. calculating the number of objects with equal attributes from the set of attributes *attrD*.

6. **Insert post-processed data into a database.** When data is transformed according to defined scheme, they are being inserted into a database. It is allowing to extract heterogeneous data from the database without additional manipulation with data.

4 Use Case

4.1 An Architecture for the Method Implementation

The proposed method was used as a data fusion method for urban management decision support system regarding traffic jams analysis and preventive management problem. We used video streams and logged files about traffic from different data sources [14]. Video streams obtained from intersections or road cameras, and logged files collected from mobile vehicles.

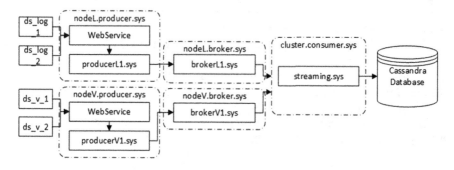

Fig. 2. Distributed architecture for data collecting and data fusion implementing proposed method. Legend: ds_*_* – data sources, WebService* – web service for data collecting, producer*.* – a component produces internal data flows, consumer* – a component for data flows consuming

The decision support system has to provide information about number of vehicles at intersection according all available data sources. This information is a result of query. The aim is to minimise user query run-time to data storage containing described types of data. Distributed architecture help to overcome the problem of big volume of data. In this study, we used distributed architecture where Spark Streaming components combining with Kafka Server. Spark Streaming allows performing data pre-processing in memory (step 5 of the proposed method). We expect to see decreasing run-time of data preprocessing. Figure 2 shows the representation of proposed distributed architecture for solution implementing the method.

The solution is based on distributed architecture containing on five layers:

- s-layer of data sources with different data sources indicated as ds_*_* in the architecture;
- p-layer of producers with components named producer*.sys;
- b-layer of brokers (broker*.sys);
- c-layer of consumers (cluster.consumer.sys);
- k-layer of data storages with SQL-based access to data (Cassandra Database).

Desired scheme is define in general keyspace in two MSDB Cassandra tables *IoTCameraTraffic* and *IoTVehicleTraffic* in the k-layer.

Here the script of keyspace and tables creating in Cassandra MSDB.

```
// create key space
CREATE KEYSPACE Traffic_Space WITH
replication = {'class':'SimpleStrategy',
'replication_factor':3};

// create table for saving iot vehicle traffic
CREATE TABLE Traffic_Space.IoTVehicleTraffic
(IdVehicle text, TypeofVehicle text, timeStamp timestamp,
recordDate text, SpeedofVehicle float,
PRIMARY KEY (IdVehicle,recordDate,TypeofVehicle));

// create table for saving iot camera traffic
CREATE TABLE Traffic_Space.IoTCameraTraffic
(IdCamera text, timeStamp timestamp,
recordDate text, totalCount bigint,
PRIMARY KEY (IdCamera,recordDate));
```

Setting of p-layer contains of data sources description and data settings of data collecting tools. Usually these settings are represented as settings files, which are loaded by p-layer components. Setting of binding of different data schemas are located in the JSON files for c-layer. c-layer contains the algorithms for real-time data transfering according the binding rules and Spark Streaming functionality. In case of high load, to avoid the missing data during transfering from p-layer to c-layer, the broker layer b-layer is used. Apache Kafka b-layer is managed by Zookeeper. In c-layer, Spark Streaming splits data streams obtained from b-layer into micro-package data streams. Micro-package data streams are uploaded in DStream for in-memory processing. After, the processed data is stored in Cassandra database in k-layer.

In updated version of implementation, image processing library OpenCV is used for video streams analysis. The main benefit of proposed architecture is reduction of data pre-processing time. Note, the number of producers should be proportional to a number of data sources. Video data processing is performed in the p-layer.

4.2   Experiment Setup

Several experiments with duration of 30 min 38 s each were made (365 completed batches, 1,318,747 records). Each experiment has three parts:

1. Part A: where logged files and video (here and after 'video & log') streams come with low intensity ~100 events per second,
2. Part B: where only logged files (here and after 'log') come with high frequency ~1,500 event per second and
3. part C where logged files and video streams come with intensity ~1,000 event per second.

Figure 3 illustrates data intensity during the experiment.

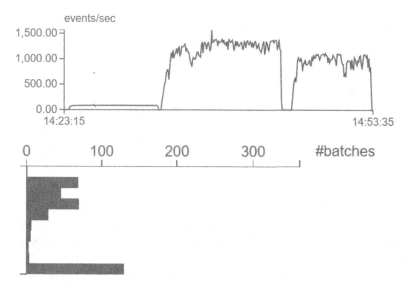

Fig. 3. Timelines (Last 365 batches, 0 active, 365 completed) of an experiment (above). Histogram show distribution of batches depending on intensity (events/sec). The chart obtained from Streaming Statistics.

There are 9 types of intensity modes were used during the experiment: 69 batches (18,9%) between 1,252 and 1,408 events/sec; 46 batches (12,6%) between 1,095.5 and 1,252 events/sec; 70 batches (19,8%) between 939 and 1,095.5 events/sec; 29 batches (7,95%) between 782.5 and 939 events/sec; 7 batches (1,92%) between 626 and 782.5 events/sec; 6 batches (1,64%) between 469.5 and 626 events/sec; 3 batches (0,82%) between 313 and 469.5 events/sec; 4 batches (1,1%) between 156.5 and 313 events/sec; 130 batches (35.62,1%) between 0 and 156.5 events/sec.

4.3   Results and Discussion

Results are obtained using the workstation with Linux OS, 8 GB RAM. Producers of two types were used: the producers for logged data sources (I type producer) and producers for video stream data sources (II type producer). Video streams were defragmented in p-layer, where we obtain images and descriptions. Table 1 contains average results for several runs (from 18 to 27 runs) according to the described profile of experiments.

Table 1. Results of experiments for defined load profile: average processing time and average delay for different configuration

Experiment's part	Brokers	Intencity (event/sec)	Average processing time, sec.	Average delay, sec.
Part A: video & log	1 broker	100	0.339 ± 0.802	0.356 ± 1.104
Part B: log	1 broker	1500	3.095 ± 17.81	3.095 ± 17.81
Part C: video & log	1 broker	1000	2.195 ± 16.290	2.195 ± 16.290
Part A: video & log	3 brokers	100	0.191 ± 0.118	0.191 ± 0.118
Part B: log	3 brokers	1500	1.519 ± 8.741	1.519 ± 8.741
Part C: video & log	3 brokers	1000	1.066 ± 6.12	1.066 ± 6.12

A number of producers is changed from 1 to 10, brokers from 1 to 3. Producers send data to Kafka Cluster brokers. Consumers based on Spark Streaming received data according to the pre-defined topics.

Figure 4 shows the delay profile for the one experiment (a number of producer is 1, the number of brokers is 1 as well). The first groups of experiments were made with a number of brokers is 1. In part A, producers I and II types generated data with low intensity, so the average latency is low as well 0.339 s.

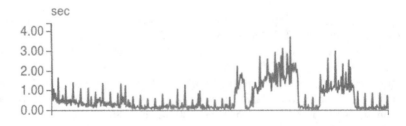

Fig. 4. Delay in one run during the experiment (in sec.): x-axes is time axis. The chart obtained from Streaming Statistics

Delay is slightly differs from processing time due to specific of computational environment. In part B, we observe peak intensity with average delay 3.095 s. Note, in the part B only logged files were processed. Finally, the average processing time in part C of the experiment is about 2.195 s. As expected, the latency increases when a number of connected to producers data sources increases as well. However, the latency peaks are more smoothly even in case of maximal intensity.

For experiments with 3 brokers, the average processing time is reducing. So for Part A: the reduction is aprx. in 1.77 times, for Part B, the reduction is about 2.6 times, and finally for Part C, the reduction is 2.05 times less. The main conclusion is the increasing of number of brokers for observed intensity is reasonable.

5   Conclusion

The paper presents advanced method for various types of data (video and log files) fusion and preprocessing in proactive decision support systems. The method is need to be implemented in distributed architecture containing five layers: s-layer of data sources, p-layer of producers, b-layer of brokers, c-layer of consumers and k-layer of data storages.

In case of heterogeneous data (like video and loged data) and functional requirements for query execution over these data, the distributed 'on-fly' preprocessing might be efficient in comparison with batch processing. As proposed method is implemented in the five-layer distributed architecture based on Apache Kafka and Spark Streaming technology, in-memory processing is more efficient.

Acknowledgments. The reported study was partially supported by RFBR research projects 16-37-60066 mol_a_dk, and project MD-6964.2016.9.

References

1. Arnott, D., Pervan, G.: A critical analysis of decision support systems research revisited: the rise of design science. J. Inf. Technol. **29**(4), 269–293 (2014)
2. Bengio, Y.: Practical recommendations for gradient-based training of deep architectures. In: Montavon, G., Orr, G.B., Müller, K.-R. (eds.) Neural Networks: Tricks of the Trade. LNCS, vol. 7700, pp. 437–478. Springer, Heidelberg (2012). doi:10. 1007/978-3-642-35289-8_26
3. Engel, Y., Etzion, O.: Towards proactive event-driven computing. In: Proceedings of the 5th ACM International Conference on Distributed Event-Based System, pp. 125–136 (2011)
4. Bemporad, A., Morari, M.: Robust model predictive control: a survey. In: Garulli, A., Tesi, A. (eds.) Robustness in Identification and Control. LNCIS, vol. 245, pp. 207–226. Springer, London (1999). doi:10.1007/BFb0109870
5. Golubev, A., Chechetkin, I., Solnushkin, K.S., Sadovnikova, N., Parygin, D., Strategway, S.M.: Web solutions for building public transportation routes using big geodata analysis. In: 17th International Conference on Information Integration and Web-Based Applications and Services (2015)

6. Golubev, A., Chechetkin, I., Parygin, D., Sokolov, A., Shcherbakov, M.: Geospatial data generation and preprocessing tools for urban computing system development. Procedia Comput. Sci. **101**, 217–226 (2016)
7. De Gooijer, J.G., Hyndman, R.J.: 25 years of time series forecasting. Int. J. Forecast. **22**(3), 443–473 (2006)
8. HIPI Hadoop image processing interface. http://hipi.cs.virginia.edu/. Accessed 5 May 2017
9. Benchmarking Apache Kafka: 2 Million Writes Per Second (On Three Cheap Machines). https://engineering.linkedin.com/kafka/. Accessed 5 May 2017
10. Maarala, A.I., Rautiainen, M., Salmi, M., Pirttikangas, S., Riekki, J.: Low latency analytics for streaming traffic data with Apache Spark. In: 2015 IEEE International Conference on Big Data, pp. 2855–2858 (2015)
11. Sage, A.: Decision support systems. http://onlinelibrary.wiley.com/doi/10.1002/9780470172339.ch4/summary. Accessed
12. Tennenhouse, D.: Proactive computing. Commun. ACM **43**(5), 43–50 (2000)
13. Tran V.P., Shcherbakov M., Nguyen T.A., Skorobogatchenko D.A.: A method for data acquisition and data fusion in intelligent proactive decision support systems. Neurocomputers (11), 40–44 (2016). (in Russian)
14. Tran, V.P., Shcherbakov, M., Nguyen, T.: A framework for event generator in proactive system design. In: 2016 7th International Conference on Information, Intelligence, Systems Applications (IISA), pp. 1–5 (2016)
15. Qin, S.J., Badgwell, T.: A survey of industrial model predictive control technology. Control Eng. Pract. **11**(7), 733–764 (2003)

Control and Safety of Operation of Duplicated Computer Systems

Vladimir Bogatyrev$^{(\boxtimes)}$ and Maria Vinokurova

ITMO University, Kronversky Pr. 49, 197101 St. Petersburg, Russia
vladimir.bogatyrev@gmail.com, gudvin606@gmail.com
http://www.ifmo.ru/ru/

Abstract. The subject of the research is a duplicated computer system, which contains means of operational and test control, and will be considered as a basic computer node for infocommunication systems of responsible meaning. There are different ways of organizing system control: testing of only one node, testing of only two nodes and testing whether one or two computer nodes. Besides the functional and the failed states of the system, there are dangerous conditions that may be not detected by control malfunction. The inbound flow of requests is heterogeneous - there are non-critical requests as well as critical, which require duplicated computations. The aim of the work is to choose a discipline of test control and define optimal intervals of testing initialization that ensure the maximum probability of the system's readiness for safe implementation of functional requests for different flows of requests. Research method is based on constructing a Markov model.

Keywords: Markov model · Control · The optimal frequency of test monitoring · Critical requests · Safety · Reliability · Aqueuing system

1 Introduction

The stability and the safety [1–3] of distributed computer systems' functioning [4–7] in conditions of failures and destructive accidental or malicious actions is achieved due to the redundancy of communication channels [8–10], computer nodes, organization of their control and operability's restoration after detection of violations.

As a basis for building computer nodes of distributed fault-tolerant systems, including clustered ones, we consider a system consisting of two computer [11], equipped with means of operational and test control. Periodic initialization of testing is carried out by timer.

The effectiveness of an operational disruptions' detection is determined by the completeness of the operational and the periodicity of the test control. The reduction of the test control leads to a decrease of system's readiness due to an increase of time costs for testing, but at the same time it increases system's safety level by reducing the probability of system's functioning in conditions of undetected failures.

© Springer International Publishing AG 2017
V.M. Vishnevskiy et al. (Eds.): DCCN 2017, CCIS 700, pp. 331–342, 2017.
DOI: 10.1007/978-3-319-66836-9_28

The choice of the best control options for redundant cluster systems is complicated, because besides an optimizing test intervals, it includes a determining the number of machines that may be tested together, depending on the nature of the requests flow. Including the heterogeneity of this flow, at which some requests require duplicated calculations on different machines (with the outputing of results only if they coincide), and some may be serviced by one of the machines in the load-sharing mode.

Depending on the homogeneity of the input flow, we distinguish the following cases:

- the flow is homogeneous, any request is serviced by one of the computer nodes (option A_1);
- the flow is homogeneous, any request requires duplicated computing by two nodes (option A_2);
- the flow is heterogeneous, herewith some requests require a duplicated execution by two computers, other requests can be serviced by one of the nodes (option A_3).

For a homogeneous flow of requests that do not require a duplication of calculations, and which are serviced in the load-sharing mode, simultaneous testing of two nodes is undesirable, cause it may lead to a decrease of system's availability probability [11]. For an input flow that provides a duplicated service of requests in two machines with a comparison of the results [11], the readiness of the system increases with simultaneous testing of both nodes, since the resources of only one node are insufficient for organizing a duplicated computational process.

For a heterogeneous flow of requests, some of which involve a duplication of calculations, the impact of the organization and the periodicity of the test control on the system's readiness for safe queries servicing requires additional research.

We will analyze the ways of control organization in which simultaneous testing of two nodes is unacceptable (option B_1), is possible (option B_2) and is always implemented (option B_3).

The following combinations of variants, depending on the homogeneity of the input flow of requests and the organization of control by the number of simultaneously initialized nodes for testing, are possible:

$$(A_1, B_1), (A_1, B_2), (A_1, B_3),$$
$$(A_2, B_1), (A_2, B_2), (A_2, B_3),$$
$$(A_3, B_1), (A_3, B_2), (A_3, B_3).$$

An option without test control B_0 can be considered as a basic variant of the control organization. Taking into account variations in the homogeneity of the input flow, the following cases are possible:

$$(A_1, B_0), (A_2, B_0), (A_3, B_0).$$

The explored protected systems, in addition to operating and failed states, can be in hazardous states, in which there are functional violations that are not detected by the monitoring means of one or two nodes.

The aim of the work is to choose a discipline of test control and define optimal intervals of testing initialization that ensure the maximum probability of the system's readiness for safe operation, depending on the nature of the flow of requests requiring execution by one or two computer nodes simultaneously.

To achieve this aim, we should:

- define the options of organizing control (service discipline);
- form a system of criteria for the effectiveness of the investigated options;
- construct Markov models for each control option that allow to determine the probabilities of different states of the system, as well as the influence of the discipline and the frequency of the test control initialization on the probability of the system's readiness for safe operation when the incoming requests are heterogeneous;
- determine the optimal periodicity of testing for the considered control options;
- choose the option of organizing the system's control.

2 Models of the Options of Test Control Organization

While constructing the model, the initialization and testing intervals, as well as the intervals before the occurrence of violations and restoration of the node's operation, will be considered as distributed according to the exponential law. It is assumed that all violations, including those that occur during testing, are detected as the result of the test control. During testing, the operational control will be considered as disabled.

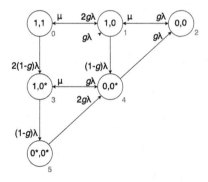

Fig. 1. Graphs of states and transitions of Markov model of a duplicated system with operational control (option B_0)

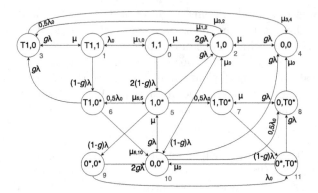

Fig. 2. Graphs of states and transitions of Markov model of a duplicated system with operational and test control (option B_1)

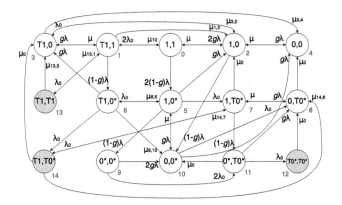

Fig. 3. Graphs of states and transitions of Markov model of a duplicated system with operational and test control (option B_2)

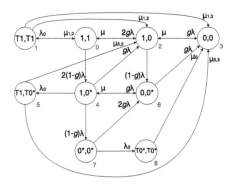

Fig. 4. Graphs of states and transitions of Markov model of a duplicated system with operational and test control (option B_3)

Graphs of states and transitions of Markov models for the basic option of the control organization B_0, which provides only operational control shown in Fig. 1. Graphs of states and transitions of Markov models for the considered options of control organization B_1, B_2, B_3 of duplicated computer systems are shown in Figs. 2, 3 and 4.

Each vertex of the graphs in Figs. 1, 2, 3 and 4 corresponds to the state of the pair of nodes (regardless of the numbering of nodes because of their identity). When coding the states of each node, the following notation is used: "1" - the node's operation; "0" - the detected violation; "0*" - undetected violation (if at least one node of the system is in this state, then the state of the entire system is considered as dangerous); "T1" - testing of an efficient node; "T0*" - testing of the node with undetected violations. The presented graphs of states and transitions introduce the following symbols: λ - a failure rate of the node; λ_0 - the intensity of initialization of the test control of the node (the value opposite to the average time between tests); g - the probability of failure's detection by the operational control (share of resources covered by the operational control); μ_0 - the test intensity (the value inverse of the average test time); μ - the intensity of the restoration of the operability of the node. It is assumed that one repairman is involved. The states that differentiate option B_2 from B_1, in Fig. 3 are darkened.

The initial state with the number 0 corresponds to the readiness of two nodes to functioning, and the states in Figs. 2 and 3 with the numbers 1, 2, 7 correspond to the readiness of only one efficient node. The states with a violation of safety are designated as 5, 6, 9–11. For the control option B_2 in states 12, 13, 14, both nodes of the system are tested. For the control option B_3 (Fig. 4), the system is ready for safe functioning, if it is possible to use one node in state 2.

In accordance with known rules, a system of Kolmogorov equations is compiled on the state and transition diagram. The solution of this system of equations can be obtained using computer mathematics, for example, Mathcad 15, based on the "Given-Find" construction.

In accordance with Fig. 1 we have the following Kolmogorov system of algebraic equations:

$$\begin{cases} -[2(1-g)\lambda + 2g\lambda]P_0 + \mu P_1 = 0, \\ -[\mu + g\lambda + (1-g)\lambda]P_1 + 2g\lambda P_0 + g\lambda P_3 + \mu P_2 = 0, \\ -\mu P_2 + g\lambda P_1 + g\lambda P_4 = 0, \\ -\mu P_4 + (1-g)\lambda P_1 + g\lambda P_3 + 2g\lambda P_5 = 0, \\ -2g\lambda P_5 + (1-g)\lambda P_3 = 0, \\ \sum_{i=0}^{5} P_i = 1, \end{cases} \qquad (1)$$

where P_i - the probability of the i-th state of the system.

According to the diagram in Fig. 2 we obtain [12] the system of equations:

$$
\begin{cases}
-[\lambda_0 + 2(1-g)\lambda + 2g\lambda]P_0 + \mu_{1,0}P_1 + \mu P_2 = 0, \\
-[g\lambda + (1-g)\lambda + \mu_{1,0} + \mu_{1,2}]P_1 + \mu P_3 + \lambda_0 P_0 = 0, \\
-[0.5\lambda_0 + \mu + (1-g)\lambda + g\lambda]P_2 + \mu_{3,2}P_3 + \mu_{1,2}P_1 + 2g\lambda P_0 + g\lambda P_5 + \\
+\mu_0 P_7 + \mu P_4 = 0, \\
-[\mu_{3,2} + \mu_{3,4} + \mu]P_3 + g\lambda P_1 + 0.5\lambda_0 P_2 + g\lambda P_6 = 0, \\
-\mu P_4 + \mu_{3,4}P_3 + g\lambda P_2 + \mu_0 P_8 + g\lambda P_{10} = 0, \\
-[0.5\lambda_0 + (1-g)\lambda + g\lambda + 0.5\lambda_0 + g\lambda]P_5 + 2(1-g)\lambda P_0 + \mu_{6,5}P_6 + \mu P_{10} = 0, \\
-[g\lambda + \mu_{6,5} + \mu_{6,10}]P_6 + (1-g)\lambda P_1 + 0.5\lambda_0 P_5 = 0, \\
-[\mu_0 + g\lambda + (1-g)\lambda]P_7 + 0.5\lambda_0 P_5 + \mu P_8 = 0, \\
-[\mu + \mu_0]P_8 + g\lambda P_7 + 0.5\lambda_0 P_{10} + g\lambda P_{11} = 0, \\
-[2g\lambda + \lambda_0]P_9 + (1-g)\lambda P_5 = 0, \\
-[\mu_0 + g\lambda]P_{11} + \lambda_0 P_9 + (1-g)\lambda P_7 = 0, \\
\sum_{i=0}^{11} P_i = 1,
\end{cases}
\tag{2}
$$

From the diagram in Fig. 3 we compose a system of algebraic equations:

$$
\begin{cases}
-[2\lambda_0 + 2(1-g)\lambda + 2g\lambda]P_0 + \mu_{1,0}P_1 + \mu P_2 = 0, \\
-[g\lambda + \lambda_0 + (1-g)\lambda + \mu_{1,0} + \mu_{1,2}]P_1 + \mu P_3 + 2\lambda_0 P_0 + \mu_{13,1}P_{13} = 0, \\
-[\lambda_0 + \mu + (1-g)\lambda + g\lambda]P_2 + \mu_{3,2}P_3 + \mu_{1,2}P_1 + 2g\lambda P_0 + g\lambda P_5 + \\
+\mu_0 P_7 + \mu P_4 = 0, \\
-[\mu_{3,2} + \mu_{3,4} + \mu]P_3 + g\lambda P_1 + \lambda_0 P_2 + g\lambda P_6 + \mu_{13,3}P_{13} + \mu_0 P_{14} = 0, \\
-\mu P_4 + \mu_{3,4}P_3 + g\lambda P_2 + \mu_0 P_8 + g\lambda P_{10} = 0, \\
-[\lambda_0 + (1-g)\lambda + g\lambda + \lambda_0 + g\lambda]P_5 + 2(1-g)\lambda P_0 + \mu_{6,5}P_6 + \mu P_{10} = 0, \\
-[g\lambda + \mu_{6,5} + \mu_{6,10} + \lambda_0]P_6 + (1-g)\lambda P_1 + \lambda_0 P_5 = 0, \\
-[\mu_0 + \lambda_0 + g\lambda + (1-g)\lambda]P_7 + \lambda_0 P_5 + \mu P_8 + \mu_{14,7}P_{14} = 0, \\
-[\mu + \mu_0]P_8 + g\lambda P_7 + \lambda_0 P_{10} + g\lambda P_{11} + \mu_0 P_{12} + \mu_{14,8}P_{14} = 0, \\
-[2g\lambda + 2\lambda_0]P_9 + (1-g)\lambda P_5 = 0, \\
-[\mu_0 + g\lambda + \lambda_0]P_{11} + 2\lambda_0 P_9 + (1-g)\lambda P_7 = 0, \\
-\mu_0 P_{12} + \lambda_0 P_{11} = 0, \\
-[\mu_{13,1} + \mu_{13,3}]P_{13} + \lambda_0 P_1 = 0, \\
-[\mu_0 + \mu_{14,7} + \mu_{14,8}]P_{14} + \lambda_0 P_6 + \lambda_0 P_7 = 0, \\
\sum_{i=0}^{14} P_i = 1,
\end{cases}
\tag{3}
$$

where P_i - the probability of the i-th state of the system;

$$\mu_{1,0} = \mu_{3,2} = \mu_{6,5} = \mu_0 e^{-\lambda/\mu_0},$$

$$\mu_{1,2} = \mu_{3,4} = \mu_{6,10} = \mu_0 (1 - e^{-\lambda/\mu_0}),$$

$$\mu_{13,1} = \mu_{14,7} = 0.5\mu_0 e^{-\lambda/\mu_0},$$

$$\mu_{13,3} = \mu_{14,8} = 0.5\mu_0 e^{-\lambda/\mu_0}.$$

For the option B_3 from the diagram in Fig. 4 we have:

$$\begin{cases}
-[\lambda_0 + 2(1-g)\lambda + 2g\lambda]P_0 + \mu_{1,0}P_1 + \mu P_2 = 0, \\
-[\mu_{1,0} + \mu_{1,2} + \mu_{1,3}]P_1 + \lambda_0 P_0 = 0, \\
-\mu P_3 + \mu_{1,3}P_1 + g\lambda P_2 + \mu_{5,3}P_5 + g\lambda P_6 + \mu_0 P_8 = 0, \\
-[\lambda_0 + g\lambda + (1-g)\lambda + g\lambda]P_4 + 2(1-g)\lambda P_0 + \mu P_6 = 0, \\
-[\mu_{5,2} + \mu_{5,3}]P_5 + \lambda_0 P_4 = 0, \\
-[\mu + g\lambda]P_6 + (1-g)\lambda P_2 + g\lambda P_4 + 2g\lambda P_7 = 0, \\
-[2g\lambda + \lambda_0]P_7 + (1-g)\lambda P_4 = 0, \\
-\mu_0 P_8 + \lambda_0 P_7 = 0, \\
\sum_{i=0}^{8} P_i = 1,
\end{cases} \qquad (4)$$

where

$$\mu_{1,0} = \mu_0 (e^{-\lambda/\mu_0})^2,$$

$$\mu_{1,2} = 2\mu_0 (1 - e^{-\lambda/\mu_0}) e^{-\lambda/\mu_0},$$

$$\mu_{1,3} = \mu_0 (1 - e^{-\lambda/\mu_0})^2,$$

$$\mu_{5,2} = \mu_0 e^{-\lambda/\mu_0},$$

$$\mu_{5,3} = \mu_0 (1 - e^{-\lambda/\mu_0}),$$

and P_i - the probability of the i-th state of the system.

3 Efficiency of the Options of Control Organization

For the option of the input flow A_1 that allows requests to be serviced by only one node - (A_1, B_0), (A_1, B_1), (A_1, B_2), (A_1, B_3) the probability of the system's readiness to perform required functions is defined as the sum of the probabilities of P_i states in which both computers of the node are in a safe state and at least one of them is ready to perform required functions. In this case, the availability factor for the options (A_1, B_1), (A_1, B_2) is calculated as:

$$K_{(1,2)} = P_0 + P_1 + P_2 + P_7,$$

while for the options B_1, B_2 the probabilities P_0, P_1, P_2, P_7 are found when solving the systems of Kolmogorov Eqs. (2) and (3).

For the option (A_1, B_3) availability factor is found as:

$$K = P_0 + P_2,$$

where the probabilities of the states P_0, P_2 are found from the system of Eqs. (4). And for the option (A_1, B_0) availability factor is calculated as:

$$K = P_0 + P_1,$$

where the probabilities of the states P_0, P_1 are found from the system of Eqs. (1). Considering the object of the research as a queuing system, we will determine the probability of its readiness for safe performing the required functions, provided the service mode is stationary. For the options (A_1, B_1) [12], (A_1, B_2), allowing the servicing of requests by one computer, the probability of system's readiness is found as:

$$F_{1(1,2)} = P_0\delta_2 + (P_1 + P_2 + P_7)\delta_1,$$

and for the option (A_1, B_3):

$$F_1 = P_0\delta_2 + P_2\delta_1,$$

where

$$\delta_1 = \begin{cases} 1, & \text{if } \Lambda v < 1, \\ 0, & \text{if } \Lambda v \geq 1, \end{cases} \quad \delta_2 = \begin{cases} 1, & \text{if } \Lambda v/2 < 1, \\ 0, & \text{if } \Lambda v/2 \geq 1. \end{cases}$$

In this case, $\rho = \Lambda v$ is the load of one node of the system, Λ is the intensity of the flow of incoming functional requests, v is the average time of their execution. For the control options B_0, B_1, B_2, B_3, the probabilities of states are found from the systems of Kolmogorov equations.

The efficiency ratio of the system, taking into account idle times during testing for the options (A_1, B_1), (A_1, B_2), is defined as:

$$R_1 = P_0\delta_2 + \frac{T_2}{T_1}(P_1 + P_2 + P_7)\delta_1 = P_0\delta_2 + \frac{1-\rho}{1-0.5\rho}(P_1 + P_2 + P_7)\delta_1.$$

For the option (A_1, B_3):

$$R_1 = P_0\delta_2 + \frac{1-\rho}{1-0.5\rho}P_2\delta_1,$$

And for the option (A_1, B_0):

$$R_1 = P_0\delta_2 + \frac{1-\rho}{1-0.5\rho}P_1\delta_1,$$

where $T_1 = v/(1-\rho)$ and $T_2 = v/(1-0.5\rho)$ - the average times of requests [13] staying in the system, provided the possibility that functional requests are distributed to be serviced in one or two workable computer nodes, are found as:

For the options of the input flow A_2, A_3, the readiness of the system is determined by its ability to service requests that require duplicated calculations, therefore, taking into account the requirement of stationarity of the service mode and the inadmissibility of a loss of requests requiring duplicated calculations, the probability of the readiness and the efficiency ratio we find, respectively, as:

$$F_2 = R_2 = P_0\delta_2, \quad F_3 = R_3 = P_0\delta_3,$$

where for the option of the input flow A_3:

$$\delta_3 = \begin{cases} 1, & \text{if } 0.5\ \Lambda v(d+1) < 1, \\ 0, & \text{if } 0.5\ \Lambda v(d+1) \geq 1, \end{cases}$$

herewith the probability of the state P_0 for the control options B_1, B_2, B_3 is respectively determined from the systems of equations.

4 Results of Calculations

We will compare the considered options of the control organization depending on the nature of the input flow of requests.

Dependencies of the readiness of the system to the safe servicing of requests F from the duration of the intervals of testing initialization λ_0 for the options of the control organizations B_1, B_2, B_3 are shown in Figs. 5, 6 and 7. This figures represent the options of the input flow A_1, A_2, A_3, respectively.

Curves 1, 2, 3 correspond to the options of the control organizations B_1, B_2, B_3. The calculations were performed with $\rho = 0.995$, $\lambda = 10^{-4}\ h^{-1}$, $g = 0.6$, $\mu = 1\ h^{-1}$, $\mu_0 = 1\ h^{-1}$.

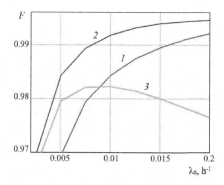

Fig. 5. The probabilities of the readiness to safe servicing of requests F for the options of the control organization B_1, B_2, B_3 and the input flows of requests A_1

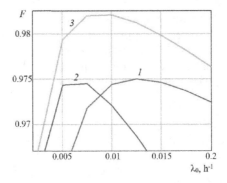

Fig. 6. The probabilities of the readiness to safe servicing of requests F for the options of the control organization B_1, B_2, B_3 and the input flows of requests A_2

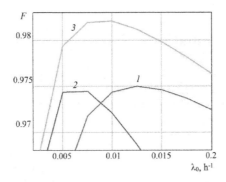

Fig. 7. The probabilities of the readiness to safe servicing of requests F for the options of the control organization B_1, B_2, B_3 and the input flows of requests A_3

From Figs. 5, 6 and 7 it can be concluded that for a homogeneous flow of requests A_1 that do not require a duplication of calculations (d = 0), it is advisable to use either the B_1 control option (testing of only one computer node), or B_2 (it is possible to test both one or two computer nodes). For these service disciplines, approximately the same efficiencies of system's functioning are achievable, by which the maximum probability of the system's readiness for safe performing the required functions is achieved. Moreover, the desired maximum of efficiency is achieved for different periodicity of testing initialization (greater for B_2).

For a heterogeneous flow of requests and a flow of requests requiring duplicated calculations (d > 0), the maximum probability of the system's readiness for safe operation is achieved with the control option B_3 (both computer nodes are tested simultaneously). It should be noted that in addition to the considered mathematical models, choosing the method of organization of duplicated computer nodes, it is also important to take into account the requirements of the applied application processes [14,15], the organization of the interconnection of computer nodes, including one based on a multipath redundant data transfer [16–19] and a priority of requests service [20,21].

5   Conclusion

Thus, the Markov models of a duplicated computer system that allow analyze the influence of the investigated options of the control organization and the periodicity of the initialization of the test control on the system's readiness for safe performing the required functions are proposed.

It is shown that there are optimal periods of initialization of testing for all of the considered options of the control organization, in which the maximum probability of the system's readiness for safe performing the required functions is achieved.

References

1. Aysan, H.: Fault-tolerance strategies and probabilistic guarantees for real-time systems. Mälardalen University, Västerås, Sweden, p. 190 (2012)
2. Sorin, D.: Fault Tolerant Computer Architecture, p. 103. Morgan & Claypool, San Rafael (2009)
3. Koren, I.: Fault Tolerant Systems, p. 378. Morgan Kaufmann Publications, San Francisco (2009)
4. Bogatyrev, V.A., Bogatyrev, S.V., Golubev, I.Y.: Optimization and the process of task distribution between computer system clusters. Autom. Control Comput. Sci. **46**(3), 103–111 (2012)
5. Bogatyrev, V.A.: Exchange of duplicated computing complexes in fault tolerant systems. Autom. Control Comput. Sci. **45**(5), 268–276 (2011)
6. Bogatyrev, V.A., Bogatyrev, A.V.: Functional reliability of a real-time redundant computational process in cluster architecture systems. Autom. Control Comput. Sci. **49**(1), 45–56 (2015)
7. Bogatyrev, V.A.: Fault tolerance of clusters configurations with direct connection of storage devices. Autom. Control Comput. Sci. **45**(6), 330–337 (2011)
8. Bogatyrev, V.A.: An interval signal method of dynamic interrupt handling with load balancing. Autom. Control Comput. Sci. **34**(6), 51–57 (2000)
9. Bogatyrev, V.A.: Protocols for dynamic distribution of requests through a bus with variable logic ring for reception authority transfer. Autom. Control Comput. Sci. **33**(1), 57–63 (1999)
10. Bogatyrev, V.A.: On interconnection control in redundancy of local network buses with limited availability. Eng. Simul. **16**(4), 463–469 (1999)
11. Arustamov, S.A., Bogatyrev, V.A., Polyakov, V.I.: Back up data transmission in real-time duplicated computer systems. Adv. Intel. Syst. Comput. **451**, 103–109 (2016)
12. Bogatyrev, V.A., Vinokurova, M.S., Petrov, P.A., Nazarova, M.L., Shabakov, R.V.: Control and safety of the duplicated computer systems. Sci. Tech. J. Inform. Technol. Mech. Opt. **17**(2), 369–373 (2017). doi:10.17586/2226-1494-2017-17-2-369-373. (in Russian)

13. Vishnevsky, V.M.: Teoreticheskie osnovy proektirovaniya komputernykh setey. Theoretical Fundamentals for Design of Computer Networks. – Tekhnosfera, Moscow (2003) (in Russian)
14. Korobeynikov, A.G., Fedorovsky, M.E., Maltseva, N.K., Baranova, O.V., Zharinov, I.O., Gurjanov, A.V., Zharinov, O.O.: Use of information technologies in design and production activities of instrument-making plants. Indian J. Sci. Technol. IET **9**(44), 104–108 (2016)
15. Kolomoitcev, V.S., Bogatyrev, V.A.: The fault-tolerant structure of multilevel secure access to the resources of the public network. Commun. Comput. Inform. Sci. **678**, 302–313 (2016)
16. Bogatyrev, V.A., Parshutina, S.A.: Efficiency of redundant multipath transmission of requests through the network to destination servers. Commun. Comput. Inform. Sci. **678**, 290–301 (2016)
17. Bogatyrev, V.A., Parshutina, S.A., Poptcova, N.A., Bogatyrev, A.V.: Efficiency of redundant service with destruction of expired andirrelevant request copies in real-time clusters. Commun. Comput. Inform. Sci. **678**, 337–348 (2016)
18. Bogatyrev, V.A., Parshutina, S.A.: Redundant distribution of requests through the network by transferring them over multiple paths. Commun. Comput. Inform. Sci. **601**, 199–207 (2016)
19. Bogatyrev, V.A.: Increasing the fault tolerance of a multi-trunk channel by means of inter-trunk packet forwarding. Autom. Control Comput. Sci. **33**(2), 70–76 (1999)
20. Aliev, T.I.: The synthesis of service discipline in systems with limits. Commun. Comput. Inform. Sci. **601**, 151–156 (2016). doi:10.1007/978-3-319-30843-2-16
21. Kalinin, I.V., Muraveva-Vitkovskaia, L.A.: Evaluation of functionality's efficiency in priority telecommunication networks with heterogeneous traffic. Commun. Comput. Inform. Sci. IET **601**, 253–259 (2016)

Fluid Limit for Switching Closed Queueing Network with Two Multi-servers

Svetlana Anulova[✉]

ICS, RAS, Profsoyuznaya, 65, 117997 Moscow, Russian Federation
anulovas@ipu.rssi.ru
http://www.ipu.ru/

Abstract. A closed network consists of two multi-servers with n customers. Service requirements of customers at a multi-server have a common cdf. State parameters of the network: for each multi-server empirical measure of the age of customers being serviced and for the queue the number of customers in it, all multiplied by n^{-1}.

Our objective: asymptotics of dynamics as $n \to \infty$. The asymptotics of dynamics of a single multi-server and its queue with an arrival process as the number of servers $n \to \infty$ is currently studied by famous scientists K. Ramanan, W. Whitt et al. In the last publications the arrival process is generalized to time-dependent. We develop our previous asymptotics results for a network also in this direction: instead of a simple time dependence a markov swithching behavior of one multi-server is introduced. For the asymptotic process we in a rough way find equilibrium and prove convergence as $n \to \infty$.

Motivation for studying such models: they represent call/contact centers, and switching expresses the changes of the system environment.

Keywords: Multi-server queues · GI/G/n queue · Fluid limits · Mean-field limits · Strong law of large numbers · Measure-valued processes · Fluid limit equilibrium and convergence · Switching networks · Call/contact centers

We have suggested in [2] a suitable model for call/contact centers—a closed network consisting of 2 multi-servers and n customers. Its dynamics has a fluid limit as $n \to \infty$ (introduced in [3]) with an equilibrium point and a partial convergence to it. Now we generalize the model to the changing environment. To understand easier our research see details of description and proofs of behavior for the model in [1,2]:

https://link.springer.com/content/pdf/10.2F978-3-319-51917-3_33.pdf
https://link.springer.com/content/pdf/10.2F978-3-319-30843-2_19.pdf

---

S. Anulova—This work was partially supported by RFBR grants No. 16-08-01285 and No. 17-01- 00633.

V.M. Vishnevskiy et al. (Eds.): DCCN 2017, CCIS 700, pp. 343–354, 2017.
DOI: 10.1007/978-3-319-66836-9_29

1   Introduction

1.1   Review of Investigated Contact Centers Models

In the last 15 years an extensive research in mathematical models for telephone call centers has been carried out, cf. References [2-6, 8–17] of [2]. The object has been expanded to more general customer contact centers (with contact also made by other means, such as fax and e-mail). In order to describe the object efficiently the state of the model must include: (1) for every customer in the queue the time that he has spent in it and (2) for every customer in the multi-server the time that he has spent after entering the service area, that is being received by one of the available servers.

The focus of research was on a multi-server with a large number of servers, because it is typical of contact centers. One of important relating questions is the dynamics of the queue of a multi-server with a large number of servers. For such queues were found fluid limits as the number of servers tends to infinity. Notice that such a limit is a deterministic fumction of time with values in a certain measure space, or in a space containing such a component.

An important particular question is the convergence of the fluid limit to a stable state as time tends to infinity. For a discrete time model W. Whitt has found equilibrium states (a multitude) of the fluid model and proved the time convergence in a special case—for a primitive arrival process and for initial condition with empty multi-server and queue, [3, Sect. 7].

1.2   A New Model for Contact Centers and Its Fluid Limit with Equilibrium Behavior

We have suggested in [2] a more suitable model for contact centers. The number of customers is fixed. Customers may be situated in two states: normal and failure. There is a multi-server which repairs customers in the failure state. The repair time/the time duration of a normal state is a random variable, independent and identically distributed for all customers. Now "the arrival process" in the multi-server does not correspond to that of the previous $G/GI/s+GI$ model. For a large number of customers and a suitable number of servers we have calculated approximately the dynamics of the normalized state of the system—its fluid limit. In [1] we explored the convergence of the fluid limit as time tends to infinity and found its steady-state (or equilibrium). Now we allow in the network random changings of the environment.

We confine ourselves to a discrete time model and in [1] we justified its importance.

In Whitt's article [3] the ideas of the convergences proofs are true and very lucid, and the proofs are clearly presented. We transfer his proof technique to our network model and also exploit his equilibrium technique.

2   Closed Multi-server Network with n Customers and Its Fluid Limit Equilibrium

In this section we describe the previous model with a constant environment and results for it. This makes the results for the present model understandable and explains the course of their proofs.

2.1   Network Description

Consider a closed network consisting of n customers. They may be situated in two states: normal and failure. A multi-server repairs customers in the failure state. The repair time (resp., the time duration of a normal state) is a random variable, independent and identically distributed for all customers. We set the problem to calculate the number of current failures as $n \to \infty$ and the number of servers grows suitably, so much as an approximation.

Now we give a rigorous description of this model.

The network consists of two multi-servers with n customers and is closed. Multi-server 1 (further denoted MS1) consists of n servers (for the customers in the normal state), the time they service a customer has distribution G^1. Multi-server 2 (further denoted MS2) consists of $s_n n$ servers with a number $s_n \in (0,1)$ (for the customers in the failure state), the time they service a customer has distribution G^2. The distributions G^1, G^2 are discrete: they are concentrated on $\{1,2,\ldots\}$. Service times are independent for both multi-servers and all customers. We investigate the behavior of the network as $n \to \infty$, namely, we shall establish a stochastic-process fluid limit. It is done only in a special case: discrete time $t = 0,1,2,\ldots$ Everywhere further we demand:

Assumption 1. *Upon arrival to a multi-server a customer*

- *enters service immediately if there is a server available;*
- *waits in queue if the servers in the multi-server are all busy.*

Customers from queue are served in order of their arrival (FCFS) by the first available server.

In MS1 no queue may arise—if all n its servers are occupied then all customers are in MS1, therefore no new customer can arrive. In MS2 a queue may arise.

Denote the number of customers at a moment $t = 0,1,\ldots$ in MS1 (resp., MS2) by $B_n^1(t)$ (resp., $B_n^2(t)$) and the number of customers in the queue $Q_n(t) = n - B_n^1(t) - B_n^2(t)$. The quantities for multi-servers must be defined more exactly, namely, $B_n^i(t) = \sum_{k=0}^{\infty} b_n^i(t,k)$, $i = 1,2$, with $b_n^i(t,k)$ being the number of customers in MSi at the moment t who have spent there time k, $i = 1,2$. $b_n^i(t,k)$ may also be interpreted as the number of busy servers at time t in the multi-server i that are serving customers that have been in service precisely for time k, $i = 1,2$.

At the same time moment $t \in \{1, 2, \ldots\}$ multiple events can take place, so we have to specify their order.

We must create a fictitious queue for the MS1—in fact this multi-server is so large (n servers) that any customer of the whole quantity n trying to enter the MS1 at once finds a free server in it.

At the time moment t the parameters b^1, b^2, Q are taken from the previous time $t - 1$ and processed to the current situation.

For both multi-servers:

- first, customers in service are served;
- second, the served customers move to another multi-server queue, to the end of it;
- third, waiting customers in queue move into service of the multi-server according to Assumption 1.

Customers enter service in MS2 whenever a server is available, so that the system is work-conserving; i.e. we assume that $Q_n(t) = 0$ whenever $B_n^2(t) < s_n n$, and that $B_n^2(t) = s_n n$ whenever $Q_n(t) > 0$, $t = 0, 1, 2, \ldots$. This condition can be summarized by the equation

$$(s_n - B_n^2(t)/n)Q_n(t) = 0 \text{ for all } t \text{ and } n.$$

2.2   Fluid Limit Dynamics

Notations

Denote for $i = 1, 2$:

- $G^{i;c}(k) := 1 - G^i(k)$ and $g^i(k) := G^i(k) - G^i(k-1), k = 1, 2, \ldots$
- E^i the expectation of the time the server in MSi services a customer (for the last equality see "Expected value" in Wikipedia.):

$$E^i := \sum_{k=1}^{\infty} k g^i(k) = 1 + \sum_{k=1}^{\infty} G^i(k) .$$

- $\sigma_n^i(t)$ the number of service completions in MSi at time moment $t = 1, 2, \ldots$.

Symbol $\Rightarrow$ means convergence of the network state characteristics to a constant in probability as the index n denoting the number of customers tends to infinity. **Fluid Limit Dynamics**

Under certain conditions, specifically $\lim_{n \to \infty} s_n = s \in (0, 1)$, the fluid limit exists and its dynamics is described below (the proof is given in [2, Theorem 1]). As $n \to \infty$, for $i = 1, 2$ and $t, k \geq 0$,

$$\frac{b_n^i(t, k)}{n} \Rightarrow b^i(t, k), \quad \frac{\sigma_n^i(t)}{n} \Rightarrow \sigma^i(t), \quad \frac{B_n^i(t)}{n} \Rightarrow B^i(t) \equiv \sum_{k=0}^{\infty} b^i(t, k),$$

$$\frac{Q_n(t)}{n} \Rightarrow Q(t), \quad \sum_{i=1}^{2} B^i(t) + Q(t) = 1, \qquad (1)$$

and $B^2(t) \leq s$ and $(s - B^2(t))Q(t) = 0$. The evolution of the vector $(b^i, \sigma^i, i = 1, \ldots, N)(t)$, $t = 0, 1 \ldots$, proceeds with steps of t:

$$b^i(t, k) = b^i(t - 1, k - 1)\frac{G^{i;c}(k)}{G^{i;c}(k - 1)}, \quad k = 1, 2, \ldots, i = 1, 2, \qquad (2)$$

$$b^2(t, 0) = \min\{s - B^2(t - 1) + \sigma^2(t), Q(t - 1) + \sigma^1(t)\}, \qquad (3)$$

$$b^1(t, 0) = \sigma^2(t), \ \sigma^i(t) = \sum_{k=1}^{\infty} b^i(t - 1, k - 1)\frac{g^i(k)}{G^{i;c}(k - 1)}, \quad i = 1, 2. \qquad (4)$$

2.3   Fluid Limit Equilibrium

Consider the discrete time fluid limit for the closed network model dynamics established in our article [2].

Definition 1. *A point in the state space of deterministic fluid processes is called "equilibrium" if fluid processes after reaching this point remain in it. Deterministic fluid processes are formally described/characterized by sets (b^1, b^2, Q) consisting of non-negative functions of $b^1(t, k)$, $b^2(t, k)$, $Q(t), t = 0, 1, \ldots, k = 0, 1, \ldots$, satisfying*

$$\sum_{k=0}^{\infty} b^2(t, k) \leq s, \ \sum_{k=0}^{\infty} (b^1 + b^2)(t, k) + Q(t) = 1, t = 0, 1, \ldots,$$

and equilibrium points are described/characterized by sets (b^{*1}, b^{*2}, Q^*) consisting of non-negative functions $b^{*1}(k), b^{*2}(k), k = 0, 1, \ldots$, and a non-negative number Q^* satisfying

$$\sum_{k=0}^{\infty} b^{*2}(k) \leq s, \ \sum_{k=0}^{\infty} (b^{*1} + b^{*2})(k) + Q^* = 1. \qquad (5)$$

If the initial condition of a fluid process is an equilibrium, then this fluid process is constant in time.

The following theorem is proved in [1]. Here it is abridged.

Theorem 1. *For the deterministic fluid processes there exists a single equilibrium point. The characteristics b^{*1}, b^{*2}, Q^* of this equilibrium point have the form*

$$b^{*i}(k) = b^{*i}(0)G^{i;c}(k), \ k = 1, 2, \ldots, i = 1, 2, \ and \ Q^* = 1 - (B^{*1} + B^{*2}) \qquad (6)$$

with the values of $b^{*1}(0)$ and $b^{*2}(0)$ being equal.

2.4   Fluid Limit Convergence to Equilibrium as $t \to \infty$

W. Whitt tried to investigate convergence of the fluid limit trajectory to equilibrium in [3, Sect. 7]. No strong result for universal convergence has been presented, only in particular case—starting from an empty multi-server and an empty queue, see [3, Theorem 7.3]. We transfer this simple theorem to our closed network model. The following theorem is proved in [1].

Theorem 2. *Suppose the fluid limit satisfies at time $t = 0$ the following conditions: $B^1(0) = 0$ and $b^2(0, \cdot) = b^{*2}$ (b^{*2} is the component of equilibrium point (b^{*1}, b^{*2}, Q^*), see Theorem 1). Then the fluid limit converges to the equilibrium point as $t \to \infty$. Namely:*

- *the state of MS2 remains equilibrium:*

$$b^2(t, \cdot) = b^{*2}, \ t = 0, 1, 2, \ldots; \tag{7}$$

- *the state of MS1 grows occupying its equilibrium state—with each time step adds the next age equilibrium parameter:*

$$b^1(0, \cdot) \equiv 0 \ and \ for \ t = 1, 2, \ldots \ b^1(t, \cdot) = b^{*1}(\cdot)I_t \quad with \ I_t = I_{\{0,1,\ldots,t-1\}}; \tag{8}$$

- *the queue decreases—with each time step loses the amount of the previous age MS1 equilibrium parameter: $Q(0) = 1 - B^{*2}$,*

$$Q(t) = Q(t-1) - b^{*1}(t-1) = 1 - B^{*2} - \sum_{l=0}^{t-1} b^{*1}(l), \ t = 1, 2, \ldots$$

3   Generalization to Changing Environment

Return to the beginning of Subsect. 2.1. Now the environment of the system is not permanent, it may change (with discrete time step, stochastically). And the time duration of a customer's normal state (namely, its distribution) changes also—it corresponds to the current environment. The multi-server which repairs customers in the failure state is irrespective of the environment—the distribution of the customer repair time does not change. We make this restriction on the repairing multi-server for the simplicity of our research. We shall describe a generalization of the fluid limit and its equilibrium in the changing environment.

There is a finite number N of changing environments, each of them is denoted by its index $i \in \{1, \ldots, N\}$. In the environment $i \in \{1, \ldots, N\}$ the time duration of the normal state is marked by the corresponding i. So, instead of the previous distribution G^1 of the time MS1 services a customer, we have distributions $G^{1|i}$, $i \in \{1, \ldots, N\}$.

The change of environments is specified by a discrete time-homogeneous markov chain $u := u(t)$, $t = 0, 1, \ldots$, with state space $\{1, \ldots, N\}$ and transition probabilities

$$\mathbb{P}(u(1) = j | u(0) = i) = p_{ij}, \ i, j \in \{1, \ldots, N\}.$$

It is obvious that the fluid limit now turns to a piecewise-deterministic Markov process (see [5, Chap. 2]). Really, denote by τ_i, $i = 1, 2, \ldots$, the moments when u receives its new value, and by τ_0 the starting moment 0:

- τ_i is the moment when the value which u had at the moment τ_{i-1} becomes changed: $\tau_i = \min\{t > \tau_{i-1} : u(t) \neq u(\tau_{i-1})\}$, $i = 1, 2, \ldots$

For τ_0 the probability space divides into N non-overlapping probability subspaces $\{u(\tau_0) = j\}$, $j \in \{1, \ldots, N\}$, and on each nonempty subspace $\{u(\tau_0) = j\}$ equations (1)-(4) hold with $G^{1|j}$, G^2 on time interval $[\tau_0, \tau_1]$—the distribution of customers ages in both multi-servers and the queue quantity converges to a fluid limit. And at time τ_1 first the evolution of the limit process makes the same way step with its distribution value and then the environment changes by virtue of switching, at once to all variants $i \in \{1, \ldots, N\}$ with $p_{ji} > 0$. Thus the fluid limit at time τ_1 divides for the future into different fluid limits, only one of which with $u(\tau_1) = j$ continues the development of the previous fluid limit (under the condition $p_{jj} > 0$). And at the moment τ_1 and further the development of the process repeats the behavior for τ_0, on each version $\{\tau_1 = l\}$, $l = 1, 2, \ldots$

And so forth, for each $i = 1, 2, \ldots$ and $j \in \{1, \ldots, N\}$ on the interval $[\tau_i, \tau_{i+1}]$ and on the probability subspace of fixed values $\{(\tau_1, u(\tau_1)), \ldots, (\tau_i, u(\tau_i))\}$ with $\{u(\tau_i) = j\}$ equations (1)-(4) hold with $G^{1|j}$, G^2. And even more—the distributions of customers ages in both multi-servers and the queue quantity converge to a fluid limit on the interval $[\tau_0, \tau_{i+1}]$ and on the described probability subspace of fixed values. At time τ_{i+1} first the limit process makes a step with its distribution value and then the environment changes, at once to all variants $i \in \{1, \ldots, N\}$ with $p_{ji} > 0$. Thus the fluid limit at time τ_{i+1} divides for the future into different fluid limits.

It is clear that the described limit process as $n \to \infty$ may be only random, not deterministic, and respectively an equilibrium for this process may be only random, not constant, as it was for the previous model with a fixed environment in [1, Subsect. 2.3].

The whole limit process is described by $(b, Q, u)(t)$, $t = 0, 1, \ldots$, where (b, Q) is a distribution identifying the state of the network (restrictions $b = (b^1, b^2)$ for the multi-servers 1 and 2, Q for the queue). To find an equilibrium for this piecewise-deterministic Markov process is complicated. Therefore we switch over to the conditional expectation of this process: $\mathbb{E}\left((b, Q)(t)|u(t)\right)$, $t = 0, 1, \ldots$ We denote for environment $i = 1, \ldots, N$ the conditional distribution in multi-servers by $b^i(t)$, its restrictions to the multi-server 1 by $b^{1|i}(t)$, to the multi-server 2 by $b^{2|i}(t)$, and the conditional expectation of the queue by $Q^i(t)$:

$$\mathbb{E}\left((b, Q)(t)|u(t) = i\right) = (b^i, Q^i)(t) = (b^{1|i}, b^{2|i}, Q^i)(t), \ t = 0, 1, \ldots$$

The equations for conditional distributions are similar to the equations for the fluid limit in [1, Subsect. 2.2], only for elementary events $b^{2|i}(\cdot, 0)$, $i \in \{1, \ldots, N\}$, they are specific:

$$\mathbb{P}(u(t+1)=i)b^{1|i}(t+1,0) = \sum_{j=1}^{N} \mathbb{P}(u(t)=j)p_{ji}\,\sigma^{2|j}(t),$$

$$\mathbb{P}(u(t+1)=i)b^{1|i}(t+1,k+1) = \sum_{j=1}^{N} \mathbb{P}(u(t)=j)p_{ji}\,b^{1|j}(t,k)\frac{G^{1|j;c}(k+1)}{G^{1|j;c}(k)},$$

$$\mathbb{P}(u(t+1)=i)b^{2|i}(t+1,k+1)$$

$$= \sum_{j=1}^{N} \mathbb{P}(u(t)=j)p_{ji}\,b^{2|j}(t,k)\frac{G^{2;c}(k+1)}{G^{2;c}(k)}, \quad t=0,1,\ldots, \ k=0,1,\ldots$$

$$(9)$$

And for $b^{2|i}(\cdot,0)$? Unfortunately, this equation exists only for specific models, because for $i=1,\ldots,N$ $b^{2|i}(t+1,0)$ has a complicated connection to the random distributions $(b,Q)(t)$ transformed with $G^{1|j}$, $j=1,\ldots,N$, and G^2, $t=0,1,\ldots$ The property of the model must be such: with a lapse of time $G^{1|j}$, $j=1,\ldots,N$, and G^2 turn the MS2 to either fully occupied with customers, or to a bounded part occupied with customers. Here are the equations for large t and $i=1,\ldots,N$: if MS2 is fully occupied with customers, then

$$\mathbb{P}(u(t+1)=i)b^{2|i}(t+1,0) = \sum_{j=1}^{N} \mathbb{P}(u(t)=j)p_{ji}\,\sigma^{2|j}(t+1), \qquad (10)$$

because in i the sum of the MS2 customers after servicing is

$$\sum_{j=1}^{N} \mathbb{P}(u(t)=j)p_{ji}\left(s - \sigma^{2|j}(t+1)\right)$$

$$= s\mathbb{P}(u(t+1)=i) - \sum_{j=1}^{N} \mathbb{P}(u(t)=j)p_{ji}\,\sigma^{2|j}(t+1); \qquad (11)$$

if MS2 is occupied with customers in a bounded part, then

$$\mathbb{P}(u(t+1)=i)b^{2|i}(t+1,0) = \sum_{j=1}^{N} \mathbb{P}(u(t)=j)p_{ji}\,\sigma^{1|j}(t+1), \qquad (12)$$

because there are no queues and all serviced customers in MS1 move to MS2.

We have generalized the distribution process describing the state of the network (it is defined in [1, Subsects. 2.1 and 2.2]) to a switching (see [6]) distribution process. We explained how this switching distribution process converges as the number of customers $n \to \infty$. Namely, it converges in a complicated way: not to the fluid limit, but to a quasi-fluid limit. We have simplified this process with the conditional expectation and found equations for the evolution of the conditional expectation process $(\mathbb{E}\left((b,Q)(t)|u(t)\right), u(t))$, $t=0,1,\ldots$ Now we intend to find an equilibrium for this process.

Definition 2. *A moment distribution of the process* $(\mathbb{E}((b, Q)|u), u)$, *that is, a distribution of u and corresponding expectations of distributions (b, Q) with respect to the value of u, is "equilibrium" if the process after reaching this distribution remains in it. In particular, u must have its stable distribution.*

Notations for Equilibriums

An equilibrium for $u = j$ is a constant distribution $(b^{*1|j} = (b^{*1|j}, b^{*2|j}), Q^{*|j})$, $j \in \{1, \ldots, N\}$, and the Markov process u has a stable distribution denoted by $P_j := \mathbb{P}(u = j)$, $j \in \{1, \ldots, N\}$. Denote $B^{*i|j} \equiv \sum_{k=0}^{\infty} b^{*i|j}(k)$, $\sigma^{*i|j}$ the number of service completions in MSi before the environment changes, $i = 1, 2$, $j = 1, \ldots, N$.

A designation in equilibrium with an added upper left symbol ^{t} means the corresponding object in time step before switching ("t" for "transient"), for example:

$$^t b^{*1|i}(k+1) = b^{*1|j}(k) \frac{G^{1|j;c}(k+1)}{G^{1|j;c}(k)}, \ k = 0, 1, \ldots$$

The following theorem describes evident connections in the equilibrium of the conditional expectation process.

Theorem 3. *The following holds for each $i \in \{1, \ldots, N\}$.*
For $k = 0, 1, \ldots$

$$P_i b^{*1|i}(k+1) = \sum_{j=1}^{N} P_j p_{ji} \, b^{*1|j}(k) \frac{G^{1|j;c}(k+1)}{G^{1|j;c}(k)},$$

$$P_i b^{*2|i}(k+1) = \sum_{j=1}^{N} P_j p_{ji} \, b^{*2|j}(k) \frac{G^{2;c}(k+1)}{G^{2;c}(k)}. \tag{13}$$

And for $b^{*1|i}(0)$ and $b^{*2|i}(0)$:

– $P_i b^{*1|i}(0) = \sum_{j=1}^{N} P_j p_{ji} \, \sigma^{*2|j}$;

– *if MS2 is fully occupied with customers, then*

$$P_i b^{*2|i}(0) = \sum_{j=1}^{N} P_j p_{ji} \, \sigma^{*2|j}; \tag{14}$$

– *if MS2 is occupied with customers in a bounded part, then*

$$P_i b^{*2|i}(0) = \sum_{j=1}^{N} P_j p_{ji} \, \sigma^{*1|j}. \tag{15}$$

Of course, for the present model it is more difficult to calculate an equilibrium than we made it without changing environment in [1, Subsect. 2.3]. But think just of bounded distributions $G^2, G^{1;i}$, $i \in \{1, \ldots, N\}$—surely in this case we can calculate the equilibrium point. We show how to find an equilibrium in this rough way:

Theorem 4. *Let* $N = 2$, $G^2(1) = G^{1|i}(1) = 1$, $i \in \{1,2\}$, *and denote* $a^i := \mathfrak{b}^{*2|i}(0)$, $i \in \{1,2\}$ *For each object* o *in equilibrium there are two constants* $c_1(o)$ *and* $c_2(o)$, *depending only on* $\{P_i, p_{ij}, G^{1|i}, G^2, i, j \in \{1,2\}\}$ *and setting the object equal to:*

- *for* $o = b^{*i|j}(k), B^{*i|j}, \sigma^{*i|j}, i, j \in \{1,2\}, \quad o = c_1(o)a^1 + c_2(o)a^2;$
- *for* $o = Q^{*|j}, i, j \in \{1,2\}, \quad o = 1 - (c_1(o)a^1 + c_2(o)a^2).$

As $a^j = \mathfrak{b}^{*1|j}(0) = \min\{s - (B^{*2|j} - \sigma^{*2|j}), Q^{*|j} + \sigma^{*1|j}\}$, $j \in \{1,2\}$, a^1 *and* a^2 *can be calculated in equations*

$$
a^j =
$$

$$
\min\{s - \left((c_1(B^{*2|j}) - c_1(\sigma^{*2|j}))a^1 + (c_2(B^{*2|j}) - c_2(\sigma^{*2|j}))a^2\right),
$$
$$
1 - \left((c_1(Q^{*|j}) - c_1(\sigma^{*1|j}))a^1 + (c_2(Q^{*|j}) - c_2(\sigma^{*1|j}))a^2\right)\}, \tag{16}
$$
$$
j = 1, 2.
$$

Proof. It is trivial to find constants $c_1(o)$ and $c_2(o)$, for example for $b^{*2|i}(0)$ and $b^{*2|i}(1)$, $i \in \{1,2\}$:

$$
P_i b^{*2|i}(0) = \sum_{j=1}^{2} P_j p_{ji} \mathfrak{b}^{*2|j}(0) = \sum_{j=1}^{2} P_j p_{ji} a^j ; \tag{17}
$$

$$
P_i b^{*2|i}(1) = \sum_{j=1}^{2} P_j p_{ji} \, b^{*2|j}(0) \frac{G^{2;c}(1)}{G^{2;c}(0)} = \frac{G^{2;c}(1)}{G^{2;c}(0)} \sum_{j=1}^{2} p_{ji} \, P_j b^{*2|j}(0),
$$

and $\displaystyle\sum_{j=1}^{2} p_{ji} \, P_j b^{*2|j}(0) = \sum_{j=1}^{2} p_{ji} \sum_{l=1}^{2} P_l p_{lj} a^l = a^1 P_1 \sum_{j=1}^{2} p_{1j} p_{ji} + a^2 P_2 \sum_{j=1}^{2} p_{2j} p_{ji} .$

$$
\tag{18}
$$

In the same way are calculated coefficients for the whole $b^{*1|i}$, $i \in \{1,2\}$, and other objects of equilibrium. And to solve the equations (16) we transfer them to 4 new simple equations systems, setting in these equations instead of minimums in turn all their components, in all possible variants. Further steps work for longer $G^2, G^{1;i}$, $i \in \{1,2\}$, too. $\qquad\blacksquare$

We generalize now Subsect. 2.4 "Fluid Limit Convergence to Equilibrium As $t \to \infty$" from [1] (its abridgement presented here) to the switching model.

Theorem 5. *If the equilibrium satisfies* $B^{*2|i} = s$, $i = 1, \dots, N$, *and a quasi-fluid limit starts with the following conditions: the distribution of* $u(0)$ *is stable;* $B^{1|i}(0) = 0$ *and* $b^{2|i}(0, \cdot) = b^{*2|i}$, $i = 1, \dots, N$, *then this quasi-fluid limit converges steadily to the equilibrium as* $t \to \infty$. *Namely, for* $i = 1, \dots, N$

- *the state of MS2 remains equilibrium:*

$$
b^{2|i}(t, \cdot) = b^{*2|i}, t = 0, 1, 2, \dots ; \tag{19}
$$

– the state of MS1 grows occupying its equilibrium state—with each time step adds the next age equilibrium parameter: $b^{1|i}(0, \cdot) \equiv 0$ and

$$\text{for } t = 1, 2, \dots \; b^{1|i}(t, \cdot) = b^{*1|i}(\cdot) I_t \quad \text{with } I_t = I_{\{0,1,\dots,t-1\}}; \qquad (20)$$

– the queue decreases—with each time step loses the amount of the previous time age MS1 equilibrium parameter:

$$Q^i(0) = 1 - B^{*2|i}, \; Q^i(t) = Q^i(t-1) - b^{*1|i}(t-1) =$$

$$1 - \left(B^{*2|i} + \sum_{l=0}^{t-1} b^{*1|i}(l) \right), t = 1, 2, \dots$$

Lemma 1. *If for equilibrium* $\mathbb{E}B^{*2|u} = s$, *then for each* $i \in \{1, \dots, N\}$ $B^{*2|i} = s$ *and* ${}^t B^{*2|i} = s$.

Proof. Since $\mathbb{E}B^{*2|u} = \sum_{i=1}^{N} P_i B^{*2|i}$, it equals s only if $B^{*2|i} = s$ for each $i \in \{1, \dots, N\}$ (we suppose that all P_i are positive; otherwise the whole problem reduces only to positive ones). Consequently ${}^t B^{*2|i} = s$ for each $i \in \{1, \dots, N\}$: the stability of u implies $P_i = \sum_{j=1}^{N} P_j p_{ji}$ and

$$P_i B^{*2|i} = \sum_{j=1}^{N} P_j p_{ji} \, {}^t B^{*2|j} = \sum_{j=1}^{N} P_j p_{ji} \left(s + ({}^t B^{*2|j} - s) \right)$$

$$= \sum_{j=1}^{N} P_j p_{ji} s + \sum_{j=1}^{N} P_j p_{ji} \left({}^t B^{*2|j} - s \right) = s P_i + \sum_{j=1}^{N} P_j p_{ji} \left({}^t B^{*2|j} - s \right).$$

If for some $j \in \{1, \dots, N\}$ ${}^t B^{*2|j} < s$ then choose $p_{ji} > 0$ and for this i the last sum is negative and consequently $B^{*2|i} < s$.

Now we prove the theorem.

Proof. Fix $i \in \{1, \dots, N\}$.
   A trivial evaluation shows:

– $b^{2|i}(1, k) = b^{*2|i}(k), \; k = 1, 2, \dots$;
– $b^{1|i}(1, \cdot)$ equals $\{b^{*1|i}(0), 0, 0, \dots\}$;
– $\sigma^{2|i}(1) = {}^t \sigma^{*2|i}$;
– $Q^i(0) \geq Q^{*i}$ because

$$Q^i(0) = 1 - (B^{1|i}(0) + B^{2|i}(0)) = 1 - B^{*2|i} = B^{*1|i} + Q^{*i}.$$

It is left to calculate $b^{2|i}(1, 0)$ and $Q^i(1)$. Obviously, ${}^b\!b^{2|i}(1, 0) = \min\{s - (B^{2|i}(0) - \sigma^{2|i}(1)), Q^i(0)\} = \min\{{}^t \sigma^{*2|i}, Q^i(0)\}$ as $B^{2|i}(0) = B^{*2|i} = s$. Since ${}^t B^{*2|i} = B^{*2|i} = s$ (cf. Lemma 1) $\min\{{}^t \sigma^{*2|i}, Q^{*i}\} = {}^t \sigma^{*2|i}$, and also $\min\{{}^t \sigma^{*2|i}, Q^i(0)\} = {}^t \sigma^{*2|i}$ as $Q^{*i} \leq Q^i(0)$. Thus ${}^b\!b^{2|i}(1, 0) = {}^b\!b^{*2|i}(0)$. It holds for all $i \in \{1, \ldots, N\}$, consequently $b^{2|i}(1,0) = b^{*2|i}(0)$. Finally $Q^i(1) = 1 - (B^{*2|i} + b^{*1|i}(0))$.

We finish the proof by induction. Suppose at time t which is not less than 1 $b^2(t) = b^{*2}$ and the statement of the theorem about $b^{1|i}(t)$ in equation (20) holds. Evidently, $Q^i(t)$ must satisfy the statement of the theorem:

$$Q^i(t) = 1 - \left(B^{*2|i} + \sum_{l=0}^{t-1} b^{*1|i}(l) \right), \ t = 1, 2, \ldots$$

Repeating the calculations from $t = 0$ to $t = 1$ for the time step from t to $t+1$ we obtain:

- $b^{2|i}(t+1, k) = b^{*2|i}(k)$, $k = 1, 2, \ldots$;
- $b^{1|i}(t+1, \cdot)$ equals $\{b^{*1|i}(0), b^{*1|i}(1), \ldots b^{*1|i}(t), 0, 0, \ldots\}$;
- ${}^tQ^i(t+1) = Q^i(t) + \sigma^{1|i}(t+1)$ and is not less than Q^{*i};

and finally $b^{2|i}(t+1, 0) = b^{*2|i}(0)$ ensuring $b^{2|i} = b^{*2|i}$, and $Q^i(t+1) = 1 - (B^{*2|i} + \sum_{l=0}^{t} b^{*1|i}(l))$. This demonstrates that the quasi-fluid limit converges steadily to the equilibrium.

4   Conclusion

The models of call/contact centers with a changing environment have an important approximate description for their simplified dynamics—quasi-fluid limit. We plan to investigate for our models with a fully general starting case the convergence in time of the quasi-fluid limit to the equilibrium.

References

1. Anulova, S.: Properties of fluid limit for closed queueing network with two multi-servers. In: Vishnevskiy, V.M., Samouylov, K.E., Kozyrev, D.V. (eds.) DCCN 2016. CCIS, vol. 678, pp. 369–380. Springer, Cham (2016). doi:10.1007/978-3-319-51917-3_33
2. Anulova, S.: Approximate description of dynamics of a closed queueing network including multi-servers. In: Vishnevsky, V., Kozyrev, D. (eds.) DCCN 2015. Communications in Computer and Information Science, pp. 177–187. Springer, Cham (2016)
3. Whitt, W.: Fluid models for multi-server queues with abandonments. Oper. Res. **54**(1), 37–54 (2006). http://pubsonline.informs.org/doi/abs/10.1287/opre.1050.0227
4. Walsh Zuñiga, A.: Fluid limits of many-server queues with abandonments, general service and continuous patience time distributions. Stochast. Process. Appl. **124**(3), 1436–1468 (2014)
5. Davis, M.: Markov models and optimization. Monographs on Statistics and Applied Probability, vol. 49. Chapman & Hall, London (1993)
6. Yin, G., Zhu, C.: Hybrid Switching Diffusions: Properties and Applications. Springer, Heidelberg (2010)

Analysis of Unreliable Open Queueing Network with Dynamic Routing

Elmira Yu. Kalimulina[(✉)]

V.A. Trapeznikov Institute of Control Sciences, Russian Academy of Sciences,
Profsoyuznaya St. 65, Moscow 11799, Russia
elmira.yu.k@gmail.com

Abstract. An open queuing network with m unreliable nodes is considered. The routing of jobs in a network is transformed depending on availability/unavailability of some nodes. If there is no direct transition from the node i to j due the node j failure, then jobs are redirecting to the node the most close to the j. In this way there is no blocking in a network. The initial probability r_{ij} in a transition matrix R is proportional divided between available nodes. The network dynamic is described by a continuous in time random process $X(t)$ taking values from the state space enlarged by the ensemble G_N. G_N contains all graphs (transition matrices), and described by the initial state R^0 and a given sequence of transition rates $\{\alpha_i\}$ and $\{\beta_i\}$, where α_i and β_i are failure and recovery rates for the node i, $i = 1, \ldots m$.

Keywords: Dynamic routing · Queueing network · Unreliable nodes · Emsemble of networks

1 Introduction

Queueing systems and networks are the most suitable mathematical tools for modelling and performance evaluation of complex systems such as modern computer systems, telecommunication networks, transport, energy and others [1–5]. The reliability is another important factor for quality assessment of these systems [6]. So models with unreliable elements are a subject of great interest last years [27]. A large number of research papers study queueing systems with unreliable servers [7–12]. The less ones consider queueing networks. In this paper we analyse the performance characteristics of an open queueing network, whose nodes are subject to failure and repair. This assumption is often missed in theoretical papers, but it's essential for applications, for example, for telecommunication, sensor, ad-hoc, mesh and other kind of networks.

This work is motivated by a practical task of modelling of modern complex networks (telecommunication, transport, distributed computing systems, etc.). We consider the mathematical model of the queueing network as a set of connected nodes that can break down and repair. We propose a modification of the classical model of an open queueing network (see e.g. [13, Chap. 2]), based on the principle of dynamic routing.

© Springer International Publishing AG 2017
V.M. Vishnevskiy et al. (Eds.): DCCN 2017, CCIS 700, pp. 355–367, 2017.
DOI: 10.1007/978-3-319-66836-9_30

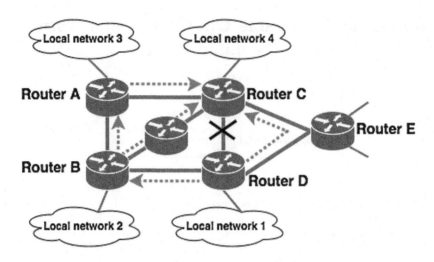

Fig. 1. A six-node internetwork with some route redundancy.

A strong mathematical definition of the term "dynamic routing" doesn't exist. It originally appeared in telecommunication industry. Dynamic routing is the technology that enables active network nodes (called routers) to perform many vital functions: detection, maintaining and modification of routes with considering of a network's topology, as well as some functions of routes calculation and their estimations. In difference from static routing technology routes are calculated dynamically using any one of a number of dynamic routing protocols. Dynamic routing is crucial technology for reliable packet transmission in case of failures. Protocols are used by routers to detect nodes availability and find routes available for packets forwarding over network. Figure 1 shows a simple example of six-node internetworking transmission. The transmission between *Local network 1* and *Local network 4* uses the link *Router D - Router C*. Assume the transmission facility between Gateway *Routers C* and *D* has failed. This renders the link between C and D unusable and the data transmission between *Local network 1* and *2* is impossible (for static routing). The dynamic routing protocols use a route redistribution scheme and dynamically define new paths for transmission between *Router D* and *Router C* via *B* and *A*, or *E*, or central router (dotted lines). Thus network availability can be significantly enhanced through dynamic routing, alternate routes of more length are redundant communications links between nodes D and C.

There are some math research papers where queueing networks with dynamic routing were considered. Queueing networks with constant routing matrix were considered in papers [14,15], each node was modelled as a multichannel system, principle of dynamic routing was a random selection of a channel at the node. Kelly in his papers (see. [16]) considered a network as a set of parallel queues with several types of requests incoming into it, and the dynamic routing principle is the selection of particular queue depending on the type of request.

We will adhere to interpret the dynamic routing as it's defined in telecoms and we will understand it as reconstruction of the route depending on the availability/unavailabiity of a specific node in the network, sending the requests (message) via alternate paths. In terms of queueing networks models it means the change the values of a routing matrix $\{r_{ij}\}$ [25, 26, 28]. A request will be rerouted to the available node in the case of failure of node j, i.e., this concept is as close as possible to this definition in telecommunication networks [18].

It is noted that other approaches can be applied to the problem of estimation of networks reliability. For example, the Erdos-Renyi random graphs model can be used to analyse the network connectivity (Erdos-Renyi graph) [19].

But as our task is the performance evaluation and analysis of traffic flows in networks taking into account reliability, we use queueing theory models and a model of open queueing network. There are some researches on unreliable queueing networks. Several algorithms for modifying the routing matrix $\{r_{ij}\}$ in the case of failures of nodes were described in [17], the common idea which they are based on is the principle of blocking of requests and repeated service after nodes recovery. The result related to the rate of convergence to the stationary distribution for unreliable network is given in [20, 21]. In this paper we give some results for unreliable networks similarly as it was done in [21], but we propose another approach to the modification of the route matrix $\{r_{ij}\}$ and consider a more general model for network nodes. There are several main classical problems in the analysis of queueing systems(networks) models that have to been considered: stability, ergodicity and probability of overflow. These problems are strongly connected with the transient behaviour of networks and the speed of convergence to stationary regime, so here we will also consider this problem for unreliable network systems.

2 Process Definition

2.1 The Classic Queueing Network

The classical model of Jackson network consists of m nodes $(m < \infty)$, $M := \{1, \ldots, m\}$. Each node is a $M/M/1$ queueing system with the "first-come-first-served" discipline of service. Incoming requests are indistinguishable. The external flow into the network is a Poisson process with the parameter $\Lambda(t)$, depending on time. Incoming request is sending for the service to the node j c with probability r_{0j}, $\sum_{j=1}^{m} r_{0j} = r \leq 1$. A request is served at the node j with intensity $\mu_j(n_j)$, where n_j - the numbers of requests at the node j. $X_j(t)$ is denoted as the number of requests at the node j at time t, so the system state at time t is characterised with a vector $X(t) = (X_1(t), X_2(t), ..., x_m(t))$. The unique stationary distribution for the process $X(t)$ exists and is defined as

$$C_i = 1 + \sum_{n=1}^{\infty} \frac{\lambda_i^n}{\prod_{y=1}^{n} \mu_i(y)}, i = 1, 2, ..., m, \tag{1}$$

if the system of equations for a network traffic

$$\lambda_i = \Lambda * r_{0i} + \sum_{j=1}^{m} \lambda_j r_{ji}, i = 1, 2, ..., m \tag{2}$$

has the unique solution [17].

Degradable Networks and Some Routing Mechanisms. If we consider the servers at the nodes to be unreliable, the network's performance is degraded because a subset of the operating nodes is not available. Different routing mechanisms were considered by Daduna [17]. He provided the most extensive research of this subject. A short survey from [17] about the methods of routing is provided here.

The first method of routing is a blocking after service (BAS). It assumes that after completing the service at node at time t a customer chooses the next destination node according to his routing instruction. If at time t the destination node is not able to accept further customers, the customer stays and blocks current server until the situation at the destination node has changed and the customer can enter the node. The current node is blocked during this waiting period, i.e., it cannot start serving another customer, who might be waiting in the waiting room. If several nodes are simultaneously blocked by the same node j, then the "first blocked - first unblocked- rule" usually determines the order in which the nodes will be unblocked when a departure occurs from the destination node j.

The second method is blocking before present service. A customer at node i selects the following destination node j according to the routing rule before he starts receiving service at node i at time t. If node j is full at time t, node i immediately becomes blocked. When a departure occurs from node j, node i becomes unblocked and service begins. However, as soon as the destination node j becomes full again during the customer's service at i, the service is interrupted and node i becomes blocked again. Depending upon whether the customer at the blocked node i is allowed to occupy the position in front of the server or not, one can distinguish the two cases BBS-server occupied and BBS-server not occupied. Of course this is only meaningful when node i has finite waiting capacity.

The third one is the repetitive service. A customer after being served at node i chooses the next destination node j according to the routing instruction. If node j is full, the customer stays at node i to obtain another service. When this additional service expires the customer either selects his destination node anew according to his routing instruction (repetitive service - random destination) or the previously chosen node j remains his fixed destination. In this case, the customer's service at node i has to be repeated until at the end of a service, node j is able to receive another customer (repetitive service - fixed destination).

The last mechanism is the skipping of unavailable set of nodes. Customers are not allowed to enter to the node from the certain set, skipped it and immediately performs the jump to the next one according to the routing matrix.

We provided the another approach called the dynamic routing [25, 26, 28].

2.2 Unreliable Queueing Network with Dynamic Routing

It is assumed now that nodes at the network are unreliable and may break down or repair.

We will refer to $M_0 = \{0, 1, 2, ..., m\}$ as the set of nodes, where "0" is the "external node" (entry and exit from the network) and to $D \subset M$ as the subset of failed nodes, $I \subset M \setminus D$ the subset of working nodes, nodes from I may break down with the intensity $\alpha_{D \cup I}^D(n_i(t))$. Nodes from $H \subset D$ may recover with the intensity $\beta_{D \setminus H}^D(n_i(t))$. It is assumed the routing matrix $R = (r_{ij})$ is given. Additionally the adjacency matrix for our network $S = (s_{ij})$ is considered:

$$s_{ij} = \begin{cases} 1, & \text{if } r_{ij} \neq 0, \\ 0, & \text{if } r_{ij} = 0. \end{cases}$$

Now we can consider all possible paths of the network graph. To find them we need to calculate the following matrix: $(s_{ij})^2$, $(s_{ij})^3$, ..., $(s_{ij})^m$, $m < \infty$, $(s_{ij})^1 = (s_{ij})$. The matrix $(s_{ij})^m$ has the following property: the element in row i and column j is the number of paths from node i in the unit j of length m (including $(m-1)$ transitional nodes) [23].

We take the following routing scheme for network nodes from the subset D (we call it as "dynamic routing without blocking"). Only transitions to $M_0 \setminus D$ are possible for nodes from D:

$$r_{ij}^D = \begin{cases} 0, & \text{if } j \in D, i \neq j, \\ r_{ij} + r_{ik}/s_{ik}^p, & \text{if } j \notin D, k \in D \\ \quad \exists\, i \to j \to i' \to j' \to ... \to i'' \to k : \underbrace{s_{ij}^1 * s_{ji'}^1 * s_{i'j'}^1 * ... * s_{i''k}^1}_{p+1} \neq 0, \\ \text{where} \quad p = \min\{2, 3, ..., m : s_{ik_{k \in D}}^p \neq 0\}, \\ r_{ii} + \sum\limits_{\substack{k \in D \\ s_{ik}^p = 0\ \forall\ 1 < p \leq m}} r_{ik}, & \text{if } i \in M_0 \setminus D, i = j, \end{cases}$$

where s_{ik}^p - element of a matrix $(s_{ij})^p$.

The routing matrix is changed according to the same way for the input flow:

$$\Lambda r_{0j}^D = \begin{cases} \Lambda r_{0j}, & \text{if } j \in M \setminus D, \\ \Lambda(r_{0j} + r_{0k}/s_{0k}^p * \underbrace{(s_{0j}^1 * s_{ji'}^1 * s_{i'j'}^1 * ... * s_{i''k}^1))}_{p+1}, & \text{if } j \notin D, k \in D \\ 0, & \text{otherwise.} \end{cases}$$

Futher we will refer to The modified routing matrix as $R^D = (r_{ij}^D)$. The intensities of failures and recoveries depend on the state of nodes and does not depend on network load and are defined as follows [17]:

$$\alpha(D, I) = \frac{\psi(D \cup I)}{\psi(D)},$$

$$\beta(D, H) = \frac{\phi(D)}{\phi(D \setminus H)},$$

where ψ, ϕ are positive functions with domain on all subsets of set of nodes and taking only finite values for finite sets ($\psi(\emptyset) := 1$, $\phi(\emptyset) := 1$).

A more general model than in [20] is considered for network nodes. It is assumed that each network node is a queueing system type $M/G/1$. The system's dynamic will be described by a continuous in time random process $X(t)$ taking values from the following enlarged state space $\mathbb{E}$:

$$\tilde{n} = ((n_1, z_1), (n_3, z_2), ..., (n_m, z_m), D) \in \{\mathbb{Z}_+ \times \{R_+ \cup 0\}\}^m \times |D| = \mathbb{E},$$

where n_i is the number of requests at the node i, z_i - may be considered as the time elapsed from the beginning of service for the current request i or as remaining service time of the customer in service, $|D|$ - the cardinality of set D. Intensity rates $\mu_i(n_i, z_i)$ depend on both the number of requests at nodes $n_i(t)$ and time $z_i(t)$. If $z_i(t)$ - the time elapsed from the beginning of service for the current request at time t, the conditional probability of the absence of any events occurring in a fixed interval of time $[t, t + \Delta t)$ (= {no new request } $\cup$ {no current service finished at all nodes } under the condition that the current value of the process $X(t)$):

$$\exp\left(-\int_t^{t+\Delta t} \left(\Lambda(s) + \sum_{i=1}^m \mu_i(n_i(t), z_i(t) + s)\right) ds\right),$$

which, if Δt is small enough, equals to [24, Chaps.2–4]

$$1 - \int_t^{t+\Delta t} \left(\Lambda(s) + \sum_{i=1}^m \mu_i(n_i(t), z_i(t) + s)\right) ds + O(\Delta t)^2,$$

if $\Delta t \to 0$ terms with $O((\Delta t)^2)$ are negligible in comparison with the terms without Δt or with Δt in a power one. The probability of any jump (finishing of service in one of the nodes or new request arriving in a network) is defined similarly:

$$\mu_j(n_j(t), z_j(t)) \Delta t \left(1 - \int_t^{t+\Delta t} \left(\Lambda(s) + \sum_{i \neq j}^m \mu_i(n_i(t), z_i(t) + s)\right) ds + O(\Delta t)^2\right), \quad (3)$$

$$\Lambda(t) \Delta t \left(1 - \int_t^{t+\Delta t} \left(\sum_{i=1}^m \mu_i(n_i(t), z_i(t) + s)\right) ds + O(\Delta t)^2\right). \quad (4)$$

The following transitions in a network are possible:

$$T_{ij}\tilde{n} := (D, n_1, \cdots, n_i - 1, \cdots, n_j + 1 \cdots, n_m),$$
$$T_{0j}\tilde{n} := (D, n_1, \cdots, n_j + 1, \cdots, n_m),$$
$$T_{i0}\tilde{n} := (D, n_1, \cdots, n_i - 1, \cdots, n_m),$$
$$T_H\tilde{n} := (D \setminus H, n_1, \cdots, n_m),$$
$$T^I\tilde{n} := (D \cup I, n_1, \cdots, n_m).$$

Definition 1. *The markov process* $\mathbf{X} = (X(t), t \geq 0)$ *is called unreliable queueing network if it's defined by the following infinitesimal generator:*

$$\tilde{\mathbf{Q}}f(\tilde{\mathbf{n}}) = \sum_{j=1}^{m} [f(T_{0j}\tilde{\mathbf{n}}) - f(\tilde{\mathbf{n}})]\Lambda(t)r_{0j}^{D}$$

$$+ \sum_{i=1}^{m}\sum_{j=1}^{m} [f(T_{ij}\tilde{\mathbf{n}}) - f(\tilde{\mathbf{n}})]\mu_i(n_i(t), z_i(t))r_{ij}^{D}$$

$$+ \sum_{I \subset M} [f(T^I\tilde{\mathbf{n}}) - f(\tilde{\mathbf{n}})]\alpha(D, I) \tag{5}$$

$$+ \sum_{H \subset M} [f(T^I\tilde{\mathbf{n}}) - f(\tilde{\mathbf{n}})]\beta(D, H)$$

$$+ \sum_{j=1}^{m} [f(T_{j0}\tilde{\mathbf{n}}) - f(\tilde{\mathbf{n}})]\mu_j(n_i(t), z_i(t))r_{j0}^{D}.$$

3   Main Results

Like the classical and the Jackson network with blocking cases [17] the existence of a stationary distribution for an unreliable network with dynamic routing may be proved.

Theorem 1. *It is assumed the following conditions for unreliable network from the Definition 1*

$$(1) \inf_{n_j, t} \mu_j(n_j, z_j) > 0 \quad \forall\, j,$$

(2) time of service and time between new arrivals are independent random variables,

(3) routing matrix R^D *is reversible,*

then the stationary distribution for unreliable networks is defined by formulae

$$\pi(\tilde{\mathbf{n}}) = \pi(D, n_1, n_2, \cdots, n_m) = \frac{1}{C}\frac{\psi(D)}{\phi(D)}\prod_{i=1}^{m}\frac{1}{C_i}\frac{\lambda_i^{n_i}}{\prod_{k=1}^{n_i}\mu_i(k)}$$

where

$$C_i = \sum_{n=0}^{\infty}\frac{\lambda_i^{n_i}}{\prod_{y=1}^{n}\mu_i(y)}, \quad \lambda_i = \sum_{j=0}^{m}\Lambda * r_{ji}.$$

The main result for the convergence rate is formulated in terms of the spectral gap for unreliable queueing network. The preliminary notations and results on the spectral gap: there is a Markov process $\mathbf{X} = (X_t, t \geq 0)$ with the matrix of transition intensities $Q = [q(\mathbf{e}, \mathbf{e}')]_{\mathbf{e}, \mathbf{e}' \in \mathbb{E}}$, with stationary distribution π and an infinitesimal generator given by

$$\mathbf{Q}f(\mathbf{e}) = \sum_{\mathbf{e}' \in \mathbb{E}}(f(\mathbf{e}') - f(\mathbf{e}))q(\mathbf{e}, \mathbf{e}').$$

The usual scalar product on $L_2(\mathbb{E}, \pi)$ is defined as

$$\langle f, g \rangle_{pi} = \sum_{e \in \mathbb{E}} f(\mathbf{e}) g(\mathbf{e}) \pi(\mathbf{e}). \tag{6}$$

The spectral gap for $\mathbf{X}$ is

$$Gap(\mathbf{Q}) = \inf\{-\langle f, \mathbf{Q}f \rangle_\pi : \|f\|_2 = 1, \langle f, \mathbf{1} \rangle_\pi = 0\}.$$

The main result for a network is formulated in the following theorems:

Theorem 2. *If* $\mathbf{X}$ *is a markov process with infinitesimal generator* $\mathbf{Q}$, *it is assumed that* $\mathbf{Q}$ *is bounded, the minimal intensity of service is strictly positive* $\inf_{n_j,t} \mu_j(n_j, z_j) > 0$ *and the routing matrix* (r_{ij}^D) *is reversible, then* $Gap(\mathbf{Q}) > 0$, *if the following condition is true: for any* $i = 1, \cdots, m$, *for the birth and death process, corresponding to the node* i *with parameters* λ_i *and* $\mu_i(n_i, z_i)$ *the spectral gap is strictly positive* $Gap_i(\mathbf{Q}_i) > 0$.

Theorem 3. *If* $\mathbf{X}$ *is a markov process with a bounded infinitesimal generator* $\mathbf{Q}$, *positive minimal intensity of service* $\inf_{n_j,t} \mu_j(n_j, z_j) > 0$ *and reversible routing matrix* (r_{ij}^D), *then* $Gap(\mathbf{Q}) > 0$ *iff for any* $i = 1, \cdots, m$, *the distribution* $\pi = (\pi_i), i \geq 0$ *has light tails, i.e. the following condition* $\inf_k \frac{\pi_i(k)}{\sum_{j>k} \pi_i(j)} > 0$.

Theorem 4 *(Corollary from [22]).* *If* $\mathbf{X}$ *is unreliable queueing network with dynamic routing from Definition 1 with infinitesimal generator* $\mathbf{Q}$ *and transition probabilities matrix* P_t. *It is assumed that routing matrix* (r_{ij}^D) *is reversible and* $(r_{ij}^D)^k > 0$ *dor* $k \geq 1$. *If the distribution* π_i *has light tails for any* $i = 1, \cdots, m$, *then the following conditions are equivalent*

– *for any* $f \in L_2(\mathbb{E}, \pi)$

$$\|P_t f - \pi(f)\|_2 \leq e^{-Gap(\mathbf{Q})t} \|f - \pi(f)\|_2, t > 0,$$

– *for any* $\mathbf{e} \in \mathbb{E}$ *the constant* $C(\mathbf{e}) > 0$ *exists such, that*

$$\|\delta_{\mathbf{e}} - \pi(f)\|_{TV} \leq C(\mathbf{e}) e^{-Gap(\mathbf{Q})t}, t > 0.$$

4   The Numerical Example

We consider two numerical examples of network state probabilities calculation:

Example 1: The network consists of three nodes, each node is a system with two servers (see Fig. 2).

Example 2: The network consists of two nodes, each node is a system with three servers (see Fig. 3).

We use the following initial data for calculations in Example 1:
the number of nodes: $m = 3$,

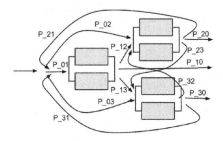

Fig. 2. Network with two-servers nodes

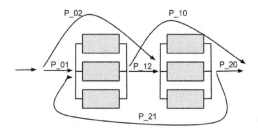

Fig. 3. Network with three-servers nodes

the transition probabilities matrix:

$$P_{ij} = \begin{pmatrix} 0.03 & 0.57 & 0.35 & 0.05 \\ 0.1 & 0.002 & 0.398 & 0.5 \\ 0.35 & 0.25 & 0.15 & 0.25 \\ 0.2 & 0.25 & 0.3 & 0.25 \end{pmatrix}$$

The matrix P_{ij} has a size of $\times(m+1)x(m+1)$. The first row of matrix (P_{0j}, $j = 0, 1, 2, 3$) defines the probabilities the application received by the network, will be sent for service to node j, the first column gives the probability that the application leaves the system.

The intensities of input flow λ (depend on the number of customers in the system):

$$\lambda(0) = 10,$$
$$\lambda(1) = 10,$$
$$\lambda(2) = 10,$$
$$\lambda(3) = 10,$$
$$\lambda(4) = 10,$$
$$\lambda(5) = 10,$$
$$\lambda(6) = 10,$$
$$\lambda(7) = 10.$$

The service rates μ (also depend on the node number and the number of customers):

$$\mu_{0,0} = 20.0,$$
$$\mu_{0,1} = 10.0,$$
$$\mu_{0,2} = 5.0,$$
$$\mu_{0,3} = 7.0,$$
$$\mu_{1,0} = 32.0,$$
$$\mu_{1,1} = 30.0,$$
$$\mu_{1,2} = 20.0,$$
$$\mu_{1,3} = 17.0.$$

Server failure rates α (depend on the node number):

$$\alpha_0 = 1.0,$$
$$\alpha_1 = 2.0,$$
$$\alpha_2 = 3.0.$$

Recovery rates β (depend on the node number):

$$\beta_0 = 2.0,$$
$$\beta_1 = 4.0,$$
$$\beta_2 = 6.0.$$

The calculation results for Example 1. The stationary probabilities for Example 1:

$$P_0(0,0,0) = 0,340448527,$$
$$P_0(0,0,1) = 0,010639016,$$
$$P_0(0,0,2) = 0,003546339,$$
$$P_0(0,1,0) = 0,048635504,$$
$$P_0(0,1,1) = 0,015198595,$$
$$P_0(0,1,2) = 0,005066198,$$
$$P_0(0,2,0) = 0,162118346,$$
$$P_0(0,2,1) = 0,050661983,$$
$$P_0(0,2,2) = 0,016887328,$$
$$P_0(1,0,0) = 0,017022426,$$
$$P_0(1,0,1) = 0,005319508,$$
$$P_0(1,0,2) = 0,001773169,$$
$$P_0(1,1,0) = 0,024317752,$$
$$P_0(1,1,1) = 0,007599297,$$

$$P_0(1,1,2) = 0,002533099,$$
$$P_0(1,2,0) = 0,081059173,$$
$$P_0(1,2,1) = 0,025330992,$$
$$P_0(1,2,2) = 0,008443664,$$
$$P_0(2,0,0) = 0,017022426,$$
$$P_0(2,0,1) = 0,005319508,$$
$$P_0(2,0,2) = 0,001773169,$$
$$P_0(2,1,0) = 0,024317752,$$
$$P_0(2,1,1) = 0,007599297,$$
$$P_0(2,1,2) = 0,002533099,$$
$$P_0(2,2,0) = 0,081059173,$$
$$P_0(2,2,1) = 0,025330992,$$
$$P_0(2,2,2) = 0,008443664.$$

Based on these results can be obtained by other characteristics of the network, such as: - The probability of denial of service (the probability that all sites are occupied) = 0.0084; - Availability factor of the system (the system is completely free) = 0.34.

References

1. Lakatos, L., Szeidl, L., Telek, M.: Introduction to Queueing Systems with Telecommunication Applications. Springer Science & Business Media, New York (2012). Mathematics- 388 pages
2. Ghosal, A., Gujaria, S.C., Ghosal, R.: Network Queueing Systems: With Industrial Applications. South Asian Publishers, New Delhi (2004). Computer network protocols - 138 pages
3. Daigle, J.: Queueing Theory with Applications to Packet Telecommunication. Springer Science & Business Media, New york (2006). Technology & Engineering - 316 pages
4. Alexander, T.: Analysis of fork/join and related queueing systems. ACM Comput. Surv. **47**(2), 71 (2014). Article 17, http://dx.doi.org/10.1145/2628913
5. Lakshmi, C., Sivakumar, A.I.: Application of queueing theory in health care: a literature review. Oper. Res. Health Care **2**(1–2), 25–39 (2013)
6. Sterbenz, J.P.G., Çietinkaya, E.K., Hameed, M.A., et al.: Evaluation of network resilience, survivability, and disruption tolerance: analysis, topology generation, simulation, and experimentation. Telecommun. Syst. **52**, 705–736 (2013). doi:10. 1007/s11235-011-9573-6
7. Hou, I.-H., Kumar, P.R.: Queueing systems with hard delay constraints: a framework for real-time communication over unreliable wireless channels. Queueing Syst. **71**(1), 151–177 (2012). doi:10.1007/s11134-012-9293-y

8. Jain, M., Sharma, G.C., Sharma, R.: Unreliable server M/G/1 queue with multi-optional services and multi-optional vacations. Int. J. Math. Oper. Res. 5(2), 145–169 (2013). doi:10.1504/IJ-MOR.2013.052458

9. Ba-Rukab, O.M., Tadj, L., Ke, J.-C.: Binomial schedule for an M/G/1 queueing system with an unreliable server. Int. J. Model. Oper. Manage. 3(3–4), 206–218 (2013). doi:10.1504/IJMOM.2013.058326 http://dx.doi.org/10.1504/IJMOM.2013.058326

10. Tadj, L., Choudhury, G., Rekab, K.: A two-phase quorum queueing system with Bernoulli vacation schedule, setup, and N-policy for an unreliable server with delaying repair. Int. J. Serv. Oper. Manage. 12(2), 139–164 (2012). doi:10.1504/IJSOM.2012.047103

11. Bama, S., Afthab Begum, M.I., Fijy Jose, P.: Unreliable Mx/G/1 queueing system with two types of repair. Int. J. Innovative Res. Dev. 4(10), 25–38 (2015)

12. Klimenok, V., Vishnevsky, V.: Unreliable queueing system with cold redundancy. In: Gaj, P., Kwiecień, A., Stera, P. (eds.) CN 2015. CCIS, vol. 522, pp. 336–346. Springer, Cham (2015). doi:10.1007/978-3-319-19419-6_32

13. Chen, H., Yao, D.D.: Fundamentals of Queueing Networks: Performance, Asymptotics, and Optimization. Stochastic Modelling and Applied Probability, vol. 46. Springer, New York (2001). doi:10.1007/978-1-4757-5301-1

14. Vvedenskaya, N.D.: Configuration of overloaded servers with dynamic routing. In: Probl. Inf. Transm. 47, 289 (2011). doi:10.1134/S0032946011030070

15. Sukhov, Y.M., Vvedenskaya, N.D.: Fast Jackson networks with dynamic routing. Probl. Inf. Transm. 38, 136–153 (2002). doi:10.1023/A:1020010710507

16. Kelly, F.P., Laws, C.N.: Dynamic routing in open queueing networks: Brownian models, cut constraints and resource pooling. Queueing Syst. 13, 47–86 (1993)

17. Sauer, C., Daduna, H.: Availability formulas and performance measures for separable degradable networks. Econ. Qual. Control 18(2), 165–194 (2003)

18. Marco, C.: Dynamic Routing in Broadband Networks Springer Science & Business Media, Computers, Berlin (2012)

19. Yavuz, F., Zhao, J., Yaan, O., Gligor, V.: Toward k -connectivity of the random graph induced by a pairwise key predistribution scheme with unreliable links. IEEE Trans. Inf. Theo. 61(11), 6251–6271 (2015). doi:10.1109/TIT.2015.2471295

20. Lorek, P., Szekli, R.: Computable bounds on the spectral gap for unreliable Jackson networks. Adv. Appl. Probab. 47, 402–424 (2015)

21. Lorek, P.: The exact asymptotic for the stationary distribution of some unreliable systems. arXiv:1102.4707 [math.PR] (2011)

22. Chen, M.F.: Eigenvalues, Inequalities, and Ergodic Theory. Springer, London (2005)

23. Cormen, T.H., Leiserson, C.E., Rivest, R.L., Stein, C.: Introduction to Algorithms, 3rd edn. The MIT Press, Cambridge (2009)

24. Saaty, T.L.: Elements of Queueing theory with applications. Dover Publications, NY (1983)

25. Kalimulina, E.Y.: Rate of convergence to stationary distribution for unreliable Jackson-type queueing network with dynamic routing. In: Vishnevskiy, V.M., Samouylov, K.E., Kozyrev, D.V. (eds.) DCCN 2016. CCIS, vol. 678, pp. 253–265. Springer, Cham (2016). doi:10.1007/978-3-319-51917-3_23

26. Kalimulina, E.Y.: Queueing system convergence rate In: Proceedings of the 19th International Conference, Distributed Computer and Communication Networks (DCCN 2016, Moscow, Russia), Vol. 3. pp. 203–211. RUDN, Moscow (2016)

27. Kalimulina, E.Y.: Analysis of system reliability with control, dependent failures, and arbitrary repair times. Int. J. Syst. Assur. Eng. Manage. **8**, 180–188 (2016). doi:10.1007/s13198-016-0520-5
28. Kalimulina, E.Y.: Analysis of Unreliable Jackson-Type Queueing Networks with Dynamic Routing, December 2016. http://dx.doi.org/10.2139/ssrn.2881956
29. Dorogovtsev, S.N., Mendes, J.F.F.: Evolution of Networks: From Biological Nets to the Internet and WWW (Physics). Oxford University Press Inc, New York, NY (2003)

Coupling Method for Backward Renewal Process and Lorden's Inequality

Galina Zverkina[1,2]([⊠])

[1] Moscow State University of Railway Engineering, Moscow, Russia
zverkina@inbox.ru
[2] V.A. Trapeznikov Institute of Control Sciences of Russian Academy of Sciences, Moscow, Russia

Abstract. A scheme of using the coupling method to obtain strong bounds for the convergence rate of the distribution of the backward renewal process in the total variation distance is described. This scheme can be applied to a wide class of regenerative processes in queuing theory.

Keywords: Backward renewal process · Renewal process · Convergence rate · Strong bounds · Total variation metric · Lorden's inequality

1 Introduction

Obviously, the behaviour of the queueing system can be described by regenerative process. Hence a study of the behaviour of regenerative processes is an important problem in the queuing theory and in the reliability theory.

Definition 1. *Recall that the stochastic process $\{X_t,\ t \geqslant 0\}$ with the measurable state space $(\mathcal{X}, \mathcal{B}(\mathcal{X}))$ and filtration $\{\mathcal{F}_t,\ t \geqslant 0\}$ defined on the probability space $(\Omega, \mathcal{F}, \mathbf{P})$ is called regenerative process if there exists a sequence of Markov moments $\{\theta_i\}_{i \in \mathbb{N}}$ with respect to the filtration $\mathcal{F}_t$ such that: 1. $X_{\theta_i} = X_{\theta_j}$ for all $i, j \in \mathbb{N}$; 2. The random elements $\Xi_i \stackrel{\text{def}}{=} \{X_t,\ t \in [\theta_i, \theta_{i+1}]\}$ $(i \in \mathbb{N})$, are mutually independent and identically distributed.* ▷

Hence, the random variables $\zeta_{i+1} \stackrel{\text{def}}{=} \theta_{i+1} - \theta_i$, $i \in \mathbb{N}$ are i.i.d.; denote

$$F(s) \stackrel{\text{def}}{=} \mathbf{P}\{\zeta_i \leqslant s\},\ i \in \mathbb{N},\ \text{and}\ G(s) \stackrel{\text{def}}{=} \mathbf{P}\{\theta_1 \leqslant s\}.$$

Also, the random variables $\{\{\zeta_i\}_{i=1}^{\infty}$ and $\theta_1\}$ are mutually independent.

If the random variables ζ_i, $i \in \mathbb{N}$, and $\zeta_1 \stackrel{\text{def}}{=} \theta_1$ have finite expectations, then the ergodicity of the process X_t follows from the Harris-Khasminsky principle, i.e. the distribution of X_t weekly converges to the unique probability (*stationary*) measure $\mathcal{P}$ on $(\mathcal{X}, \mathcal{B}(\mathcal{X}))$.

G. Zverkina is supported by the RFBR, project No 17-01-00633 A.

V.M. Vishnevskiy et al. (Eds.): DCCN 2017, CCIS 700, pp. 368–379, 2017.
DOI: 10.1007/978-3-319-66836-9_31

Moreover, it is well-known, that if for some $\kappa > 1$ the condition

$$\{\mathbf{E}\,\theta_1^\kappa < \infty, \; \mathbf{E}\,\zeta_2^\kappa < \infty\}$$

is satisfied, then for all $\alpha \leqslant \kappa - 1$ there exists the (unknown) constant $K(\alpha)$ such that for all $S \in \mathcal{B}(\mathcal{X})$ and all $t > 0$ the inequality

$$|\mathbf{P}\{X_t \in S\} - \mathcal{P}(S)| < \frac{K(\alpha)}{t^\alpha}$$

is true [1].

The knowledge of the convergence rate (i.e. the constants $K(\alpha)$) is very important in applications. Often in the analysis of the behavior of the queueing system and in the management of such queueing system it is necessary to know the distribution of some characteristics of this queueing system. Usually, if this distribution is unknown, then it is replaced by the stationary distribution. Therefore, if the convergence rate of the distribution of queueing system is known, then they can be sure that after some time T this replacement is correct.

The bounds for the constant $K(\alpha)$ was defined in some particular cases – see e.g. [7–9,11] et al.; our goal is to give some procedure to obtain such bounds for a sufficiently wide class of regenerative processes.

Firstly we remark that the Markov moments $\{\theta_i\}_{i\in\mathbb{N}}$ form an embedded renewal process $R_t \stackrel{\text{def}}{=} \sum_{i=1}^{\infty} \mathbf{1}(\theta_i \leqslant t)$. If $\theta_1 = 0$, then R_t is a renewal process without delay. If $\theta_1 > 0$, then we can interpret the time interval $(0, \theta_1)$ as the residual time of the regeneration period of X_t (or the renewal period of R_t), where the process X_t began its regeneration period at the some fixed time $(-a)$.

The random variable $D_t \stackrel{\text{def}}{=} (\theta_{R_t+1} - t)$ is called *forward renewal time*. For $t > \theta_1$, *Lorden's inequality* [2] is true:

$$\mathbf{E}\,D_t \leqslant \Theta \stackrel{\text{def}}{=} \frac{\mathbf{E}\,\zeta_2^2}{\mathbf{E}\,\zeta_2}. \tag{1}$$

Also denote $B_t \stackrel{\text{def}}{=} t - \theta_{R_t}$ for $t \geqslant \theta_1$; B_t is a time from the last renewal $(\theta_{R_t} \leqslant t)$ to the time t, it is called *backward renewal time* of the renewal process R_t. The state space of the process B_t is $(\mathbb{R}_+, \mathcal{B}(\mathbb{R}_+))$; if $\theta_1 = 0$ then the process B_t starts from the state $B_0 = 0$ (non-delay process). If $\theta_1 \neq 0$ then we consider the random variable θ_1 as the residual time of the first renewal period of the renewal process R_t started at some fixed time $(-a)$; consequently we put $G(s) = F_a(s) \stackrel{\text{def}}{=} \frac{F(a+s) - F(a)}{1 - F(a)}$; *we suppose that $F(a) < 1$ for all $a \in \mathbb{R}$. So, for $t \in [0, \theta_1]$ we put $B_t = a + t$, in the assumption that $G(s) = F_a(s)$.*

It is easy to see that the process B_t is Markov. If $\mathbf{E}\,\zeta_2 < \infty$, and consequently $\mathbf{E}\,\theta_1 < \infty$, then B_t is ergodic, i.e. its distribution weekly converges to the stationary distribution as $t \to \infty$. Moreover, if the distribution of B_t converges to the stationary distribution, then the distribution of X_t also converges to the stationary distribution as $t \to \infty$.

And if we know the bounds of the convergence rate of distribution of the process B_t (in some sense), then we know the bounds of the convergence rate of distribution of the process X_t (in the same sense), because the distribution of X_t is determined by the value of B_t.

So, our goal is to obtain the bounds for the convergence rate of the backward renewal process, and for this aim we will use the *coupling method*.

2   Coupling Method

Let X_t' and X_t'' be the homogeneous independent Markov processes with the same state space $(\mathcal{X}, \mathcal{B}(\mathcal{X}))$ and the same transition function, but with different initial states: $X_0' = x' \neq X_0'' = x''$. Denote the distribution of the process X_t with the initial state x at the time t by $\mathcal{P}_t^x$, i.e. $\mathcal{P}_t^x(S) = \mathbf{P}\{X_t \in S | X_0 = x\}$; and for all $x \in \mathcal{X}$, $\mathcal{P}_t^x \Longrightarrow \mathcal{P}$ as $t \to \infty$.

Let the paired process $\mathcal{Z}_t = (U_t', U_t'')$ created on some probability space satisfies the following conditions (i)–(iii):

(i) For all $t \geqslant 0$ and $S \in \mathcal{B}(\mathcal{X})$, $\mathbf{P}\{U_t' \in S\} = \mathbf{P}\{X_t' \in S\}$, and $\mathbf{P}\{U_t'' \in S\} = \mathbf{P}\{X_t'' \in S\}$, therefore, $U_0' = X_0' = x'$ and $U_0'' = X_0'' = x''$.
(ii) For all $t \geqslant \tau(x', x'') \stackrel{\text{def}}{=} \inf\{t \geqslant 0 : U_t' = U_t''\}$, the equality $U_t' = U_t''$ is true.
(iii) For all x', $x'' \in \mathcal{X}$, $\mathbf{P}\{\tau(x', '') < \infty\} = 1$.

The paired process $\mathcal{Z}_t$ satisfying conditions (i)–(iii) is called *successful coupling* of the processes X_t' and X_t'', and for them, the based coupling inequality can be written so: for all $S \in \mathcal{B}(\mathcal{X})$

$$
\begin{aligned}
\left| \mathcal{P}_t^{x'}(S) - \mathcal{P}_t^{x''}(S) \right| &\\
= \left| \mathbf{P}\{X_t' \in S\} - \mathbf{P}\{X_t'' \in S\} \right| &= \left| \mathbf{P}\{U_t' \in S\} - \mathbf{P}\{U_t'' \in S\} \right| \\
= \big| \mathbf{P}\{U_t' \in S \,\&\, \tau(x', x'') \leqslant t\} &+ \mathbf{P}\{U_t' \in S \,\&\, \tau(x', x'') > t\} \\
- (\mathbf{P}\{U_t'' \in S \,\&\, \tau(x', x'') \leqslant t\} &+ \mathbf{P}\{U_t'' \in S \,\&\, \tau(x', x'') > t\}) \big| \\
\leqslant \left| \mathbf{P}\{U_t' \in S \,\&\, \tau(x', x'') > t\} &- \mathbf{P}\{U_t'' \in S \,\&\, \tau(x', x'') > t\} \right| \\
\leqslant \mathbf{P}\{\tau(x', x'') > t\}. &
\end{aligned}
$$

Hence, if we can find an estimate $\mathbf{P}\{\tau(x, y) > t\} \leqslant \varphi(x, y, t)$, and if $\widehat{\varphi}(x, t) \stackrel{\text{def}}{=} \mathbf{E}\varphi\left(x, \widetilde{\zeta}, t\right) < \infty$ for the random variable $\widetilde{\zeta}$ with the stationary distribution $\mathcal{P}$, then for all $S \in \mathcal{B}(\mathcal{X})$

$$
|\mathcal{P}_t^x(S) - \mathcal{P}(S)| \leqslant \int_{\mathcal{X}} \varphi(x, u, t)\mathcal{P}(\mathrm{d}u) = \widehat{\varphi}(x, t); \tag{2}
$$

$$
\|\mathcal{P}_t^x - \mathcal{P}\|_{TV} \stackrel{\text{def}}{=} 2 \sup_{S \in \mathcal{B}(\mathcal{X})} \left| \mathcal{P}_t^{X_0}(S) - \mathcal{P}(S) \right| \leqslant 2\widehat{\varphi}(x, t). \tag{3}
$$

For more details see [3, 5, 10].

3   Auxiliary Considerations

Definition 2. *The common part of the distributions of the random variables* ξ_1 *and* ξ_2 *with distribution functions* $\Psi_j(s) = \mathbf{P}\{\xi_j \leqslant s\}$, $s \in \mathbb{R}$ *is*

$$\varkappa \overset{\text{def}}{=\!=} \varkappa(\Psi_1(s), \Psi_2(s)) \overset{\text{def}}{=\!=} \int\limits_{-\infty}^{\infty} \min(\psi_1(u), \psi_2(u))\, \mathrm{d}\, u,$$

where

$$\psi_j(s) \overset{\text{def}}{=\!=} \begin{cases} \Psi_j'(s), & \text{if } \Psi_j'(s) \text{ exists}, \\[2mm] 0, & \text{otherwise} \end{cases}$$

(here and hereafter $j = 1, 2$).   ▷

Proposition 1. *1. If $\varkappa > 0$, then the function*

$$\widetilde{\Psi}(s) \overset{\text{def}}{=\!=} \varkappa^{-1} \int\limits_{-\infty}^{s} \min(\psi_1(u), \psi_2(u))\, \mathrm{d}\, u$$

is a distribution function, and if $\varkappa < 1$, then the functions

$$\widetilde{\Psi}_j(s) \overset{\text{def}}{=\!=} \frac{\Psi_j(s) - \varkappa\widetilde{\Psi}(s)}{1 - \varkappa}$$

are a distribution functions ($j = 1, 2$).

2. Let $\mathcal{U}'$, $\mathcal{U}''$ and $\mathcal{U}'''$ be independent random variables with uniform distribution on $[0, 1]$. Then

$$\widetilde{\xi}_j \overset{\text{def}}{=\!=} \widetilde{\Psi}^{-1}(\mathcal{U}'')\mathbf{1}(\mathcal{U}' < \varkappa) + \widetilde{\Psi}_j^{-1}(\mathcal{U}''')\mathbf{1}(\mathcal{U}' \geqslant \varkappa) \overset{\mathcal{D}}{=\!=} \xi_j,$$

and $\mathbf{P}\left\{\widetilde{\xi}_1 = \widetilde{\xi}_2\right\} \geqslant \varkappa$.   ▷

Here and hereafter we put $h^{-1}(s) \overset{\text{def}}{=\!=} \inf\{x \in \mathbb{R} : h(x) \geqslant s\}$ for non-decreasing function $h(x)$.

Remark 1. Proposition 1 is a simplified variant of the *Coupling Lemma* or *"Lemma about three random variables"* (see, e.g., [6]).   ▷

Proof.

1. It is easy to see that $\widetilde{\Psi}(s) \geqslant 0$, it is not decreasing, and $\lim\limits_{s\to\infty} \widetilde{\Psi}(s) = \dfrac{\varkappa}{\varkappa} = 1$ as $\varkappa > 0$.

Now, $\widetilde{\Psi}_j(s) = \Psi_j(s) - \int\limits_{0}^{s} \min(\psi_1(u), \psi_2(u))\, \mathrm{d}\, u$, and again $\widetilde{\Psi}_j(s) \geqslant 0$, it is not

decreasing, and $\lim\limits_{s\to\infty} \widetilde{\Psi}_j(s) = \dfrac{1 - \varkappa}{1 - \varkappa} = 1$ as $\varkappa < 1$.

2. By the formula of total probability:

$$\mathbf{P}\{\widetilde{\xi}_j \leqslant s\} = \mathbf{P}\{\widetilde{\xi}_j \leqslant s | \mathcal{U}' < \varkappa\}\mathbf{P}\{\mathcal{U}' < \varkappa\} + \mathbf{P}\{\widetilde{\xi}_j \leqslant s | \mathcal{U}' \geqslant \varkappa\}\mathbf{P}\{\mathcal{U}' \geqslant \varkappa\}$$
$$= \widetilde{\Psi}(s)\varkappa + \widetilde{\Psi}_j(s)(1 - \varkappa) = \Psi_j(s) = \mathbf{P}\{\xi_j \leqslant s\};$$

and

$$\mathbf{P}\{\widetilde{\xi}_1 = \widetilde{\xi}_1\} = \mathbf{P}\{\widetilde{\xi}_1 = \widetilde{\xi}_2 | \mathcal{U}' < \varkappa\}\mathbf{P}\{\mathcal{U}' < \varkappa\} + \mathbf{P}\{\widetilde{\xi}_1 = \widetilde{\xi}_2 | \mathcal{U}' \geqslant \varkappa\}\mathbf{P}\{\mathcal{U}' \geqslant \varkappa\}$$
$$= 1 \times \varkappa + \mathbf{P}\{\widetilde{\xi}_1 = \widetilde{\xi}_2 | \mathcal{U}' \geqslant \varkappa\}\mathbf{P}\{\mathcal{U}' \geqslant \varkappa\} \geqslant \varkappa.$$

Proposition 1 is proved.

4   Successful Coupling for Backward Renewal Process

Now we consider the backward renewal process B_t; its state space is $(\mathbb{R}_+, \mathcal{B}(\mathbb{R}_+))$; its stationary distribution has a distribution function

$$\widetilde{F}(s) = (\mathbf{E}\,\zeta_1)^{-1} \int_0^s (1 - F(u))\,\mathrm{d}\,u; \qquad \mathcal{P}(S) = (\mathbf{E}\,\zeta_1)^{-1} \int_S (1 - F(u))\,\mathrm{d}\,u. \quad (4)$$

Remark 2. Remind that B_t is the embedded backward renewal process of the regenerative process X_t (see Sect. 1). So, if at the some time τ the distributions of two embedded (for the processes $X_t^{(1)}$ and $X_t^{(2)}$) backward renewal processes $B_t^{(1)}$ and $B_t^{(2)}$ coinside, then at this time τ the distributions of the regenerative processes coinside too. $\triangleright$

4.1   Basic Assumption

Here and hereafter we suppose: $\kappa \geqslant 2$, and for some $A > 0$ for all $t > A$ and for all $\varepsilon > 0$,

$$\int_t^{t+\varepsilon} f(s)\,\mathrm{d}\,s > 0, \text{ where } f(s) \overset{\mathrm{def}}{=} \begin{cases} F'(s), & \text{if } \exists F'(s); \\ 0, & \text{otherwise,} \end{cases} \quad (*)$$

and $\mathbf{E}\,\zeta_2^\kappa < \infty$.

We will construct the successful coupling for two versions of the process B_t started from different initial states b_1 and b_2; denote them $B_t^{(1)}$ and $B_t^{(2)}$.

4.2   Construction of Renewal Process

Recall the method of construction of the renewal process R_t with delay θ_1 having the distribution function $G(s)$ (see, e.g., [4]).

For simplicity, we suppose that $G(s) = F_a(s)$ for some $a \geqslant 0$; i.e. we observe the renewal process which starts at the time $-a$, and its first renewal time is positive; $B_0 = a$.

Let $\{\mathcal{U}_n\}$ be a sequence of independent random variables with uniform distribution on $[0, 1]$.

The construction of the renewal times for the renewal process R_t, i.e. $\theta_1, \theta_2, \ldots, \theta_n, \ldots$ is follow:

$$\theta_1 \overset{\text{def}}{=} G^{-1}(\mathcal{U}_1) = F_a^{-1}(\mathcal{U}_1) = \zeta_1; \theta_2 \overset{\text{def}}{=} \theta_1^a + F^{-1}(\mathcal{U}_2) = \theta_1 + \zeta_2;$$

$$\ldots \theta_n \overset{\text{def}}{=} \theta_{n-1} + F^{-1}(\mathcal{U}_n) = \theta_{n-1} + \zeta_n; \ldots$$

So, for construction of two independent backward renewal processes with initial states $B_0^{(j)} = b_j \geqslant 0$ we need two sequences of independent random variables with uniform distribution on $[0, 1]$ – let they be $\{\mathcal{U}_{n,1}\}$ and $\{\mathcal{U}_{n,2}\}$. Now we denote for $j = 1, 2$

$$\theta_1^{(j)} \overset{\text{def}}{=} F_{b_j}^{-1}(\mathcal{U}_{1,j}) = \zeta_1^{(j)}; \theta_2^{(j)} \overset{\text{def}}{=} \theta_1^{(j)} + F^{-1}(\mathcal{U}_{2,j}) = \theta_1^{(j)} + \zeta_2^{(j)};$$

$$\ldots \theta_n^{(j)} \overset{\text{def}}{=} \theta_{n-1}^{(j)} + F^{-1}(\mathcal{U}_{n,j}) = \theta_{n-1}^{(j)} + +\zeta_n^{(j)}; \ldots,$$

and

$$B_t^{(j)} = \begin{cases} b_j + t, & \text{if } t < \theta_1^{(j)}; \\ t - \sum_{i=1}^{\infty}\left(1\left(\theta_i^{(j)} \leqslant t\right)\zeta_i^{(j)}\right), & \text{otherwise}; \end{cases} \tag{5}$$

$$R_t^{(j)} = \sum_{i=1}^{\infty} 1\left(\theta_i^{(j)} \leqslant t\right); \qquad D_t^{(j)} = \theta_{R_t^{(j)}+1}^{(j)} - t.$$

From (5) we see that the processes $B_t^{(1)}$ and $B_t^{(2)}$ are piecewise linear, and they can begin to be equal only at the time when both of them are equal to zero. But for independent processes $B_t^{(1)}$ and $B_t^{(2)}$ the probability of they coincidence is zero because the distribution of residual time of any renewal periods of corresponding renewal process has a continuous component – see $(*)$.

Therefore we will construct the successful coupling concerning of two *dependent* processes $\widetilde{B}_t^{(1)}$ and $\widetilde{B}_t^{(2)}$ such that $\widetilde{B}_t^{(j)} \overset{\mathcal{D}}{=} B_t^{(j)}$; for this aim we need an additional sequences of independent random variables $\{\mathcal{U}_n^i\}$ with uniform distribution on $[0, 1]$.

4.3   Construction of the Successful Coupling $\mathcal{Z}_t = \left(\widetilde{B}_t^{(1)}, \widetilde{B}_t^{(2)}\right)$

We suppose that the first renewal time of the process $\widetilde{B}_t^{(j)}$ has distribution function $F_{a_j}(s)$.

We will construct the paired process $\mathcal{Z}_t$ by following algorithm.

Step 1. We begin to construct independent processes $\widetilde{B}_t^{(1)}$ and $\widetilde{B}_t^{(2)}$ according to the scheme Sect. 4.2 – i.e. we construct the renewal times $\theta_i^{(j)}$ for the processes $\widetilde{B}_t^{(j)}$. Put $T_1 \stackrel{\text{def}}{=} \max\left(\theta_1^{(1)}, \theta_1^{(2)}\right)$.

Note, that $T_1 \leqslant \theta_1^{(1)} + \theta_1^{(2)}$. For simplicity, here we suppose that $T_1 = \theta_1^{(1)}$. If $\theta_1^{(1)} = \theta_1^{(2)}$ then we go to the Step 3 $\left(\text{but } \mathbf{P}\left\{\theta_1^{(1)} = \theta_1^{(2)}\right\} = 0\right)$. Otherwise, we go to the Step 2.

Step 2. At the time $T_k = \theta_k^{(1)} \in \left(\theta_{\nu_k}^{(2)}, \theta_{\nu_k+1}^{(2)}\right)$, $\mathbf{E}\, D_{T_k}^{(2)} \leqslant \Theta$ (see (1)), and $\widetilde{B}_{T_k}^{(1)} = 0 \neq \widetilde{B}_{T_k}^{(2)}$.

By Markov inequality, for some $R > \Theta$,

$$\mathbf{P}\left\{D_{T_k}^{(2)} \leqslant R\right\} \geqslant \pi_R \stackrel{\text{def}}{=} 1 - \frac{\Theta}{R}.$$

And $\mathbf{P}\left\{\theta_{k+1}^{(1)} - T_k = \zeta_k > R\right\} = 1 - F(R)$.

So, at the time $\theta_{\nu_k+1}^{(2)} = T_k + D_{T_k}^{(2)}$

$$\mathbf{P}\left\{\widetilde{B}_{\theta_{\nu_k+1}^{(2)}}^{(2)} = 0 \,\&\, \beta \stackrel{\text{def}}{=} \widetilde{B}_{\theta_{\nu_k+1}^{(2)}}^{(1)} = D_{T_k}^{(2)} \leqslant R\right\} \geqslant P_R \stackrel{\text{def}}{=} \pi_R(1 - F(R)).$$

If $D_{T_k}^{(2)} > R$ or $\zeta_k^{(1)} < R$ then we move on to the next time $\theta_{k+1}^{(1)}$, i.e. we replace k by $k+1$ and return to the Step 2.

If $D_{T_k}^{(2)} \leqslant R$ and $\zeta_k^{(1)} \geqslant R$ then we stop both processes at the time $\theta_{\nu_k+1}^{(2)}$. Then we prolong these processes (i.e. their residual times with distributions $F_\beta(s)$ and $F(s)$) using the additional random variables $\mathcal{U}_k^1, \mathcal{U}_k^2, \mathcal{U}_k^3$—by such a way that with probability $\varkappa(F_\beta(s), F(s))$ the next renewal times of both processes coincide (see Proposition 1).

Note that the condition $(*)$ implies that the common part of distribution of the forward renewal times of both processes

$$\varkappa(F_\beta(s), F(s)) \geqslant \varkappa_R \stackrel{\text{def}}{=} \inf_{a \in [0, R]}\{\varkappa(F_a(s), F(s))\} > 0$$

(see Definition 2).

Hence, at the time $\theta_{k+1}^{(1)}$ the constructed processes $\widetilde{B}_t^{(1)}$ and $\widetilde{B}_t^{(2)}$ coincide with probability $p_k \geqslant \varkappa_R P_R$ – denote this event by $\mathcal{E}_k$; $\mathbf{P}(\mathcal{E}_k | \overline{\mathcal{E}}_{k-1}) \geqslant \varkappa_R P_R$.

If $\widetilde{B}_{\theta_{k+1}^{(1)}}^{(1)} = \widetilde{B}_{\theta_{k+1}^{(1)}}^{(2)} (= 0)$ the we go to the Step 3.

Otherwise, we move on to the next time $\theta_{k+1}^{(1)}$: we replace k by $k+1$ and return to the Step 2.

Step 3. After the time $\tau \stackrel{\text{def}}{=} \theta_{k+1}^{(1)}$ such that $\widetilde{B}_\tau^{(1)} = \widetilde{B}_\tau^{(2)} = 0$ we prolong the construction of the processes $\widetilde{B}_t^{(1)}$ and $\widetilde{B}_t^{(2)}$ identically by the scheme (Sect. 4.2).

Denote

$$\mathfrak{E}_k \overset{\text{def}}{=} \left\{ \widetilde{B}^{(1)}_{\theta^{(1)}_{k+1}} = \widetilde{B}^{(2)}_{\theta^{(1)}_{k+1}} \ \& \ \widetilde{B}^{(1)}_{\theta^{(1)}_{k}} \neq \widetilde{B}^{(2)}_{\theta^{(1)}_{k}} \right\}.$$

Then

$$\mathfrak{E}_k = \left(\bigcap_{m=2}^{k} \overline{\mathcal{E}}_m \right) \cap \mathcal{E}_{k+1};$$

$$\mathbf{P}(\mathfrak{E}_k) = p_k \prod_{m=2}^{k} (1 - p_m) \leqslant (1 - \varkappa_R P_R)^{k-1} \overset{\text{def}}{=} q_R^{k-1}.$$

Proposition 2. *The paired process* $\mathcal{Z}_t = \left(\widetilde{B}^{(1)}_t, \widetilde{B}^{(2)}_t \right)$ *is a successful coupling.*   ▷

Proof.

(i) By construction, the distribution function of the random variables $\theta^{(1)}_1$ and $\theta^{(2)}_1$ of processes $B^{(1)}_t$ and $B^{(2)}_t$ are correspondingly $F_{a_1}(s)$ and $F_{a_2}(s)$.

Similarly, the random variables $\theta^{(j)}_{i+1} - \theta^{(j)}_i$ ($i \in \mathbb{N}$) also have the distribution function $F(s)$. Moreover, the random variables $\{\theta^{(j)}_1; \theta^{(j)}_{i+1} - \theta^{(j)}_i, \ i \in \mathbb{N}\}$ are mutually independent for fixed j.

So, for all time $t \geqslant 0$, $\mathbf{P}\left\{\widetilde{B}^{(j)}_t \in S\right\} = \mathbf{P}\left\{B^{(j)}_t \in S\right\}$.

(ii) By construction, for all $t \geqslant \tau(b_1, b_2) \overset{\text{def}}{=} \inf\left\{t \geqslant 0 : \widetilde{B}^{(1)}_t = \widetilde{B}^{(2)}_t\right\}$, the equality $\widetilde{B}^{(1)}_t = \widetilde{B}^{(2)}_t$ is true.

(iii) By construction, $\tau(b_1, b_2) \leqslant T_1 + \sum_{k=1}^{\nu} \zeta_{i+1}$, where $\mathbf{P}\{\nu > k\} \leqslant (1 - \varkappa P_R)$.

As $\mathbf{E}\,\zeta_i < \infty$, $\mathbf{E}\,\tau(b_1, b_2) \leqslant \infty$.

Proposition 2 is proved.

5 Strong Bounds for Convergence Rate

Now, for $\phi(s) = s^\alpha$ we can find an upper bound for $\mathbf{E}\,\phi(\tau(b_1, b_2)) = \mathbf{E}\left(\tau(b_1, b_2)^\alpha\right)$, $\alpha \geqslant 1$ using Jensen's inequality for positive a_i in the form

$$\left(\sum_{k=1}^{n} a_i \right)^\alpha \leqslant n^{\alpha-1} \left(\sum_{k=1}^{n} a_i^\alpha \right), \tag{6}$$

namely:

$$
\mathbf{E}\left(\tau(b_1, b_2)\right)^{\alpha} \leqslant \sum_{n=1}^{\infty} \mathbf{E}\left(\left(T_1 + \sum_{m=2}^{n+1} \zeta_m^{(1)}\right) \mathbf{1}(\mathfrak{E}_n)\right)^{\alpha}
$$

$$
\leqslant \sum_{n=1}^{\infty} (n+2)^{\alpha-1} \mathbf{E}\left(\left(\left(\theta_1^{(1)}\right)^{\alpha} + \left(\theta_1^{(2)}\right)^{\alpha} + \sum_{m=2}^{n+1}\left(\zeta_m^{(1)}\right)^{\alpha}\right) \mathbf{1}(\mathfrak{E}_n)\right)
$$

$$
\leqslant K_1(\alpha)\left(\mathbf{E}\left(\theta_1^{(1)}\right)^{\alpha} + \mathbf{E}\left(\theta_1^{(2)}\right)^{\alpha}\right)
$$

$$
+ \sum_{n=1}^{\infty} (n+2)^{\alpha-1}\left(\sum_{m=2}^{n} \mathbf{E}\left(\left(\zeta_m^{(1)}\right)^{\alpha} \mathbf{1}(\overline{\mathscr{E}}_m)\right) q_R^{n-2}\right.
$$

$$
\left. + \mathbf{E}\left(\left(\zeta_{n+1}^{(1)}\right)^{\alpha} \mathbf{1}(\mathscr{E}_{n+1})\right) q_R^{n-1}\right)
$$

$$
\leqslant K_1(\alpha)\left(\mathbf{E}\left(\theta_1^{(1)}\right)^{\alpha} + \mathbf{E}\left(\theta_1^{(2)}\right)^{\alpha}\right) + K_2(\alpha)\mathbf{E}\,\zeta_2^{\alpha}
$$

$$
= \varpi_{\alpha}(b_1, b_2, t),
$$

where $K_1(\alpha) = \sum_{n=1}^{\infty} (n+2)^{\alpha-1} q_R^{n-1}$, $K_2(\alpha) = \sum_{n=1}^{\infty} (n+2)^{\alpha} q_R^{n-1}$, $\sum_{m=2}^{1} \overset{\text{def}}{=} 0$.

Now, by Markov inequality,

$$
\mathbf{P}\{\tau(b_1, b_2) > t\} \leqslant \frac{\mathbf{E}\left(\tau(b_1, b_2)\right)^{\alpha}}{t^{\alpha}} \leqslant \varphi_{\alpha}(b_1, b_2, t) \overset{\text{def}}{=} \frac{\varpi_{\alpha}(b_1, b_2, t)}{t^{\alpha}}.
$$

Then, for $\alpha \leqslant \kappa - 1$ we have (taking into consideration (4))

$$
\int_0^{\infty} \mathbf{E}\left(\tau(b_1, b_2)\right)^{\alpha} \mathrm{d}\widetilde{F}(b_2) \leqslant K(\alpha, b_1) \overset{\text{def}}{=} \int_0^{\infty} \varpi_{\alpha}(b_1, b_2, t)\,\mathrm{d}\widetilde{F}(b_2)
$$

$$
= K_1(\alpha)\mathbf{E}\left(\theta_1^{(1)}\right)^{\alpha} + K_2(\alpha)\mathbf{E}\,\zeta_2^{\alpha} + \int_0^{\infty} K_1(\alpha)\mathbf{E}\left(\theta_1^{(2)}\right)^{\alpha} \mathrm{d}\widetilde{F}(b_2) \tag{7}
$$

$$
= K_1(\alpha)\mathbf{E}\left(\theta_1^{(1)}\right)^{\alpha} + K_2(\alpha)\mathbf{E}\,\zeta_2^{\alpha} + \frac{K_1(\alpha)}{\alpha+1}\mathbf{E}\,\zeta_2^{\alpha+1};
$$

and $\widehat{\varphi}_{\alpha}(b_1, t) = \dfrac{K(\alpha, b_1)}{t^{\alpha}}$ for $\alpha \in [1, \kappa - 1]$.

Therefore we have the bounds (2) and (3) for backward renewal process with $\dfrac{K(\alpha, b_1)}{t^{\alpha}}$ in right hand side.

The estimate (7) can be improved by the choice of R. It is not optimal, and it can be done better by use of the properties of the distribution F, and by more accurate estimation of the series.

The bounds (2) and (3) for the convergence rate of backward renewal processes are applicable to regenerative processes described in Sect. 1.

Remark 3. Bounds (7) are not optimal for two reasons:

(1) all possible coupling epochs are not considered,
(2) the Jensen's inequality (6) gives a large inaccuracy.   ▷

For example, for the distribution function $F(s) = 1 - \left(1 + \dfrac{x}{n-1}\right)^{-n}$ of ζ_2 we have:

1. For $n = 3.001$ $p \stackrel{\text{def}}{=} \sup\limits_{R>0} \varkappa_R P_R \approx 3.106 \times 10^{-3}$, and $K(1,0) \approx 1.03 \times 10^5$, $K(2,0) \approx 1.08 \times 10^9$;
2. For $n = 5.001$ $p \stackrel{\text{def}}{=} \sup\limits_{R>0} \varkappa_R P_R \approx 8.268 \times 10^{-3}$, and $K(2,0) \approx 1.74 \times 10^8$, $K(4,0) \approx 8.36 \times 10^{16}$;
3. For $n = 10.001$ $p \stackrel{\text{def}}{=} \sup\limits_{R>0} \varkappa_R P_R \approx 0.0127$, and $K(2,0) \approx 1.46 \times 10^8$, $K(5,0) \approx 5.88 \times 10^{19}$, $K(9,0) \approx 6.22 \times 10^{39}$;
4. for $n = 20.001$ $p \stackrel{\text{def}}{=} \sup\limits_{R>0} \varkappa_R P_R \approx 0.015$, and $K(2,0) \approx 1.37 \times 10^8$, $K(5,0) \approx 2.67 \times 10^{19}$, $K(10,0) \approx 8.19 \times 10^{41}$, $K(19,0) \approx 1.13 \times 10^{92}$.

For comparing, the simulation modeling gives the empirical bounds for Kolmogorov metric

$$K^*(\mathcal{P}_t, \mathcal{P}) = \|\mathcal{P}_t - \mathcal{P}\|_K \stackrel{\text{def}}{=} \sup\limits_{x \in \mathbb{R}} |\mathcal{P}_t((-\infty, x)) - \mathcal{P}((-\infty, x))| \leqslant \|\mathcal{P}_t - \mathcal{P}\|_{TV}$$

(in the simulation on the time period $[0, 300]$):

1. For $n = 3.001$

$$K^*(1,0) \approx 296.82, \qquad K^*(2,0) \approx 89046;$$

2. For $n = 5.001$

$$K^*(2,0) \approx 89172, \qquad K^*(4,0) \approx 8.03 \times 10^9;$$

3. For $n = 10.001$

$$K^*(2,0) = 89153, \qquad K^*(5,0) \approx 2.41 \times 10^{12}, \qquad K^*(9,0) \approx 1.95 \times 10^{22};$$

4. For $n = 20.001$

$$K^*(2,0) \approx 88669, \qquad K(5,0)^* \approx 2.39 \times 10^{12},$$

$$K^*(10,0) \approx 5.83 \times 10^{24}, \qquad K^*(19,0) \approx 1.14 \times 10^{47}.$$

Remark 4. In applications, the existence of any bounds of the convergence rate is better than their absence. The bounds (7) can be improved using the properties of a distribution of ζ and a more accurate calculation.   ▷

Remark 5. Bounds for the constants $K(\cdot, \cdot)$ are not an end in itself. They need to answer the question: "When you can replace the unknown distribution $\mathcal{P}_t$ by a stationary distribution $\mathcal{P}_t$?"

Hence, we have a new question about how to use these bounds to estimate the time when $\|\mathcal{P}_t^x - \mathcal{P}\|_{TV} < \varepsilon$ for given $\varepsilon > 0$.

Because $K(\alpha, X_0) \to \infty$ as $\alpha \uparrow \sup\{a : \mathbf{E}\,\zeta_2^a < \infty\}$, it must find

$$T(X_0, a, \varepsilon) \stackrel{\text{def}}{=} \inf_{\alpha > 0} \mathfrak{M}(X_0, a, \varepsilon, \alpha), \text{ where } \mathfrak{M}(X_0, a, \varepsilon, \alpha) \stackrel{\text{def}}{=} \left\{ \left(\frac{K(\alpha, X_0)}{\varepsilon} \right)^{\frac{1}{\alpha}} \right\};$$

denote $k(X_0, a, \varepsilon) \stackrel{\text{def}}{=} \operatorname{argmin}(\mathfrak{M}(X_0, a, \varepsilon, \alpha))$.

Calculation of the constants $K(\alpha, 0)$ and $T(X_0, a, \varepsilon)$ ($\alpha \in \mathbb{N}$) for $F(s) = 1 - \left(1 + \dfrac{x}{a-1}\right)^{-a}$ make the results given in the Table 1.

Table 1. Constants $K(\alpha, 0)$ and $T(X_0, a, \varepsilon)$ for $F(s) = 1 - \left(1 + \dfrac{x}{a-1}\right)^{-a}$

a	ε	$T(0, a, \varepsilon)$	$k(0, a, \varepsilon)$	a	ε	$T(0, a, \varepsilon)$	$k(0, a, \varepsilon)$
3.001	10^{-1}	$1.038 \cdot 10^5$	2	5.001	10^{-1}	$2.151 \cdot 10^4$	3
3.001	10^{-2}	$3.281 \cdot 10^5$	2	5.001	10^{-2}	$4.635 \cdot 10^4$	3
3.001	10^{-3}	$1.038 \cdot 10^6$	2	5.001	10^{-3}	$9.561 \cdot 10^4$	4
3.001	10^{-4}	$3.281 \cdot 10^6$	2	5.001	10^{-4}	$1.700 \cdot 10^5$	4
3.001	10^{-5}	$1.038 \cdot 10^7$	2	5.001	10^{-5}	$3.024 \cdot 10^5$	4
10.001	10^{-1}	$1.424 \cdot 10^4$	4	20.001	10^{-1}	$1.217 \cdot 10^4$	5
10.001	10^{-2}	$2.259 \cdot 10^4$	5	20.001	10^{-2}	$1.861 \cdot 10^4$	6
10.001	10^{-3}	$3.389 \cdot 10^4$	6	20.001	10^{-3}	$2.661 \cdot 10^4$	7
10.001	10^{-4}	$4.931 \cdot 10^4$	7	20.001	10^{-4}	$3.635 \cdot 10^4$	8
20.001	10^{-5}	$6.852 \cdot 10^4$	7	20.001	10^{-5}	$4.801 \cdot 10^4$	9

So, we see that (using the obtained bounds for the constants $K(\cdot, \cdot)$) the time before the distribution becomes close to the stationary distribution is almost independent of the number of finite moments of the distribution, and this time is estimated by the value of about $10^5 \times \mathbb{E}\,\zeta_2$ – because for the random variable with the distribution function defined in the Remark 3 has the mean equal to 1.

Moreover, for calculate the value of $T(0, a, \varepsilon)$, it is enough calculate the constants $K(\alpha, \cdot)$ only about for $\alpha = 1, \ldots, 6$.

Comparison of the results of the Table 1 with the results of the simulation seems impossible, because the simulation computes the characteristics of a random model only in discrete time, and in this situation it is very difficult to evaluate the cumulative error of the computations. $\triangleright$

Remark 6. The schema described in this paper can be applied for more complicated processes, for example for the queueing processes which have an embedded alternating renewal process (where there are the idle and busy alternate independent periods).   ▷

Acknowledgments. The author is grateful to L.G. Afanasieva and A.Yu. Veretennikov for very useful consultations.

References

1. Borovkov, A.A.: Stochastic Processes in Queueing Theory. Springer, New York (1976)
2. Lorden, G.: On excess over the boundary. Ann. Math. Stat. **41**(2), 520–527 (1970)
3. Griffeath, D.: A maximal coupling for Markov chains. Zeitschrift für Wahrscheinlichkeitstheorie und Verwandte Gebiete **31**(2), 95–106 (1975)
4. Kalashnikov, V.V.: Some properties of piecewise linear Markov processes. Teor. Veroyatnost. i Primenen. **20**(3), 571–583 (1975)
5. Lindvall, T.: Lectures on the Coupling Method. Wiley, New York (1992)
6. Veretennikov, A.: Ergodic Markov processes and Poisson equations (lecture notes). arXiv:1610.09661 [math.PR] (2017)
7. Veretennikov, A., Zverkina, G.: On polynomial convergence rate of the availability factor to its stationary value. In: Proceedings of the Eighteenth International Scientific Conference on Distributed Computer and Communication Networks: Control. Computation, Communications (DCCN-2015), pp. 168–175. ICS RAS, Moscow (2015)
8. Veretennikov, A.Yu.: On the rate of convergence for infinite server Erlang-Sevastyanov's problem. Queueing Syst. **76**(2), 181–203 (2014)
9. Veretennikov, A.Yu.: On the rate of convergence to the stationary distribution in the single-server queuing system. Autom. Remote Control **74**(10), 1620–1629 (2013)
10. Veretennikov, A.: Coupling method for Markov chains under integral Doeblin type condition. Theory Stoch. Process. **8**(24), 383–390 (2002). No. 3–4
11. Veretennikov, A.Yu., Zverkina, G.A.: Simple proof of Dynkin's formula for single-server systems and polynomial convergence rates. Markov Process. Relat. Fields **20**(3), 479–504 (2014)

Modified Cramer-Lundberg Models with On/Off Control and Hyperexponential Distribution of Demands Purchases Values

Anatoly Nazarov[1,2]([✉]) and Valentina Broner[1,2]

[1] RUDN University, 6 Miklukho-Maklaya st, Moscow 117198, Russia
nazarov.tsu@gmail.com, valsubbotina@mail.ru
[2] Tomsk State University, Lenina avenue 36, Tomsk 634050, Russia

Abstract. The paper contains research of modified Cramer-Lundberg models of inventory management with On/Off control and hyperexponential distribution of demands purchases values. In first model we assume that input product flow has piecewise-constant rate, which depends on some threshold value, the random part of demand is modeled as Poisson process with constant intensity. Hereby system control is to switch the rate of the input product flow if the stock level in the system is more than threshold. Second model is investigated under following conditions: the rate of input product flow is a constant, the random part of demand is modeled as Poisson process with piecewise-constant intensity. In this case the idea of control the system is similar to the first model. We find explicit expressions for the stationary distribution of the inventory level for each models. The results are discussed with illustrative numerical example.

Keywords: Cramer-Lundberg model · Inventory management · On/Off control · Hyperexponential distribution · Mathematical modelling

1 Introduction

The inventory management problem is widely known. In an increasingly competitive environment to ensure safe and stable operation commercial companies should to meeting the consumer demand and the reduction of trade costs. Accordingly inventory policy should be improve based on control of input and output product flows using mathematical modelling.

Well-known Newsvendor problem is considered in Arrow et al. [1]; Gallego and Moon [2]; Silver et al. [8]; Qin et al. [9]; Kitaeva et al. [3]. For example, there is the classical single-period problem describes in Khouja [4]. In the study of such

This paper was financially supported by the Ministry of Education and Science of the Russian Federation on the program to improve the competitiveness of Peoples' Friendship University among the world's leading research and education centers in the 2016–2020.

© Springer International Publishing AG 2017
V.M. Vishnevskiy et al. (Eds.): DCCN 2017, CCIS 700, pp. 380–394, 2017.
DOI: 10.1007/978-3-319-66836-9_32

models it is assumed remained stocks at the period end have deteriorated, and the seller should make a discount or dispose of stocks.

Multi-period inventory management models are considered in Zhang et al. [10], Mousavi et al. [5]. Multi-period models with On/Off control are discussed in Nazarov and Broner [6,7]. In [6] explicit expression for the stationary distribution of the inventory level is found under following condition: Demand occurs according to a Poisson process with piecewise-constant intensity, the purchase values of demand have Erlang distribution. Approximation of probability density function of inventory level is provided in [7] for similarly mathematical model with arbitrary distribution of purchase values of demand.

In this paper we consider modified Cramer-Lundberg models with On/Off control as multi-period inventory management models.

2   Mathematical Model of System with the Piecewise-Constant Rate of Input Product Flow

Let us consider a modified Cramer-Lundberg model of inventory management system (Fig. 1).

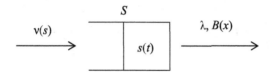

Fig. 1. Inventory management model

We denote the inventory level accumulated in the system at the time t as $s(t)$. Lets inventory arrive in the system with piecewise constant rate $\nu(s)$, where s is a value of the process $s(t)$, and S is some fixed threshold value. If S is reached, then a change rate of input product flow takes place

$$\nu(s) = \begin{cases} \nu_1, s < S, \\ \nu_2, s \geq S. \end{cases}$$

If the inventory level $s(t)$ accumulated in the system is below a threshold value S, then rate of input product flow is ν_1. If the inventory level is above threshold value S we try to avoid the overflow and we begin to deliver the product to outlets with rate $\nu_2, \nu_1 > \nu_1$.

Let us assume that demand occurs according to a Poisson process with constant intensity λ. The purchases values are independent, identically distributed random variables, having the distribution function $B(x)$. We assume that $s(t)$ can take negative values, i.e. the customer waits for the arrival the required amount of stocks. The condition of existence of a stationary regime for the system it can be determined from

$$\nu_1 > \lambda b > \nu_2, \tag{1}$$

where b is the first moment of function $B(x)$.

2.1   Problem Statement

According to the mathematical model process $s(t)$ is Markovian with continuous time t and a continuous set of values $-\infty < s < \infty$.

Denotes

$$P(s) = \frac{\partial P\{s(t) < s\}}{\partial s},$$

then we write

$$P(s + \nu(s)\Delta t) = P(s)(1 - \lambda\Delta t) + \lambda\Delta t \int\limits_0^\infty P(s + x)dB(x).$$

Using series expansion, we write the Kolmogorov equation for the stationary probability density function $P(s)$ of inventory level $s(t)$

$$\nu(s)P'(s) + \lambda P(s) = \lambda \int\limits_0^\infty P(s + x)dB(x), s \neq S, -\infty < s < \infty. \tag{2}$$

The solution $P(s)$ of Eq. (2) satisfied following boundary conditions

$$P(-\infty) = P(\infty) = 0. \tag{3}$$

The plan of this section is to identify some property of the solution $P(s), s > S$ for an arbitrary distribution $B(x)$ and to find the function P(s), $s < S$ for hyperexponential distribution B(x).

2.2   The Probabilities R_1 and R_2

Let us denote

$$\int\limits_{-\infty}^S P(s)ds = R_1, \int\limits_S^\infty P(s)ds = R_2.$$

We try to find an expression for determining the probabilities R_1, R_2.

Proposition 1. *The probabilities R_1 and R_2 are determined by following expression*

$$R_1 = \frac{\nu_1 - \lambda b}{\nu_1 - \nu_2}, R_2 = \frac{\lambda b - \nu_2}{\nu_1 - \nu_2}. \tag{4}$$

Proof. We multiply Eq. (2) by s, and integrate the obtained equality

$$\int\limits_{-\infty}^{\infty} s\nu(s)P'(s)ds + \lambda \int\limits_{-\infty}^{\infty} sP(s)ds = \lambda \int\limits_{-\infty}^{\infty} s \int\limits_{0}^{\infty} P(s+x)dB(x)ds. \qquad (5)$$

Let us consider the right side of expression (5)

$$\lambda \int\limits_{-\infty}^{\infty} s \int\limits_{0}^{\infty} P(s+x)dB(x)ds = \lambda \int\limits_{0}^{\infty} \int\limits_{-\infty}^{\infty} sP(s+x)dsdB(x)$$

$$= [s+x=y] = \lambda \int\limits_{0}^{\infty} \int\limits_{-\infty}^{\infty} (y-x)P(y)dydB(x) = \lambda \int\limits_{-\infty}^{\infty} \int\limits_{0}^{\infty} (y-x)dB(x)P(y)dy$$

$$= \lambda \int\limits_{-\infty}^{\infty} (y-b)P(y)dy = \lambda \int\limits_{-\infty}^{\infty} yP(y)dy - \lambda b \int\limits_{-\infty}^{\infty} P(y)dy.$$

Substituting this expression into (5), we obtain the equality

$$\int\limits_{-\infty}^{\infty} \nu(s)sP'(s)ds + \lambda \int\limits_{-\infty}^{\infty} sP(s)ds = \lambda \int\limits_{-\infty}^{\infty} yP(y)dy - \lambda b \int\limits_{-\infty}^{\infty} P(y)dy.$$

Using condition

$$\int\limits_{-\infty}^{\infty} P(y)dy = 1,$$

we obtain

$$\int\limits_{-\infty}^{\infty} \nu(s)sP'(s)ds = -\lambda b.$$

We write the integral in the left side of equation as two integral

$$\int\limits_{-\infty}^{S} \nu_1 sP'(s)ds + \int\limits_{S}^{\infty} \nu_2 sP'(s)ds = -\lambda b.$$

Using integration by parts we get

$$\nu_1 SP_1(S) - \nu_2 SP_2(S) - \nu_1 \int\limits_{-\infty}^{S} P(s)ds - \nu_2 \int\limits_{S}^{\infty} sP_2(s)ds = -\lambda b.$$

It is possible to prove that

$$\nu_1 SP_1(S) - \nu_2 SP_2(S) = 0,$$

then

$$\nu_1 \int_{-\infty}^{S} P(s)ds + \nu_2 \int_{S}^{\infty} sP_2(s)ds = \lambda b.$$

We get

$$\nu_1 R_1 + \nu_2 R_2 = \lambda b.$$

Tacking into account expression

$$R_1 + R_2 = 1,$$

we find the probabilities

$$R_1 = \frac{\nu_1 - \lambda b}{\nu_1 - \nu_2}, R_2 = \frac{\lambda b - \nu_2}{\nu_1 - \nu_2}.$$

2.3 The Solution $P(s)$ of (2) for $s > S$

For $s > S$ Eq. (2) has the form

$$\nu_2 P'(s) + \lambda P(s) = \lambda \int_{0}^{\infty} P(s+x)dB(x). \tag{6}$$

We will find the solution of this equation in the form of exponential function

$$P(s) = Ce^{-\gamma(s-S)}, s \geq S. \tag{7}$$

Substituting (7) into (6), we obtain

$$\lambda - \nu_2\gamma = \lambda \int_{0}^{\infty} e^{-\gamma x}dB(x), \tag{8}$$

which is a nonlinear equation for γ.

By virtue of condition (1), Eq. (8) has a unique positive solution. Obviously other solutions of Eq. (8) are extraneous. Substituting (7) into the expression for the probability R_2

$$R_2 = \int_{S}^{\infty} P(s)ds,$$

we obtain

$$R_2 = \int_{S}^{\infty} Ce^{-\gamma(s-S)}ds = \int_{S}^{\infty} e^{-\gamma x}dx = \frac{C}{\gamma},$$

then we can write expression that defines the value of the parameter C of the function $P(s)$

$$C = \gamma R_2 = \gamma\frac{\lambda b - \nu_2}{\nu_1 - \nu_2}, \tag{9}$$

Thus, we can write the solution $P(s)$ of Eq. (2) in the following form

$$P(s) = \begin{cases} P_1(s), s < S, \\ Ce^{-\gamma(s-S)}, s > S, \end{cases} \quad (10)$$

where the function $P_1(x)$ for $x < 0$ will be defined below.

2.4 The Solution $P_1(s)$ of (2) for $s < S$

The main objective of this section is to try to find the solution $P(s)$ of Eq. (2). Equation (2) for $s < S$ has the form

$$\nu_1 P_1'(s) + \lambda P_1(s) = \lambda \int_0^{S-s} P_1(s+x)dB(x) + \lambda \int_{S-s}^{\infty} P_1(s+x)dB(x). \quad (11)$$

Lets the values of purchases are independent and identically distributed random variables having m-th order hyperexponential distribution

$$B(x) = \sum_{k=1}^{m} q_k \left(1 - e^{-\mu_k x}\right), \quad (12)$$

with positive parameters $\mu_k > 0$ and $q_k > 0$

$$\sum_{k=1}^{m} q_k = 1. \quad (13)$$

Taking into account (7), rewrite (11)

$$\nu_1 P_1{}'(s) + \lambda P_1(s) = \lambda \int_0^{S-s} P_1(s+x)dB(x) + \lambda Ce^{-\gamma(s-S)} \int_{S-s}^{\infty} e^{-\gamma x}dB(x). \quad (14)$$

Using (12), we get

$$\int_{S-s}^{\infty} e^{-\gamma x}dB(x) = \int_{S-s}^{\infty} e^{-\gamma x} \sum_{k=1}^{m} q_k \mu_k e^{-\mu_k x}dx = \sum_{k=1}^{m} q_k \mu_k \int_{S-s}^{\infty} e^{-(\mu_k+\gamma)x}dx$$

$$= \sum_{k=1}^{m} q_k \frac{\mu_k}{\mu_k + \gamma} e^{-(\mu_k+\gamma)(S-s)},$$

then Eq. (14) has form

$$\nu_1 P_1{}'(s) + \lambda P_1(s)$$

$$= \sum_{k=1}^{m} q_k \mu_k \left\{ \lambda_1 \int_0^{S-s} P_1(s+x)e^{-\mu_k x}dx + Ce^{\mu_k s} \frac{\lambda_2}{\mu_k + \gamma} e^{-\mu_k S} \right\}. \quad (15)$$

It is necessary to find the solution $P_1(s)$ of (15) which will be defined in the Theorem 1.

Before formulating the theorem about the function $P_1(s)$ we consider the equation

$$\nu_1 z + \lambda = \lambda \sum_{k=1}^{m} q_k \frac{\mu_k}{\mu_k - z}. \tag{16}$$

Equation (16) can be transformed to the algebraic equation of degree $m + 1$, consequently Eq. (16) has $m + 1$ roots. Obviously $z = 0$ is a root of this equation.

For the other roots $z = z_k, k = \overline{1, m}$ of the Eq. (16) we prove the following lemma.

Lemma 1. *If the condition (1) is satisfied*

$$\nu_1 > \lambda b$$

all the roots $z = z_k, k = \overline{1, m}$ of Eq. (16) are real and positive.

Proof. We assume that μ_k arranged by increasing, i.e.

$$\mu_1 < \mu_2 < ... < \mu_m.$$

Consider the function (Fig. 2) from right side of the Eq. (16)

$$f(z) = \lambda \sum_{k=1}^{m} q_k \frac{\mu_k}{\mu_k - z}.$$

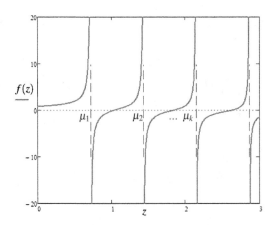

Fig. 2. Function $f(z)$

Differentiating the function $f(z)$, we obtain

$$f'(z) = \lambda \sum_{k=1}^{m} q_k \frac{\mu_k}{(\mu_k - z)^2} > 0.$$

Hence function $f(z)$ is continuous, monotone increasing and takes values $\lambda < f(z) < \infty$ on the interval $0 < z < \mu_1$. When the condition $\nu_1 > \lambda b$ is satisfied, on the interval $0 < z < \mu_1$ Eq. (16) has at least one root.

On the intervals $\mu_{n-1} < z < \mu_n$, $n \geq 2$ the function $f(z)$ is continuous and monotonically increases and takes a values $\infty < f(z) < \infty$, so Eq. (16) also has at least one root $z_n > 0$ on each interval.

Number m of the interval equals the number of positive roots $z = z_m$.

Lemma is proved.

Theorem 1. *The solution $P_1(s)$ of Eq. (11) have form*

$$P_1(s) = C \sum_{n=1}^{m} x_n e^{z_n(s-S)}, s < S, \tag{17}$$

where z_n are a positive roots of Eq. (16), x_ν are components of the vector $\mathbf{X}$. This vector is a solution to a system of linear algebraic equations

$$\mathbf{AX} = \mathbf{h}, \tag{18}$$

where A_{kn} are elements of the matrix $\mathbf{A}$, h_k are elements of the vector $\mathbf{h}$. The elements A_{kn} and h_k have following form

$$A_{kn} = \frac{1}{\mu_k - z_n}, h_k = \frac{1}{\mu_k + \gamma}, k = \overline{1,m}, n = \overline{1,m} \tag{19}$$

normalizing constant C is determined by (9).

Proof. The solution $P_1(s)$ of the Eq. (11) we will to find in the form (17). Substituting expression (17) into (11) we obtain the equation

$$\sum_{n=1}^{m} x_n e^{z_n(s-S)} \left\{ \nu_1 z_n + \lambda + \lambda \sum_{k=1}^{m} q_k \frac{\mu_k}{z_n - \mu_k} \right\}$$

$$= \lambda \sum_{k=1}^{m} q_k \mu_k e^{\mu_k(s-S)} \left\{ \sum_{n=1}^{m} x_n \frac{1}{z_n - \mu_k} + \frac{1}{\mu_k + \gamma} \right\}.$$

Equating to zero the coefficients in the linear combination of exponents $e^{z_n(s-S)}$ in this expression, we obtain the equality

$$\nu_1 z_n + \lambda + \lambda \sum_{k=1}^{m} q_k \frac{\mu_k}{z_n - \mu_k} = 0, n = \overline{1,m}.$$

Obviously, that this expression matches with (16) for $n = \overline{1,m}$. Therefore, z_n are the roots of the Eq. (16).

Analogically, for linear combination of exponents $e^{\mu_k s}$ we get

$$\sum_{n=1}^{m} x_n \frac{1}{\mu_k - z_n} = \frac{1}{\mu_k + \gamma}, k = \overline{1,m}.$$

These equations are non-homogeneous system of linear algebraic equations for the x_n, and the same as the system (18), the elements A_{kn} of matrix $\mathbf{A}$ is defined by (19). The theorem is proved.

Taking into account (7) and (17), probability density function $P(s)$ is given by

$$
P(s) = C \begin{cases} \sum\limits_{n=1}^{m} x_n e^{z_n(s-S)}, s \leq S, \\ e^{-\gamma(s-S)}, s \geq S, \end{cases} \tag{20}
$$

where z_n are a nonzero roots of Eq. (16), γ is unique positive root of Eq. (8), x_n are components of the vector $\mathbf{X}$, where $\mathbf{X}$ is a solution to a system of linear algebraic equations (18).

Using $\nu_1 S P_1(S) - \nu_2 S P_2(S) = 0$, we have following expression

$$
P_1(S) = \frac{\nu_2}{\nu_1} P_2(S),
$$

thus the probability density function $P(s)$ of process $s(t)$ is continuous for all values of s, but at the point $s = S$ it has a discontinuity, i.e. $P_1(S) \neq P_2(S)$.

3   Mathematical Model of System with the Constant Rate of Input Product Flow

Lets consider second mathematical model of inventory management (Fig. 3). It's assumed that the product flow be continuous with fixed rate $\nu = 1$. Demand occurs according to a Poisson process with piecewise-constant intensity $\lambda(s)$

$$
\lambda(s) = \begin{cases} \lambda_1, s < S, \\ \lambda_2, s \geq S, \end{cases}
$$

where S is the threshold inventory level of $s(t)$.

The values of purchases are independent and identically distributed random variables having $m-$th order hyperexponential distribution (12) with the first moment equals 1.

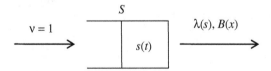

Fig. 3. Inventory management model

The condition of existence of a stationary regime for this model has form

$$
\lambda_1 b < 1 < \lambda_2 b. \tag{21}
$$

Denotes stationary probability density function of process $s(t)$

$$P(s) = \frac{\partial P\{s(t) < s\}}{\partial s}.$$

We obtain Kolmogorov equation for the distribution $P(s)$

$$P'(s) + \lambda(s)P(s) = \int_0^\infty \lambda(s+x)P(s+x)dB(x), \qquad (22)$$

where $P(s)$ satisfies the boundary conditions

$$P(-\infty) = P(\infty) = 0. \qquad (23)$$

The plan of this section is to determine the function $P(s)$.
Similarly Sect. 1, denotes

$$\int_{-\infty}^S P(s)ds = R_1, \int_S^\infty P(s)ds = R_2.$$

Proposition about form of probabilities R_1 and R_2 are formulated below.

Proposition 2. *Probabilities R_1 and R_2 is defined by*

$$R_1 = \frac{\lambda_2 b_1 - 1}{\lambda_2 b_1 - \lambda_1 b_1}, R_2 = \frac{1 - \lambda_1 b_1}{\lambda_2 b_1 - \lambda_1 b_1}.$$

The proof of this proposition is analogous to Subsect. 2.2.

3.1   Form of the Solution $P(s)$ for $s \geq S$

For $s \geq S$ Eq. (22) has the form

$$P'(s) + \lambda_2 P(s) = \lambda_2 \int_0^\infty P(s+x)dB(x), s \geq S. \qquad (24)$$

We seek the solution of this equation in the form of an exponential function

$$P(s) = Ce^{-\gamma(s-S)}, s \geq S. \qquad (25)$$

Substituting expression (25) into (24), we obtain the equality

$$\lambda_2 - \gamma = \lambda_2 \int_0^\infty e^{-\gamma x}dB(x), \qquad (26)$$

which is a nonlinear equation from γ.

Obviously Eq. (26) has a unique positive solution.

Substituting expression (26) into the expression for the probability R_2

$$R_2 = \int_S^\infty P(s)ds,$$

we obtain the equality

$$R_2 = \int_S^\infty Ce^{-\gamma(s-S)}ds = \int_S^\infty e^{-\gamma x}dx = \frac{C}{\gamma},$$

from which we get

$$C = \gamma R_2 = \gamma \frac{\lambda_2 b_1 - 1}{\lambda_2 b_1 - \lambda_1 b_1}. \qquad (27)$$

3.2   The Solution $P_1(s)$ of Eq. (22) for $s < S$

Lets the values of purchases are independent and identically distributed random variables having m-th order hyperexponential distribution (12).

Using (25), from (22) we get

$$P_1'(s) + \lambda_1 P_1(s) = \lambda_1 \int_0^{S-s} P_1(s+x)dB(x) + \lambda_2 Ce^{-\gamma(s-S)} \int_{S-s}^\infty e^{-\gamma x}dB(x). \quad (28)$$

Taking into account (12), we obtain

$$P_1'(s) + \lambda_1 P_1(s) = \sum_{k=1}^m b_k \mu_k \left\{ \lambda_1 \int_0^{S-s} P_1(s+x)e^{-\mu_k x}dx + Ce^{\mu_k s} \frac{\lambda_2}{\mu_k + \gamma}e^{-\mu_k S} \right\}.$$

It is necessary to find the solution $P_1(s)$ of (22) which will be defined in the Theorem 2.

Firstly we formulate Lemma and proposition about roots of equation

$$z + \lambda_1 = \lambda_1 \sum_{k=1}^m b_k \frac{\mu_k}{\mu_k - z}. \qquad (29)$$

Lemma 2. *If condition (21) is satisfied*

$$\lambda_1 b < 1,$$

then all roots $z = z_k, k = \overline{1,m}$ of Eq. (29) are real and positive.

Proposition 3. *For roots of Eq. (29) following conditions*

$$0 < z_1 < \mu_1,$$

$\mu_n - 1 < z_n < \mu_n, n \geq 2$ *is satisfied.*

Formulated proposition greatly simplifies the numerical computation of positive roots of the Eq. (29).

Theorem 2. *If $B(x)$ is hyperexponential distribution, then the solution $P_1(s)$ of Eq. (22) have form*

$$P_1(s) = C \sum_{n=1}^{m} x_n e^{z_n(s-S)}, s < S, \tag{30}$$

where z_n are a positive roots of Eq. (29), x_ν are components of the vector $\mathbf{X}$. This vector is a solution to a system of linear algebraic equations

$$\mathbf{AX} = \mathbf{h}, \tag{31}$$

where A_{kn} are elements of the matrix $\mathbf{A}$, h_k are elements of the vector $\mathbf{h}$. The elements A_{kn} and h_k have following form

$$A_{kn} = \frac{\lambda_1}{\mu_k - z_n}, h_k = \frac{\lambda_2}{\mu_k + \gamma}, k = \overline{1,m}, n = \overline{1,m}, \tag{32}$$

normalizing constant C is determined by the expression (27).

Proof. Solution $P_1(s)$ of the Eq. (22) will be find in the form (30). Substituting (30) into (22) we obtain the equation

$$\sum_{n=1}^{m} x_n e^{z_n(s-S)} \left\{ z_n + \lambda_1 + \lambda_1 \sum_{k=1}^{m} q_k \frac{\mu_k}{z_n - \mu_k} \right\}$$

$$= \sum_{k=1}^{m} q_k \mu_k e^{\mu_k(s-S)} \left\{ \sum_{n=1}^{m} x_n \frac{\lambda_1}{z_n - \mu_k} + \frac{\lambda_2}{\mu_k + \gamma} \right\}.$$

Equating to zero the coefficients in the linear combination of exponents $e^{z_n(s-S)}$ in this expression, we obtain the equality

$$z_n + \lambda_1 + \lambda_1 \sum_{k=1}^{m} q_k \frac{\mu_k}{z_n - \mu_k} = 0, n = \overline{1,m},$$

It is easy to see that this expression matches with (29) for $n = \overline{1,m}$. Therefore, z_n are the roots of the Eq. (29).

Similarly, for $e^{\mu_k s}$ we obtain

$$\sum_{n=1}^{m} x_n \frac{\lambda_1}{\mu_k - z_n} = \frac{\lambda_2}{\mu_k + \gamma}, k = \overline{1,m}.$$

These equations are non-homogeneous system of linear algebraic equations for the x_n, and the same as the system (31) in which the elements A_{kn} matrix $\mathbf{A}$ defined by (32). The theorem is proved.

From (25) and (30), probability density function $P(s)$ is given by

$$P(s) = C \begin{cases} \sum_{n=1}^{m} x_n e^{z_n(s-S)}, s \leq S, \\ e^{-\gamma(s-S)}, s \geq S, \end{cases} \tag{33}$$

where z_n are a nonzero roots of Eq. (29), γ is unique positive root of Eq. (26), x_n are components of the vector $\mathbf{X}$, where $\mathbf{X}$ is the solution to the system of linear algebraic equations (31).

4   Numerical Results

In this section, we discuss the numerical results of inventory management models with On/Off control. For simplicity, we will assume that the random demand has third-order hyperexponential distribution $B(x)$ with first moments $b = 1$, therefore we choose following value of parameters

$$\mu_1 = 0.4, \mu_2 = 1, \mu_3 = 10, q_1 = 0.3, q_2 = 0.2, q_3 = 0.5. \tag{34}$$

4.1   Case of a Piecewise-Constant Rate of Input Product Flow

We use following values of the parameters: rate of input product flow $\nu_1 = 1.2, s < S$ and $\nu_2 = 0.8, s > S$; intensity of Poisson process $\lambda = 1$; threshold value $S = 40$.
    We found the roots of Eqs. (8) and (16)

$$\gamma = 0.124, z_1 = 0.078, z_2 = 0.886, z_3 = 9.602.$$

Note that the Eq. (8) has only one positive root γ, Eq. (16) has three positive roots $z_n, n = \overline{1,3}$. Let us find probability density function $P(s)$ for the given parameters. The parameters x_n, of (18) have following value $x_1 = 0.63; x_2 = 0.024; x_3 = 0.013$, and normalizing constant $C = 0.062$.
    The numerical results are illustrated in Fig. 4.

4.2   Case of Piecewise-Constant Intensity of Output Product Flow

In second model we consider following values of parameters: rate of input product flow $\nu = 1$; intensity of Poisson process $\lambda_1 = 0.8, s < S$ and $\lambda_2 = 1.2, s > S$; threshold value $S = 40$. The positive roots of Eqs. (26) and (29) are found

$$\gamma = 0.099; z_1 = 0.094; z_2 = 0.899; z_3 = 9.617.$$

Thus, the Eq. (26) has a unique solution, and the Eq. (29) has three positive roots. Both equations have zero extraneous roots. The parameters $x_n, n = \overline{1,3}$ of (31) $x_1 = 0.945; x_2 = 0.036; x_3 = 0.019$, normalizing constant $C = 0.049$.
    The numerical results are illustrated in Fig. 5.

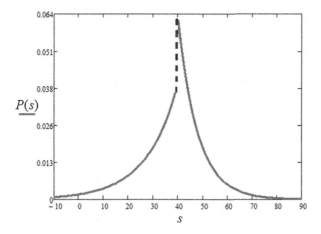

Fig. 4. Solution $P(s)$ of Eq. (2)

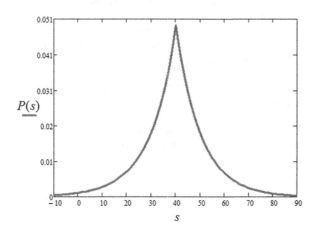

Fig. 5. Solution $P(s)$ of Eq. (22)

5 Conclusions

In this work we consider modified Cramer-Lundberg models of the inventory management system with On/Off control. In the first model we manage the rate of input product flow. In the second model intensity of output product flow is changed. For each models we find some properties stationary distributions of inventory level accumulated in the system for arbitrary distribution of demands purchase values. We obtain the explicit expressions of the stationary distributions of inventory level for hyperexponential distribution of purchase values of demand.

References

1. Arrow, K.J., Harris, T.E., Marschak, J.: Optimal inventory policy. Econometrica **19**(3), 205–272 (1951)
2. Gallego, G., Moon, I.: The distribution free newsboy problem: review and extensions. J. Oper. Res. Soc. **44**, 825–834 (1993)
3. Kitaeva, A., Subbotina, V., Zmeev, O.: The newsvendor problem with fast moving items and a compound poisson price dependent demand. In: 15th IFAC Symposium on Information Control Problems in Manufacturing INCOM 2015 (IFAC-PapersOnLine), vol. 48, pp. 1375–1379. Elsevier (2015)
4. Khouja, M.: The single-period (news-vendor) problem: literature review and suggestionsfor future research. OMEGA-INT J **27**(5), 537–553 (1999)
5. Mousavia, S.M., Hajipoura, V., Niakib, S.T.A., Alikar, N.: Optimizing multi-item multi-period inventory control system with discounted cash flow and inflation: two calibrated meta-heuristic algorithms. Appl. Math. Model. **37**(4), 2241–2256 (2013)
6. Nazarov, A., Broner, V.: Inventory management system with erlang distribution of batch sizes. In: Dudin, A., Gortsev, A., Nazarov, A., Yakupov, R. (eds.) ITMM 2016. CCIS, vol. 638, pp. 273–280. Springer, Cham (2016). doi:10.1007/978-3-319-44615-8_24
7. Nazarov, A.A., Broner, V.I.: Resource control for physical experiments in the cramer-lundberg model. Russ. Phys. J. **59**(7), 1024–1036 (2016)
8. Silver, E.A., Pyke, D.F., Peterson, R.: Inventory Management and Production Planning and Scheduling. Wiley, New York (1998)
9. Qin, Y., Wang, R., Vakharia, A., Chen, Y., Hanna-Seref, M.: The newsvendor problem: review and directions for future research. Eur. J. Oper. Res. **213**, 361–374 (2011)
10. Zhang, D., Xu, H., Wu, Y.: Single and multi-period optimal inventory control models with risk-averse constraints. Eur. J. Oper. Res. **199**, 420–434 (2009)

Dobrushin Mean-Field Approach for Queueing Large-Scale Networks with a Small Parameter

Galina O. Tsareva and Sergey A. Vasilyev$^{(\boxtimes)}$

Department of Applied Probability and Informatics, RUDN University,
Miklukho-Maklaya St. 6, Moscow 117198, Russia
gotsareva@gmail.com, svasilyev@sci.pfu.edu.ru

Abstract. In this paper it is considered a system that consists of infinite number of servers with a Poisson input flow of requests of intensity $N\lambda$. Each requests arriving to the system randomly selects two servers and is instantly sent to the one with the shorter queue. In this case a share $u_k(t)$ of the servers that have the queues lengths with not less than k can be described using an infinite system of differential equations. It is possible to investigate Tikhonov type Cauchy problem for this system with small parameter μ for building the solutions $u_k(t)$. The evolution analysis of $u_k(t)$ $(k = 1, 2, \ldots)$ be applied to application in queueing large-scale networks analysis.

Keywords: Analytical methods in probability theory · Systems of differential equations of infinite order · Singular perturbated systems of differential equations · Small parameter · Countable Markov chains · Large network modeling

1 Introduction

The recent research of service networks with complex routing discipline in [14,22–24] transport networks [1,6,7] and the asymptotic behavior of Jackson networks [18] faced with the problem of proving the global convergence of the solutions of certain infinite systems of ordinary differential equations to a time-independent solution. Scattered results of these studies, however, allow a common approach to their justification. This approach will be expounded here. In work [15] the countable systems of differential equations with bounded Jacobi operators are studied and the sufficient conditions of global stability and global asymptotic stability are obtained. In [13] it was considered finite closed Jackson networks with N first come, first serve nodes and M customers. In the limit $M \to \infty$, $N \to \infty$, $M/N \to \lambda > 0$, it was got conditions when mean queue lengths are uniformly bounded and when there exists a node where the mean queue length tends to ∞ under the above limit (condensation phenomena, traffic jams), in terms

S.A. Vasilyev—The reported study was funded within the Agreement 02.03.21.0008 dated 24.11.2016 between the Ministry of Education and Science of the Russian Federation and RUDN University.

V.M. Vishnevskiy et al. (Eds.): DCCN 2017, CCIS 700, pp. 395–405, 2017.
DOI: 10.1007/978-3-319-66836-9_33

of the limit distribution of the relative utilizations of the nodes. It was deriven asymptotics of the partition function and of correlation functions. Cauchy problems for the systems of ordinary differential equations of infinite order was investigated A.N. Tihonov [19], K.P. Persidsky [16], O.A. Zhautykov [25,26], J. Korobeinik [9], A.M. Samoilenko, Yu V. Teplinskii [17] other researchers. For example, Markus Kreer, Aye Kzlers and Anthony W. Thomas [11] investigated fractional Poisson processes, a rapidly growing area of non-Markovian stochastic processes, that are useful in statistics to describe data from counting processes when waiting times are not exponentially distributed. They showed that the fractional KolmogorovFeller equations for the probabilities at time t could be represented by an infinite linear system of ordinary differential equations of first order in a transformed time variable. These new equations resemble a linear version of the discrete coagulation fragmentation equations, well-known from the non-equilibrium theory of gelation, cluster-dynamics and phase transitions in physics and chemistry.

It was studied the singular perturbed systems of ordinary differential equations by A.N. Tihonov [20], A.B. Vasil'eva [21], S.A. Lomov [12] other researchers.

In paper [2] we investigated the singular perturbed systems of ordinary differential equations of infinite order of Tikhonov-type $\mu\dot{x} = F(x(t, g_x), y(t, g_y), t)$, $\dot{y} = f(x(t, g_x), y(t, g_y), t)$ with the initial conditions $x(t_0) = g_x$, $y(t_0) = g_y$, where $x, g_x \in X$, $X \subset l_1$ and $y, g_y \in Y$, $Y \in \mathbf{R^n}$, $t \in [t_0, t_1]$ $(t_0 < t_1)$, $t_0, t_1 \in T$, $T \in \mathbf{R}$, g_x and g_y are given vectors, $\mu > 0$ is a small real parameter.

In this paper we apply Dobrushin mean-field approachs from [22] for the singular perturbed systems of ordinary differential equations of infinite order of Tikhonov type. We considered a system that consists of infinite number of servers with a Poisson input flow of requests of intensity $N\lambda$. Each requests arriving to the system randomly selects two servers and is instantly sent to the one with the shorter queue. In this case a share $u_k(t)$ of the servers that have the queues lengths with not less than k can be described using an infinite system of differential equations. It is possible to investigate Tikhonov type Cauchy problem for this system with small parameter μ and initial conditions. It is studying the singular perturbed Tikhonov systems of ordinary differential equations of infinite order $\dot{u} = f(u(t, g_u), U(t, g_U), t)$, $\mu\dot{U} = F(u(t, g_u), U(t, g_U), t)$ with the initial conditions $u(t_0, g_u) = g_u$, $U(t_0, g_U) = g_U$, where $u, f \in X$, $X \in \mathbf{R^n}$ are n-dimensional functions; $U, F \in Y$, $Y \subset l_1$ are infinite-dimensional functions and $t \in [0, t_1]$ $(0 < t_1 \leq \infty)$, $t \in T$, $T \in \mathbf{R}$; $g_u \in X$ and $g_U \in Y$ are given vectors, $\mu > 0$ is a small real parameter. The evolution analysis of $u_k(t)$ $(k = 1, 2, \ldots)$ be applied to application in queueing large-scale networks analysis.

2   General Concepts of the Theory of Infinite Systems

Consider a system of differential equations of the form [17,25,26]

$$\frac{du_i}{dt} = F(u_1, u_2, \ldots, u_n, \ldots), \quad i = 1, 2, \ldots. \tag{1}$$

Definition 1. A function $F(u_1, u_2, \ldots, u_n, \ldots)$ is called strongly continuous if for any $\varepsilon > 0$, there exist N_0 and $\delta > 0$ such that the inequality $|u_i' - u_i''| < \delta$, $i = 1, 2, \ldots, N_0$, implies the estimate

$$\left| F\left(u_1', u_2', \ldots \right) - F\left(u_1'', u_2'', \ldots \right) \right| < \varepsilon. \tag{2}$$

Theorem 1. *Assume that the right-hand sides of the system of equations* (1)

1. *are defined for any* $u_i \in R^1, i = 1, 2, \ldots$, *and all* $t \in T_0 = [t_0, t_0 + \Delta t] \subset R^1$;
2. *are strongly continuous in* $u_1, u_2, \ldots$ *for fixed* t *and measurable in* t *for fixed* $u_i, i = 1, 2, \ldots$;
3. *satisfy the inequalities*

$$|F_i(t, u_1, u_2, \ldots)| < M_i(t) \tag{3}$$

for all $i = 1, 2, \ldots$, *where* $M_i(t)$ *are functions summable on the segment* T_0.

Then, for any vector $(u_1^0, u_2^0, \ldots)$ with real coordinates, there exists at least one solution $(u_1(t), u_2(t), \ldots)$ of the system of equations (1) such that $u_i(t_0) = u_i^0, i = 1, 2, \ldots$.

Proof. We replace the system of equations (1) by the following system of integral equations:

$$u_i(t) = u_i^0 + \int_{t_0}^{t} F_i(t, u_1(t), u_2(t), \ldots) \, dt, i = 1, 2, \ldots, \tag{4}$$

and consider a mapping (A)

$$z_i(t) = u_i^0 + \int_{t_0}^{t} F_i(t, u_1(t), u_2(t), \ldots) \, dt, i = 1, 2, \ldots, \tag{5}$$

which establishes a correspondence between an arbitrary countable system of continuous functions $\{u_i(t)\}_{i=1}^{\infty}$ and another system of this sort $\{z_i(t)\}_{i=1}^{\infty}$. Note that if $F(t, u_1, \ldots, u_n)$ is a continuous function of finitely many variables $\{u_i(t)\}_{i=1}^{n}$ measurable with respect to t for fixed $u_i, i = \overline{1, n}$, then the function

$$\Phi(t) = F(t, \phi_1(t), \ldots, \phi_n(t))$$

is measurable if $\phi_i(t), i = \overline{1, n}$, are measurable.

Thus, the function

$$\Psi_n(t) = F(t, \phi_1(t), \ldots, \phi_n(t), 0, 0, \ldots)$$

is measurable and, therefore, the function

$$F(t, \phi_1(t), \phi_2(t), \ldots) = \Psi(t)$$

is also measurable because

$$\Psi(t) = \lim_{n \to \infty} \Psi_n(t), \tag{6}$$

which readily follows from the condition of strong continuity. The requirement of summability follows from condition 3 of Theorem 1. We consider a system of functions $\{u_i(t)\}_{i=1}^\infty$ as a point P of an abstract space R. If there exists a point P invariant under mapping (A) (5), then it specifies a solution of the system of equations (4) and, hence, of system (1).

Consider a set M_0 formed by three points P for which $\{u_i(t)\}_{i=1}^\infty$ satisfy the conditions

1. $|u_k(t) - u_k^0| \leq \int\limits_{t_0}^{t} M_k(t)dt,$

2. $|u_k(t') - u_k(t'')| \leq \int\limits_{t'}^{t''} M_k(t)dt, \; k = 1, 2, \ldots.$

It is easy to see that mapping (A) (5) maps the set M_0 into itself. We now introduce mapping (B) by putting every point P in correspondence with a set of numbers

$$\frac{a_1^1}{N_1}, \ldots, \frac{a_1^n}{N_1}, \ldots,$$

$$\cdots\cdots, \tag{7}$$

$$\frac{a_n^1}{nN_n}, \ldots, \frac{a_n^n}{nN_n}, \ldots,$$

$$\cdots\cdots,$$

where $N_i = u_i^0 + \int\limits_{t_0}^{t_0+\Delta t} M_i(t)\, dt$ and the numbers $\{a_n^r\}_{n,r=1}^\infty$ $(a_n^1, \ldots, a_n^n, \ldots)$ are the coefficients of the Fourier expansion of a function $u_n(t)$ in a certain complete orthogonal system of functions on the segment T_0. By ordering the set of numbers (7), we obtain a numerical sequence $b_1, b_2, \ldots, b_n, \ldots$. Moreover, we have

$$\sum_{k=1}^\infty \left(a_n^k\right)^2 = \int\limits_{t_0}^{t_0+\Delta t} (u_n(t))^2\, dt \leq \int\limits_{t_0}^{t_0+\Delta t} \left(u_n^0 + \int\limits_{t_0}^{t} M_k(t)dt\right)^2 dt$$

$$\leq \int\limits_{t_0}^{t_0+\Delta t} N_n^2 dt = aN_n^2, \tag{8}$$

whence it follows that

$$\sum_{i=1}^\infty b_i^2 = \sum_{n=1}^\infty \sum_{k=1}^\infty \left(\frac{a_n^k}{nN_n}\right)^2 \leq a\sum_{n=1}^\infty \frac{1}{n^2} = \frac{a\pi^2}{6}. \tag{9}$$

Thus, mapping (B) maps the set M_0 into a subset M_0^* of the Hilbert space l_2. Therefore, mapping (A) induces a mapping (A^*) of the set M_0^* into itself. Further, if mapping (A^*) has a fixed point $P^* \in M_0^*$, then the corresponding point $P^* \in M_0$ determines the solution of Eq. (4) and, hence, (1). To use the Schauder theorem, it suffices to show that the set M_0^* is compact and convex. If $P'^* = (b_1', \ldots, b_n', \ldots)$ and $P''^* = (b_1'', \ldots, b_n'', \ldots)$ are points from M_0^*, then the point

$$\alpha P'^* + \beta P''^* = (\alpha b_1' + \beta b_1'', \alpha b_2' + \beta b_2'', \ldots), \alpha + \beta = 1, \alpha > 0, \beta > 0, \quad (10)$$

belongs to M_0^* because it corresponds to the system of functions

$$\alpha u_1'(t) + \beta u_1''(t), \alpha u_2'(t) + \beta u_2''(t), \ldots. \quad (11)$$

specifying a point from the set M_0. Indeed,

$$\left| \alpha u_k'(t) + \beta u_k''(t) - u_k^0 \right| = \left| \alpha(u_k'(t) - u_k^0) + \beta(u_k''(t) - u_k^0) \right|$$

$$\leq (\alpha + \beta) \int_{t_0}^{t} M_k(t)dt = \int_{t_0}^{t} M_k(t)dt, \quad (12)$$

i.e., condition 1 is satisfied. Similarly, the inequality

$$\left| \alpha u_k'(t') + \beta u_k''(t') - \alpha u_k'(t'') - \beta u_k''(t'') \right| \leq \alpha + \beta) \int_{t_0}^{t} M_k(t)dt \quad (13)$$

implies condition 2. Hence, the set M_0^* is convex. In this set, we choose an arbitrary sequence of points P_i^*. This sequence corresponds to the sequence of points $P_i \left(u_1^{(i)}(t), u_2^{(i)}(t), \ldots \right)$ in the set M_0. According to conditions 1 and 2, the sequence $u_1^{(i)}(t), i = 1, 2, \ldots$, is uniformly bounded and equicontinuous and, consequently, it contains a subsequence $u_1^{(\alpha_1)}(t), u_1^{(\alpha_1)}(t), \ldots, u_1^{(\alpha_s)}(t), \ldots$ that converges uniformly in $t \in T_0$. However, the sequence $u_2^{(\alpha_s)}(t), s \to \infty$, is also uniformly bounded and equicontinuous and, hence, it also contains a convergent subsequence

$$u_2^{(\beta_1)}(t), u_2^{(\beta_2)}(t), \ldots, u_2^{(\beta_s)}(t), \ldots. \quad (14)$$

This process can be continued infinitely.

We compose the table

$$u_1^{(\alpha_1)}(t) u_1^{(\alpha_2)}(t) u_1^{(\alpha_3)}(t) \ldots$$

$$u_2^{(\beta_1)}(t) u_2^{(\beta_2)}(t) u_2^{(\beta_3)}(t) \ldots \quad (15)$$

$$u_3^{(\gamma_1)}(t) u_3^{(\gamma_2)}(t) u_3^{(\gamma_3)}(t) \ldots$$

$$\ldots \ldots \ldots \ldots$$

and rewrite the set of sequences row by row

$$u_1^{(\alpha_1)}(t)u_1^{(\beta_2)}(t)u_1^{(\gamma_3)}(t)\ldots$$

$$u_2^{(\alpha_1)}(t)u_2^{(\beta_2)}(t)u_2^{(\gamma_3)}(t)\ldots \tag{16}$$

$$u_3^{(\alpha_1)}(t)u_3^{(\beta_2)}(t)u_3^{(\gamma_3)}(t)\ldots$$

$$\ldots\ldots\ldots\ldots$$

Each of these sequences converges as a subsequence of a convergent sequence supplemented by finitely many elements. Thus, the sequence of points

$$P_{\alpha_1}, P_{\beta_2}, P_{\gamma_3}, \ldots \subset M_0 \tag{17}$$

converges weakly (coordinatewise) to a point $P_0 \in M_0$ (uniformly in $t \in T_0$). For the sake of convenience, we rewrite sequence (17) as

$$P_1, P_2, P_3, \ldots, P_n, \ldots \tag{18}$$

Let us show that the sequence of the corresponding points $P_1^*, P_2^*, P_3^*, \ldots$ from the set M_0^* converges to the point $P_0^* \in M_0^*$ in tne norm of the Hilbert space l_2. Indeed, the distance between the points P'^* and P''^* from M_0^* is given by the formula

$$\rho(P'^*, P''^*) = \sqrt{\sum_{i=1}^{\infty}(b_i' - b_i'')^2} = \sqrt{\sum_{n=1}^{\infty}\frac{1}{n^2 N_n^2}\int_{t_0}^{t_0+\Delta t}(u_n' - u_n'')^2 dt}, \tag{19}$$

whence it follows that

$$\rho(P_0^*, P_k^*) \le \sqrt{\sum_{n=1}^{n_0}\frac{1}{n^2 N_n^2}\int_{t_0}^{t_0+\Delta t}(u_n^0 - u_n^k)^2 dt + \Delta t \sum_{n=n_0}^{\infty}\frac{1}{n^2}} \tag{20}$$

is arbitrarily small for sufficiently large n_0 and k. This means that the set M_0^* is compact. Note that one can easily prove that mapping (B) is a homeomorphism, i.e., the sets M_0 and M_0^* are topologically equivalent. Theorem 1 is proved.

3   Large-Scale Network Model

Let's consider a system that consists of N servers with a Poisson input flow of requests of intensity $N\lambda$. Each request arriving to the system randomly selects two servers and is instantly sent to the one with the shorter queue. The service time is distributed exponentially with mean $\bar{t} = 1$. Let $u_k(t)$ be a share servers that have the queues lengths with not less than k. It is possible to investigate the asymptotic distribution of the queue lengths as $N \to \infty$ and $\lambda < 1$ [22]. The considered system of the servers is described by ergodic Markov chain. There is a stationary probability distribution for the states of the system and if $N \to \infty$ the evolution of the values $u_k(t)$ becomes deterministic and the Markov chain asymptotically converges to a dynamic system the evolution of which is described by infinite system of differential-difference equations

$$\begin{cases} \dot{u}_k(t) = u_{k+1}(t) - u_k(t) + \lambda \left((u_{k-1}(t))^2 - (u_k(t))^2 \right), \\ u_k(0) = g_k \geq 0, \ k = 0, 1, 2, \ldots, \end{cases} \tag{21}$$

where $g = \{g_k\}_{k=1}^{\infty}$ is a numerical sequence $(1 = g_0 \geq g_1 \geq g_2, \ldots)$ [22]. This system of differential equations satisfies the requirements of Theorem 1.

We can investigate infinite system of differential-difference equations with small parameter such form

$$\begin{cases} \dot{u}_k(t) = u_{k+1}(t) - u_k(t) + \lambda \left((u_{k-1}(t))^2 - (u_k(t))^2 \right), \\ k = 0, 1, \ldots, n-1, \\ \dot{u}_n(t) = U_{n+1}(t) - u_n(t) + \lambda \left((u_{n-1}(t))^2 - (u_n(t))^2 \right), \\ \mu \dot{U}_k(t) = U_{k+1}(t) - U_k(t) + \lambda \left((U_{k-1}(t))^2 - (U_k(t))^2 \right), \\ k = n+1, n+1, \ldots, \\ u_k(0) = g_k \geq 0, \ k = 0, 1, 2, \ldots, \end{cases} \tag{22}$$

where μ is a small parameter that bring a singular perturbation to the system (21) which allows us to describe the processes of rapid change of the systems.

Using (22) we can write Tikhonov problems for systems of ordinary differential equations of infinite order with a small parameter μ and initial conditions

$$\begin{cases} \dot{u} = f(u(t, \lambda, g_u), U(t, \lambda, g_U), t), \\ \mu \dot{U} = F(U(t, \lambda, g_U), t); \\ u(0, \lambda, g_u) = g_u, \ U(0, \lambda, g_U) = g_U, \end{cases} \tag{23}$$

where $u, f \in X$, $X \in \mathbf{R}^n$ are n-dimensional functions; $U, F \in Y, Y \subset l_1$ are infinite-dimensional functions and $t \in [0, t_1]$ $(t_0 < t_1 \leq \infty)$, $t \in T$, $T \in \mathbf{R}$; $g_u \in X$ and $g_U \in Y$ are given vectors $(g_u = \{g_k\}_{k=1}^n, g_U = \{g_k\}_{k=n+1}^{\infty})$, $\mu > 0$ is a small real parameter; $u(t, g_u) = g_u = \{u_k\}_{k=1}^n$ and $U(t, g_U) = \{u_k\}_{k=n+1}^{\infty}$ are solutions of (23). Given functions $f(u(t, \lambda, g_u), U(t, \lambda, g_U), t)$ and $F(U(t, \lambda, g_U), t)$ are continuous functions for all variables

$$\begin{cases} f_k(u(t, \lambda, g_u), t) = u_{k+1}(t) - u_k(t) \\ \quad + \lambda \left((u_{k-1}(t))^2 - (u_k(t))^2 \right), k = 0, 1, \ldots, n-1, \\ f_n(u(t, \lambda, g_u), U(t, \lambda, g_U), t) = U_{n+1}(t) - u_n(t) \\ \quad + \lambda \left((u_{n-1}(t))^2 - (u_n(t))^2 \right), \\ F_k(U(t, \lambda, g_U), t) = U_{k+1}(t) - U_k(t) \\ \quad + \lambda \left((U_{k-1}(t))^2 - (U_k(t))^2 \right), k = n+1, n+1, \ldots \end{cases} \tag{24}$$

Let S is an integral manifold of the system (23) in $X \times Y \times T$. If any point $t^* \in [t_0, t_1]$ $(u(t^*), U(t^*), t^*) \in S$ of trajectory of this system has at least one common point on S this trajectory $(u(t, G), U(t, g), t) \in S$ belongs the integral manifold S totally.

If we assume in (23) that $\mu = 0$ than we have a degenerate system of the ordinary differential equations and a problem of singular perturbations

$$\begin{cases} \dot{u} = f(u(t, \lambda, g_x), U(t), t), \\ 0 = F(u(t, \lambda, g_x), U(t), t); \\ u(t_0, \lambda, g_u) = g_u, \end{cases} \tag{25}$$

where the dimension of this system is less than the dimension of the system (23), since the relations $F(u(t, \lambda), U(t, \lambda), \lambda, t) = 0$ in the system (25) are the algebraic equations (not differential equations). Thus for the system (24) we can use limited number of the initial conditions then for system (23). Most natural for this case we can use the initial conditions $u(t_0, \lambda, g_u) = g_u$ for the system (25) and the initial conditions $U(t_0, \lambda, U_y) = g_U$ disregard otherwise we get the overdefined system. We can solve the system (25) if the equation $F(u(t, \lambda), U(t, \lambda), \lambda, t) = 0$ has roots. If it is possible to solve we can find a finite set or countable set of the roots $U_q(t, \lambda, g_u) = u_q(u(t, \lambda, g_u), t)$ where $q \in \mathbf{N}$. If the implicit function $F(u(t, \lambda), U(t, \lambda), \lambda, t) = 0$ has not simple structure we must investigate the question about the choice of roots. Hence we can use the roots $U_q(t, \lambda, g_u) = u_q(u(t, \lambda, g_u), t)$ $(q \in \mathbf{N})$ in (25) and solve the degenerate system

$$\begin{cases} \dot{u}_d = f(u_d(t, \lambda, g_u), u_q(u_d(t, \lambda, g_u), t), \lambda, t); \\ U_d(t_0, \lambda, g_u) = g_u. \end{cases} \tag{26}$$

Since it is not assumed that the roots $U_q(t, \lambda, g_u) = u_q(u(t, \lambda, g_u), \lambda, t)$ satisfy the initial conditions of the Cauchy problem (23) $(U_q(t_0) \neq g_u, q \in \mathbf{N})$, the solutions $U(t, \lambda, g_U)$ (23) and $U_q(t, \lambda, g_u)$ do not close to each other at the initial moments of time $t > 0$. Also there is a very interesting question about behaviors of the solutions $u(t, \lambda, g_u)$ of the singular perturbed problem (23) and the solutions $u_d(t, \lambda, g_u)$ of the degenerate problem (25). When $t = 0$ we have $u(t_0, \lambda, g_u) = u_d(t_0, \lambda, g_u)$. Do these solutions close to each other when $t \in (0, t_1]$? The answer to this question depends on using roots $U_q(t, \lambda, g_u) = u_q(u(t, \lambda, g_u), t)$ and the initial conditions which we apply for the systems (23) and (26).

4   Truncation Large-Scale Network Model and Numerical Analysis

Using (22) we can write the truncation system of differential-difference equations

$$\begin{cases} \dot{u}_k(t) = u_{k+1}(t) - u_k(t) + \lambda \left((u_{k-1}(t))^2 - (u_k(t))^2 \right), \\ k = 0, 1, \ldots, n, \\ \mu \dot{u}_k(t) = u_{k+1}(t) - u_k(t) + \lambda \left((u_{k-1}(t))^2 - (u_k(t))^2 \right), \\ k = n+1, \ldots, N, \\ u_k(0) = g_k \geq 0, \ k = 0, 1, 2, \ldots, N, \\ u_{N+1} = g_{N+1} \geq 0. \end{cases} \tag{27}$$

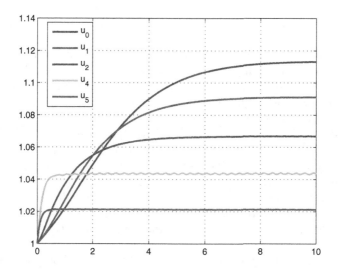

Fig. 1. Evolution analysis of u_k ($\mu = 0.1$).

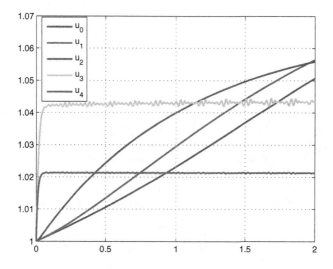

Fig. 2. Evolution analysis of u_k ($\mu = 0.01$).

The numerical example is presented in the figure (see Figs. 1 and 2) where $n = 2$, $N = 4$, $l = 0.2$, $g_k = 1$, $k = 0,\overline{5}$ and a small parameter $\mu = 0.1$ (Fig. 1), $\mu = 0.01$ (Fig. 2). In this numerical example it was shown the existence of stady state conditions for evolutions $u_i(t)$, $i = 0, 1, 2$ and quasi-periodic conditions with boundary layers for evolutions $u_i(t)$, $i = 3, 4$.

5   Conclusions

We investigated the large-scale network model that consists of infinite number of servers with a Poisson input flow of requests of intensity $N\lambda$. Each requests arriving to the system randomly selects two servers and is instantly sent to the one with the shorter queue. In this case a share $u_k(t)$ of the servers that have the queues lengths with not less than k can be described using an infinite system of differential equations. For Tikhonov problem for infinite system of differential equations with small parameter μ and initial conditions we applied the truncation method and studied a finite system of differential equations with small parameter μ order N. The numerical analysis of $u_k(t)$ $(k = 1, 2, \ldots, 5)$ be applied for queueing large-scale networks evolution conditions. It was shown the existence of stady state conditions for evolutions $u_i(t)$, $i = 0, 1, 2$ and quasi-periodic conditions with boundary layers for evolutions $u_i(t)$, $i = 3, 4$.

Acknowledments. The reported study was funded within the Agreement 02.03.21.0008 dated 24.11.2016 between the Ministry of Education and Science of the Russian Federation and RUDN University.

References

1. Afanassieva, L.G., Fayolle, G., Yu, P.S.: Models for transportation networks. J. Math. Sci. **84**(3), 1092–1103 (1997)
2. Bolotova, Galina, Vasilyev, S.A., Udin, D.N.: Systems of differential equations of infinite order with small parameter and countable Markov chains. In: Vishnevskiy, V.M., Samouylov, K.E., Kozyrev, D.V. (eds.) DCCN 2016. CCIS, vol. 678, pp. 565–576. Springer, Cham (2016). doi:10.1007/978-3-319-51917-3_48
3. McDonald, D.R., Reynier, J.: A mean-field model for multiple TCP connections through a buffer implementing RED. Perform. Eval. **49**(14), 77–97 (2002)
4. Daletsky, Y.L., Krein, M.G.: Stability of Solutions of Differential Equations in Banach Space. Science Publishers, Moscow (1970)
5. Henry, Daniel: Geometric Theory of Semilinear Parabolic Equations. LNM, vol. 840. Springer, Heidelberg (1981)
6. Khmelev, D.V., Oseledets, V.I.: Mean-field approximation for stochastic transportation network and stability of dynamical system. Preprint No. 434 of University of Bremen (1999)
7. Khmelev, D.V.: Limit theorems for nonsymmetric transportation networks. Fundamentalnaya i Priklladnaya Matematika **7**(4), 1259–1266 (2001)
8. Kirstein, B.M., Franken, D.E., Stoian, D.: Comparability and monotonicity of Markov processes. Theor. Prob. Appl. **22**(1), 43–54 (1977)
9. Korobeinik, Ju: Differential equations of infinite order and infinite systems of differential equations. Izv. Akad. Nauk SSSR Ser. Mat. **34**, 881–922 (1970)
10. Krasnoselsky, M.A., Zabreyko, P.P.: Geometrical Methods of Nonlinear Analysis. Springer, Berlin (1984)
11. Kreer, M., Kzlersu, A., Thomas, A.W.: Fractional Poisson processes and their representation by infinite systems of ordinary differential equations. Stat. Prob. Lett. **84**, 27–32 (2014)

12. Lomov, S.A.: The construction of asymptotic solutions of certain problems with parameters. Izv. Akad. Nauk SSSR Ser. Mat. **32**, 884–913 (1968)
13. Malyshev, V., Yakovlev, A.: Condensation in large closed Jackson networks. Ann. Appl. Probab. **6**(1), 92–115 (1996)
14. Mitzenmacher, M.: The Power of Two Choices in Randomized Load Balancing. Ph.D. Thesis. University of California at Berkley (1996)
15. Oseledets, V.I., Khmelev, D.V.: Global stability of infinite systems of nonlinear differential equations, and nonhomogeneous countable Markov chains. Problemy Peredachi Informatsii (Russian) **36**(1), 60–76 (2000)
16. Persidsky K.P.: Izv AN KazSSR. Ser. Mat. Mach. (2), pp. 3–34 (1948)
17. Samoilenko, A.M., Teplinskii, Y.V.: Countable Systems of Differential Equations. Brill, Utrecht (2003)
18. Scherbakov, V.V.: Time scales hierarchy in large closed Jackson networks: Preprint No. 4. French-Russian A.M. Liapunov Institute of Moscow State University, Moscow (1997)
19. Tihonov, A.N.: Über unendliche Systeme von Differentialgleichungen. Rec. Math. **41**(4), 551–555 (1934)
20. Tihonov, A.N.: Systems of differential equations containing small parameters in the derivatives. Mat. Sbornik N. S. **31**(73), 575–586 (1952)
21. Vasil'eva, A.B.: Asymptotic behaviour of solutions of certain problems for ordinary non-linear differential equations with a small parameter multiplying the highest derivatives. Uspehi Mat. Nauk. **18**(3), 15–86 (1963). Issie 111
22. Vvedenskaya, N.D., Dobrushin, R.L., Kharpelevich, F.I.: Queueing system with a choice of the lesser of two queues the asymptotic approach. Probl. inform. **32**(1), 15–27 (1996)
23. Vvedenskaya, N.D., Suhov, Y.M.: Dobrushin's mean-field approximation for a queue with dynamic routing. Markov Process. Relat. Fields **3**, 493–526 (1997)
24. Vvedenskaya, N.D.: A large queueing system with message transmission along several routes. Problemy Peredachi Informatsii. **34**(2), 98–108 (1998)
25. Zhautykov, O.A.: On a countable system of differential equations with variable parameters. Mat. Sb. N. S. **49**(91), 317–330 (1959)
26. Zhautykov, O.A.: Extension of the Hamilton-Jacobi theorems to an infinite canonical system of equations. Mat. Sb. N. S. **53**(95), 313–328 (1961)

Retrial Queue M/M/1 with Negative Calls Under Heavy Load Condition

Mais Farkhadov[1(✉)] and Ekaterina Fedorova[2,3]

[1] V.A. Trapeznikov Institute of Control Sciences of Russian Academy of Sciences,
65 Profsoyuznaya St., Moscow 117997, Russian Federation
mais.farhadov@gmail.com
[2] Peoples' Friendship University of Russia, 6 Miklukho-Maklaya St.,
Moscow 117198, Russian Federation
moiskate@mail.ru
[3] Tomsk State University, 36 Lenina Ave., Tomsk 634050, Russian Federation

Abstract. In the paper, the retrial queueing system of M/M/1 type with negative calls is considered. The system of Kolmogorov equations for the system states process is derived. The method of asymptotic analysis is proposed for the system solving under the heavy load condition. The theorem about the gamma form of the asymptotic characteristic function of the number of calls in the orbit is formulated and proved. During the study, the expression for the system throughput is obtained. Also the exact characteristic function is derived. Numerical examples of comparison asymptotic and exact distributions are presented. The conclusion about the asymptotic method application area is made.

Keywords: Retrial queue · Negative calls · Asymptotic analysis · Heavy load

1 Introduction

Retrial queue as a mathematical model of queueing theory is characterized by the feature that an arriving call finding a server busy does not joint a queue and does not leave the system immediately, but goes to some virtual place (orbit), then it tries to get service again after some random time. The comprehensive description, the comparison of classical queueing systems and retrial queues and detailed overviews are contained in books of G. Falin and J. Artalejo [1–3].

Retrial queues are widely used for many practical problems in telecommunication networks, mobile networks, computer systems and various daily life situations [4–9].

Nowadays, the majority of the studies devoted to retrial queueing systems are performed by matrix methods [10–12] and numerical analysis or computer simulation [1, 13–15].

Performance characteristics for retrial queueing systems under heavy, light loads and long delay conditions were studied by Falin, Aissani, etc. [2, 16, 17].

© Springer International Publishing AG 2017
V.M. Vishnevskiy et al. (Eds.): DCCN 2017, CCIS 700, pp. 406–416, 2017.
DOI: 10.1007/978-3-319-66836-9_34

But for the asymptotic analysis, they use explicit formulas, so they study only Poisson arrivals system.

Queueing models with negative calls or G-queues were introduced by E. Gelenbe [18,19]. A negative customer does not receive service, but it has the effect the system behaviour such that deleting the queue, breaking the server, etc. In the simplest situation, a negative customer delete a positive customer from the system, according to some strategy (for example, it eliminates all customers in the system, or it removes the customer from the head of queue or on the server, etc.). Negative arrivals have been interpreted as virus, computer attacks or disasters. The G-queues overview is presented in [20,21].

Retrial queues with negative customers are also investigated [22,23]. But it is known explicit formulas for stationary distributions of retrial queues only with Poisson arrival process. Usually for more complex retrial queues exact solutions cannot be obtained. But some approximations or asymptotic solutions can be proposed.

In paper [24], Markov and non-Markov probabilistic models of annihilating particles flows interact were suggested as infinite-server queuing systems with positive and negative applications. For the Markov model, the probability distribution of the number of positive applications in the model was found. Asymptotic results were submitted in case of high intensity incoming flows. The system with non-exponential service were studied using asymptotic analysis. It was shown that the probability distribution becomes Gaussian as the intensity of incoming flows grows; also the values of the distribution parameters were found.

In this paper, we propose the asymptotic analysis method under the heavy load condition [25,26], which gets approximate solutions for different types of retrial queues. Firstly, we apply it to the simplest model for the comparison asymptotic and exact distributions and making conclusions about its applicability area, which helps us to provide the research of more complex models, e.g. with non-Poisson arrivals.

The rest of the paper is organized as follows. In Sect. 2, the considered mathematical model is described and the process under study is introduced. In Sect. 3, the retrial queue is studied under a limit condition of heavy load. The theorem about the gamma distribution of the asymptotic characteristic function is proved. In addition, the formula for the system throughput is obtained. In Sect. 4, the exact formula for the characteristic function is derived and some numerical examples of the comparison of the asymptotic distributions with exact ones are presented. Conclusions about the asymptotic method applicability area are made.

2   Model Description

Consider a single server retrial queueing system $M/M/1$. Primary calls arrive from outside at the system according to Poisson Process with a rate λ. We say that this calls are positive. If a positive call finds the server free, it stays here with service time distributed exponentially with rate μ. Otherwise, the call goes to an orbit, where it stays during random time distributed by the exponential law with rate σ. After the delay, the call makes an attempt to reach the server again. If the server is free, the call gets the service, otherwise, the call instantly returns to the orbit. Also there are negative calls arrived according to Poisson Process with a rate γ. If a negative call finds the server busy (by a primary or repeated call), it cancels the served positive call, and both call leave the system. Otherwise, the negative call leaves the system. The system structure is presented in Fig. 1.

The arrival processes, the service times and the retrial times are assumed to be mutually independent.

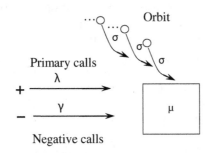

Fig. 1. Retrial queueing system $M/M/1$ with negative calls

Let $i(t)$ be the random process of the number of calls in the orbit and $k(t)$ be the random process that defines the server states as follows

$$k(t) = \begin{cases} 0, \text{if server are free at the moment } t, \\ 1, \text{if server is busy at the moment } t. \end{cases}$$

The aim of the research is to find the probability distribution of the number of calls in the orbit.

The process $i(t)$ is not Markovian, therefore we consider the two-dimensional continuous time Markov chain $\{k(t), i(t)\}$.

By $P_k(i) = P\{k(t) = k, i(t) = i\}$ denote the stationary probability distribution of the system states $\{k(t), i(t)\}$, where $k = 0 \ldots N$, $i = 0 \ldots \infty$. The considered process is Markovian, so, the following system of Kolmogorov equations for $P_n(i)$ can be written as

$$\begin{cases} -(\lambda + i\sigma)P_0(i) + (\mu + \gamma)P_1(i) = 0, \\ -(\lambda + \mu + \gamma)P_1(i) + \lambda P_0(i) + (i + 1)\sigma P_0(i + 1) + \lambda P_1(i - 1) = 0. \end{cases} \tag{1}$$

Let us introduce the partial characteristic functions

$$H_k(u) = \sum_{i=0}^{\infty} e^{jui} P_k(i), \tag{2}$$

where $j = \sqrt{-1}$ is an imaginary unit.

Substituting functions (2) into Eq. (1), the following system of equations is obtained

$$\begin{cases} -\lambda H_0(u) + j\sigma H_0'(u) + (\mu + \gamma)H_1(u) = 0, \\ -(\lambda + \mu + \gamma)H_1(u) - j\sigma e^{-ju}H_0'(u) + \lambda H_0(u) + \lambda e^{ju}H_1(u) = 0. \end{cases} \quad (3)$$

3   Asymptotic Analysis Under Heavy Load

By $\rho = \lambda/\mu$ denote the system load. The stationary regime of the retrial queue exists if $\rho < S$, where S is called as the system throughput.

Let us consider the system under the limit condition of the heavy load $\rho \to S$. The following theorem holds.

Theorem. The asymptotic characteristic function of the probability distribution of the number of calls in the orbit in the retrial queueing system of M/M/1 type with negative calls under the heavy load condition has the gamma distribution form

$$\lim_{\rho \to S} M e^{jw(S-\rho)i(t)} = \left(1 - \frac{jw}{\beta}\right)^{-s}, \quad (4)$$

where $S = 1 + \dfrac{\gamma}{\mu}$ is the system throughput, $\beta = \dfrac{1}{S}$, $s = 1 + \dfrac{\mu}{\sigma}\beta$.

To prove the theorem, we introduce the following notation

$$\begin{aligned} \lambda = (1 - \varepsilon)\mu, \quad u = \varepsilon w, \\ H_0(u) = \varepsilon F_0(w, \varepsilon), \quad H_1(u) = F_1(w, \varepsilon). \end{aligned} \quad (5)$$

Substitute expressions (5) into system (3).

$$\begin{cases} -(S - \varepsilon)\varepsilon F_0(w, \varepsilon) + j\dfrac{\sigma}{\mu}\dfrac{\partial F_0(w, \varepsilon)}{\partial w} + \left(1 + \dfrac{\gamma}{\mu}\right)F_1(w, \varepsilon) = 0, \\ \left[(S - \varepsilon)\left(1 - e^{jw\varepsilon}\right) + 1 + \dfrac{\gamma}{\mu}\right]F_1(w, \varepsilon) + (S - \varepsilon)\varepsilon F_0(w, \varepsilon) \\ \qquad - j\dfrac{\sigma}{\mu}e^{-j\varepsilon w}\dfrac{\partial F_0(w, \varepsilon)}{\partial w} = 0. \end{cases} \quad (6)$$

Let us consider expansions of the functions $F_k(w, \varepsilon)$ in the form

$$F_k(w, \varepsilon) = F_k(w) + \varepsilon f_k(w) + O(\varepsilon^2), \quad (7)$$

where $O(\varepsilon^2)$ is an infinitesimal value of order ε^2.

Substituting (7) into system (6), and writing equalities for members with equal powers of ε, we obtain the following system of equations

$$\begin{cases} j\dfrac{\sigma}{\mu}F_0'(w) + \left(1 + \dfrac{\gamma}{\mu}\right)F_1(w) = 0, \\ -\left(1 + \dfrac{\gamma}{\mu}\right)F_1(w) - j\dfrac{\sigma}{\mu}F_0'(w) = 0, \\ -SF_0(w) + j\dfrac{\sigma}{\mu}f_0'(w) + \left(1 + \dfrac{\gamma}{\mu}\right)f_1(w) = 0, \\ SjwF_1(w) - \left(1 + \dfrac{\gamma}{\mu}\right)f_1(w) + SF_0(w) + j\dfrac{\sigma}{\mu}jwF_0'(w) - j\dfrac{\sigma}{\mu}f_0'(w) = 0. \end{cases} \quad (8)$$

Summing all equations of system (6), we get the following equation

$$j\frac{\sigma}{\mu}\left(1 - e^{-jw\varepsilon}\right)\frac{\partial F_0(w,\varepsilon)}{\partial w} - \left(1 - e^{jw\varepsilon}\right)(S - \varepsilon)F_1(w,\varepsilon) = 0.$$

$$j\frac{\sigma}{\mu}\frac{\partial F_0(w,\varepsilon)}{\partial w} + e^{jw\varepsilon}(S - \varepsilon)F_1(w,\varepsilon) = 0.$$

Substitute expansions (7), we obtain two additional equations

$$\begin{cases} j\dfrac{\sigma}{\mu}F_0'(w) + SF_1(w) = 0, \\ j\dfrac{\sigma}{\mu}f_0'(w) + (jwS - 1)F_1(w) + Sf_1(w) = 0. \end{cases} \tag{9}$$

Thus, we have the system of six linearly dependent equations (8) and (9).

The characteristic function of the number of calls in the orbit in considered retrial queue is calculated as

$$H(u) = Me^{jui(t)} = H_0(u) + H_1(u).$$

Under the heavy load condition, $H(u)$ can be presented in the form

$$h(u) = \lim_{\rho \to S} Me^{jw(S-\rho)i(t)} = F_1(w) + O(\varepsilon). \tag{10}$$

Therefore, it is necessary to obtain the function $F_1(w)$ from Eqs. (8) and (9). The derivation is in three steps.

Step 1. From the first equation of (8), we get

$$j\frac{\sigma}{\mu}F_0'(w) = -\left(1 + \frac{\gamma}{\mu}\right)F_1(w). \tag{11}$$

Substitute (11) into the first equation of system (9).

$$-\left(1 + \frac{\gamma}{\mu}\right)F_1(w) + SF_1(w) = 0.$$

In this way we have the value of the system throughput.

$$S = 1 + \frac{\gamma}{\mu},$$

so the stationary regime of the retrial queue exists if $\lambda < \mu + \gamma$.

Step 2. From the second equation of (8), we get

$$j\frac{\sigma}{\mu}f_0'(w) = SF_0(w) - \left(1 + \frac{\gamma}{\mu}\right)f_1(w). \tag{12}$$

Substituting expressions (11), (12) into the last equation of system (9), we obtain that

$$SF_0(w) - (jwS - 1)F_1(w) = 0. \tag{13}$$

Step 3. Differentiating (13) and substituting (11), we have the following equation

$$(1 - jwS)F_1'(w) = jSF_1(w)\left(1 + S\frac{\mu}{\sigma}\right). \tag{14}$$

Obviously, the solution of (14) has the form

$$F_1(w) = C\left(1 - jwS\right)^{-1-S\mu/\sigma},$$

where $C = 1$ from the normalization requirement.

Substituting the last expression into (10), we obtain

$$h(u) = \left(1 - \frac{ju}{(S - \rho)\beta}\right)^{-s},$$

where $S = 1 + \dfrac{\gamma}{\mu}$ is the system throughput, $\beta = \dfrac{1}{S}$, $s = 1 + \dfrac{\mu}{\sigma}\beta$.

This completes the prove.

4   Numerical Analysis of Asymptotic Results

Here we present some numerical examples to demonstrate the applicability area of the obtained results. Using the inverse Fourier transformations, we compare probability distribution obtained by asymptotic analysis and exact one.

The exact formula for the characteristic function of the number of calls in the orbit in this system can be obtained as follows.

From the first equation of (3), we have

$$(\mu + \gamma)H_1(u) = \lambda H_0(u) - j\sigma H_0'(u), \tag{15}$$

Substitute (15) to the second equation of (3)

$$H_0(u)\left(-\lambda\frac{\lambda + \mu + \gamma}{\mu + \gamma} + \lambda + \frac{\lambda^2}{\mu + \gamma}e^{ju}\right)$$
$$-j\sigma H_0'(u)\left(-\frac{\lambda + \mu + \gamma}{\mu + \gamma} + e^{-ju} + \frac{\lambda}{\mu + \gamma}e^{ju}\right) = 0,$$

and perform some transformations

$$\frac{\lambda}{\mu + \gamma}H_0(u)\left(-\lambda - \mu - \gamma + \mu + \gamma + \lambda e^{ju}\right)$$
$$= \frac{j\sigma}{\mu + \gamma}H_0'(u)\left(-\lambda - \mu - \gamma + e^{-ju}(\mu + \gamma) + \lambda e^{ju}\right),$$

$$\lambda^2 H_0(u) = j\sigma H_0'(u)\left(-e^{-ju}(\mu + \gamma) + \lambda\right).$$

Integrating the last equation

$$\ln H_0(u) = -\frac{\lambda^2}{j\sigma} \int \frac{e^{ju} du}{(\mu + \gamma) - \lambda e^{ju}}.$$

Finally, the following solution is obtained

$$H_0(u) = C \left(\mu + \gamma - \lambda e^{ju}\right)^{-\frac{\lambda}{\sigma}}, \qquad (16)$$

where $C = const$.

Substitute (16) into (15).

$$H_1(u) = \frac{C\lambda}{\mu + \gamma} \left(\mu + \gamma - \lambda e^{ju}\right)^{-\frac{\lambda}{\sigma}} \left(1 + \frac{\lambda e^{ju}}{\mu + \gamma - \lambda e^{ju}}\right),$$

$$H_1(u) = \frac{C\lambda}{\mu + \gamma} \left(\mu + \gamma - \lambda e^{ju}\right)^{-\frac{\lambda}{\sigma} - 1}.$$

Then the characteristic function of the number of calls in the orbit has the following form

$$H(u) = H_0(u) + H_1(u) = C \left(\mu + \gamma - \lambda e^{ju}\right)^{-\frac{\lambda}{\sigma}} \left(1 + \frac{\lambda}{\mu + \gamma - \lambda e^{ju}}\right),$$

$$H(u) = C \left(\mu + \gamma - \lambda e^{ju}\right)^{-\frac{\lambda}{\sigma} - 1} \left(\mu + \gamma + \lambda(1 - e^{ju})\right). \qquad (17)$$

From the normalization requirement $H(0) = 1$, it is easy to show that

$$C = \frac{(\mu + \gamma - \lambda)^{\frac{\lambda}{\sigma} + 1}}{\mu + \gamma}.$$

Thus, we finally have

$$H(u) = \frac{\mu + \gamma + \lambda(1 - e^{ju})}{\mu + \gamma} \left(\frac{\mu + \gamma - \lambda e^{ju}}{\mu + \gamma - \lambda}\right)^{-\frac{\lambda}{\sigma} - 1}. \qquad (18)$$

For the comparison of the asymptotic and the exact distributions, we use Kolmogorov distance between respective distribution functions

$$d = \max_{i \geq 0} \left| \sum_{l=0}^{i} [\tilde{p}(l) - p(l)] \right|.$$

Here, $p(l)$ is a probability distribution calculated using the asymptotic formula and $\tilde{p}(l)$ is an exact distribution of the number of calls in the orbit.

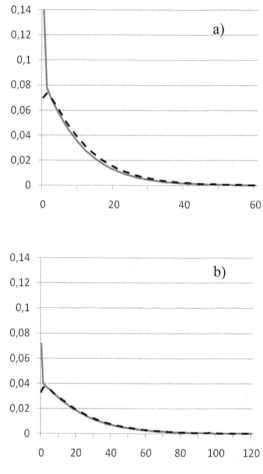

Fig. 2. Comparisons of the asymptotic (dashed line) and the exact (solid line) distributions for $\sigma = 10$ and: (a) $\rho = 0.9 \cdot S$; (b) $\rho = 0.95 \cdot S$

Let the service rate be $\mu = 1$, the arrival positive rate be $\lambda = \mu\rho$, the arrival negative rate be $\gamma = 1$. We variate parameters ρ and σ for the asymptotic analysis applicability area demonstrating. The comparison of the distributions is shown in Figs. 2 and 3. Values of Kolmogorov distance are presented in Tables 1 and 2.

Note, we obtain the same results of the numerical comparison for different values of service, arrival and impatient parameters.

We assume that values $d \leq 0.05$ are sufficient for good accuracy of approximations. Thus, the proposed asymptotic method can be applied for $\rho \geq 0.95 \cdot S$.

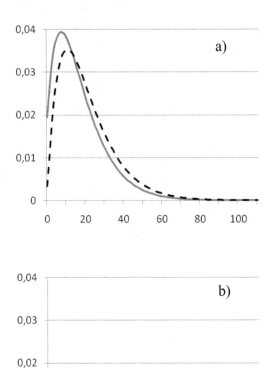

Fig. 3. Comparisons of the asymptotic (dashed line) and the exact (solid line) distributions for $\sigma = 1$ and: (a) $\rho = 0.9 \cdot S$; (b) $\rho = 0.95 \cdot S$

Table 1. Kolmogorov distances d for various values of the parameter ρ for $\sigma = 10$

ρ	$0.9 \cdot S$	$0.95 \cdot S$	$0.97 \cdot S$
d	0.086	0.043	0.026

Table 2. Kolmogorov distances d for various values of the parameter ρ for $\sigma = 1$

ρ	$0.9 \cdot S$	$0.95 \cdot S$	$0.97 \cdot S$
d	0.108	0.050	0.033

5   Conclusions

In the paper, we study the retrial queueing system of $M/M/1$ type with negative calls. We have obtained the expression for the characteristic function of the number of calls in the orbit. Moreover, we have proposed the asymptotic method for the retrial queue with negative calls research under the heavy load condition. We have proved that the asymptotic characteristic function has the gamma distribution (4). Also we have obtained the value of the system throughput. Finally, we have provided numerical analysis of comparison asymptotic and exact distributions that allows to make the conclusion about the asymptotic method applicability area.

In previous papers [25, 26], it was proved that the distribution of the number of calls in the orbit under the heavy load condition has the gamma form for different retrial queues without negative calls, i.e. retrial queues with non-Poisson arrivals. In future, we will consider different problems in retrial queus with negative calls. We plan to study multi-server retrial queues with negative calls, where there are additional problems (such as which device will the call come up, if there is a choice between an empty server and one with a negative call?). In addition, the retrial queues with negative calls with MMPP arrivals will be investigated. Also we plan to categorize types of negative calls according to kinds of atacks and disasters, and analyze possibilities of using negative calls models to cyberthreats protection (with examples of application).

Acknowledgments. The publication was financially supported by the Ministry of Education and Science of the Russian Federation (the Agreement number 02.a03.21.0008) and by RFBR according to the research project No.16-31-00292mol-a.

References

1. Artalejo, J.R., Gómez-Corral, A.: Retrial Queueing Systems, A Computational Approach. Springer, Heidelberg (2008)
2. Falin, G.I., Templeton, J.G.C.: Retrial Queues. Chapman & Hall, London (1997)
3. Artalejo, J.R., Falin, G.I.: Standard and retrial queueing systems: a comparative analysis. Rev. Mat. Complut. **15**, 101–129 (2002)
4. Wilkinson, R.I.: Theories for toll traffic engineering in the USA. Bell Syst. Tech. J. **35**(2), 421–507 (1956)
5. Elldin, A., Lind, G.: Elementary Telephone Trafic Theory. Ericsson Public Telecommunications, Stockholm (1971)
6. Gosztony, G.: Repeated call attempts and their efect on trafic engineering. Budavox Telecommun. Rev. **2**, 16–26 (1976)
7. Kuznetsov, D.Y., Nazarov, A.A.: Analysis of non-markovian models of communication networks with adaptive protocols of multiple random access. Avtomatika i Telemekhanika **5**, 124–146 (2001)
8. Roszik, J., Sztrik, J., Kim, C.S.: Retrial queues in the performance modelling of cellular mobile networks using MOSEL. Int. J. Simul. **6**, 38–47 (2005)
9. Tran-Gia, P., Mandjes, M.: Modeling of customer retrial phenomenon in cellular mobile networks. IEEE J. Sel. Areas Commun. **15**, 1406–1414 (1997)

10. Dudin, A.N., Klimenok, V.I.: Queueing system $BMAP/G/1$ with repeated calls. Math. Comput. Modell. **30**(3–4), 115–128 (1999)
11. Gómez-Corral, A.: A bibliographical guide to the analysis of retrial queues through matrix analytic techniques. Ann. Oper. Res. **141**, 163–191 (2006)
12. Diamond, J.E., Alfa, A.S.: Matrix analytical methods for $M/PH/1$ retrial queues. Stoch. Models **11**, 447–470 (1995)
13. Artalejo, J.R., Gómez-Corral, A., Neuts, M.F.: Analysis of multiserver queues with constant retrial rate. Eur. J. Oper. Res. **135**, 569–581 (2001)
14. Ridder, A.: Fast simulation of retrial queues. In: Third Workshop on Rare Event Simulation and Related Combinatorial Optimization Problems, pp. 1–5, Pisa, Italy (2000)
15. Artalejo, J.R., Pozo, M.: Numerical calculation of the stationary distribution of the main multiserver retrial queue. Ann. Oper. Res. **116**, 41–56 (2002)
16. Aissani, A.: Heavy loading approximation of the unreliable queue with repeated orders. In: Actes du Colloque Methodes et Outils d'Aide 'a la Decision (MOAD 1992), Bejaa, pp. 97–102 (1992)
17. Anisimov, V.V.: Asymptotic analysis of reliability for switching systems in light and heavy traffic conditions. In: Recent Advances in Reliability Theory: Methodology, Practice, and Inference, pp. 119–133 (2000)
18. Gelenbe, E.: Random neural networks with positive and negative signals and product form solution. Neural Comput. **1**(4), 502–510 (1989)
19. Gelenbe, E.: Queueing networks with negative and positive customers. J. Appl. Probab. **28**, 656–663 (1991)
20. Gelenbe, E.: The first decade of G-networks. Eur. J. Oper. Res. **126**, 231–232 (2000)
21. Do, T.V.: Bibliography on G-networks, negative customers and applications. Math. Comput. Modell. **53**(1–2), 205–212 (2011)
22. Anisimov, V.V., Artalejo, J.R.: Analysis of Markov multiserver retrial queues with negative arrivals. Queueing Syst. Theo. Appl. **39**(2/3), 157–182 (2001)
23. Berdjoudj, L., Aissani, D.: Martingale methods for analyzing the M/M/1 retrial queue with negative arrivals. J. Math. Sci. **131**(3), 5595–5599 (2005)
24. Nazarov, A.A., Farkhadov, M.P., Gelenbe, E.: Markov and non-Markov probabilistic models of interacting flows of annihilating particles. In: Dudin, A., Gortsev, A., Nazarov, A., Yakupov, R. (eds.) ITMM 2016. CCIS, vol. 638. Springer, Cham (2016). doi:10.1007/978-3-319-44615-8_25
25. Moiseeva, E., Nazarov, A.: Asymptotic analysis of RQ-systems M/M/1 on heavy load condition. In: Proceedings of the IV International Conference Problems of Cybernetics and Informatics, Baku, Azerbaijan, pp. 164–166 (2012)
26. Fedorova, E.: The second order asymptotic analysis under heavy load condition for retrial queueing system MMPP/M/1. In: Dudin, A., Nazarov, A., Yakupov, R. (eds.) ITMM 2015. CCIS, vol. 564, pp. 344–357. Springer, Cham (2015). doi:10.1007/978-3-319-25861-4_29

The Survey on Markov-Modulated Arrival Processes and Their Application to the Analysis of Active Queue Management Algorithms

Ivan Zaryadov[1,2], Anna Korolkova[1], Dmitriy Kulyabov[1,3(✉)],
Tatiana Milovanova[1], and Vladimir Tsurlukov[1]

[1] Department of Applied Probability and Informatics, RUDN University,
6 Miklukho-Maklaya Street, Moscow 117198, Russia
izaryadov@gmail.com, {akorolkova,dharma}@sci.pfu.edu.ru,
milovanova_ta@rudn.university, dober.vvt@gmail.com
[2] Institute of Informatics Problems of the Federal Research Center
"Computer Science and Control", Russian Academy of Sciences,
44-2 Vavilova Street, Moscow 119333, Russia
[3] Laboratory of Information Technologies, Joint Institute for Nuclear Research,
Joliot-Curie 6, Dubna, Moscow Region 141980, Russia

Abstract. The article is devoted to the application of Markov modulated arrival processes (Markov modulated Poisson process — MMPP,
Markov modulated Bernoulli process — MMBP and Markov modulated
fluid flow — MMFF) models to the analysis of Active Queue Management (AQM) algorithms (Random Early Detection (RED) family, for
example). The main ideas and properties of Markov modulated arrival
processes (MMAP) are presented as the brief description of RED-type
AQM algorithms. A review of the main results obtained with the help
of MMAP processes in the analysis of AQM algorithms models is made.
The authors formulated problems that also can be solved with the help of
MMAP processes when analysing the systems with RED-like algorithms.

Keywords: Markov modulated arrival processes · Markov modulated
Poisson process · Markov modulated Bernoulli process · Markov modu-
lated fluid flow · Active queue management · Random Early Detection
(RED)

1 Introduction

This article introduces a brief overview of Markov modulated arrival processes
(Markov modulated Poisson process (MMPP), Markov modulated Bernoulli
process (MMBP) (the discrete analogue of MMPP) and Markov modulated fluid
flow) and their application to active queue management (AQM) algorithms.

The structure of the paper is follows: the Sect. 1 is Introduction, the Sect. 2
is a brief overview of Markov modulated arrival processes, the Sect. 3 is devoted
to active queue management algorithms. The Sect. 4 introduces the review of

© Springer International Publishing AG 2017
V.M. Vishnevskiy et al. (Eds.): DCCN 2017, CCIS 700, pp. 417–430, 2017.
DOI: 10.1007/978-3-319-66836-9_35

the main results obtained with the help of MMAP processes in the analysis of AQM algorithms models. And the Sect. 5 — Conclusions — a brief summary of the content and purposes of the paper.

2 Markov Modulated Arrival Processes

If one wants to construct a mathematical model considering a heterogeneous character of incoming traffic (for example, burstiness) or service (failures, break-downs, different modes of service) the point processes whose arrival (service) rates vary randomly over time are needed. These processes with heterogeneous rate arise are appliable for mathematical modelling of broadband networks, some asynchronous transfer mode (ATM) services, hybrid wireless communication channels operation, or (as will be shown later in this article) active queue management algorithms.

The first works concerning queueing models with heterogeneous arrivals and service are the works of M. Eisen and M. Tainiter [1], U. Yechiali and P. Naor [2], M.F. Neuts [3] and P. Purdue [4]. In [1,2] the same M/M/1 queueing system (with heterogeneous arrivals and service in terms of [2]) was under investigation (we should note that the term "Markovian environment" and its general form — "random environment" are still used and relevant [5–7]). The rates of arrival and service change their values due to transition of external Markov process from one state to another.

In [1,2] the similar probability generation functions (PGF) for steady-state probability distribution of two-dimensional Markov process, describing the of customers in the system, as well as mean queue length and mean waiting times were obtained. But in [1] the exact solution for PGF and mean waiting time were obtained only for infinite service rate case. In [2] the level independent customers probability distribution was derived when ratios of incoming and service rates are the same for states of governing Markov process. In [3] (the term "extraneous phase changes" is used for random environment) besides M/M/1 queueing system the general case — M/G/1 system with extraneous irreducible Markov chain with m states was considered. Two approaches to the time dependent study of this queue were presented. One generalizes the imbedded semi-Markov process obtained by considering the queue immediately following departure points; the other approach exploits the relationship between this queue and branching processes. The equilibrium condition of the queue is obtained with the help of Perron-Frobenius theory of positive matrices. As a result the recurrence relations which yield the joint distribution of the phase state at time t, the queue length, the total number served and the virtual waiting time at t were obtained.

In [4] (the term "Markovian environment") for M/M/1 system with m-state irreducible continuous time governing Markov chain the busy period, equilibrium conditions and probabilities of the system being empty (for each state of Markovian environment) were obtained with the help of analytic matrix function.

The further works are [8–13]. In [11,13] the matrix analytical method for such system investigation was suggested.

In this chapter we will study the case when the only arrival process is under Markov modulation.

Markov modulated Poisson process (MMPP). The term "Markov modulated Poisson process" was introduced by M.F. Neuts in [12].

The MMPP is the doubly stochastic Poisson process whose arrival rate is defined by an m-state irreducible Markov process. Equivalently, a Markov-modulated Poisson process can be constructed by varying the arrival rate of a Poisson process according to an m-state irreducible continuous time Markov chain which is independent of the arrival process. When the Markov chain is in state i, arrivals occur according to a Poisson process of rate λ_i. The MMPP is parameterized by the m-state continuous-time Markov chain with infinitesimal generator Q and the m Poisson arrival rates $\lambda_1, \lambda_2, \ldots, \lambda_m$.

The work [14] is an excellent survey about Markov-modulated Poisson processes and queues with Markov-modulated input and gives a lot of references for more detailed information. Also the queueing systems with MMPP are described in [15].

Markov modulated Bernoulli process (MMBP) — is a discrete time analogue of MMPP, where a time is discretized in slots (with fixed length) and the number of time slots spent by arrival process in each state of m-state irreducible Markov chain follows geometric distribution. The MMBP is defined by Bernoulli process $X = \{X_i; i = 1, 2, \ldots\}$ where each X_i is a binary random variable which arrival probability p depends on Y_n state of Markov chain (Y_n, $n = \overline{1, m}$, m — number of states of Markov chain) with transition matrix P (the elements of P define the change of Bernoulli process arrival probability) on a discrete state space E. The theory and analysis of MMBP are presented in [16,17], the estimation of MMBP parameters is given in [17]. Also the theory of MMBP may be found in [18–21] as well as examples of application.

Markov modulated fluid flow — the rate of incoming flow (represents the incoming traffic as a stream with a finite rate) is modulated by external continuous time Markov process [19]. The main difference with the previous two approaches is that Markov modulated fluid flow approach based on theory of stochastic differential equations (SDE). The fluid flow models are consider to be appropriate approximation to systems (real-life models) where the number of arrival customers is relatively large (the arrival traffic has to be heavy with system load greater than 90%).

One of the first works concerning MMFF is [23] with superposition of several incoming flows with arrival rates governed by two-state Markov chains with application to computer network analysis. The work [24] suggests for steady-state analysis of MMFF the matrix-analytical approach, based on the theory of quasi birth-death processes, for which efficient solution methods are well known [29,30]

In [25] the algorithmic approach to numerical solution of steady state distribution for queueing flow models (a buffer of infinite (finite) size) with underlying time continuous Markov is suggested.

Also worth noting is that the fluid flow modulated by diffusion process or Brownian noise is usually called as second order fluid queue [22].

The MMFF queue approach may be applied for modelling of the IEEE 802.11 protocol [29], peer-to-peer file sharing [26], optical burst switching [31], ad hoc networks [27, 28], Internet of Things (IoT) [32].

3   Active Queue Management, RED-type Algorithms

According to RFC 7567 [34] active queue management (AQM) is considered as a best practice of network congestion avoidance (reducing) in Internet routers. The active queue management is a based on some rules (algorithms such as random early detection (RED) [35], Explicit Congestion Notification (ECN) [36], or controlled delay (CoDel) [33]) technique of intelligent drop of network packets inside a buffer associated with a network interface controller (NIC), when that buffer becomes full or gets close to becoming full.

We will consider only the case of RED algorithm and some its modifications. RED has the ability to absorb bursts and also is simple, robust and quite effective at reducing persistent queues.

The classic RED (random early detection or random early discard or random early drop) is a queueing discipline with two thresholds (Q_{min} and Q_{max}) and a low-pass filter to calculate the average queue size $\hat{Q}$ [35, 76]:

$$\hat{Q}_{k+1} = (1 - w_q)\hat{Q}_k + w_q\hat{Q}_k, \quad k = 0, 1, 2, \ldots, \tag{1}$$

where w_q, $0 < w_q < 1$ is a weight coefficient of the exponentially weighted moving-average and determines the time constant of the low-pass filter. As said in [35] RED monitors the average queue size and drops (or marks when used in conjunction with ECN) packets based on statistical probabilities $p(\hat{Q})$:

$$p(\hat{Q}) = \begin{cases} 0, & 0 \le \hat{Q} \le Q_{min}, \\ \dfrac{\hat{Q} - Q_{min}}{Q_{max} - Q_{min}}p_{max}, & Q_{min} < \hat{Q} \le Q_{max}, \\ 1, & \hat{Q} > Q_{max}, \end{cases} \tag{2}$$

p_{max} is the maximum level of packages to be dropped (marked or reset).

RED is more fair than tail drop when the incoming packet is dropped only if the buffer is full. Also RED does not possess a bias against bursty traffic that uses only a small portion of the bandwidth. But, as shown in [38], RED has a number of problems, one of which is that it need tuning and has a little guidance on how to set configuration parameters.

The RED modifications are well presented in [37], but we will specify some of them.

Weighted RED (WRED) [39, 48] — in this algorithm different probabilities for different types of traffic with different priorities (IP precedence, DSCP) and/or queues may be defined. The modification of WRED is Distributed Weighted RED (DWRED)[40].

Adaptive RED or active RED (ARED) [41, 48] — was designed in order to make RED algorithm (based on the observation of the average queue length) more or less aggressive. If the average queue length $\hat{Q}$ oscillates around Q_{min} minimum threshold then early detection considers to be aggressive. If the average queue length $\hat{Q}$ oscillates around Q_{max} threshold then early detection is being too conservative. The drop probability is changed by the algorithm according to how aggressively it senses it has been discarding traffic.

Robust RED (RRED) [42] — is proposed for TCP throughput improvement against DoS (Denial-of-Service) attacks, especially LDoS (Low-rate Denial-of-Service) attacks. The basic idea behind the RRED is to detect and filter out LDoS attack packets from incoming flows before they feed to the RED algorithm. When loss of a sent packet is detected by the source then there will be a transmit delay, so a packet which was sent within a short-range after a loss detection will be suspected to be an attacking packet. This is the basic idea of the detection algorithm of Robust RED (RRED).

EASY RED [43] — is a simpler variant of RED. the drop probability is defined not by average queue length but by instantaneous queue length. The reason is to inform the sender about congestion as soon as possible. The EASY RED parameters are the minimum threshold Q_{min} and the drop probability p_{drop}, which is a constant and used only when the instantaneous queue length is greater or equal to Q_{min}.

Stabilized RED (SRED) [44] — aims at stabilizing buffer occupation by estimating the number of active connections in order to set the drop probability as a function of the number of the active flows and of the instantaneous queue length.

Flow RED (FRED) [45, 48] — uses per-active-flow accounting (based on minimum and maximum limits on the packets that a each flow may have in the queue) to impose on each incoming flow a loss rate (depends on the degree of buffer usage by a flow), it also uses a more aggressive drop against the flows that violates the maximum bound. The state information about active connections also needs to be maintained in the routers

Balanced RED (BRED) [46, 48] — is proposed to regulate the bandwidth of a flow also by doing per-active-flow accounting for the buffer, similar to FRED but with a different approach: in BRED, two variables (the measures of the packet number for one flow in the buffer and the packet number accepted from this flow since the previous packet dropping) for each flow having packets in the buffer are maintained, which are. As a result the decision of packet drop or acceptance is based on before mentioned two flow state variables.

Dynamic RED (DRED) [47] — is proposed to discard packets with a load dependent probability. The drop probability is updated by employing an integral controller (the input of the controller is the difference between the average queue length and the target buffer level, the output is the drop probability).

The more information about RED and its modifications (Gentle RED (GRED), RED with In and Out (RIO), WRED with thresholds (WRT), Exponential RED (EXPRED), Double Slope RED (DSRED), Random Early Dynamic

Detection (REDD)), as other AQM algorithms (Random Exponential Marking (REM), Blue [53] and stochastic fair Blue (SFB), Adaptive virtual queue algorithm) is available at Sally Floyd webpage [49] (up to 2008 year), or in [48,50–52], new approaches to AQM [54].

4   The Application of Markov Modulated Arrival Processes Theory to the Active Queue Management

In this part the brief overview of articles with Markov modulated arrival processes application (since Internet traffic is known to be bursty and exhibits Long Range Dependence (LRD) characteristic, Poisson and Bernoulli processes failed to model bursty and correlated traffic in adequate manner) to AQM algorithms will be presented.

Markov modulated Poisson process approach. Markov Modulated Poisson Process (MMPP) is used for bursty and correlated traffic modelling due to the simplicity of its mathematical model [55,57,59,60]. It was shown that superposition of MMPP can be used to model variable packet traffic with Long Range Dependence [55,57]. In [55] superpositions of two-state Markovian sources is suggested as a very versatile tool for the modelling of variable packet traffic with Long Range Dependence (LRD). The article [57] gives a short survey of Internet traffic modelling and introduces MMPP model for traffic with LRD. One of the most recent works on this item is [58], where the queueing behavior of the Internet router employing priority based partial buffer sharing mechanism under synchronous self-similar traffic input is investigated with Markov modulated Poisson process modelling of incoming traffic. The authors consider the output port of the router as a multi-server queueing system with deterministic service times (service time is a packet length) — $MMPP/D/c/K$ (according to the Kendalls notation), where K is the number of wavelength channels for each fiber line of the router (which consists of N input fiber lines and N output fiber lines), c – the size of a wavelength converter pool ($0 \leq c \leq K$), dedicated to each output fiber line. By using the matrix-geometric methods and approximate Markovian model authors computed high priority and low priority packet loss probabilities, and mean length of non-critical and critical periods against the system parameters and traffic parameters.

In [56] the MMPP model was applied for modelling of exogenous stream, representing the superposition of all incoming UDP connections into a queue, for the problem of the stability and performance analysis of a system involving several TCP connections (TCP Tahoe and the TCP Reno) passing through a tandem of RED controlled queues (each of which has an incoming exogenous stream) was studied. As results the conditions for stability of the system and closed form expressions for the throughput of the TCP connections also as the mean sojourn times of the TCP and the exogenous streams were obtained.

In [61] the problem of service rate control (the optimal service rate) was considered ans it was shown that the optimal service rate was nondecreasing in the number of customers in the system and higher congestion levels warrant higher service rates. Also the authors shown that the optimal service rate is not necessarily monotone in the current arrival rate. The full version of this work is [62].

In [59] the queueing model for GRED-I (Gentle RED with instantaneous queue length) algorithm with different classes of bursty traffic is investigated. The aggregate traffic in proposed queueing model is defined by the superposition of 2-state MMPP. For each traffic class two individual thresholds are assigned in order to differentially control traffic injection rate. In [60] the analytical model for priority-based AQM with heterogeneous bursty traffic (Markov-Modulated Poisson Process) and non-bursty traffic (Poisson process) is discussed. The each traffic class has individual thresholds. For traffic injection rate control the Preemptive Resume (PR) priority scheduling mechanism is adopted to control. The expressions of the aggregate and marginal performance characteristics of the priority-based AQM system are obtained.

Even MMPP approach is regarded to be less complex, some scientists [68] consider MMPP models to disagree with the digitalised communication world, so Markov modulated Bernoulli process approach is more appropriate.

Markov modulated Bernoulli process approach. Internet traffic consists of traffic flows from various network applications and these flows may be classified at edge routers into different traffic classes in order to enable differentiated service to be applied (based on the traffic classes defined by router).

The methodology of bursty traffic sources modelling by the Markov Modulated Bernoulli Process with two states (MMBP-2) is presented in [63]. The proposed technique can be extended to an MMBP with m states. In this article the following queueing models were under study: the infinite buffer system with MMBP-2 batch arrivals, the infinite buffer system with a group of two identical MMBP-2 sources, finite buffer system with two MMBP-2 sources with different parameters. For all cases the queue length distribution in the framework of Markov theory was obtained.

In works [64–68] the Markov modulated Bernoulli process is applied to AQM modelling and analysis. The [64–67] consider analytical model proposed for a single MMBP-2 arrival process. In [64] a discrete-time stochastic queueing model for the performance analysis of congestion control mechanism based on Random Early Detection (RED) methodology with bursty and correlated traffic by using a two-state Markov-Modulated Bernoulli arrival process (MMBP-2) as the traffic source was presented. Two traffic classes were modelled by two-dimensional discrete-time Markov chain (each Markov chain dimension corresponds to a traffic class with its own RED parameters). The authors computed such performance characteristics as the mean system occupancy and mean packet delay, system throughput and the probability of packet loss (functions of the thresholds and maximum drop probability). In [65,66] the same model as in [64] but for multiple-class traffic with short range dependent traffic characteristics was considered. The results of this work — the analytical expressions for various performance metrics, the demonstration of the effect of input parameters on derived performance metrics.

Since the analytical models proposed in [64–67] were not able to represent traffic flows with different characteristics and precedence, the model based on superposition of multiple MMBP-2 sources (for aggregated Internet traffic from multiple traffic classes) for RED and WRED queue management schemes performance evaluation was introduced in [68].

Markov modulated fluid flow approach. It is commonly used to apply fluid models to AQM analysis, for example for the first time the fluid models for TCP and RED analysis were applied in [69–71]. In [69] the simple fluid analysis of TCP and other TCP-like congestion control mechanism was described. The authors modelled the window size behaviour as a Poisson Counter driven Stochastic Differential Equation and performed analysis. In [70] jump process driven Stochastic Differential Equations to model the interactions of a set of TCP flows and Active Queue Management routers in a network setting were used. As an application, the system with RED as the AQM policy was modelled and solved. By using fluid flow approach the role played by the RED configuration parameters on the behaviour of the algorithm in a network was explained. And in [71] the authors designed an AQM control system using the random early detection (RED) scheme by relating its free parameters such as the low-pass filter break point and loss probability profile to the network parameters. The guidelines for designing linearly stable systems subject to network parameters like propagation delay and load level were presented.

The fluid flow model approach for AQM analysis was also used in [72,73] and developed for Wiener fluid process governed by Poisson process (Poisson counter). The method of stochastization (randomisation) of one-step processes (birth-death processes) [74–79] is proposed for active queue management analysis in before mentioned model. In [75] the methodology of combinatorial and operator approaches of the method of one-step processes stochastization (randomisation) to RED modelling is introduced and discussed.

In [80] the wireless TCP fluid model (WTFM) (based on cross layers) was proposed for stability analysis of wireless network. The active queue management, abnormality of wireless channels and packets collisions were taken into consideration. According to obtained in [80] results the proposed WTFM model performs better than other schemes (the classical fluid model and the convex optimisation model) in comprehensive aspects on capturing the characteristic of the wireless network and computing complexity.

5   Conclusions

The brief history and the theory of Markov modulated arrival process and their applications to the modelling and analysis of RED-like active queue management algorithms were presented in the article.

The authors consider that not only Markov modulated arrival processes may be applied to the AQM modelling and analysis (as shown before), but the queueing systems with renovation (general renovation) [81–86] may be also applied (as shown, for example, in [87]) coupled with Markov modulated arrival or service processes.

Authors plan to expand the queueing model of RED algorithm with batch arrivals (it was considered in [88]) by Markov modulated arrival and service processes.

At the end, it is worthy of mention that the other approach to the analysis of behaviour of networks with burst traffic (for overload (congestion) control) may be applied — the method based on hysteretic thresholds load control [89, 90].

Acknowledgments. The work was financially supported by the Ministry of Education and Science of the Russian Federation (the Agreement No. 02.A03.21.0008) and partially supported by RFBR grants No. 15-07-03007, No. 15-07-03406 and No. 14-07-00090.

References

1. Eisen, M., Tainiter, M.: Stochastic variations in queuing processes. Oper. Res. **11**(6), 922–927 (1963)
2. Yechiali, U., Naor, P.: Queueing problems with heterogeneous arrivals and service. Oper. Res. **19**(3), 722–734 (1971)
3. Neuts, M.F.: A queue subject to extraneous phase changes. Adv. Appl. Probability **3**(1), 78–119 (1971)
4. Purdue, P.: The M/M/1 queue in a Markovian environment. Oper. Res. **22**(3), 562–569 (1974)
5. Falin, G.: The M/M/∞ queue in a random environment. Queueing Syst. **58**, 65–76 (2008)
6. Rykov, V., Tran, A.N.: On Markov reliability model of a system, operating in random environment. In: XXXI International Seminar on Stability Problems for Stochastic Models, pp. 114–116. IIP, Moscow (2013)
7. Andronov, A.M., Vishnevsky, V.M.: Markov-modulated continuous time finite Markov chain as the model of hybrid wireless communication channels operation. Autom. Control Comput. Sci. **50**(3), 125–132 (2016)
8. Neuts, M.F.: The M/M/1 queue with randomly varying arrival and service rates. Technical report No./77, Department of Statistics and Computer Science, University of Delaware, Newark DE, U.S.A. (1977)
9. Neuts, M.F.: Further results on the M/M/1 queue with randomly varying rates. Technical report No./78-4, Department of Statistics and Computer Science, University of Delaware, Newark DE, U.S.A. (1978)
10. Neuts, M.F.: Further results on the M/M/1 queue with randomly varying rates. OPSEARCH **15**(4), 158–168 (1978)
11. Neuts, M.F.: Matrix Geometric Solutions in Stochastic Models: An Algorithmic Approach. Johns Hopkins University Press, Baltimore (1981)
12. Neuts, M.F.: A versatile Markovian point process. J. Appl. Probability **16**(4), 764–779 (1979)
13. Neuts, M.F.: Structured Stochastic Matrices of M/G/1 Type and Their Applications. Marcel Dekker Inc., New York (1989)
14. Fisher, W., Meier-Hellstern, K.S.: The Markov-Modulated Poisson Process (MMPP) cookbook. Perform. Eval. **18**(2), 149–171 (1993)
15. Prabhu, N.U., Zhu, Y.: Markov-modulated queueing systems. Queueing Syst. **5**(1–3), 215–245 (1989)

16. Özekici, S.: Markov Modulated Bernoulli process. Math. Methods Oper. Res. **45**(3), 311–324 (1997)
17. Özekici, S., Soyer, R.: Bayesian analysis of Markov Modulated Bernoulli processes. Math. Methods Oper. Res. **57**(1), 125–140 (2003)
18. Perros, H.G.: An Introduction to ATM Networks. Wiley, New York (2001)
19. Ng, P.C.H., Boon-Hee, P.S.: Queueing Modelling Fundamentals: With Applications in Communication Networks. Wiley, New York (2008)
20. Ibe, O.: Markov Processes for Stochastic Modeling. Elsevier Science (2013)
21. Trivedi, K.S.: Probability and Statistics with Reliability, Queuing, and Computer Science Applications. Wiley, Hoboken (2016)
22. Asmussen, S.: Stationary distributions for Fluid Flow Models with or without Brownian Noise. Commun. Stat. Stochast. Models **11**(1), 21–49 (1995)
23. Anick, D., Mitra, D., Sondhi, M.M.: Stochastic theory of a data-handling system with multiple sources. Bell Syst. Tech. J. **61**(8), 1871–1894 (1982)
24. Ramaswami, V.: Matrix analytic methods for stochastic fluid flows. In: Teletraffic Engineering in a Competitive World. ITC – 16: International Teletraffic Congress, Edinburgh, 3a&3b, pp. 1019–1030 (1999)
25. Akar, N., Sohraby, K.: Infinite and finite buffer Markov fluid queues: a unified analysis. J. Appl. Probability **41**(2), 557–569 (2004)
26. Gaeta, R., Gribaudo, M., Manini, D., Sereno, M.: Analysis of resource transfers in peer-to-peer file sharing applications using fluid models. Perform. Eval. **63**(3), 149–174 (2006)
27. Bekker, R., Mandjes, M.: A fluid model for a relay node in an ad hoc network: the case of heavy-tailed input. Math. Methods Oper. Res. **70**(2), 357–384 (2009)
28. Latouche, G., Taylor, P.G.: A stochastic fluid model for an ad hoc mobile network. Queueing Syst. **63**, 109–129 (2009)
29. Arunachalam, V., Gupta, V., Dharmaraja, S.: A fluid queue modulated by two independent birthdeath processes. Comput. Math. Appl. **60**(8), 2433–4444 (2010)
30. Govorun, M., Latouche, G., Remiche, M.A.: Stability for fluid queues: characteristic inequalities. Stoch. Model **29**, 64–88 (2013)
31. Yazici, M.A., Akar, N.: Analysis of continuous feedback markov fluid queues and its applications to modeling optical burst switching. In: Proceedings of the 25th International Teletraffic Congress (ITC), pp. 1–8 (2013)
32. Tunc, C., Akar, N.: Markov Fluid Queue Model of an energy harvesting IoT device with adaptive sensing. Perform. Eval. **111**, 1–16 (2017)
33. Nichols, K., Jacobson, V.: Controlling queue delay. Commun. ACM **55**(7), 42–50 (2012)
34. Baker, F., Fairhurst, G.: IETF Recommendations Regarding Active Queue Management. RFC 7567, Internet Engineering Task Force (2015). https://tools.ietf.org/html/rfc7567
35. Floyd, S., Jacobson, V.: Random early detection gateways for congestion avoidance. IEEE/ACM Trans. Networking **4**(1), 397–413 (1993)
36. Ramakrishnan, K., Floyd, S., Black, D.: The Addition of Explicit Congestion Notification (ECN) to IP. RFC 3168, Internet Engineering Task Force (2001). https://tools.ietf.org/html/rfc3168
37. Korolkova, A.V., Kulyabov, D.S., Chernoivanov, A.I.: On the classification of RED Algorithms. Math. Inf. Sci. Phys. **3**, 34–46 (2009). Bulletin of Peoples' Friendship University of Russia
38. Jacobson, V., Nichols, K., Poduri, K.: RED in a Different Light. http://citeseerx.ist.psu.edu/viewdoc/summary?doi=10.1.1.22.9406

39. Class-Based Weighted Fair Queueing and Weighted Random Early Detection. http://www.cisco.com/c/en/us/td/docs/ios/12_0s/feature/guide/fswfq26.html

40. Cisco IOS Quality of Service Solutions Configuration Guide, Release 12.2. http://www.cisco.com/c/en/us/td/docs/ios/12_2/qos/configuration/guide/fqos_c.html

41. Floyd, S., Gummadi, R., Shenker, S.: Adaptive RED: an algorithm for increasing the robustness of RED's active queue management (2001). http://www.icir.org/floyd/papers/adaptiveRed.pdf

42. Changwang, Z., Jianping, Y., Zhiping, C., Weifeng, C.: RRED: Robust RED algorithm to counter low-rate denial-of-service attacks. IEEE Commun. Lett. **14**(5), 489–491 (2010)

43. Grieco, L.A., Mascolo, S.: TCP westwood and easy RED to improve fairness in high-speed networks. In: Carle, G., Zitterbart, M. (eds.) PfHSN 2002. LNCS, vol. 2334, pp. 130–146. Springer, Heidelberg (2002). doi:10.1007/3-540-47828-0_9

44. Ott, T.J., Lakshman, T.V., Wong, L.H.: SRED: Stabilized RED. In: Proceedings IEEE INFOCOM 1999, vol. 3, pp. 1346–1355. IEEE (1999)

45. Lin, D., Morris, R.: Dynamics of random early detection. Comput. Commun. Rev. **27**(4), 127–137 (1997)

46. Anjum, F.M., Tassiulas, L.: Balanced RED: an algorithm to achieve fairness in the internet. Technical Research Report (1999). http://www.dtic.mil/dtic/tr/fulltext/u2/a439654.pdf

47. Aweya, J., Ouellette, M., Montuno, D.Y.: A control theoretic approach to active queue management. Comput. Netw. **36**, 203–235 (2001)

48. Jun, H.X.: Variants of RED. http://www.ee.ust.hk/~heixj/publication/thesis/node37.html

49. Sally Floyd Website. http://www.icir.org/floyd/

50. Chrysostomoua, C., Pitsillidesa, A., Rossidesa, L., Polycarpoub, M., Sekercioglu, A.: Congestion control in differentiated services networks using Fuzzy-RED. Control Eng. Pract. **11**, 1153–1170 (2003)

51. Feng, W.-C.: Improving internet congestion control and queue management algorithms. http://thefengs.com/wuchang/umich_diss.html

52. Al-Raddady, F., Woodward, M.: A new adaptive congestion control mechanism for the internet based on RED. In: 21st International Conference on Advanced Information Networking and Applications, AINAW 2007 Workshops (2007)

53. Feng, W., Kandlur, D.D., Saha, D., Shin, K.G.: BLUE: a new class of active queue management algorithms. UM CSE-TR-387-99 (1999). https://www.cse.umich.edu/techreports/cse/99/CSE-TR-387-99.pdf

54. Baldi, S., Kosmatopoulos, E.B., Pitsillides, A., Lestas, M., Ioannou, P.A., Wan, Y.: Adaptive optimization for active queue management supporting TCP flows. In: 2016 American Control Conference (ACC), pp. 751–756 (2016)

55. Andersen, A., Nielsen, B.: A Markovian approach for modelling packet traffic with long-range dependence. IEEE J. Sel. Areas Commun. **16**(5), 719–732 (1998)

56. Sharma, V., Purkayastha, P.: Performance analysis of TCP connections with RED control and exogenous traffic. Queueing Syst. **48**(3), 193–235 (2004)

57. Muscariello, L., Mellia, M., Meo, M., Marsan, M.A., Cigno, R.L.: Markov Models of internet traffic and a new hierarchical MMPP model. Comput. Commun. **28**(16), 1835–1852 (2005)

58. Gudimalla, R.K., Perati, M.R.: Loss behavior of internet router with priority based self-similar synchronous traffic-multi server queueing system with Markovian input. OPSEARCH **54**, 283–305 (2017)

59. Wang, L., Min, G., Awan, I.: Stochastic modeling and analysis of GRED-I congestion control for differentiated bursty traffic. In: 21st International Conference on Advanced Information Networking and Applications (AINA 2007), pp. 1022–1030 (2007)
60. Wang, L., Min, G., Awan, I.: An Analytical model for priority based AQM in the presence of heterogeneous network traffic. In: 22nd International Conference on Advanced Information Networking and Applications (AINA 2008), pp. 93-99 (2008)
61. Kumar, R., Lewis, M.E., Topaloglu, H.: Dynamic service rate control for a single server queue with Markov modulated arrivals. Naval Logistics Res. **60**(8), 661–677 (2013)
62. Kumar, R., Lewis, M.E., Topaloglu, H.: Dynamic service rate control for a single server queue with Markov modulated arrivals (2013). https://arxiv.org/pdf/1307.2601.pdf
63. Ng, C., Yuan, L., Fu, W., Zhang, L.: Methodology for traffic modeling using two-state Markov-Modulated Bernoulli Process. Comput. Commun. **22**(13), 1266–1273 (1999)
64. Guan, L., Woodward, M.E., Awan, I.U.: Stochastic approach for modeling multi-class congestion control mechanisms based on RED in TCP/IP networks. In: The 2nd International Conference on the Performance Modelling and Evaluation of Heterogeneous Networks (HER-NETs 2004), pp. 361–369 (2004)
65. Guan, L., Awan, I.U., Woodward, M.E.: Stochastic modelling of random early detection based congestion control mechanism for bursty and correlated traffic. IEE Proc. Softw. **151**(5), 240–247 (2004)
66. Guan, L., Woodward, M.E., Awan, I.U.: Performance analysis of active queue management scheme for bursty and correlated multi-class traffic. In: The 19th International Teletraffic Congress (ITC 19, China), pp. 1001–1010 (2005)
67. Guan, L., Awan, I.U., Woodward, M.E., Wang, X.: Discrete-time performance analysis of a congestion control mechanism based on RED under multi-class bursty and correlated traffic. J. Syst. Softw. **80**(10), 1716–1725 (2007)
68. Lim, L.B., Guan, L., Grigg, A., Phillips, I.W., Wang, X.G., Awan, I.U.: RED and WRED performance analysis based on superposition of N MMBP arrival proccess. In: 24th IEEE International Conference on Advanced Information Networking and Applications (AINA), pp. 66–73 (2010)
69. Misra, V., Gong, W.-B., Towsley, D.: Stochastic differential equation modeling and analysis of TCP-window size behavior. In: Proceedings of Performance, pp. 42–50 (1999)
70. Misra, V., Gong, W.-B., Towsley, D.: Fluid-based analysis of a network of AQM routers supporting TCP flows with an application to RED. ACM SIGCOMM Comput. Commun. Rev. **30**(4), 151–160 (2000)
71. Hollot, C.V., Misra, V., Towsley, D., Gong, W.-B.: A control theoretic analysis of RED. In: Proceedings of IEEE Infocom (2001)
72. Korolkova, A.V., Kulyabov, D.S.: Mathematical model of the dynamic behavior of RED-like system parameters. Math. Inf. Sci. Phys. **1**, 54–64 (2010). Bulletin of Peoples' Friendship University of Russia
73. Velieva, T.R., Korolkova, A.V., Kulyabov, D.S., Dos Santos, B.A.: Model queue management on routers. Math. Inf. Sci. Phys. **2**, 81–92 (2014). Bulletin of Peoples' Friendship University of Russia

74. Velieva, T.R., Korolkova, A.V., Kulyabov, D.S.: Designing installations for verification of the model of active queue management discipline RED in the GNS3. In: 6th International Congress on Ultra Modern Telecommunications and Control Systems and Workshops (ICUMT), pp. 570–577. IEEE Computer Society (2015)

75. Korolkova, A.V., Kulyabov, D.S., Sevastianov, L.A.: Combinatorial and operator approaches to RED modeling. Math. Model. Geom. **3**, 1–18 (2015)

76. Korolkova, A.V., Velieva, T.R., Abaev, P.A., Sevastianov, L.A., Kulyabov, D.S.: Hybrid simulation of active traffic management. In: Proceedings 30th European Conference on Modelling and Simulation, pp. 685–691. ECMS, Regensburg, Germany (2016)

77. Hnatič, M., Eferina, E.G., Korolkova, A.V., Kulyabov, D.S., Sevastyanov, L.A.: Operator approach to the master equation for the one-step process. EPJ Web Conf. **108**, 58–59 (2015)

78. Eferina, E.G., Hnatich, M., Korolkova, A.V., Kulyabov, D.S., Sevastianov, L.A., Velieva, T.R.: Diagram representation for the stochastization of single-step processes. In: Vishnevskiy, V.M., Samouylov, K.E., Kozyrev, D.V. (eds.) DCCN 2016. CCIS, vol. 678, pp. 483–497. Springer, Cham (2016). doi:10.1007/978-3-319-51917-3_42

79. Korolkova, A.V., Eferina, E.G., Laneev, E.B., Gudkova, I.A., Sevastianov, L.A., Kulyabov, D.S.: Stochastization of one-step processes in the occupations number representation. In: Proceedings 30th European Conference on Modelling and Simulation, pp. 698–704. ECMS, Regensburg, Germany (2016)

80. Zhou, Z., Xiao, Y., Wang, D.: Stability analysis of wireless network with improved fluid model. J. Syst. Eng. Electron. **26**(6), 1149–1158 (2015)

81. Kreinin, A.: Queueing systems with renovation. J. Appl. Math. Stoch. Anal. **10**(4), 431–443 (1997)

82. Bocharov, P.P., Zaryadov, I.S.: Probability distribution in queueing systems with renovation. Math. Inf. Sci. Phys. **1–2**, 15–25 (2007). Bulletin of Peoples' Friendship University of Russia

83. Zaryadov, I.S., Pechinkin, A.V.: Stationary time characteristics of the $GI/M/n/\infty$ system with some variants of the generalized renovation discipline. Autom. Remote Control **70**(12), 2085–2097 (2009)

84. Zaryadov, I.S.: Queueing systems with general renovation. In: ICUMT 2009 – International Conference on Ultra Modern Telecommunications, pp. 1–6. IEEE, St.-Petersburg (2009)

85. Zaryadov, I.S.: The $GI/M/n/\infty$ queuing system with generalized renovation. Autom. Remote Control **71**(4), 663–671 (2010)

86. Zaryadov, I., Razumchik, R., Milovanova, T.: Stationary waiting time distribution in $G/M/n/r$ with random renovation policy. In: Vishnevskiy, V.M., Samouylov, K.E., Kozyrev, D.V. (eds.) DCCN 2016. CCIS, vol. 678, pp. 349–360. Springer, Cham (2016). doi:10.1007/978-3-319-51917-3_31

87. Zaryadov, I.S., Korolkova, A.V.: The application of model with general renovation to the analysis of characteristics of active queue management with Random Early Detection (RED). T-Comm: Telecommun. Transport **7**, 84–88 (2011)

88. Korolkova, A.V., Zaryadov, I.S.: The mathematical model of the traffic transfer process with a rate adjustable by RED. In: International Congress on Ultra Modern Telecommunications and Control Systems and Workshops (ICUMT), pp. 1046–1050. IEEE. Moscow, Russia (2010)

89. Abaev, P., Gaidamaka, Y., Samouylov, K., Pechinkin, A., Razumchik, R., Shorgin, S.: Hysteretic control technique for overload problem solution in network of SIP servers. Comput. Inform. **33**(1), 218–236 (2014)
90. Gaidamaka, Y., Pechinkin, A., Razumchik, R., Samouylov, K., Sopin, E.: Analysis of an M/G/1/R queue with batch arrivals and two hysteretic overload control policies. Int. J. Appl. Math. Comput. Sci. **24**(3), 519–534 (2014)

Control of System Dynamics and Constraints Stabilization

Robert Garabshevich Mukharlyamov[1,2]
and Marat Idrisovich Tleubergenov[1,2(✉)]

[1] Department of Theoretical Physics and Mechanics,
Peoples' Friendship University of Russia, Miklukho-Maklaya Str. 6,
Moscow 117198, Russia
{robgar,marat207}@mail.ru
[2] Institute of Mathematics and Mathematical Modeling,
Pushkina Str. 125, Almaty 050010, Kazakhstan

Abstract. The equations of classical mechanics used for describing a dynamical process of controlled systems containing different elements. The method of constructing differential equations of known partial integrals is used to stabilize the constraints imposed on the mechanical system dynamics which is described by Lagrange equations and Hamilton equations. The problem of constructing the dynamics equations with known properties of motion in the class of Ito stochastic differential equations was investigated by Tleubergenov M.I., Azhymbaev D.T. Assuming that some of the properties of the motion are known and the random perturbing forces belong to the class of processes with independent increments, Lagrange functions, Hamilton functions and Birkhoff functions can be constructed. Stability conditions for solutions of equations of dynamics with respect to the constraint equations are obtained, and an algorithm for constructing equations of constraint perturbations that guarantees the stabilization of constraints in the course of numerical solution is proposed. The problem of controlling the rectilinear motion of a cart with inverted pendulum is solved.

Keywords: Modeling · System · Differential · Equation · Dynamics · Stability · Construction · Solution · Control

1 Introduction

Modern methods for modeling the dynamics of complex systems imply that the required operating properties satisfied at the stage of setting up the equations of dynamics. To set up these equations and to investigate the kinematic and dynamic properties of controlled systems equations methods of classical mechanics [1–5] are used. The analogy of dynamic processes in a simple economic object with

This study was supported by the Russian Foundation for Basic Research, project 16-08-00558 A and the state task 3.1939.2014/A; design part.

the motion of a point of variable mass [3] makes it possible to solve problems of scheduling and managing the dynamics of production systems using the methods of analytical dynamics of systems with variable mass [6]. The motions of mechanical systems are described by differential-algebraic equations composed of equations of dynamics and constraint equations. A considerable problem in the numerical solution of such systems is the stabilization of constraints [7,8], which is formulated as a problem of limitation of deviations due to additional forces or adequate modification of constraint forces.

2   Statement of the Problem

We consider the problem of modeling the dynamics of a system with its kinematic properties and control objectives given by constraint equations. The control of dynamics should ensure that the constraint equations are satisfied for appropriate initial conditions and the constraints are stabilized under perturbations caused by deviations of the initial conditions and errors in the numerical solution of dynamics equations. This problem can be solved only if the constraint equations involve the partial integrals of the corresponding equations of system dynamics. In this case, the solution of the equations of dynamics must be asymptotically stable with respect to the functions that estimate the deviations from the constraint equations. The existing methods of constrain stabilization [7–10] are based on a linear combination of equations of constraints and their derivatives. The control constraint forces can be determined using the generic approach based on the construction of systems of differential equations given known partial integrals [11–13].

3   Construction of Dynamics Equations

Due to the dynamic analogies, methods of classical mechanics can be used for the analysis and synthesis of control systems [12]. The necessary kinematic properties and dynamics of the controlled system are described by the system of differential-algebraic equations, presented by constraints equations and, for example, the Lagrange equations

$$f(q,t) = 0, \quad f_q v + f_t = 0, \quad \varphi(q,v,t) = 0,$$
$$\frac{dq}{dt} = v, \frac{d}{dt}\left(\frac{\partial L}{\partial v}\right) - \frac{\partial L}{\partial q} = Q - \frac{\partial D}{\partial v} + Bu,$$
$$q(t_0) = q^0, \qquad v(t_0) = v^0,$$
$$q,v \in R_n, \qquad f \in R_m, \qquad \varphi \in R_r, \qquad u \in R_k,$$
$$B = B(q,v,t), m + r \leqslant k \leqslant n.$$

Here q, v are the vectors of generalized coordinates and velocities, Q, u are generalized forces and controlling, L is Lagrange function, D is dissipative function. The general solution of problem of control synthesis that provides constraints stabilization is proposed. For the purpose of constraint stabilization excessive coordinates $\tilde{q} = f(q,t)$, $\tilde{v} = f_q v + f$, $v' = \varphi(q,v,t)$ are introduced and consider the extended system to which correspond Lagrange function $\tilde{L} = \tilde{L}(x,y,t)$ and dissipative function $\tilde{D} = \tilde{D}(x,y,t)$, $x = (q,v)$, $y = (\tilde{q},\tilde{v},v')$, which satisfy the conditions $\tilde{L}(x,0,T) = L(x,t)$, $\tilde{D}(x,0,t) = D(x,t)$. For definition of constraint stabilization conditions we will present the expanded system equations in the form.

$$\frac{dx}{dt} = w(x,t) + G(x,t)y, \tag{1}$$

$$\frac{dy}{dt} = K(x,t)y, \tag{2}$$

$$y - g(x,t) = 0, \tag{3}$$

$$x(t_0) = x^0, \qquad y^0 = g(x^0,t). \tag{4}$$

4   Stability and Stabilization

The stability with respect to constraints equations we shall consider in sense of the following definitions.

Definition 1. The solution of the expanded system equations is stable with respect to constraints equations, if for any ε there exists such δ, that for any initial conditions $x(t_0) = x^0$, that satisfy the inequalities $\|y(t_0)\| \leqslant \delta$, for all $t > t_0$ the inequality $\|y(t)\| \leqslant \varepsilon$ fulfills.

Definition 2. The solution of the expanded system equations is asymptotically stable with respect to constraints equations, if it is stable and the condition $\lim_{t \to \infty} \|y(t)\| = 0$ is fulfilled.

The solution of the system of Eqs. (1) and (3) is asymptotically stable with respect to constraint equations $g(x,t) = 0$, if the trivial solution $y = 0$ of the system of constraint perturbation Eq. (2) has this property. The matrix K of the coefficients of Eq. (2) of constraint perturbations is determined by the choice of functions $\tilde{L}$, $\tilde{D}$, and it generally depends on the phase coordinates q,v of the original system. If matrix K is constant, then the trivial solution is asymptotically stable if the roots of the characteristic equation have negative real parts. In the general case, the stability conditions are determined directly by the method of Lyapunovs functions [14].

For the successful numerical simulation of the dynamics of controlled systems, the asymptotic stability of the solution of Eqs. (1) and (3) with respect to constraint equations $g(x,t) = 0$ is insufficient. The constraints will be stabilized only if the deviations of the numerical solution of the equations of dynamics satisfy the constraint equations with certain accuracy. Obviously, the estimation for deviations of the solution to the equations of dynamics depends on the numerical solution method.

Let differential Eq. (1) with (3) and the initial conditions (4) be solved using the simple difference scheme

$$x^{k+1} = x^k + \tau \tilde{v}(x^k, t_k),$$
$$\tilde{v}(x, t) = w(x, t) + G(x, t)y,$$
$$x^k = x(t_k),$$
$$\tau = t_{k+1} - t_k,$$
$$k = 0, 1, 2 \ldots$$

Then, expanding the function $y^{k+1} = g(x^{k+1}, t_{k+1})$ in powers of τ with regard to (1), (3) as

$$y^{k+1} = y^k + \left(\frac{\partial g}{\partial x}\right)^k \triangle x^k + \tau \left(\frac{\partial g}{\partial t}\right)^k + \frac{\tau^2}{2} g^{k(2)},$$

we obtain the bound

$$\|y^{k+1}\| \leqslant \|I_{m+r} + \tau K(x^k, t_k)\| \|y^k\| + \frac{\tau^2}{2} \|g^{k(2)}\|.$$

Hence, if $\|y^k\| \leqslant \varepsilon$, $\|I_{m+r} + \tau K(x^k, t_k)\| \leqslant \alpha \leqslant 1$, $\tau^2 \|g^{k(2)}\| \leqslant 2\varepsilon(1 - \alpha)$, then we have the inequality $\|y^{k+1}\| \leqslant \varepsilon$.

If Eqs. (1) and (3) are solved using the second order difference scheme

$$x^{k+1} = x^k + \triangle x^k, \qquad \triangle x^k = \tau(1 - \sigma)\tilde{v}^k + \tau \sigma \tilde{\tilde{v}}^k,$$
$$\tilde{v}^k = \tilde{v}(x^k, t_k + \alpha \tau), \qquad \tilde{x}_k = x_k + \alpha \tau \tilde{v}^k, \qquad (5)$$
$$\sigma > 0, \qquad \alpha > 0, \qquad k = 0, 1, 2, \ldots$$

where α and σ are constants and the conditions

$$2\alpha\sigma = 1, \qquad \|y^k\| \leqslant \varepsilon, \qquad \tau^3 \|g^{k(3)}\| \leqslant 6\varepsilon(1 - \beta),$$
$$\|I_{m+r} + \tau K_1(x^k, t_k) + \frac{1}{2}\tau^2 K_2(x^k, t_k)\| \leqslant \beta \leqslant 1, \qquad (6)$$
$$K_2(x^k, t_k) = K^2(x^k, t_k) + \frac{dK(x^k, t_k)}{dt},$$

hold for all $k = 0, 1, 2, \ldots$, then we have the inequality $\|y^{k+1}\| \leqslant \varepsilon$ (see [15]). If is a constant matrix, inequality (6) is simplified:

$$\|I_{m+r} + \tau K + \frac{1}{2}\tau^2 K\| \leqslant \beta \leqslant 1. \qquad (7)$$

The constraint stabilization conditions to be imposed on the coefficients of constraint perturbation Eq. (2) with the help of higher order finite difference schemes were obtained in [16]. Specifically, the solution of Eqs. (1) and (3) by a fourth order RungeKutta method can provide the stabilization of constraints if the following conditions are satisfied:

$$\tau^5 \|g^{k(5)}\| \leqslant (5!)\varepsilon(1 - \beta)\ddot{x},$$

$$\|I_{m+r} + \sum_{s=1}^{4} \frac{\tau^s}{s!}(K_s)^k\| \leqslant \beta < 1,$$

$$K_1 = K,$$

$$K_2 = \frac{dK}{dt} + K^2,$$

$$K_3 = \frac{d^2K}{dt^2} + 3\frac{dK}{dt} + K^3,$$

$$K_4 = \frac{d^3K}{dt^3} + 4\frac{d^2K}{dt^2} + 3\left(\frac{dK}{dt}\right)^2 + 6\frac{dK}{dt}K^2 + K^4.$$

At each computing step, the constraint stabilization conditions ensure that the constraint equations are satisfied with a given accuracy and the deviation errors are not accumulated. This makes it possible to use simple numerical methods for solving the equations of the controlled system dynamics.

Example. The solution of the system of first order differential equations

$$\frac{dx}{dt} = X(x, y), \qquad \frac{dy}{dt} = Y(x, y),$$

$$X(x, y) = -4cy - \frac{x}{x^2 + 16y^2}u(x, y),$$

$$Y(x, y) = cx - \frac{4y}{x^2 + 16y^2}u(x, y),$$

$$u(x, y) = k(x^2 + 4y^2 - 4),$$

satisfies the constraint equation $g(x, y) = (x^2 + 4y^2 - 4)/2 = 0$. The equation of constraint perturbations $dg/dt = /kg$ has an asymptotically stable trivial solution $g = 0$ for all $k > 0$. The system is numerically solved using the difference scheme

$$x^{k+1} = x^k + \tau X^k, \qquad y^{k+1} = y^k + \tau Y^k, \qquad t_{k+1} = t_k + \tau.$$

We assume that

$$\varepsilon = 0.001, \tau = 0.001, |x| \leqslant 2.1, |y| \leqslant 1.1, c = 1,$$

$$t_0 = 0, x^0 = 2, y^0 = 0, |g(x^0, y^0)| = 0 < \varepsilon.$$

Inequalities for α and k take the form $\alpha \leqslant 0.815$, $0 < k < 38217$. Assuming that $\alpha = 0.8$, we obtain the conditions for k: $200 < k < 1800$. The direct calculations show that, for $k = 50$ and $k = 2050$, the condition for the numerical solution stability is violated at $t = 0.023$: $|g^{23}| = 0.00011 > \varepsilon$ and $t = 0.069$: $|g^{69}| = 0.00011 > \varepsilon$, respectively. For $k = 300$, the stabilization conditions are satisfied: $|g| \leqslant 2.66E - 05$. The graphs of changes in the deviation of $g(t)$ from the constraint equation $g(x, y) = 0$, corresponding to the values $k = 50, 300, 2050$, are shown in Fig. 1.

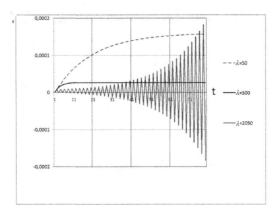

Fig. 1. The deviations from the constraint equation

5   Construction of Stochastic Differential Equations

The problem of modeling control dynamics systems in general formulation has a broader interpretation and relates to inverse problems of dynamics. The problem of reducing the system of differential equations to the structure of the Lagrange equations and systems of Helmholtz is studied in [17,18]. A method of constructing dynamics equations of systems with constraints in the form of Lagrange equations is proposed in [19]. Inverse problems of the Helmholtz dynamics systems are investigated in [20,21]. In [22,23] methods for solving inverse problems of dynamics are used to construct the equations of the dynamics of systems with constraints in generalized coordinates, in the canonical variables and equations in the form of Birkoff.

The task of constructing stochastic equations in the form of Lagrange equations is as follows. From the known constraint equations

$$g(q, v, t) = 0, \qquad g \in R_m, \qquad q \in R_n, \qquad v = \frac{dq}{dt}, \tag{8}$$

required to construct a stochastic differential equation in the form of Lagrange equations.

$$\frac{dq}{dt} = v, \qquad \frac{d}{dt}\left(\frac{\partial L}{\partial v_\nu}\right) - \frac{\partial L}{\partial q_\nu} = \sigma'_{\nu j}(q, v, t)\dot{\xi}^j, \tag{9}$$

$$\xi \in R_k, k \geqslant m,$$

so that the set (8) was a system of partial integrals of the system (9). Here, $\xi(t, \omega)$ is a system of random processes with independent increments.

We represent the random process as the sum [24] $\xi = \xi_0 + \int c(y)P^0(t, dy)$, where ξ_0 - Wiener process, P^0 - Poisson process, $P^0(t, dy)$ - the number of jumps of the process P^0 at the interval $[0, t]$, entering into a set of dy, $c(y)$ - vector function mapping space R_{2n} in the space values of $\xi(t)$ for all t. To solve the problem in the first phase build a system of Ito differential equations of the second order

$$\frac{dq}{dt} = v, \qquad \frac{dv}{dt} = f(q, v, t) + \sigma(q, v, t)\dot{\xi}, \tag{10}$$

for which the equality (8) is a set of partial integrals. Next, construct stochastic equations of Lagrangian, Hamiltonian or Birkoffian structure equivalent to the Eq. (10).

Previously, by the rule of stochastic Ito differentiation derivative is calculated:

$$\frac{dg}{dt} = \frac{\partial g}{\partial q}v + \frac{\partial g}{\partial v}f + \frac{\partial g}{\partial t} + S_1 + S_2 + S_3 + \frac{\partial g}{\partial v}\sigma\dot{\xi},$$

$$S_1 = \frac{1}{2}\sigma^T\frac{\partial^2 g}{\partial v^T \partial v},$$

$$S_2 = \left\{ g(q, v + \sigma c(y), t) - g(q, v, t) + \frac{\partial g}{\partial v}\sigma c(y) \right\} dy, \tag{11}$$

$$S_3 = \int [g(q, v + \sigma c(y), t) - g(q, v, t)] P^0(t, dy).$$

Equating the right-hand side of Eq. (3) expressions

$$\frac{dg}{dt} = A(g, q, v, t) + B(g, q, v, t)\dot{\xi},$$
$$A(0, q, v, t) = 0, \qquad B(0, q, v, t) = 0, \tag{12}$$

form the equations of perturbation constraints. Comparing Eqs. (11) and (12) we arrive at the relations

$$\begin{cases} \dfrac{\partial g}{\partial v}f = A - \dfrac{\partial g}{\partial t} - \dfrac{\partial g}{\partial q}\dot{x} - S_1 - S_2 - S_3, \\[2mm] \dfrac{\partial g}{\partial v}\sigma = B. \end{cases} \tag{13}$$

The vector f and the matrix σ is defined as the total solution of the Eq. (13)

$$f = s_0\left[\frac{\partial g}{\partial v}C\right] + \left(\frac{\partial g}{\partial v}\right)^{+}\left(A - \frac{\partial g}{\partial t} - \frac{\partial g}{\partial q}v - S_1 - S_2 - S_3\right),$$

$$\sigma = S_0\left[\frac{\partial g}{\partial v}C\right] + \left(\frac{\partial g}{\partial v}\right)^{+}B,$$

where s_0 - arbitrary scalar value, S_0 is an arbitrary diagonal matrix.

Thus the set (10) Ito differential equations of the second order, allowing partial integrals (8) has the form

$$\frac{dq}{dt} = v, \quad \frac{dv}{dt} = s_0\left[\frac{\partial g}{\partial v}C\right] + \left(\frac{\partial g}{\partial v}\right)^{+}\left(A - \frac{\partial g}{\partial t} - \frac{\partial g}{\partial q}v - S_1 - S_2 - S_3\right)$$

$$+ \left(S_0\left[\frac{\partial g}{\partial v}C\right] + \left(\frac{\partial g}{\partial v}\right)^{+}B\right)\dot{\xi} \tag{14}$$

After similar calculations for $\dfrac{d}{dt}\left(\dfrac{\partial L}{\partial v_\nu}\right)$ by the rule of Ito stochastic differentiation

$$\frac{d}{dt}\left(\frac{\partial L}{\partial v_\nu}\right) = \frac{\partial^2 L}{\partial v_\nu \partial t} + \frac{\partial^2 L}{\partial v_\nu \partial q_k}v_k + \frac{\partial^2 L}{\partial v_\nu \partial v_k}\frac{dv_k}{dt} + \tilde{S}_{1\nu} + \tilde{S}_{2\nu} + \tilde{S}_{3\nu},$$

arrive at the system

$$\frac{dq}{dt} = v, \quad \frac{\partial^2 L}{\partial v_\nu \partial t} + \frac{\partial^2 L}{\partial v_\nu \partial q_k}v_k + \frac{\partial^2 L}{\partial v_\nu \partial v_k}\frac{dv_k}{dt} - \frac{\partial L}{\partial q_\nu}$$
$$+ \tilde{S}_{1\nu} + \tilde{S}_{2\nu} + \tilde{S}_{3\nu} = \sigma'_{\nu j}(q,v,t)\dot{\xi}^j, \tag{15}$$

From a comparison of the expressions (14) and (15) the conditions that must be imposed on the Lagrangian L and the matrix σ':

$$\frac{\partial^2 L}{\partial v_\nu \partial v_k} = \delta^k_\nu,$$

$$\frac{\partial L}{\partial q_\nu} + \frac{\partial^2 L}{\partial v_\nu \partial t} + \frac{\partial^2 L}{\partial v_\nu \partial q_k}v_k + \tilde{S}_{1\nu} + \tilde{S}_{2\nu} + \tilde{S}_{3\nu} = f_\nu,$$

$$\sigma' = S_0\left[\frac{\partial g}{\partial v}C\right] + \left(\frac{\partial g}{\partial v}\right)^+ B.$$

6   Applications

The proposed methods are used for solving problems of control of production, logistics and technical systems.

A. *The control problem for discrete adaptive optical systems element.*

Mirrors element is simulated by a rigid body with six degrees of freedom and controlled by three parallel forces applied to the points of rigid body at which the mirror is attached. The mirror moves along the guides.

B. *Modeling of electromechanical system.*

In the control problem of electromechanical system, consisting of power supply unit and direct current motor, which controls crank mechanism, the variable voltage is used for control. The dynamics equations and the constraint equations are formulated. The solution of differential algebraic equations and phase portraits plotting are performed using the integrative system of computing symbol mathematics Maple.

C. *A control problem of a wheel systems motion.*

A control problem of a wheel systems movement along a given trajectory $x = -k_1 t$, $y = 0$ with avoidance of moving bodies is resolved. Three-wheel system is controlled by torque moments applied to back axle wheels. In order to formulate the trajectory set qualitative theorys inverse problem of differential equations is used. System dynamics is described by Voronets equations. The system is moving with avoidance two moving obstacles, constrained curves

$$(q_1 - 2 + k_3 t)^2 + 4q_2^2 = 1,$$

$$\frac{1}{4}(q_1 + 1 - k_4 t)^2 + \frac{16}{9}(q_2 + \frac{3}{2})^2 = 1.$$

The trajectory is obtained as a result of solution of the dynamics equations at given initial conditions (see Fig. 2).

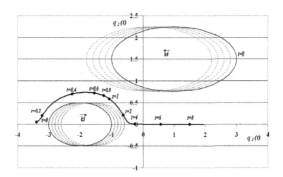

Fig. 2. The trajectory of the center of the system.

D. Management of tire enterprise consisting of two plants.

The management problem of enterprise for the production of tires, which consists of two factories: the factory of truck tire and of car tire factory. We investigate the dynamics of the necessary resources. As control functions considers the cash inflow on acquisition for the necessary equipment.

E. The problem of control for inverted pendulum on a movable base.

In a uniform gravitational field, a cart of mass m_1 can move linearly along the horizontal axis Ox of the Cartesian coordinate system under the action of force **F**. The position of the cart on the axis Ox is determined by the coordinate x of point O_1, which has a hinged uniform rod O_1A of length $2l$ and mass m_2 (Fig. 3). It is required to determine the magnitude F of force **F** and torque M applied to the rod needed to move the end A of the rod along a predetermined curve. The system dynamics is described by the equations

$$\frac{dx}{dt} = v, \qquad \frac{d\varphi}{dt} = \omega,$$

$$\frac{dv}{dt} = \frac{m_2 \cos \varphi (4l\omega^2 - 3g \sin \varphi)}{lN(\varphi)} + \frac{4}{N(\varphi)} F + \frac{3 \sin \varphi}{lN(\varphi)} M,$$

$$\frac{d\omega}{dt} = \frac{3 \cos \varphi}{lN(\varphi)} (lm_2\omega^2 \sin \varphi - (m_1 + m_2)g) + \frac{3 \sin \varphi}{lN(\varphi)} F + \frac{3(m_1 + m_2)}{m_2 l^2 N(\varphi)} M,$$

$$N(\varphi) = 4m_1 + m_2(1 + 3\cos^2 \varphi),$$

where φ is the rod slope angle with respect to the axis Ox and g is the acceleration due to gravity. Constraints corresponding to the motion of a point A on the curve described by the equations

$$x + 2l \cos \varphi - a(t) = 0, \qquad 2l \sin \varphi - b(t) = 0.$$

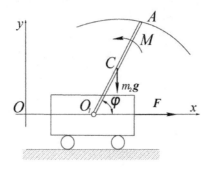

Fig. 3. Control for inverted pendulum on a movable base.

Believing

$$f_1 = x + 2l \cos \varphi - a(t), \qquad f_2 = 2l \sin \varphi - b(t).$$

the constraint perturbation equations can be represented by a linear system

$$\frac{df_i}{dt} = \dot{f}_i, \qquad \frac{\dot{f}_1}{dt} = -k_{11} f_1 - k_{12} \dot{f}_1, \qquad \frac{\dot{f}_2}{dt} = -k_{21} f_2 - k_{22} \dot{f}_2,$$
$$k_{ij} > 0, \qquad i,j = 1,2.$$

The problem of describing the point of the text is composed from straight segments and arcs of ellipses is solved (Fig. 4).

Fig. 4. Movement of a cart with inverted pendulum along a curve.

7   Conclusions

The results of theoretical research and numerical experiments confirm the efficiency of the proposed methods of stabilization of constraints that describe the system dynamics and the constraints that restrict the changes of the coordinates and velocities.

References

1. Olson, H.F.: Dynamical Analogies. Van Nostrand, New York (1943)
2. Layton, R.A.: Principles of Analytical System Dynamics. Springer, New York (1997)
3. Sirazetdinov, T.K.: A dynamical predictive model and optimal control of an economic object. Izv. Vuzov, Aviatsionnaya Tekhnika **4**, 3–8 (1972)
4. Meiser, P., Enge, O., Freudenberg, H., Kielau, G.: Electromechanical interactions in multibody systems containing electromechanical drives. Multibody Sys.Dyn. **1**, 281–302 (1997)
5. Mukharlyamov, R.G.: Modeling dynamic processes of various nature. In: Problems of Analytical Mechanics and Stability Theory (Nauka, Moscow, 2009), pp. 310–324 (2009)
6. Mukharlyamov, R.G.: Modeling the dynamics od simple economic objects as systems with program constraints. Ser. Fiz.-Mat. Nauki **1**, 25–34 (2007). Bulletin of Peoples' Friendship University of Russia
7. Baumgarte, J.: Stabilization of constraints and integrals of motion in dynamical systems. Comput. Math. Appl. Mech. Eng. **1**, 1–16 (1972)
8. Ascher, U.M., Chin, H., Petzold, L.R., Reich, S.: Stabilization of constrained mechanical systems with DAEs and invariant manifolds. J. Mech. Struct. Mach. **23**, 135–158 (1995)
9. Rentrop, P., Strehmel, K., Weiner, R.: Ein Uberblick uber Einschritten zur numerischen Integration in der technischen Simulation. GAMM Mitt. Berlin **19**(1), 9–43 (1996)
10. Amirouche, F.: Fundamentals of Multibody Dynamics: Theory and Applications. Springer, Birkhaser, Boston (2005)
11. Erugin, N.P.: Construction of the entire set of systems of differential equations that have a given integral curve. J. Appl. Math. Mech. **21**, 659–670 (1952)
12. Mukharlyamov, R.G.: On the construction of differential equations of optimal motion on the given manifold. Differ. Equat. **7**, 1825–1834 (1971)
13. Galiullin, A.S.: Methods for Solving Inverse Problems of Dynamics. Nauka, Moscow (1986)
14. Mukharlyamov, R.G.: On the construction of the set of differential equations of stable motion given an integral manifold. Differ. Equ. **5**, 688–699 (1969)
15. Mukharlyamov, R.G.: On the numerical solution of differential algebraic equations. Ser. Appl. Mat. Inform. **1**, 20–24 (1999). Bulletin of Peoples' Friendship University of Russia
16. Mukharlyamov, R.G., Beshaw, A.W.: Solving differential equation of motion for constrained mechanical systems. Ser. Mat. Inform. Fiz. **3**, 81–92 (2013). Bulletin of Peoples' Friendship University of Russia
17. Santilli, R.M.: Foundations of Theoretical Mechanics I: The Inverse Problem in Newtonian Mechanics, 266 pages. Springer, Heidelberg (1978)

18. Santilli, R.M.: Foundations of Theoretical Mechanics II: Birkhoffian Generalization of Hamiltonian Mechanics, 370 pages. Springer, Heidelberg (1982)
19. Mukharlyamov, R.G.: Reduction of dynamical equations for the systems with constraints to given structure. J. Appl. Math. Mech. **71**(3), 401–410 (2007). Elsevier
20. Galiullin, A.S., Tuladhar, B.M.: An Introduction to the Theory of Stability of Motion, 94 pages. Katmandu University (2000)
21. Galliulin, A.S., Gafarov, G.G., Malaishka, R.P., Hvan, A.M.: Gelmgolce, Bircgoff, Nambu systems analytical dynamics Moscow, 384 pages. Editorial Office Advances of Physical Sciences (1997). (in Russian)
22. Tleubergenov, M.I., Ashimbayev, D.T.: On the construction of the set of stochastic differential equations from a given integral manifold, independent of the velocity. Ukrainian Math. J. **62**(7), 1002–1009 (2010)
23. Tleubergenov, M.I., Ashimbayev, D.T.: The solution of the recovery with degenerate diffusion method of separation. J. Appl. Math. Mech. Ser. Mat. Inform. Fiz. **2**, 81–92 (2011)
24. Pugachev, V.S., Sinitsyn, I.N.: Stochastic Differential Systems: Analysis and Filtration, 632 pages (1990)

Multi-state Diagnostics for Distributed Radio Direction Finding System

Dmitry Aminev[1], Alexander Zhurkov[2], and Dmitry Kozyrev[1,3(✉)]

[1] V.A. Trapeznikov Institute of Control Sciences of Russian Academy of Sciences,
65, Profsoyuznaya Street, Moscow 117997, Russia
`aminev.d.a@ya.ru`
[2] National Research University "Higher School of Economics", Moscow, Russia
`petrovyc@gmail.com`
[3] RUDN University, Moscow, Russia
`kozyrevdv@gmail.com`

Abstract. We consider a distributed Radio Direction Finding System (RDFS) and the structure of its hardware. A classical diagnostic model of the distributed RDFS according to the binary criterion is presented, which includes diagnostic graphs and a set of tests. We propose a theoretical basis for diagnosing of the distributed RDFS according to the n-dimensional criterion. An example of implementation of the diagnostic approach to the distributed RDFS for states "failure", "deterioration", "normal" is presented.

Keywords: Technical condition monitoring · Reliability · Diagnostics · Multicriteriality · Radiotechnical system · Distributed communication network · Direction finding · Topology

1 Introduction

Despite the rapid development of global navigation systems for satellite positioning, radio direction finding is widely used nowadays, since in remote and underdeveloped areas, the undetectable automatic direction finders (ADF) are one of the main means for providing flights [1].

The distributed Radio Direction Finding System (RDFS) [2,3] which network topology type is a multilevel star, consists of the equipment of the local dispatching center (LDC), communication channels and unattended radio technical terminals (URT), which can be remoted from the LDC at distances up to hundreds kilometers (see Fig. 1a). The general configuration of the equipment of the URT and the LDC is shown in Fig. 1b.

This work has been financially supported by the Russian Science Foundation and the Department of Science and Technology (India) via grant 16-49-02021 for the joint research project by the V.A. Trapeznikov Institute of Control Sciences and the CMS College Kottayam.

© Springer International Publishing AG 2017
V.M. Vishnevskiy et al. (Eds.): DCCN 2017, CCIS 700, pp. 443–452, 2017.
DOI: 10.1007/978-3-319-66836-9_37

In general, the LDC equipment consists of the following main units: the data processing center (DPC), the channeling equipment (CE) for communication with the URT, the remote dispatching controller (RDC), the storage system (SS) for storing the direction-finding and service information, and the secondary power system (SPS). The URT equipment includes an automatic radio direction finder, a digital signal processing (DSP) and signal control unit, CE for communication with the LDC, and a SPS [4].

Each main unit can have several printed-board assemblies (PAs): DPC and DSP can consist of the main processor, backup, interface and controller circuit cards; CE can contain PAs for routers, distributors, converters for different types of communication lines; RDC includes PA of indication, input, display, and displays; SPS can have main and backup PAs for various feeding voltages; SS can have main and backup solid-state and HDD drives, buffering PAs; ADF can have a PA for antenna control, filtering and ADC.

Since the importance of operability of RDFS is high, and remote terminals are unattended, the problem of ensuring its reliable operation becomes vital [5], which is ensured, among others, by periodic and systematic diagnosis of components and modules of RDFS that can have several technical states (TSs).

The works of Russian scientists Parkhomenko, P.P., Vedeshenkov, V.A., Uvaysov, S.U. as well as foreign scientists such as Brule, D.D., Russell, J.D., Kime, C.R. etc. are devoted to the problems of diagnostic modeling of distributed RDFSs and radio-electronic equipment [4]. Despite the significant contribution of these scientists, the problems of diagnostics of the distributed RDFS are not sufficiently investigated. In particular, the problems of diagnosing complex distributed radio-technical systems with many TSs at various levels of their hierarchy (from the network topology to individual electronic components) have not been considered. Therefore, the task is urgent to develop new concepts for diagnosing RDFS, taking into account many technical conditions.

2   Diagnosis of a Distributed Radio Direction Finding System Based on Classical Model

The classical diagnostic model [6] based on topology and structure of a distributed RDFS synthesizes diagnostic graphs $G_1(U, W)$, $G_2(U, W)$, $G_3(U, W)$, and sets of tests $T = \{t_1, t_2, ..., t_p\}$ which form the values of the branches w (diagnostic links) as a function of the nodes $U : \{w_1, ..., w_N\} = T(U_1...U_m)$.

Diagnostic graphs for the RDFS of extended topology and generalized structure (see Fig. 1), consisting of a local dispatching center and N unattended terminals, some of which are connected directly to the LDC, and the rest via terminals, are shown in Fig. 2, with explanation presented in Table 1. The nodes of the topology graph $G_1(U, W)$ correspond to URT and LDC, and the arcs between the nodes correspond to the verification links and communication links. Here u_i is the verifier, $u(j)$ is the component being verified, "–" is the absence of diagnosability, "w" is the diagnostic link. In accordance with the structure of the equipment of URT and LDC (Fig. 1b), their diagnostic graphs $G_2(U, W)$

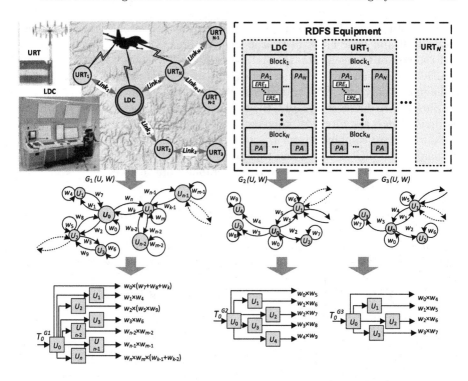

Fig. 1. Distributed RDFS, diagnostic graphs and diagnostic schemes for the generalized topology and structure

Table 1. Diagnostic links of distributed RDFS components

$G_1(U,W)$	u_i/u_j	u_0	u_1	u_2	u_3	u_{n-2}	u_{n-1}	u_n
LDC	u_0	w_0	w_1	w_2	w_3			w_n
URT$_1$	u_1	w_7	w_4					
URT$_2$	u_2	w_8		w_5				
URT$_3$	u_3	w_9			w_6			
URT$_n$	u_{n-2}					w_{m-2}		w_{n-2}
URT$_{n-1}$	u_{n-1}						w_{m-1}	w_{k-1}
URT$_{n-2}$	u_n	w_k				w_{k-2}	w_{n-1}	w_m

$G_2(U,W)$	u_i/u_j	u_0	u_1	u_2	u_3	u_4
DPC	u_0	w_0	w_1	w_2	w_3	w_4
CE	u_1		w_5	w_6		
SPS	u_2			w_7		
SD	u_3				w_8	
RDC	u_4					w_9

$G_3(U,W)$	u_i/u_j	u_0	u_1	u_2	u_3
DSP	u_0				
CE	u_1				
SPS	u_2				
ADF	u_3				

and $G_3(U,W)$, are constructed, which are subgraphs of the graph $G_1(U,W)$. The nodes of the graphs $G_2(U,W)$ and $G_3(U,W)$ correspond to the components of URT and LDC equipment, and the arcs correspond to verification links and interface links.

As it can be seen from the graphs and the table, each node of the distributed RDFS diagnoses itself and its neighbors, similarly to the Preparata-Metze-Chien model [7]. In the URT, the power supply system diagnoses only itself, CE diagnoses itself and DSP block in transit from the link, DSP unit diagnoses itself and the rest of the elements, the ADF does not self-diagnose. All the elements of LDC equipment have self-diagnostics and are controlled by a processor that is diagnosed only by CE by transit over the communication link.

Similarly to diagnostic graphs, the test complex T for a distributed RDFS can be divided into tests for T^{G1} topology, LDC equipment T^{G2}, URT equipment T^{G3}, which are combined from the tests for each of its nodes. At that, tests for analog and digital nodes, and topologies have their own characteristics. The result of the test is the diagnosis of the operability of the object.

3   Diagnostics According to Binary Criterion

The diagnostic test for the topology of the distributed RDFS in general form and the nodes of its equipment is determined by the following expression:

$$W^{G1} = \prod_{i=0}^{i=m} \mathrm{w}_i^{G1} = W^{G2} \times W_1^{G3} \times \dots \times W_m^{G3};$$

(1)

$$W^{G2} = \prod_{i=0}^{i=7} \mathrm{w}_i^{G2}; \quad W^{G3N} = \sum_{i=0}^{i=9} \mathrm{w}_i^{G3N};$$

$$\mathrm{w}_i^{G2} = f(PA_1^{\mathrm{w}_i^{G2}}, \dots, PA_N^{\mathrm{w}_i^{G2}}); \quad \mathrm{w}_i^{G3} = f(PA_1^{\mathrm{w}_i^{G3}}, \dots, PA_N^{\mathrm{w}_i^{G3}});$$

where $w_i \in \{0; 1\}$, $W^{G1} \in \{0; 1\}$—a final technical state assessment for the whole RDFS, $W^{G2} \in \{0; 1\}$—assessment of TS of LDC, $W^{G3N} \in \{0; 1\}$—N-th URT. Such test W^{G1} indicates that a failure of a LDC or at least one of URTs results in the failure of the entire RDFS.

Practical implementation of the tests consists in transferring the equipment to the testing mode, generating test effects T for the blocks and the PAs, and analyzing the parameters of the electronic analog $w_a^{G2,G3}$ and computational $w_d^{G2,G3}$ modules registered by the automatic system for technical diagnostics. In addition, temperature, humidity, smoke, etc. can be recorded in the room of an unattended terminal.

Tests $w_a^{G2,G3}$ and $w_d^{G2,G3}$ have their own peculiarities: for the power supply system, it is important to match the output voltage levels to the permissible ranges of values; for ADF—phase shifts and frequencies of converted radio signals matter; for DSP, the time duration for the execution of operations, the correct operation of the algorithm, the correspondence of the structures of the data being processed are important; CE can be diagnosed by test transactions via communication links. Measuring the voltage levels or currents of certain signals at control points can also show the operational status of both analog and digital PAs. Conditionally, the PA tests of the types $w_a^{G2,G3}$ and $w_d^{G2,G3}$ can be described by the following expressions:

$$w_a^{G2,G3} = f\left(PA_a^{w_i^{G2,G3}}\right) = \begin{cases} 0 \text{ if } P > P^a_{max} \cup P < P^a_{min} \ ; \\ 1 \text{ if } P^a_{min} < P < P^a_{max} \end{cases}$$

$$(2)$$

$$w_d^{G2,G3} = f\left(PA_d^{w_i^{G2,G3}}\right) = \begin{cases} 0 \text{ if } P^d \sim \ni \{\text{data}\} \ , \\ 1 \text{ if } P^d \ni \{\text{data}\} \end{cases}$$

where P is some parameter (voltage, current, frequency, phase, etc.), data is the reference data set for 0 (failure) and 1 (normal operation).

4 Diagnosing a Distributed RDFS with a Variety of Technical States

Let RDFS has l^{G1} technical states (TS), then, in accordance with its topology, l^{G2} and l^{G3}—are the numbers of TS of LDC and URT, and in accordance with the structure of the equipment, $l^{U0}, ..., l^{U4}$—is the number of TS of its constituent elements [8]. Since the method provides a depth of control up to a removable PA, and the completeness is up to 100%, then each PA in the general case can also have a set of TS l^{PA_i}. Then $w_i \in \{0; ...; l-1\}$, i.e. $W^{G1} \in \{0; ...; l^{G1}-1\}$ is the final estimate of the TS of the whole RDFS, $W^{G2} \in \{0; ...; l^{G2}-1\}$—the TS estimate of the LDC, $W^{G3_N} \in \{0; ...; l^{G3}-1\}$—TS estimate of the N-th URT.

Obviously, the criterion for evaluating the operational capability of the RDFS (its TS) will depend on the TS of its PAs. There are 3 reasonable approaches to determining its TS—the worst-case TS method, the band averaging method, and the priority-setting method.

Let us consider the principle of the worst-case method, in which the TS of the RDFS is determined by the normalized to l^{G1} values of l^{PA_i}. Mathematically, this diagnosis can be represented by the following expressions:

$$W^{G1} = \left\| l^{G1} \times min\left(\frac{w^{G2}}{l^{G2}}, \frac{w^{G3_1}}{l^{G3}}, \ ... \ , \frac{w^{G3_N}}{l^{G3}}\right) \right\|;$$

$$W^{G2} = \left\| l^{G2} \times min\left(\frac{w^{U0}}{l^{U0}}, \ ... \ , \frac{w^{U4}}{l^{U4}},\right) \right\|;$$

$$(3)$$

$$W^{G3} = \left\| l^{G3} \times min\left(\frac{w^{U0}}{l^{U0}}, \ ... \ , \frac{w^{U3}}{l^{U3}},\right) \right\|;$$

$$w^{Ui} = \left\| l^{Ui} \times min\left(\frac{w^{PA_1^{Ui}}}{l^{PA_1^{Ui}}}, \ ... \ , \frac{w^{PA_N^{Ui}}}{l^{PA_N^{Ui}}}\right) \right\|,$$

where $i = \overline{0,4}$ for LDC units and $i = \overline{0,3}$ for URT equipment blocks, $\| \ \|$ is the function of rounding to the nearest integer.

The principle of the method of averaged ranges is as follows: the final diagnosis of operability is calculated as the arithmetic mean of normalized diagnoses of operability of constituents:

$$W^{G1} = \left\| \frac{l^{G1}}{N+1} \times \left(\frac{\mathrm{w}^{G2}}{l^{G2}} + \sum_{i=1}^{N} \frac{\mathrm{w}^{G3_i}}{l^{G3}} \right) \right\| ;$$

$$W^{G2} = \left\| l^{G2} \times f\frac{l^{G2}}{5} \times \left(\sum_{i=0}^{4} \frac{\mathrm{w}^{Ui}}{l^{Ui}} \right) \right\| ;$$

$$W^{G3} = \left\| \frac{l^{G3}}{4} \times \left(\sum_{i=0}^{3} \frac{\mathrm{w}^{Ui}}{l^{Ui}} \right) \right\| ;$$

$$\mathrm{w}^{Ui} = \left\| \frac{l^{Ui}}{N} \times \left(\sum_{k=0}^{n} \frac{\mathrm{w}^{PA_k^{Ui}}}{l^{PA_k^{Ui}}} \right) \right\| ,$$

(4)

where $(N+1)$—number of all URT and LDC.

The principle of the method with setting priorities is to set the weighting coefficients for each block and the PA equipment. Weights k are defined for each element at each level of the hierarchy, with $\sum k_i = 1$ for each level.

In the graph form in accordance with the diagnostic model, the expressions for the technical states will look as follows:

$$W^{G1} = \left\| l^{G1} \times f\left(k^{G2} \times \frac{\mathrm{w}^{G2}}{l^{G2}}, \ k^{G3_1} \times \frac{\mathrm{w}^{G3_1}}{l^{G3}}, \ \dots, k^{G3_N} \times \frac{\mathrm{w}^{G3_N}}{l^{G3}} \right) \right\| ;$$

$$W^{G2} = \left\| l^{G2} \times f\left(k^{U0} \times \frac{\mathrm{w}^{U0}}{n^{U0}}, \ \dots, \ k^{U4} \times \frac{\mathrm{w}^{U4}}{n^{U4}} \right) \right\| ;$$

$$W^{G3} = \left\| l^{G3} \times f\left(k^{U0} \times \frac{\mathrm{w}^{U0}}{l^{U0}}, \ \dots, \ k^{U3} \times \frac{\mathrm{w}^{U3}}{l^{U3}} \right) \right\| ;$$

$$\mathrm{w}^{Ui} = \left\| l^{Ui} \times f\left(k^{PA_1^{Ui}} \times \frac{\mathrm{w}^{PA_1^{Ui}}}{l^{PA_1^{Ui}}}, \ \dots, \ k^{U4} \times \frac{\mathrm{w}^{PA_N^{Ui}}}{l^{PA_N^{Ui}}} \right) \right\| ,$$

(5)

where k—weighting factors for the distribution of priorities (for example, taking into account the achieved level of reliability of elements).

For all methods with a set of TS, the following criteria for evaluating the operability of a PA can be formulated:

$$
\mathrm{w}_a^{G2,G3} = f(PA_a^{\mathrm{w}_i^{G2,G3}}) = \begin{cases} 0 \text{ if } P > P^a{}_{max} \cup P < P^a{}_{min} \\ 1 \text{ if } P^a{}_{min1} < P < P^a{}_{max1} \\ \quad \cdots \\ l^{PA_N^{Ui}} - 1 \text{ if } P^a{}_{min^{l^{PA_{N-1}^{Ui}}}} < P < P^a{}_{max^{l^{PA_{N-1}^{Ui}}}} \end{cases}
$$

$$(6)$$

$$
\mathrm{w}_d^{G2,G3} = f(PA_d^{\mathrm{w}_i^{G2,G3}}) = \begin{cases} 0 \text{ if } P^d \sim \,\ni \{\mathrm{data}_0\} \\ 1 \text{ if } P^d \ni \mathrm{data}_1 \\ \quad \cdots \\ l^{PA_N^{Ui}} - 1 \text{ if } P^d \ni \{\mathrm{data}_{l^{PA_{N-1}^{Ui}}}\}, \end{cases}
$$

where $P^a{}_{min^{l^{PA_{N-1}^{Ui}}}}$ and $P^a{}_{max^{l^{PA_{N-1}^{Ui}}}}$ —limits of the range of values of the TC parameters for analogue PUs; $\mathrm{data}_{l^{PA_{N-1}^{Ui}}}$ — array of values for the states of digital PU.

The combined evaluation criterions principle is in using any of the above methods (minimal, medium-arithmetic and prioritizing) at various levels of the RDFS hierarchy. For example, the topology can be calculated by the arithmetic mean method, the PA—by the minimum one, and blocks—using priorities method.

5 Example of Diagnosing of a RDFS with Three TS

In the practical implementation [9], the distributed RDFS and all its elements up to PA have three TSa (deterioration, accident, normal) $l^{G1} = l^{G2} = l^{G3} = l^{Ui} = 3$. The integral criterion of the TS is determined by the worst-case method. The topology, the composition of the equipment, the diagnostic graphs, and the diagnostic schemes for such a RDFS are shown in Fig. 2, and its explanation is given in Table 2.

The considered RDFS consists of a LDC and 12 URTs. The structure of the LDC equipment includes the following PAs: a microcontroller (MC), a central processing unit (CPU), an input-output device (I/O); a central control unit (CCU), read-only memory (ROM); 12 channel interface units (CIU), map information input hardware (MIIH), interface hardware (IH), distributor (Dist.); air condition indicator (ACI), timer indicator, image forming equipment, main and backup remote dispatching controllers (RDC$_1$ and RDC$_p$), technician's console (TC); SPS; information documenting equipment (IDE), storage device (SD); adapter board, recorder.

The URT equipment includes: a peripheral control unit (PCU) with input devices (ID) and error-protecting devices (EPD), a matching device (MD), a remote control device (RC) and a remote signaling device (RS); Channel interface unit (CIU), secondary compression equipment (SCE), a modem; SPS with an

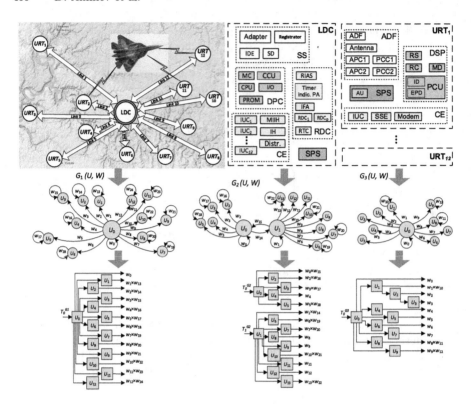

Fig. 2. Equipment, diagnostic graphs and diagnostic schemes for the existing RDFS

automation unit (AU); ADF, antenna, analog-phase converters (APC), phase-code (PC).

In accordance with the composition, there is the following distribution of PAs by blocks: for LDC - 5 in DPC, 15 in CE; 1 in SPS, 6 in the PD, 4 in the SS; For URT - 6 in DSP, 3 in CE; 2 in SSS, 6 in the ADF. The final criteria for determining the TS of such RDFS are determined by the expressions:

$$W^{G1} = \left\| 3 \times min\left(\frac{w^{G2}}{3}, \frac{w^{G3_1}}{3}, \dots, \frac{w^{G3_{12}}}{3}\right) \right\| = min(w^{G2}, w^{G3_1}, \dots, w^{G3_{12}});$$

$$W^{G2} = \left\| 3 \times min\left(\frac{w^{U0}}{3}, \dots, \frac{w^{U4}}{3}\right) \right\| = min(w^{U0}, \dots, w^{U4});$$

$$W^{G3} = \left\| 3 \times min\left(\frac{w^{U0}}{3}, \dots, \frac{w^{U4}}{3}\right) \right\| = min(w^{U0}, \dots, w^{U4}); \qquad (7)$$

$$w^{Ui} = \left\| 3 \times min\left(\frac{w^{PA_1^{Ui}}}{3}, \dots, \frac{w^{PA_N^{Ui}}}{3}\right) \right\| = min(w^{PA_1^{Ui}}, \dots, w^{PA_N^{Ui}}).$$

Table 2. Diagnostic connections of components of a distributed RDFS

$G_1(U,W)$	u_i/u_j	u_0	u_1	u_2	u_3	$u...$	u_{12}
LDC	u_0	w_0	w_1	w_2	w_3	$w...$	w_{12}
URT$_1$	u_1	w_7	w_{13}				
URT$_2$	u_2	w_8		w_{14}			
URT$_3$	u_3	w_9			w_{15}		
URT...	$u...$					$w...$	
URT$_{12}$	u_{12}						w_{24}

$G_2(U,W)$	u_i/u_j	u_0	u_1	u_2	u_3	u_4	u_5	u_6	u_7	u_8	u_9
CCU	u_0	w_0	w_1	w_2	w_3	w_4	w_5	w_6	w_7	w_8	w_9
ADF	u_1		w_{10}								
APC	u_2										
PCC	u_3										
RC	u_4										
RS	u_5										
MD	u_6										
IUC	u_7										
SSE	u_8								w_{11}		
SPS	u_9									w_{12}	

$G_3(U,W)$	u_i/u_j	u_0	u_1	u_2	u_3	u_4	u_5	u_6	u_7	u_8	u_9	u_{10}	u_{11}	u_{12}	u_{13}
CCU	u_0	w_0	w_{14}	w_2	w_3	w_4	w_5								
CPU	u_1	w_{15}	w_1					w_6	w_7	w_8	w_9	w_{10}	w_{11}	w_{12}	w_{13}
IDB	u_2			w_{16}											
PROM	u_3				w_{17}										
RTC	u_4														
IU	u_5						w_{18}								
IUC	u_6						w_{19}								
Distr.	u_7							w_{20}							
IFA	u_8														
Indic.	u_9														
SD	u_{10}										w_{21}				
RDC	u_{11}														
RIAS	u_{12}														
SPS	u_{13}														w_{22}

In all cases, TS can take one of 3 values: 0 – accident, 1 – deterioration, 2 – normal state. Completeness of control of 100% means diagnosability of all PAs of the system. Thus, given the values, a simple in implementation mechanism for determining performance criteria is obtained.

6   Conclusions

Thus, the proposed mathematical apparatus, which provides for the selection of various options for assessing the final technical state of a distributed RDFS, is universal and allows performing calculations for any number of technical states at any level of its hierarchy.

When choosing an approach, it is recommended to take into account the redundancy and the achieved level of reliability of the system and its components at each level of the hierarchy.

On the basis of the proposed worst-case method, diagnostic of distributed RDFS for the three technical states is realized—deterioration, accident, normal state.

References

1. Aminev, D.A., Zhurkov, A.P., Silaev, V.M.: Overview of U.S. patents for radio direction finding. In: Ivanov, I.A., Uvaysov, S.U. (eds.) Innovations on the Basis of Information and Communication Technologies: Materials of the International Scientific and Technical Conference, pp. 321–324. MIEM NRU HSE, Moscow (2015)
2. Aminev, D.A., Zhurkov, A.P., Kozyrev, A.A., Uvaysov, S.U.: Algorithms of software operation for microprocessor control systems for the direction-finding equipment. In: Proceedings of NIIR, Moscow, vol. 4, pp. 11–17 (2014)
3. Aminev, D.A., Zhurkov, A.P., Kozyrev, A.A.: Algorithms for controlling the equipment of the local dispatching station of the ground-based local radio direction-finding system. In: Proceedings of NIIR, Moscow, vol. 4, pp. 11–17 (2014)
4. Zhurkov, A.P., Aminev, D.A., Guseva, P.A., Miroshnichenko, S.S., Petrosjan, P.A.: Analysis of the possibilities of self-diagnosis approaches to distributed electronic surveillance system. Syst. Control Commun. Secur. 4, 114–122 (2015). http://sccs. intelgr.com/archive/2015-04/06-Zhurkov.pdf. Accessed 20 Apr 2017 (in Russian)
5. Aminev, D.A., Zhurkov, A.P., Polesskij, S.N., Kulygin, V.N., Kozyrev, D.V.: Comparative analysis of reliability prediction models for a distributed radio direction finding telecommunication system. In: Vishnevskiy, V.M., Samouylov, K.E., Kozyrev, D.V. (eds.) DCCN 2016. CCIS, vol. 678. Springer, Cham (2016)
6. Aminev, D.A., Zhurkov, A.P., Kozyrev, D.V.: Diagnostic graphs for distributed radio direction finding system. In: 19th International Conference Distributed Computer and Communication Networks: Control, Computation, Communications (DCCN-2016), vol. 1. pp. 5–15. RUDN University, Moscow (2016)
7. Preparata, F.P., Metze, G., Chien, R.T.: On the connection assignment problem of diagnosable systems. IEEE Trans. Comput. C–16, 848–854 (1967)
8. Aminev, D.A., Zhurkov, A.P., Krotkova, K.G., Ohlomenko, I.V.: Research offers for diagnosing of distributed radio direction finding systems with a some set of technical states. Syst. Control Commun. Secur. 3, 282–291 (2016). http://sccs.intelgr.com/ archive/2016-03/10-Aminev.pdf. Accessed 11 Oct 2016 (in Russian)
9. Zhurkov, A.P., Aminev, D.A.: Radio direction finding system "NIVA" and requirements to ensure its diagnostic control. In: Ivanov, I.A., Uvaysov, S.U. (eds.) Innovations on the Basis of Information and Communication Technologies: Materials of the International Scientific and Technical Conference, MIEM NRU HSE, Moscow, pp. 453–455 (2014)

On Optimal Placement of Monotype Network Functions in a Distributed Operator Network

Ekaterina Svikhnushina[1(✉)] and Andrey Larionov[2]

[1] Moscow Institute of Physics and Technology (State University), 9 Institutskiy per.,
Dolgoprudny, Moscow Region 141701, Russia
ekaterina.svikhnushina@frtk.ru
[2] V.A. Trapeznikov Institute of Control Sciences of Russian Academy of Sciences,
65 Profsoyuznaya Street, Moscow 117997, Russia
larioandr@gmail.com

Abstract. Many network operators use a large number of intermediate devices like firewalls or antiviruses implemented on the proprietary hardware. Installation and maintenance of this equipment are very expensive. Therefore the network function virtualization technology allowing flexible remote services management through a software is a promising option for organizing operator network architecture. Switching to software appliances instead of specialized hardware can optimize the administration of the network functions, significantly reducing its cost. However, a problem of determining a number of virtual network functions and their placement in a distributed network that optimizes operating costs and meets service level agreement is a complex mathematical problem. The paper deals with a problem of efficient monotype network functions placement in a distributed network in order to minimize the total cost, with restrictions on channel delays, throughput and node performance. NP-completeness of the problem is proved, the statement is given in terms of integer linear programming. A heuristic algorithm is proposed and its efficiency is shown on typical network topologies.

Keywords: Virtual network functions · NP-complete problems · Heuristic algorithms

1 Introduction

Increasing demands on flexibility and network performance have caused the onset of discussions among operators about the necessity of switching to new virtualization technologies. As a result, in 2012 the Industry Specification Group of the European Telecommunications Standardization Institute (ETSI) proposed a concept of network functions virtualization [1]. The document describes the main purpose of the new technology as the unification of various network elements using virtualization, providing a possibility of their operation on the standard servers and running virtual machines in them. Such virtual network functions (VNFs) can be located in a set of distributed nodes under a centralized control.

© Springer International Publishing AG 2017
V.M. Vishnevskiy et al. (Eds.): DCCN 2017, CCIS 700, pp. 453–466, 2017.
DOI: 10.1007/978-3-319-66836-9_38

Network functions virtualization is designed to solve many problems associated with the dedicated hardware usage. Remote control of the intermediate network functions allows to reduce the capital and operating costs, optimize power consumption and respond more quickly to end-user demands for the new services [2]. However, in order to maximize the positive economic effect, it is necessary to determine the optimal configuration of the virtual functions in the operator network that satisfies both the infrastructure capabilities and the service level agreement requirements. The problem of optimal placement of network functions is not trivial since it depends on a large number of parameters, but this problem is of a great research interest because of its relevance and mathematical complexity. Numerous authors have worked on its formalization in various statements and searched for effective algorithms to solve the problem.

In practice, network operators often face a problem of determining the locations of VNFs of the same type to provide end users with some basic services when implementing the network virtualization. An example of such a virtual network function is a firewall that monitors and filters network traffic passing between a local end-user network and the Internet. Another example is the Content Delivery Network (CDN) which can also be implemented using virtual network functions [3]. In these and other cases it is important to consider both the infrastructure capabilities of the core network, distribution of end users and their requirements for quality of service.

The main purpose of this paper is to determine the cost-optimizing placement of the monotype VNFs in a distributed operator network, subject to the constraints on the nodes and VNF performance as well as the requirements of service level agreements expressed in terms of throughput and maximum latency.

The paper is organized as follows. The second section provides a brief overview of the related work in the field of the network functions virtualization. In the third section the problem of the optimal placement of monotype VNFs is stated. Then, in the fourth section the NP-completeness of the problem is proved. The fifth section gives a formulation of the problem in terms of integer linear programming. In the sixth section a description of a heuristic algorithm is given and then in the seventh section numerical results are provided. The eighth section concludes the paper and describes the plans for the future work.

2   Related Work

Optimal VNF placement problem resembles a problem of placing the virtual machines (VM) since virtual functions per se are virtual machines providing a certain service. Many scientific papers have been devoted to the investigation of VM and VNF placement problems under various conditions.

A number of papers deal with the issue of VNF optimal placement within a single data center (DC). Adamuthe et al. [4] state a multiobjective optimization problem of the VM deployment in a cloud. The proposed objective function minimizes the inefficient use of the network resources, maximizes the profit and provides load balancing among the physical servers on which VMs reside. Shi et al. [5]

consider the maximization of the total profit from VMs usage. The paper takes into account the restrictions on the available capacity budget and the quality of service (QoS) limitations provided by the service license agreements (SLA). In contrast to the current paper, the listed papers deal with the VMs location within a single data center rather than in a distributed network.

Other authors study the VM and VNF placement in a distributed network. Chen et al. [6] investigate a problem of VMs placement in a geo-distributed network of data centers aiming at the minimization of the total operating costs for electricity and information transmission in the WAN. However, this paper does not consider the requirements determined by the quality of service agreed with the end-users.

Jemaa et al. [7] discuss an approach for optimal VNF placement in the edge-central carrier cloud infrastructure. The authors aim at optimizing network resources usage in compliance with the restrictions on the boundary cloud platforms and the terms of the user agreement. Bouet et al. [8] study a multi-objective VNFs deployment problem where each VNF implements virtual deep packets inspection (vDPI). The objective function proposed minimizes the total cost of the solution and is determined by the number of virtual functions and the network load. The approach proposed in [8] takes into account the infrastructure constraints without explicit consideration of user requirements in contrast to the current paper.

3 Problem Definition

Let us consider a network (see Fig. 1) in which the nodes can host virtual network functions (virtual machines that provide a certain set of services). One of the nodes is connected to an Internet gateway. Users access the network through the nodes defined in their agreement with a provider. Network channels are characterized by a delay value, nodes are characterized by a maximum number of VNFs the node can host and the usage cost. Each user must be provided with an Internet access, the traffic has to pass through a dedicated VNF placed in one of the nodes and reach the gateway within a time delay limited by a value specified in the user service level agreement (SLA). In addition, each VNF has limitations on the number of users it can serve and the traffic bandwidth it can process. The problem is to find a VNF deployment minimizing the total cost so that each user gets an Internet access through a dedicated VNF, subject to the delay and VNF performance constraints.

In order to give a mathematical problem statement individual components of the studied system should be described.

Physical network

Let a simple undirected graph $G(V, E)$ represent a backbone infrastructure network, nodes V represent the operator sites (sets of servers where VNFs can be placed) and edges E represent the network channels. The total resources of one site are expressed as the maximum number of VNFs $\overline{m}_v$ which can be placed at this site. Each site is also characterized by the overall cost for its usage

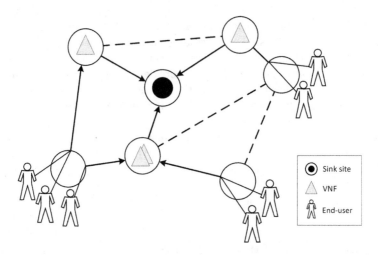

Fig. 1. Sample problem model

$c_v^{(0)}$ which may include potential costs for electricity, operating costs and other possible expenditures. To consider a VNF licensing model a price $c_v^{(1)}$ for placing one VNF is also given.

Each channel connecting two sites $(i, j) \in E$ is specified by the delay value δ_{ij}. In this paper we assume that channels have enough bandwidth to serve traffic from all users.

We denote sites though which the users get an access to the network as *boundary* and the site connected to an Internet gateway as *a sink*. The total number of sites in a network is $N_V = |V| - 1$, the sink is denoted as v_0. In the following we assume that the delay between a user and the boundary node the user is connected to is zero.

Virtual network functions

The maximum number of users a VNF can serve is denoted as η and the maximum traffic bandwidth it can process as β. We can express the total number of VNFs that can be placed in the network as $M = \sum_V \overline{m}_v$.

End Users

Let S represent a set of all users, where $N_S = |S|$. Parameter γ_{sv} is used to indicate a connectivity between a user and a boundary node: $\gamma_{sv} = 1$ if the user s has an access to the network through the boundary node v, otherwise $\gamma_{sv} = 0$. User traffic is characterized by the maximum delay d_s and required bandwidth b_s.

VNF placement problem

The VNF placement problem consists in finding optimal VNF locations at the operator network sites and connecting users to them, ensuring the minimal total cost of the solution. Meanwhile, it is necessary to comply with the restrictions imposed by the user requirements and the capacity of the infrastructure.

We need to introduce the following functions in order to give the problem formal statement.

Consider a user s and assume that he enters a network through a boundary site u. Since the delay between the user s and the site u is zero as mentioned above, we can compute a delay between the user and any other network site as the delay along the shortest path (the path with a minimal delay). Then we can define a function $\Delta : S, V \to \mathbb{R}$ giving a delay from a user s to a site v.

Let $f : S \to V$ be a function that defines a mapping between the users and nodes with VNFs, i.e. $f(s) = v$ if the user s uses a VNF placed in the node v. Let us also define an indicator function $I_f(s, v)$ such that:

$$I_f(s, v) = \begin{cases} 1, & f(s) = v \\ 0, & \text{otherwise} \end{cases}$$

We can define a number of VNFs in a node v for a given mapping f as a function $m(v)$ such that:

$$m(v) = max(\lceil \frac{\sum_S I_f(s, v)}{\eta} \rceil; \lceil \frac{\sum_S I_f(s, v) b_s}{\beta} \rceil),$$

the cost of VNFs placement in the node v as $c(v|f)$:

$$c(v|f) = \begin{cases} 0, & \nexists s \in S : f(s) = v \\ c_v^{(0)} + c_v^{(1)} m(v), & \text{otherwise} \end{cases}$$

and the total cost as $c(f) = \sum_V c(v|f)$.

Assuming $\mathfrak{F}$ be a set of all functions $f : S \to V$, we can give the VNF placement problem formulation.

Problem 1. Find a function

$$f = \arg\min_{\mathfrak{F}} c(f),$$

subject to

$$\begin{cases} m(v) \leq \overline{m}_v \\ \Delta(s, f(s)) + \delta_{f(s)v_0} \leq d_s \end{cases}$$

4   NP-Completeness

To prove the NP-completeness of the problem, a restriction method [9] is used. It consists in establishing that a problem considered includes a well-known NP-complete problem as a particular case. Let us consider the following subproblem (see Fig. 2). Let the set of nodes V be divided into two subsets $V = V_1 + V_2$, where the subset V_2 represents all boundary nodes and V_1 – all other nodes (internal), including the sink. Let us also assume the following:

- $c_v^{(1)} = 0$ for any node $v \in V$;
- $c_v^{(0)} = 0$ for all internal nodes $v \in V_1$;
- $c_v^{(0)} \neq 0$ for all boundary nodes $v \in V_2$.

This means that placing VNFs in non-boundary nodes from V_2 is free of charge while placing them in one of the boundary nodes will requires a node usage cost.

Let the delays between the nodes be arranged so that the delay for any two boundary nodes is equal to infinity and all other delays are equal to zero. In this way, a VNF for a user can be located either in a boundary node through which the user in question has an access to the network, or in one of the nodes with a free placement from V_1.

VNFs considered are characterized by the infinite bandwidth $\beta = \infty$ and can serve one user at most ($\eta = 1$).

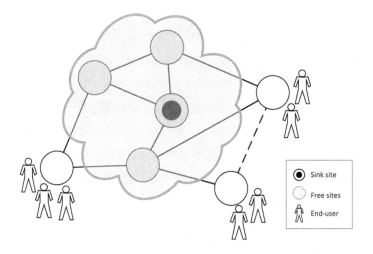

Fig. 2. Knapsack-like partial problem. The nodes in the blue cloud represent a V_1 subset and edge-nodes represent a V_2 subset. Solid lines between the nodes are channels with zero-delay while the dashed line is a channel with a delay equal to infinity. (Color figure online)

With such constraints provided, the number of VNFs in a node v $m(v)$ can be defined as $m(v) = \sum_S I_f(s, v)$. It should be noted that under the constraints described this value is the same as the number of users using any VNF placed at the node v.

The resulting partial problem can be formulated as follows:

Problem 2. Find a function

$$f = \arg\max_{\mathfrak{F}} \sum_{V_2} (-c(v|f)),$$

subject to

$$\sum_{V_2} I_f(s,v) \le \sum_{V_2} \overline{m}_v = \tilde{M}$$

This problem is a well-known NP-complete knapsack problem. Combinations of the boundary nodes and related users correspond to things in the knapsack problem: a negated cost for using a boundary node corresponds to a cost of a thing, and a number of users connected to a node corresponds to a weight of a thing. The total number of users that can theoretically be connected to VNFs placed in free of charge nodes from the subset V_1 corresponds to capacity of a knapsack. Since the VNFs are free when being placed at the nodes from V_1, the task of minimizing the total cost corresponds to the task of maximizing the cost of a set of charged nodes V_2 taken with the opposite sign.

5   Integer Linear Programming Problem Formulation

The problem of the monotype VNF placement in a distributed operator network can be formulated in terms of the integer linear programming (ILP). The following independent variables are required for the program:

- $x_{kv} = \begin{cases} 1, & \text{if VNF } k \text{ is placed at a site } v \\ 0, & \text{otherwise} \end{cases}$, $\forall k = \overline{1, M}, v = \overline{0, N_V};$

- $y_{sk} = \begin{cases} 1, & \text{if a user } s \text{ is assigned to a VNF } k \\ 0, & \text{otherwise} \end{cases}$, $\forall s = \overline{1, N_S}, k = \overline{0, M}$

For the sake of simplicity let us introduce several dependent variables:

- $z_{skv} = z_{skv}(\mathbf{x}, \mathbf{y}) = y_{sk} x_{kv}$ is a boolean variable which equals 1 if and only if a user s is assigned to a VNF k which is placed at a site v, $s = \overline{1, N_S}$, $k = \overline{1, M}$, $v = \overline{0, N_V};$

- $u_v = u_v(\mathbf{x}) = 1 - \prod_{k=1}^{M}(1 - x_{kv})$ is a boolean variable which equals 1 if and only if there is at least one VNF at a site v, $v = \overline{0, N_V};$

- $w_m = w_m(\mathbf{x}) = \sum_{v=0}^{N_V} x_{kv}$ is a number of sites over which a VNF k is placed, $k = \overline{1, M}.$

The VNF placement is constrained by the site capacity (the number of VNF the site can host) limitations and the fact that a single VNF can be placed at one site at most.

- **P1:** $\forall v = \overline{0, N_V} : \sum_{k=1}^{M} x_{kv} \le \overline{m}_v$ no more than $\overline{m}_v$ VNFs can be placed at a site;

– **P2:** $\forall k = \overline{1, M} : w_k \leq 1$ one VNF can be placed no more than at one site;

Any valid assignment between the users and the sites with VNFs is specified by the following constraints:

– **A1:** $\forall s = \overline{1, N_S} : \sum_{k=1}^{M} y_{sk} = 1$ each user must be assigned to one VNF;

– **A2:** $\forall k = \overline{1, M} : \sum_{s=1}^{N_S} y_{sk} b_s \leq \beta w_k$ a VNF with assigned users must be placed at some site and have enough bandwidth to serve the users;

– **A3:** $\forall k = \overline{1, M} : \sum_{s=1}^{S} y_{sk} \leq \eta w_k$ a VNF with assigned users must be placed at some site and a number of users assigned cannot exceed the maximum possible;

– **A4:** $\forall i = \overline{0, N_V}, \forall v = \overline{0, N_V}, \forall s = \overline{1, N_S} : (\delta_{iv} + \delta_{v0}) c_{si} z_{sv} \leq d_s$ a condition for compliance with the constraint on the delay from the network entry point of the user s to the sink through the site with VNF assigned.

Variables z_{skv} and u_v are expressed in the terms of the product of boolean variables. To avoid explicit multiplication they can be expressed using the equivalent systems of linear constraints [10]:

– **L1.1:** $\forall s = \overline{1, N_S}, \forall v = \overline{0, N_V} \; \exists k : y_{sk} + x_{kv} - z_{skv} \leq 1$;
– **L1.2:** $\forall s = \overline{1, N_S}, \forall v = \overline{0, N_V} \; \exists k : -y_{sk} - x_{kv} + 2z_{skv} \leq 0$;
– **L2.1:** $\forall v = \overline{0, N_V} : \sum_{k=1}^{M} x_{kv} \leq M u_v$;
– **L2.2:** $\forall v = \overline{0, N_V} : \sum_{k=1}^{M} x_{kv} \geq u_v$.

Using the variables defined above the total cost of the VNFs deployment can be defined using the following function:

$$c(\mathbf{x}) = \sum_{v=0}^{N_V} (c_v^{(0)} u_v(\mathbf{x}) + c_v^{(1)} \sum_{k=1}^{M} x_{kv})$$

Finally, the problem of optimal VNF placement can be formulated as the ILP:

Problem 3. Find $\mathbf{x} = (x_{kv} : 1 \leq k \leq M, 1 \leq v \leq N_V)$ and $\mathbf{y} = (y_{sk} : 1 \leq s \leq N_S, 1 \leq k \leq M)$, such that

$$c(\mathbf{x}) \rightarrow \min_{\substack{P_1 - P_2 \\ A_1 - A_2 \\ L_1 - L_2}}$$

6   Heuristic Algorithm

Since the VNF placement problem is NP-complete, an efficient greedy algorithm was developed. The algorithm operates as follows:

1. search for the nodes satisfying the delay constraints considering users one by one;
2. iteratively try to connect the most priority user in order to maximize the overall profit. The user priority is determined by the number of connection options and the gap between the most advantageous option and the next after it.

Before defining an algorithm let us introduce several auxiliary functions: *DiscoverNodes* for identifying the nodes, suitable for connecting the users, *NodesForConnection* for compiling an exact list of suitable nodes and *UserToConnect* to determine a user to be connected next.

The *DiscoverNodes* function determines nodes where the user can use VNF without violating delay requirements. This function takes a user, for which it is necessary to define the nodes possible for connection as an input. The delay constraints are sequentially checked for each node, a list of nodes is generated as a result:

DiscoverNodes(user)

```
FOR every node:
    IF available number of VNFs for placement > 0:
        IF delay from network entry point to sink through the
        node is less or equal to user delay requirement:
            add node to the result;
```

The *NodesForConnection* function identifies nodes in which the user can be assigned to a VNF considering the current load of nodes and VNFs in them, and determines the connection cost:

NodesForConnection(UnconnectedUsers)

```
FOR user in UnconnectedUsers:
    FOR node in the list of possible nodes:
        IF load of the node allows to connect a user to it:
            add node, connection type and cost to the result;
        ELSE:
            delete node from the list of possible nodes;
```

The function takes a list of unconnected users and forms a list of the nodes suitable for connecting these users. The nodes found here have sufficient resources for connecting the user. For each node from the generated list, a cost of connection and its type (connection to a new VNF in a node, to an existing VNF or to a new node) are also fixed. If the connection to the node is impossible, it is excluded from the list of possible nodes.

Finally, the *UserToConnect* function determines a user with the most profitable connection.

UserToConnect(UnconnectedUsers)

```
FOR user in UnconnectedUsers:
     IF user has no options for connection:
          add user to the list of users with no options for
          connection;
     ELSE IF user has one option for connection:
          add user to the list of users with one option for
          connection;
     ELSE: add user to the list of users for sorting;
     IF list of users with one option for connection is not
     empty:
          RETURN the first user from the list of users with one
          option for connection, list of users with no options
          for connection;
     ELSE: sort users from the list of users for sorting in
          decreasing order by the difference between the cheapest
          and the next connection options;
     RETURN the first user from the sorted list, list of users
     with no options for connection;
```

Now the main algorithm can be formulated.

At the beginning of the algorithm nodes suitable for connection are identified for each user. At the second step for each unconnected user the exact search for the suitable for connection nodes and the cost of each option is made. Further, in the *UserToConnect* function a user for which only one connection option is available or the difference between the two lowest-cost connection options is maximal is searched for. As a result, the user is connected and the algorithm returns to the second step. If there are no unconnected users the algorithm completes its execution, returning the solution in a form of user to VNF and VNF to node correspondence.

Algorithm

```
UnconnectedUsers = list of all users;
FOR user in list of all users:
     DiscoverNodes;
WHILE UnconnectedUsers is not empty:
     FOR user in UnconnectedUsers:
          NodesForConnection(UnconnectedUsers);
     FOR user in UnconnectedUsers:
          user, impossibletoconnect =
          UserToConnect(UnconnectedUsers);
     connect user to the cheapest variant;
     update UnconnectedUsers;
```

7   Experiment Results

To evaluate the quality of the proposed heuristic algorithm a number of experiments were performed on synthetic data. When running the experiments a random network was repeatedly generated (see Fig. 3). The generated network had a star-topology with the following properties:

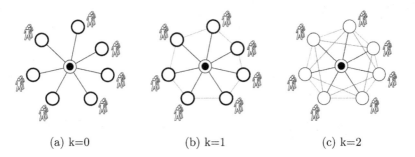

(a) k=0   (b) k=1   (c) k=2

Fig. 3. Model variants for different values of k parameter and $N = 7$

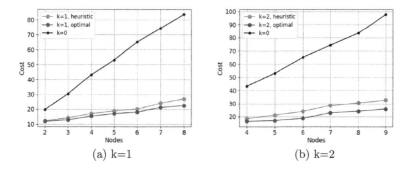

(a) k=1   (b) k=2

Fig. 4. Cost dependencies on the number of surrounding nodes N

- the central node was a sink and N boundary nodes were connected to it, where N was a variable parameter;
- users were connected to the boundary nodes, each user had $k + 1$ options of nodes for connection, where k was also a variable parameter ranging from 0 to $\lfloor \frac{N}{2} \rfloor$. Given $k = 0$, each user could connect to a VNF placed only in the user's network entry node. For $k = 1$, the two closest neighbours were added to the possible variants of nodes for connection to VNF, and so on. Additional options for connecting users to VNFs were specified by setting appropriate links between the nodes and necessary delays.

At the first stage the results retrieved from the heuristic algorithm were compared with the exact solutions. The heuristic algorithm was implemented in Python3 programming language, and the ILP program was written in Math-Prog and solved using GLPK suite [11]. During the experiments the k parameter ranged from 0 to 2 and N from 2 to 9. With a further increase of the N parameter it became impossible to get the exact solution in an appropriate time due to a lack of memory for the ILP. Figure 4 shows the results of the modelling. Each point in the charts corresponds to the average result obtained from ten simulations.

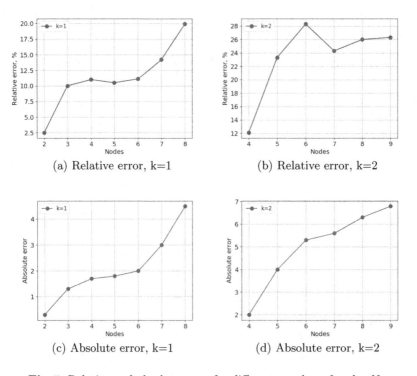

(a) Relative error, k=1   (b) Relative error, k=2

(c) Absolute error, k=1   (d) Absolute error, k=2

Fig. 5. Relative and absolute error for different number of nodes N

To confirm the correctness of the heuristic, measurements for the case $k = 0$ were made. In this case the problem becomes polynomial and convex and the results of the heuristic algorithm and the exact solution must coincide, which is evident from the method of forming the test model. As expected, the results were the same.

For other values of k, it was found that the dependence between the total cost and the number of nodes N obtained by the heuristic algorithm is correlated with the exact solution. However, the cost found by the heuristic deviates from the optimal from 2.5% to 28% for different input parameters (see Fig. 5).

Next, the heuristic performance for large N values from 2 to 20 was studied, while k was changed from 0 to 2 (see Fig. 6). The results are consistent with expectations: for $k = 0$ the values of the total cost turn out to be maximum and

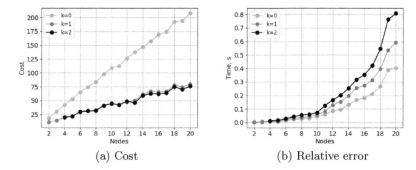

(a) Cost   (b) Relative error

Fig. 6. VNF placements cost and relative error obtained from the heuristic algorithm

they decrease with increasing of k. The heuristic time grows with the increase of the network graph size, but the order of the algorithm execution time under the considered conditions remained in the limit of a second.

The obtained results indicate that the algorithm can be used for rough estimates, when the upper bound of the total cost must be obtained in the shortest possible time.

8 Conclusion

In this paper the problem of optimal monotype VNF placement in a distributed operator network was formulated as a the cost minimization problem subject to the constraints on throughput, channel delays and node and VNF performance. It was proved that the problem defined is NP-complete and its formulation was given in terms of the integer linear programming. Next, an effective heuristic algorithm was proposed to obtain a solution in polinomial time. Estimation of its accuracy and results comparison with an optimal solution were made.

It was shown that while the algorithm estimates the desired placement in reasonable time, the relative error increases along with the growth of the number of stations. The focus of the future work is to modify the developed heuristic algorithm with the aim of improving the obtained results and to consider a more complicated problem of placing several types of virtual network functions.

References

1. Network Functions Virtualisation: Introductory White Paper. SDN and OpenFlow World Congress, Darmstadt-Germany (2012)
2. Stallings, W.: Foundations of Modern Networking. Pearson Education, Inc., Indianapolis (2016)
3. ETSI GS NFV 001 V1.1.1: Network Function Virtualisation (NFV); Use Cases (2013)

4. Adamuthe, A.C., Pandharpatte, R.M., Thampi, G.T.: Multiobjective virtual machine placement in cloud environment. In: IEEE International Conference on Cloud and Ubiquitous Computing and Emerging Technologies (CUBE), pp. 8–13 (2013)
5. Shi, W., Hong, B.: Towards profitable virtual machine placement in the data center. In: Fourth IEEE International Conference on Utility and Cloud Computing (UCC), pp. 138–145 (2011)
6. Chen, K.Y., Xu, Y., Xi, K., Chao, H.J.: Intelligent virtual machine placement for cost efficiency in geo-distributed cloud systems. In: IEEE International Conference on Communications (ICC), pp. 3498–3503 (2013)
7. Jemaa, F.B., Pujolle, G., Pariente, M.: QoS-Aware VNF placement optimization in edge-central carrier cloud architecture. In: IEEE Global Communications Conference (GLOBECOM) (2016)
8. Bouet, M., Leguay, J., Conan, V.: Cost-based placement of vDPI functions in NFV infrastructures. In: 1st IEEE Conference on Network Softwarization (NetSoft) (2015)
9. Garey, M., Johnson, D.: Computers and Intractability: A Guide to the Theory of NP-Completeness, pp. 85–88 (1982). Translate from English (in Russian)
10. Vishnevskiy, V., Lyakhov, I., Portnoy, S., Shakhnovich, I.: Broadband wireless communication networks. M. Technosphera, pp. 435–436 (2005). (In Russian)
11. GLPK. https://www.gnu.org/software/glpk/

SDN Approach to Control Internet of Thing Medical Applications Traffic

Artem Volkov[1]([✉]), Ammar Muhathanna[1], Rustam Pirmagomedov[1], and Ruslan Kirichek[1,2]

[1] State University of Telecommunication, St. Petersburg, Russia
v.artem.nikolaevich@yandex.ru, ammarexpress@gmail.com,
prya.spb@gmail.com, kirichek@sut.ru
[2] RUDN University, Moscow, Russian Federation

Abstract. The article is devoted to study the processing of traffic generated by nanonetwork of real network and built on the basis of software defined network concepts. For a more realistic experiment is considered a model that describes interactions between a group of individuals that have a nanonetwork in their bodies and the medical services through the local network of a medical institution, built based on software defined network concepts.

Keywords: Medical networks · Body area networks · Traffic emulation · e-Health · Nanonetworks · Software-defined networking · Nano machines · Lost packets · Jitter

1 Introduction

The development of Internet of Things in recent years has been moving with leaps and bounds, all over again and again dissuading the initially skeptical people to this technology on the edge of its inception. Today, Internet of Things applications are: Monitoring of the environmental situation, both locally considered areas and the level of the city, monitoring and automatic data collection of various sensors both in the electricity and water sectors. The development of networks and corresponding interaction protocols between nodes generated a whole class of networks, such as: ad hoc network, flying sensor networks, etc. These technologies are initially set to resolve the issues of human existence. For example, a scattered sensor network with autonomous nodes allows monitoring of entire forest areas far from civilization, but important for the safety of people. Early warning allows avoiding many disasters in advance, caused by human factor, or due to non-standard phenomena of nature (Floods, hurricanes, earthquakes, etc.) However, in the spectrum of all these applications related to the concept of the Internet of things, realizing the so-called "Smart Systems", recently appeared a new direction in IoT development. The development towards the nano world. In this direction of Comprehensive Internet of things development concept, it is necessary to talk about such concepts as: nanomachines,

© Springer International Publishing AG 2017
V.M. Vishnevskiy et al. (Eds.): DCCN 2017, CCIS 700, pp. 467–476, 2017.
DOI: 10.1007/978-3-319-66836-9_39

nanomachines aggregation gateways, etc. And still, what is in the world of nano Internet of things, based on which laws to act? How are nanomachines arranged? And the most important question is how to ensure the integration of the two worlds? Scientific groups work in this area of development, and have to integrate knowledges of medicine and organisms function with information and communication networks algorithms, protocols, and interaction models. Of course, in this concept it is necessary to consider a different network infrastructure and a different approach to the transmission of this type of traffic, taking into account the heterogeneity of modern networks. Medical networks [6] and nano-networks applications [2,3] are primarily aimed to analysing the organism and localizing possible anomalies (problems) in the body by the interaction of the general medical management system of human organism with the corresponding actuators realized as elements of nanomachines. Premature detection of a threat to the body is possible when implementing such systems, disease prevention and as a result prevention of all kinds of epidemics. Of course, since this concept is still at its early stage level, there are many questions about the organization of the security of these networks, and observance of the inviolability principle of the individual's personal life. Certain works carried out by the scientific group of SPbGUT with partners engaged in research in this field has already yielded certain results, by developing, modeling processes with the appropriate initial constraints that allow to develop the topic of nano networks all the way to the next level [7,8]. The network infrastructure level and algorithms and protocols level for the interaction of nano Things with a real-world Things. In general, the development of this concept can be divided into several abstract levels of interactions, defining the interfaces between them. This article is devoted specifically to the study of the possibility of processing the traffic generated by nanomachines in conventional networks, both with low-traffic networks and heavy network load.

2   The Interaction of Nano Networks with the External Network Based on the SDN Technology

The integration of information-communication technologies and medicine is called e-health [4,5]. One of applications of this integration is telemedicine, i.e. physical and psychological diagnostics, as well as distance treatment, which includes distance monitoring of indicators and providing them to medical personnel.

Telemedicine applications are implemented mainly by wireless sensor networks deployed in close proximity to the patient on the surface of his body, as well as implanted directly into his body.

The following nodes of this network can be distinguished:

- sensor nodes (measurement of required indicators)
- actuator nodes (implement effect on the organism)
- body area gateway collects information from nanosensors, if necessary performs pre-processing and data transmission to a remote information processing server.

At the same time in the field of network technologies suggested new approach to building and managing networks Software-configurable approach [9,11]. Due to the fact that it is planned to use software-Defined networks (SDN) as the transport of IMT-2020 networks, it becomes possible to dynamically organize QoS for Internet of Things applications. ITU-T TD 208 document describes the interaction of IMT-2020 + SDN.

SDN + IMT-2020 architecture is also proposed in the ONF (Open Networking Foundation) TR-526 document.

The use of SDN as a network infrastructure allows flexible configuration of the network for various services, including for the Internet of things [10]. Organization of the dynamic prioritization and control of Internet of Things applications traffic in the conditions of heterogeneity of networks can be useful for such applications as:

- tactile internet connection
- augmented reality
- medical networks, etc.

We decided to consider software defined network as a network infrastructure representing the core network of a medical institution, within the framework interaction of the model under study.

3   Body Area Gateway Interaction with External Medical Server: Description of the Experiment

Within the framework, the characteristics of traffic generated by a medical network (nano-network) and through a local body area gateway to an external network were investigated.

To reach the aim of research imitation model was realized. This model presupposes the presence of the user within the walls of a medical institution, whose data transmission network is based on SDN technologies.

The user's body area gateways are connected to the local area network of the medical institution and transmit data to a specialized medical server, also located in the telecommunications network of the institution.

Description of the experiment:

1. *goal*. It is required to conduct an experiment by transferring data from a body area gateway(data generated by a nanonetwork), and use software defined network (SDN) as the network infrastructure.
2. *technical resources*:
   - Laboratory stand of a software defined network infrastructure
   - Iperf3 traffic generator (client/server).
3. *Characteristics of the used model*

According to early studies, the result of which is presented in [1], the following initial conditions were adopted, which were incorporated into the parameters of the used traffic generator:

- The packet length is 32 bytes from one sensor, taking into account the IP header;
- Generating data every second;
- Transport Protocol UDP

In the research simulation model, the following parameters are specified:

- There are 10 clients in the local network at the same time
- Each client has 128 sensors (nanomachines)
- The time to stay in the network for all clients is 20 min. (Test Time)

For a visual illustration of the model, the structural scheme of the interaction is shown in Fig. 1.

In Fig. 2, which shows the interaction model in our experiment, it is shown that in this case the clients are connected to the SDN network through one aggregation switch, and iperf3 with fine tuning was used as the traffic generator in order to provide the necessary characteristics of the traffic from the nano

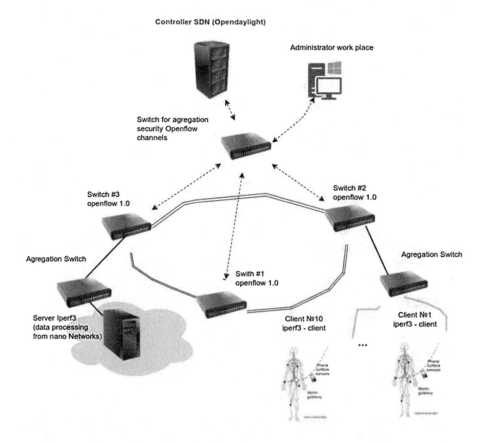

Fig. 1. Structural diagram of the interaction in this study

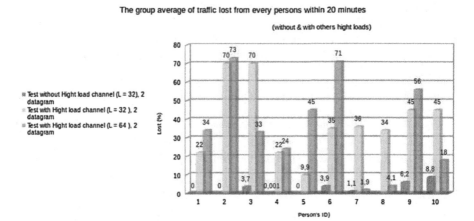

Fig. 2. The percentage value of losses for each individual client under consideration, for different values of the channel load

networks. At the same time, since the UDP protocol is considered as the transport layer protocol, in this model we considered traffic duplication transmission. That means that, data from one sensor (nanomachines) was duplicated on the gateway before sending the second packet with the same data to the network.

At the same time, within our considered model, we monitored the presence of 10 individuals at the same time in the building of a medical institution. The time of simultaneous discovery was set at 20 min, further it can be extrapolated to longer period of time.

4   Model Testing Results

The experimental stage of this work consisted of three stages that were aimed at obtaining a comparative analysis of the transmission of this type of traffic through a network with different channel load characteristics.
The following tests were performed:

1. Testing when - packet length L = 32 bytes, low load channels;
2. Testing when - packet length L = 32 bytes on the On high-loaded network by rival traffic (traffic encountered to the one under consideration);
3. In the interaction model, according to the results of the second test, it was suggested to increase the packet length L = 64, while maintaining the same intensity of the flow. This testing was also carried out on a high loaded channel with rival traffic of the same type as in the second case (traffic encountered to the considered one).

Rival traffic was also created using the iperf3 traffic generator, while passing through the TCP layer, and occupying the network link at 78–80 Mbit/s with the maximum possible through the given link of 100 Mbit/s.

The test results according to the first test are displayed in the Table 1.

Table 1. The test results

Person ID	Lost datagrams, %	Lost/Total datagrams	Jitter, ms	Bandwidth, Kbit/s
Per.1	0	0/300032	0.183	64.0
Per.2	0	0/300032	0.144	64.0
Per.3	3.7	11081/300032	0.508	64.0
Per.4	0.001	3/300032	0.253	64.0
Per.5	0	0/300032	0.209	64.0
Per.6	3.9	11732/300032	0.382	64.0
Per.7	1.1	3339/300032	0.303	64.0
Per.8	0.25	738/300032	0.307	64.0
Per.9	6.2	18554/300032	0.301	64.0
Per.10	8.8	26315/300012	0.369	64.0
Average	2.3951	-	0.2959	64.0

The test results according to the second test are displayed in the Table 2.

Table 2. The test results

Person ID	Lost datagrams, %	Lost/Total datagrams	Jitter, ms	Bandwidth, Kbit/s
1	22	66837/300032	2.793	64.0
2	70	210937/299709	2.957	64.0
3	70	208476/299844	1.424	64.0
4	22	66102/300021	3.942	64.0
5	9.9	29738/300032	0.716	64.0
6	35	104346/299927	1.308	64.0
7	36	108067/300032	3.774	64.0
8	34	103355/300032	1.288	64.0
9	45	135928/299808	2.716	64.0
10	45	135002/300020	2.055	64.0
Average	38.89	-	2.2973	64.0

The test results according to the third test are displayed in the Table 3.

Table 3. The test results

Person ID	Lost datagrams, %	Lost/Total datagrams	Jitter, ms	Bandwidth, Kbit/s
1	34	103091/300025	5.501	128.0
2	73	218964/299726	4.349	128.0
3	33	100350/299766	1.517	128.0
4	24	70936/300032	2.77	128.0
5	45	135139/299883	1.81	128.0
6	71	214064/299632	4.456	128.0
7	1.9	5817/300032	2.087	128.0
8	4.1	12158/300032	1.058	128.0
9	56	168085/299896	0.749	128.0
10	18	53683/299918	3.699	128.0
Average	36	-	2.7996	128.0

To perform a comparative analysis of the results obtained, we display the same type of data on one diagram.

A diagram (Fig. 2) shows the percentage of losses from the total number of all of transmitted packets by each individual in all three test settings.

According to the data shown in Fig. 3, we can conclude that with a high loaded network, most of the packets are lost, when having the same priority with rival traffic, which in fact occupied the bulk of link. Although, the total traffic from 10 clients did not exceed 640 kbit/s.

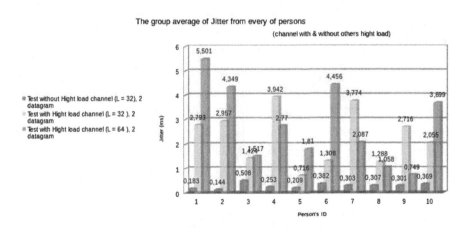

Fig. 3. Jitter value for different channel load values.

It is also noticeable that with an increase in the packet length by half, this approach does not yield significant results in reducing losses in the link with high rival traffic.

A diagram showing the value of Jitter parameter for all the cases is show in Fig. 3.

5 The Proposed Model of Interaction

According to the models studied and the analysis of the interaction of medical traffic and networks with a high peak load, we came to the conclusion that it is possible to implement an interaction model that will also dynamically or statically change the traffic prioritization settings, taking into account the running IoT services and the network load.

The proposed architecture is shown in Fig. 4.

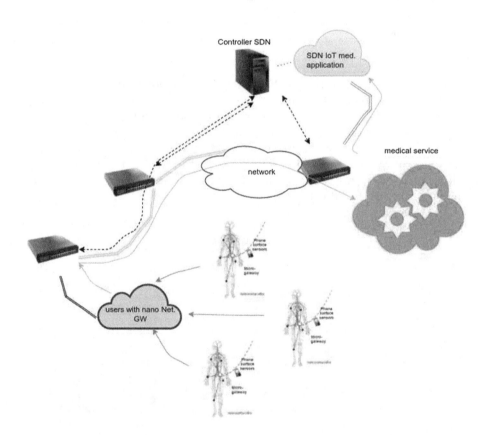

Fig. 4. The proposed traffic prioritization architecture

The interaction on this model can be described by the following algorithm:

1. The gateway generates a registration message, which, in addition to its identity, also transmits the type of Internet of Thing applications (nano-network), the required QoS characteristics, as well as the possible range of their deviations and the number of connected nano machines Their characteristics
2. The gateway logs into its service and wait for the service to respond
3. After the request for registration is received, the medical service refers to the application of the network controller via a secured, previously allocated link, which is responsible for the management of medical traffic.
4. The service on SDN application layer looks up, in the existing database, for certain identifiers of the gateway for data verification (statistics). Then it addresses to the appropriate modules of the controller that are responsible for QoS regulation policy on the network and, if possible or impossible to allocate the required resources, it responds with a message to the medical service about the solution (with a successful solution, for example, a 200 OK message). If it is impossible to allocate the required resources, the service as a response message sends possible characteristics (parameters) that it can allocate at the moment and the medical service checks the received possible characteristics with a possible range that was specified in the Internet of Things registration message.
5. After that, the service sends simultaneously a permission/denial message to the service provider and the client gateway about the established parameters, otherwise the gateway gets a failure in this working algorithm.

Acknowledgments. The publication was financially supported by the Ministry of Education and Science of the Russian Federation (the Agreement number 02.a03.21.0008) and project No. 16-37-00215 Biodriver.

References

1. Pirmagomedov, R., Hudoev, I., Shangina, D.: Simulation of medical sensor nanonetwork applications traffic. In: Vishnevskiy, V.M., Samouylov, K.E., Kozyrev, D.V. (eds.) DCCN 2016. CCIS, vol. 678, pp. 430–441. Springer, Cham (2016). doi:10.1007/978-3-319-51917-3_38
2. Akyildiz, I.F., Pierobon, M., Balasubramaniam, S., Koucheryavy, Y.: Internet of Bio-Nano things. IEEE Commun. Mag. **53**(3), 32–40 (2015)
3. Akyildiz, I.F., Jornet, J.M.: The Internet of Nano-Things. IEEE Wirel. Commun. **17**(6) (2010)
4. International Telecommunication Union: Implementing e-Health in Developing Countries: Guidance and Principles. Accessed 15 Apr 2012
5. Donoghue, M., Balasubramaniam, S., Jennings, B., Jornet, J.M.: Powering in-body Nanosensors with ultrasounds. IEEE Trans. Nanotechnol. **15**(2), 151–154 (2016)
6. Ivanov, S., Foley, C., Balasubramaniam, S., Botvich, D.: Virtual groups for patient WBAN monitoring in medical environments. IEEE Trans. Biomed. Eng. **59**(11 (Part 2)), 3238–3246 (2012)

7. Kirichek, R., Pirmagomedov, R., Glushakov, R., Koucheryavy, A.: Live substance in cyberspace - Biodriver system. In: 2016 18th International Conference on Advanced Communication Technology (ICACT), Pyeongchang, pp. 274–278 (2016). doi:10.1109/ICACT.2016.7423358
8. Pirmagomedov, R., Hudoev, I., Kirichek, R., Koucheryavy, A., Glushakov, R.: Analysis of delays in medical applications of nanonetworks. In: 8th International Congress on Ultra Modern Telecommunications and Control Systems and Workshops (ICUMT), Lisbon, Portugal, pp. 49–55 (2016). doi:10.1109/ICUMT.2016.7765231
9. Muhizi, S., Shamshin, G., Muthanna, A., Kirichek, R., Vladyko, A., Koucheryavy, A.: Analysis and performance evaluation of SDN queue model. In: Koucheryavy, Y., Mamatas, L., Matta, I., Ometov, A., Papadimitriou, P. (eds.) WWIC 2017. LNCS, vol. 10372, pp. 26–37. Springer, Cham (2017). doi:10.1007/978-3-319-61382-6_3
10. Volkov, A., Khakimov, A., Muthanna, A., Kirichek, R., Vladyko, A., Koucheryavy, A.: Interaction of the IoT traffic generated by a smart city segment with SDN core network. In: Koucheryavy, Y., Mamatas, L., Matta, I., Ometov, A., Papadimitriou, P. (eds.) WWIC 2017. LNCS, vol. 10372, pp. 115–126. Springer, Cham (2017). doi:10.1007/978-3-319-61382-6_10
11. Vladyko, A., Muthanna, A., Kirichek, R.: Comprehensive SDN testing based on model network. In: Galinina, O., Balandin, S., Koucheryavy, Y. (eds.) NEW2AN/ruSMART -2016. LNCS, vol. 9870, pp. 539–549. Springer, Cham (2016). doi:10.1007/978-3-319-46301-8_45

Author Index